$f'''(x) = D_x^3 f(x) = \dfrac{d^3f}{dx^3}$	third derivative of $f(x)$
dy	differential y
cv	critical value
$\frown$	relative maximum
$\smile$	relative minimum
P_{max}	maximum value of P
C_{min}	minimum value of C
MR	marginal revenue
MQ	marginal cost
e	$\lim_{h \to 0} (1 + h)^{1/h} = 2.718282$
e^x	exponential function
$\log x$	common logarithm of x or $\log_{10} x$
$\ln x$	natural logarithm of x or $\log_e x$
$f_x = \dfrac{\partial f}{\partial x} = f_1$	first partial derivative of f with respect to x
$f_y = \dfrac{\partial f}{\partial y} = f_2$	first partial derivative of f with respect to y
$f_{xx} = \dfrac{\partial^2 f}{\partial x^2}$	second partial derivative of f with respect to x
$f_{xy} = \dfrac{\partial^2 f}{\partial y\, \partial x}$	partial derivative of f with respect to x first, then y
l	Lagrange multiplier
$\int$	integral sign
$\int f(x)\, dx$	indefinite integral of $f(x)$ or antiderivative of $f(x)$
$\int_a^b f(x)\, dx$	definite integral from a to b of $f(x)$
$F(x) \Big]_a^b$	$F(x)$ evaluated at b minus $F(x)$ evaluated at a
$\|\Delta\|$	largest Δx_i, or mesh of a subdivision
$\sum$	summation
$\sum_{i=1}^{n} a_i$	summation of a_i from $i = 1$ to $i = n$
S_n	sum of the first n terms of a sequence
CS	consumers' surplus
PS	producers' surplus
$(\bar{q}, \bar{p})$	point of market equilibrium
$\bar{y}$	average value of y
$P(E)$	probability of event E

CALCULUS

CALCULUS

A Tool for Analysis and Decision

MICHAEL L. KOVACIC
COLORADO STATE UNIVERSITY

PRINDLE, WEBER & SCHMIDT, INC.
BOSTON, MASSACHUSETTS

Third printing: January 1979

Library of Congress Cataloging in Publication Data

Kovacic, Michael L
Calculus: a tool for analysis and decision.

Includes index.
1. Calculus. I. Title.
QA303.K77 515 76-30455
ISBN 0-87150-233-X

Prindle, Weber & Schmidt is a subsidiary of Wadsworth, Inc.

Designed by David Chelton and the production staff of Prindle, Weber & Schmidt.
Composed in Monophoto Times Roman and Helvetica by Syntax International.
Technical art by Phil Carver & Friends.
Printed and bound by Halliday Lithograph Corp.

PREFACE

Calculus: A Tool for Analysis and Decision is for courses dealing with the practical applications of calculus. The text emphasizes an intuitive approach and presents techniques for analyzing and solving problems. I believe that by keeping rigor and abstraction to a minimum, the power of the calculus becomes more evident to the student. The student using this text should have some background in basic algebra. I have provided material in Chapter 1 for students who need a review before jumping into the calculus material.

The applications in the book deal with business, economics, and related areas, as well as the life and social sciences. The text presents numerous examples and over 2,000 problems and exercises in an ascending order of difficulty. In general, theorems are simply stated and then applied to practical situations. Each chapter also contains a section of Review Problems and a section of Enrichment Topics and Problems. The enrichment sections contain more challenging exercises which should appeal to the motivated student and to the instructor who wishes to present a more rigorous course.

Chapter 1 serves a two-fold purpose: a review of functions and a short review of algebra. Taking the concept of function as its central theme, the chapter includes graphing and the construction of new functions, and also introduces factors and zeros of functions. It has been kept purposefully simple and to the point, and should be more than sufficient to prepare the average student for the calculus.

Applications to business situations are woven carefully into Chapter 2 without ignoring classic applications such as slope of a tangent line or velocity. The depth of coverage aims at helping the student cope effectively with the contents of the next chapter.

Chapter 3 presents optimizing techniques, finding extreme values both in purely mathematical and in applied situations. Mathematical examples include finding the high and low points of curves, and "real world" problems include finding optimum lot size, maximum profit, minimum cost, and other results. Chapter 4 permits students to do practical numerical work with logarithms and exponentials.

In Chapter 5 I have replaced the usual symbol for the Lagrange multiplier, λ, in favor of the letter l to reduce the number of unfamiliar symbols. The topics of Lagrange multipliers and least squares may be omitted by the instructor without affecting the treatment of partial derivatives or the continuity of the text as a whole.

In Chapter 6 the treatment of the indefinite integral precedes discussion of the definite integral, an approach I greatly favor. Areas are first found by approximating methods, which serves as a suitable background for finding exact areas by applications of the definite integral. The summation process, basic to understanding the definite integral, is denoted by sigma notation and is given expanded treatment in a separate section.

Chapter 7 abounds with applications of the definite integral. Included are areas between curves, consumers' and producers' surplus, integration by parts, improper integrals, and probability density functions.

Chapter 8 consists of a short but practical treatment of elementary differential equations to help make the text a more complete presentation of the calculus.

I believe that the content of this calculus course at this particular level should appeal to a wide variety of colleges and universities. I sincerely hope that the use of this text will kindle the student's interest in mathematics and in its application to the world around him.

ACKNOWLEDGMENTS

I would like to express my thanks and appreciation to the staff of Prindle, Weber and Schmidt, Inc., for the high standards which it has maintained not only in the publishing of this text but also many others. Finally, I must offer sincere thanks to my wife, Georgia, for her constancy during this endeavor.

Michael L. Kovacic
Fort Collins, Colorado

SUGGESTED COURSE OUTLINE

I offer the following suggested course outline, containing a core curriculum which I believe should lead to an understanding and appreciation of the basic elements of the calculus. The various options that follow depend on the type of class and the time allowed for the course.

Core Curriculum: Chapter 2, Differential Calculus (omit sections 2.5–2.7)
Chapter 3, Optimization Analysis
Chapter 5, Multivariable Differentiation (sections 5.1–5.4 only)
Chapter 6, Integral Calculus (omit section 6.6)
Chapter 7, Other Applications and Techniques of Integration (sections 7.1 and 7.4 only)

Option 1, for classes with minimal background in mathematics:
Core Curriculum plus Chapter 1, Chapter 4, and Section 7.2

Option 2, for classes with average background in mathematics:
Same as Option 1 plus sections 2.5–2.7, 7.3, 7.6, and 7.8, plus some selected problems from the Review and Enrichment Problem sets at the end of each chapter.

Option 3, for classes with better-than-average background in mathematics:
Skip Chapter 1, take Chapters 2–7 and selected sections from Chapter 8, plus many problems from the Review and Enrichment Problem sets and perhaps a few sessions on the trigonometry in Appendix A.

Option 4, for classes with exceptionally good mathematical background:
Same as Option 3 except that more challenging problems from the problem sets, especially the Enrichment Topics and Problems, should be assigned. Some developmental theory could be introduced and a few of the more important theorems proved.

CONTENTS

INTRODUCTION 1

1 ALGEBRAIC FUNCTIONS 4

A Short Review

1.1 Fundamentals 5
1.2 The Function Concept 12
1.3 Graphing Functions 13
1.4 Forming New Functions 22
1.5 Review Problems 31
1.6 Enrichment Topics and Problems 33

2 DIFFERENTIAL CALCULUS 40

The Role of Derivatives in Decision-Making

2.1 Limits of Functions 41
2.2 Continuity of Functions 48
2.3 Derivatives 54
2.4 Differentiation 60
2.4.1 Techniques of Differentiation 60
2.4.2 Additional Techniques. The Chain Rule 63
2.4.3 Derivatives of Higher Order 68
2.5 Differentials 72
2.6 The Law of the Mean 75

2.7 Newton's Method 78
2.8 Review Problems 82
2.9 Enrichment Topics and Problems 85

3 OPTIMIZATION ANALYSIS 90

Decision-Making for Optimality

3.1 Extrema 91
3.2 Applications 97
3.3 Optimum Lot Size 104
3.4 Other Applications 108
3.5 Review Problems 115
3.6 Enrichment Topics and Problems 116

4 EXPONENTIAL AND LOGARITHMIC FUNCTIONS 119

Applications of e^u and $\ln |u|$

4.1 Exponential Functions 120
4.2 Logarithmic Functions 127
4.3 The Differentiation of $\ln |u(x)|$ 135
4.4 The Differentiation of $e^{u(x)}$ 139
4.5 Review Problems 142
4.6 Enrichment Topics and Problems 145

5 MULTIVARIABLE DIFFERENTIATION 146

Applications of Multivariable Functions

5.1 Graphs 147
5.2 Partial Derivatives 151
5.3 Partial Derivatives of Higher Orders. The Total Derivative 154
5.4 Implicit Differentiation 158
5.5 Extrema for Functions of Two Variables 161
5.6 Lagrange Multipliers (Single Constraint) 166
5.7 Least Squares (Linear Regression) 172
5.8 Review Problems 177
5.9 Enrichment Topics and Problems 179

6 INTEGRAL CALCULUS 183

Antidifferentiation as a Practical Tool

6.1 The Indefinite Integral 184
6.2 Simple Differential Equations 190
6.3 More Applications 195
6.4 Other Substitutions 198
6.5 Area 202
6.5.1 Approximating Area by Rectangles 202
6.5.2 Finding Exact Areas by Calculus. The Definite Integral 209
6.6 Sigma Notation Applied 217
6.7 Review Problems 228
6.8 Enrichment Topics and Problems 231

7 OTHER APPLICATIONS AND TECHNIQUES OF INTEGRATION 235

More Tools for the Decision-Maker

7.1 Area between Two Curves 236
7.2 Consumers' and Producers' Surplus 244
7.3 Average Value of a Function 248
7.4 Integration by Parts 251
7.5 Partial Fractions: Linear Factors 255
7.6 Improper Integrals 260
7.7 Probability Density Functions 269
7.8 Integration by Use of Tables 278
7.9 Review Problems 280
7.10 Enrichment Topics and Problems 282

8 DIFFERENTIAL EQUATIONS 287

Additional Calculus Tools

8.1 Introduction 288
8.2 Solutions 289
8.3 Separation of Variables 291
8.4 Separation of Variables: Other Applications 293
8.5 Linear Equations 297
8.6 Homogeneous Equations 303
8.7 Review Problems 306
8.8 Enrichment Topics and Problems 307

APPENDIX A A-1

A Short Review of Trigonometry and Its Applications to Calculus

A.1 Basic Topics A-2
A.2 Graphs A-14
A.3 Applications A-20
A.4 Derivatives A-23
A.5 Integrals A-27

APPENDIX B B-1

Some Examples of the Use of Logarithms

Table I: Common Logarithms B-4
Table II: Exponential Functions B-6
Table III: Natural Logarithms B-7
Table IV: Powers and Roots B-8
Reference List of Mathematical Equations B-9

APPENDIX C C-1

Answers to Odd-Numbered Problems

INDEX I-1

INTRODUCTION

The calculus did not emerge suddenly upon the world's scene. Instead, it was the culmination of centuries of laborious development.

The arithmetic in use today had its origins among the Hindus and Mesopotamians. Methods for constructing buildings can be traced to the geometric techniques employed by the Egyptians. In early 800 A.D. the decimal system (the base 10 system), including the number 0 for the first time, was promoted by a mathematician in Arabia. The conquering Greeks inherited a wealth of mathematical knowledge from scholars of the Middle East and proceeded to make improvements through their powerful tools of abstraction and proof. The noteworthy result was a great advance in the field of geometry, epitomized by the works of Euclid. Other significant contributions were made by Pythagoras, Archimedes, and Diophantus. In southern Europe the Italian scholars Fibonacci, Tartaglia, Cardano, and Galileo did much for the advancement of mathematics. Kepler of Germany discovered laws concerning the orbits of planets about the sun. Copernicus of Poland devised a theory that placed the sun at the center of the solar system, a theory then considered by many as heresy. Fermat of France and Barrow of England (Isaac Newton's teacher) developed new ideas which proved to be basic to the calculus.

About 1650, the great French mathematician and philosopher René Descartes discovered what is now known as the rectangular (or cartesian) coordinate system. This revolutionary theory propounded that algebraic equations could be represented by sequences of points which, when connected, resulted in geometric figures. Conversely, these figures could be represented by sequences of numbers (points) which in turn can be expressed as equations. The introduction of coordinates was the birth of analytic geometry. This new mathematical tool added breadth to the existing mathematics and encouraged the development of two very important concepts: variable and function.

It was not long after Descartes' discovery that two mathematical giants made their contributions. These two geniuses, Isaac Newton and Gottfried Leibniz, independently invented the formal mathematical system called calculus, which is considered by some to be the most useful discovery of all time. Newton's method came in 1665 and was called the calculus of fluxions; however, he delayed publication until 1704. Leibniz' discovery occurred in 1684, also the year of publication. Long arguments followed over which one discovered the calculus first. Newton used the dot (i.e., $\dot{y}$) to denote the first derivative, whereas Leibniz used the symbol dy/dx. The latter notation, used today, seemed more conducive to the increased use of the calculus on the European continent. This development was enhanced by Lagrange of France and by the Swiss brothers Johann and Jacob Bernoulli, the latter using the power of calculus in mechanics and in the "calculus of variations." Laplace of France used the calculus to advance his theories of celestial mechanics. He also made significant contributions in the field of probability, as did Méré, Fermat, Pascal, de Moivre, and Gauss. Euler of Switzerland was still another prolific innovator and writer of mathematical ideas.

It was the German genius Carl Friedrich Gauss who rallied mathematicians after many had become confused by the proliferation of new ideas and new symbology. Gauss was said to be the last person able to master all of contemporary mathematics. One of his many areas of interest and accomplishment was that of non-Euclidean geometry, an area which the Russian, Lobachevsky, developed further. A pupil of Gauss', Bernhard Riemann, did much to extend the boundaries of mathematics, especially in higher geometry. The Riemann integral is an important concept in the integral calculus. Riemann's work also found expression in the development of Einstein's Theory of Relativity. Cauchy of France was the first to show that the concept of limit is the basic feature underlying the differential calculus. It was Dirichlet who gave the function its meaning as we know it today. There followed a long line of major contributors such as Abel, Jacobi, Hamilton, Weirstrass, Poincaré, Schrödinger, and von Neumann, to name only some. In more recent times mathematics has advanced through the work of Boole, Bourbaki, Eilenberg, Erdös, Gödel, Klein, Milnor, Nirenberg, Smale, Tate, Tukey, and Ulam, among others.

Calculus is the mathematics of motion and change. It is the tool that can "freeze" a movement so that we may analyze the process of change at a particular instant. It unifies the pre-Descartes world of *static* numbers with Descartes' static figures to produce a systematic method which gives us insight into the problems of *movement* and change. Its foundation consists of such basic ideas as function, rate of change, convergence, approximation, and integration.

Up to the time of Descartes, it may be said that mathematics was "a science without order." This inordinancy was not a new problem; it is peculiar to the nature of man. The search for order, although never-ending, at least reached a turning point with the discovery of calculus. People were slowly learning to accept the limits of their nature and their power. Within these limits they applied their energies to the development of systematic laws. Mathematicians do not express opinions; they interpret laws, using the accumulation of earlier discoveries as a guide. Systematic rigor and clarity are marks of the mathematician's technique. Such intellectual precision has become a standard for making various tests; a focus for peoples' interests and a boundary for their energies. Great mathematicians are indeed an immensely energetic lot. The impact on history of personalities such as Descartes, Newton, Leibniz, Gauss, and Einstein has been truly remarkable.

It seems that the containment of intellectual energy within the framework of systematic laws makes its use more possible. Contained energy can be harnessed and utilized indefinitely as a power. Uncontrolled energy often explodes destructively and/or dissipates uselessly.

The calculus acted as a system of controls, setting limits and providing boundaries. Within these boundaries human energy was free to release its potential force, and this force was constructive only within the system of controls. Without controls actions would be unrestrained, so that effective action would not be possible.

Thus, with peoples' ordered energy acting in direct proportion to the discipline which contained it, mathematics continued to make great progress. Of course, there were periods of disorder which affected family stability, national economies, and peoples' views of religion, science, philosophy, and other areas of life. But fortunately, chaos does not last forever. The false thinking that science can bring continual progress is on the wane. Recent protests by ecological movements against uncontrolled pollution of the environment are examples of peoples' belief in limits, boundaries, and order.

The pioneers of mathematics contributed immeasurably to the progress of man and womankind. They laid the foundations of a great discipline which has benefited the world's inhabitants in many ways. Even as we are grateful inheritors, so also should we be responsible trustees. It is obligatory that we try to inspire our youth with the great traditions of our heritage.

1 ALGEBRAIC FUNCTIONS

A Short Review

IMPORTANT EQUATIONS

$$\{x | x \in \mathbf{R}\}$$

$$P_1P_2 = \sqrt{(x_2 - x_1)^2 + (y_2 - y_1)^2}$$

$$\{(x, y) | y = mx + b\}$$

$$y - y_1 = m(x - x_1)$$

$$[fg](x) = f(x) \cdot g(x)$$

$$[f \circ g](x) = f[g(x)]$$

$$P(x) = a_0 + a_1x + a_2x^2 + \cdots + a_nx^n, \quad n \geq 0$$

1.1 FUNDAMENTALS

In this first chapter we introduce certain basic ideas which will be useful in our study of the calculus. Some students may find little that is new in this chapter, but the review will probably be helpful to most students.

A **set** is any collection of objects. These objects are called **elements**, or members of the set. A very important collection is the set of **real numbers**, which we denote by **R**. We represent the **natural numbers** (or counting numbers) by

$$\mathbf{N} = \{1, 2, 3, 4, \ldots\}$$

where the particular brackets { } are used to identify a set. The numbers 1, 2, 3, 4, ... are members of the set. The notation ... means that the numbers continue according to the pattern already established. For example, we say that 2 is an element of **N** and write this idea as $2 \in \mathbf{N}$. We use the notation $\frac{2}{3} \notin \mathbf{N}$, read "two-thirds is not an element of **N**," since 2/3 is not a natural number. Every element of **N** is an element of **R**. Thus, **N** is a **subset** of **R**, written $\mathbf{N} \subseteq \mathbf{R}$. The set composed of the natural numbers, their negatives, and the number zero is called the set of **integers**. It is denoted by

$$\mathbf{I} = \{\ldots, -3, -2, -1, 0, 1, 2, 3, \ldots\}.$$

The **rational numbers**, symbolized by **Q**, are those that can be expressed as the ratio of any two integers c and d, thus:

$$\frac{c}{d}, \quad d \neq 0.$$

Examples are 2, $-3/4$, 5/6. The **irrational numbers**, denoted by $\bar{\mathbf{Q}}$, are those real numbers which are nonterminating and nonrepeating when occurring in decimal form. Some examples are $-\sqrt{3}$, $\sqrt{2}$, π. This set can be written in so-called **set-builder notation** as

$$\bar{\mathbf{Q}} = \{x | x \in \mathbf{R} \quad \text{and} \quad x \notin \mathbf{Q}\},$$

read "$\bar{\mathbf{Q}}$ is the set of all x such that x belongs to the set of real numbers and x is not an element of the set of rational numbers."

The real numbers can be put in one-to-one correspondence with points on a horizontal straight line. That is, each real number can be paired with a point on the line (see Figure 1.1). This line L is referred to as the **number line**. All real numbers paired with points to the right of 0 are called positive numbers; those paired with points to the left of 0 are called negative numbers.

The line L provides us with a method of ordering numbers. By this method we have $0 < 1, 2 > 3/2, 0 > -1, -2 < -1$, etc., where the symbols $<$ and $>$ are called inequality signs. Expressions such as $4 > 3$ and

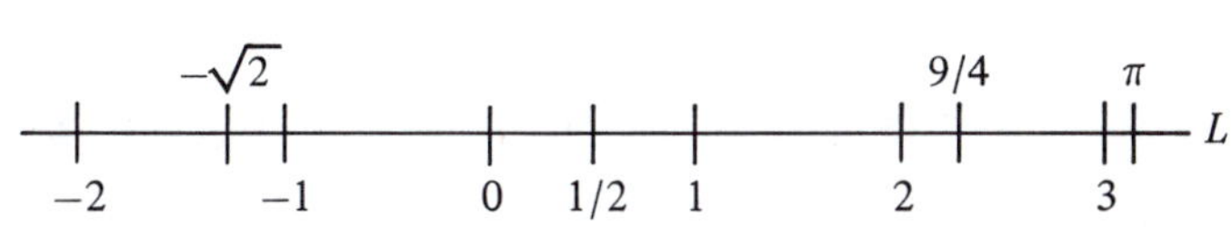

Figure 1.1

$2 < 5$ are called **inequalities**. It follows that if x and y are any two real numbers on the number line, then only one of the following is true:

(a) $x < y$, i.e., x is less than y

(b) $x > y$, i.e., x is greater than y

(c) $x = y$, i.e., x equals y.

The negation of inequalities is symbolized by $\nless$ meaning "is not less than" and by $\ngtr$ meaning "is not greater than."

The distance between any point b and the point 0 on L is the **absolute value** of b, denoted by $|b|$. By definition,

$$|b| = \begin{cases} b & \text{if } b \text{ is positive} \\ 0 & \text{if } b \text{ is } 0 \\ -b & \text{if } b \text{ is negative} \end{cases}$$

Thus, if b is 3, then $|3| = 3$; if $b = -2$, then $|-2| = 2$. In general, $|b - c|$ is the distance between b and c. To illustrate, $|2 - 5| = |-3| = 3$, the distance from 2 to 5. Similarly, $|3 - (-2)| = |5| = 5$, the distance from 3 to -2. We can identify a particular line segment on the number line by the use of set-builder notation and inequalities; thus, $S = \{y \mid 2 < y \leq 8\}$. This is read "$S$ is the set of all numbers y such that 2 is less than y and y is less than or equal to 8." This set can be written in **interval form** thus: $(2, 8]$. We call this a half-open, half-closed interval which includes the endpoint 8 but not the endpoint 2. A closed interval such as $[1, 5]$ includes both endpoints 1 and 5, whereas the open interval $(0, 2)$ does not include the endpoints 0 and 2. Sets can represent infinite intervals such as $\{x \mid x > 0\}$, which is equivalent to the infinite interval $(0, \infty)$, and $\{z \mid z \leq 9\} = (-\infty, 9]$. The symbol ∞ is read "infinity" and *is not a number*; $+\infty$ and $-\infty$ are used to indicate unboundedness in the positive and negative directions, respectively. Graphs of certain intervals are shown in Figure 1.2.

The statement:

$$\text{If} \quad a < b \quad \text{and} \quad c < 0, \qquad \text{then} \quad ac > bc$$

can be written:

$$a < b, c < 0 \rightarrow ac > bc,$$

where the arrow $\rightarrow$ means "implies that." The double arrow $\leftrightarrow$ means "if and only if." We illustrate its use thus: $a > b$ if and only if $b < a$ can be written: $a > b \leftrightarrow b < a$. The "if and only if" is usually abbreviated "iff."

We have shown how the real numbers relate to points on the number line. Now let us see how real numbers relate to points in the plane. The number line is a one-dimensional coordinate system referred to as E^1,

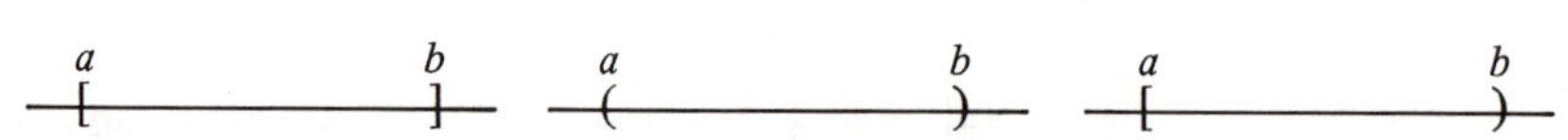

Closed interval $[a, b]$ Open interval (a, b) Half-closed, half-open interval $[a, b)$

Figure 1.2

whose points are identified by single numbers. We extend this idea to a two-dimensional system E^2, called the **cartesian** (or **rectangular**) **coordinate system**, whose points are in the plane and consist of **ordered pairs** of numbers of the form (x, y). We first draw in the plane two perpendicular lines, one vertical and one horizontal. They divide the plane into four **quadrants** with the signs of the coordinates as shown in Figure 1.3. The perpendicular lines intersect at 0 (called the **origin**), the horizontal line representing the x-axis and the vertical line representing the y-axis. Positive numbers on the vertical y-axis correspond to points above the x-axis, and negative numbers correspond to points below the x-axis. The x, y-axes are referred to as the axes of the cartesian coordinate system (or cc system). Of course, other letters besides x and y can be used. It is sometimes convenient to use different units of length on the two axes.

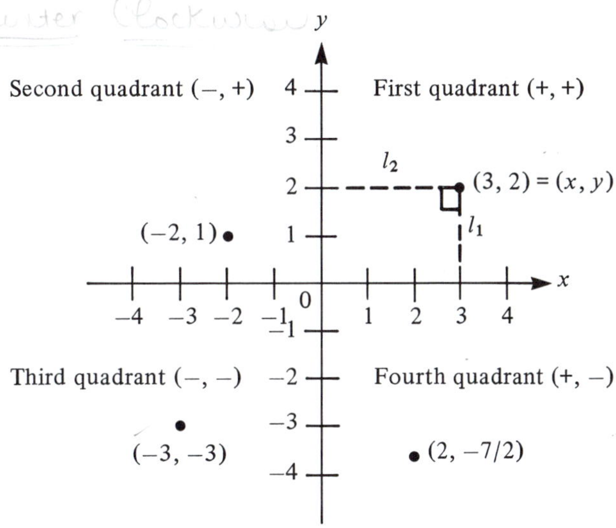

Figure 1.3

A point in a plane surface region can be uniquely identified by an ordered pair of real numbers denoted by (x, y) and referred to as the point (x, y). The location of a point represented by (3, 2), for example, is found by simply starting from the origin and going to the right a distance of 3 on the x-axis, then going up vertically a distance of 2. More precisely, we locate the point corresponding to 3 on the horizontal axis, then locate the point corresponding to 2 on the vertical axis. We draw lines l_1 and l_2 through these points (3, 0) and (0, 2) parallel to the coordinate axes (and perpendicular to each other). The intersection of l_1 and l_2 determines the position of the point (3, 2) (see Figure 1.3). Conversely, for any point in the plane there is a unique ordered pair which corresponds to that point. Accordingly, there is a one-to-one correspondence between the set of all ordered pairs of real numbers and the set of all points in the plane. This correspondence enables us to picture algebraic analysis in a geometric form, thus encouraging our intuitive understanding. In the notation (x, y) the x is called the first coordinate (or **abscissa**) and the y is called the second coordinate (or **ordinate**). Much of our analysis will depend on the relationship between these two coordinates of a point. A relationship between x and y can be expressed as a set of ordered pairs which can be plotted in the plane. This set of points constitutes the **graph** of the relationship.

Consider any two points $P_1(x_1, y_1)$ and $P_2(x_2, y_2)$, as in Figure 1.4. Draw lines through P_1 and P_2 perpendicular to each other and intersecting at the point $Q(x_2, y_1)$. Triangle P_1P_2Q is a right triangle with hypotenuse the undirected line segment P_1P_2. By the Pythagorean theorem from plane geometry, the length of P_1P_2 is given as

$$\begin{aligned}(P_1P_2)^2 &= (P_1Q)^2 + (QP_2)^2 \\ &= (x_2 - x_1)^2 + (y_2 - y_1)^2\end{aligned}$$

or

$$\boxed{P_1P_2 = \sqrt{(x_2 - x_1)^2 + (y_2 - y_1)^2}}$$

This formula represents the **distance between two points** in the plane. It is valid for all locations of the points P_1 and P_2 in all four quadrants.

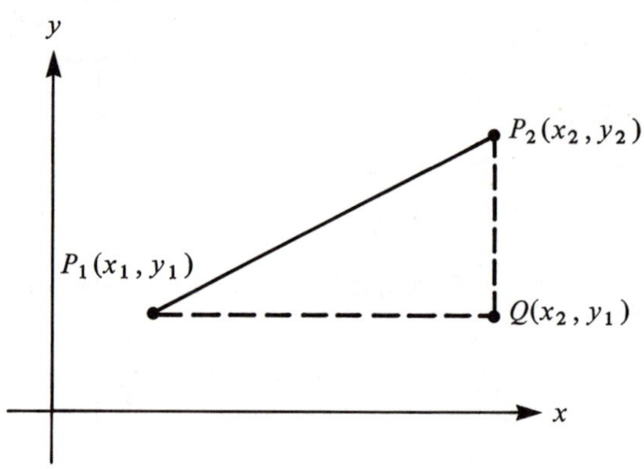

Figure 1.4

EXAMPLE The perimeter of the triangle ABC with vertices $A(-2, 1)$, $B(3, -1)$, and $C(2, 2)$ is found as follows:

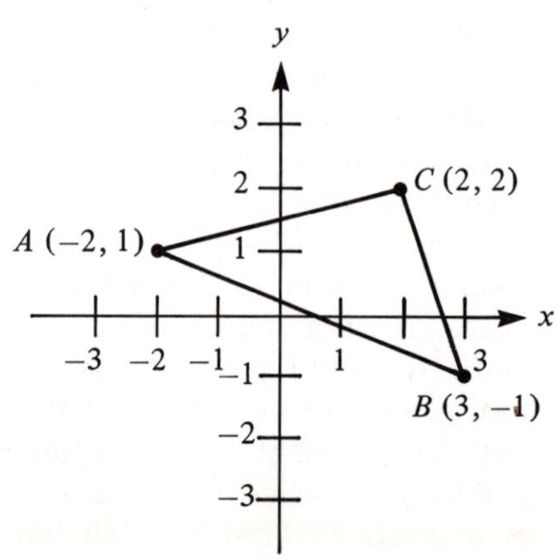

Figure 1.5

$$AB = \sqrt{(3+2)^2 + (-1-1)^2} = \sqrt{25+4} = \sqrt{29}$$
$$BC = \sqrt{(2-3)^2 + (2+1)^2} = \sqrt{1+9} = \sqrt{10}$$
$$AC = \sqrt{(2+2)^2 + (2-1)^2} = \sqrt{16+1} = \sqrt{17}$$

Therefore
$$\text{Perimeter} = \sqrt{29} + \sqrt{10} + \sqrt{17}.$$

The term **straight line** refers to a line which is either vertical, horizontal, or oblique. Graphs of these three types of lines are shown in Figure 1.6. Of course, not all oblique lines go through the origin.

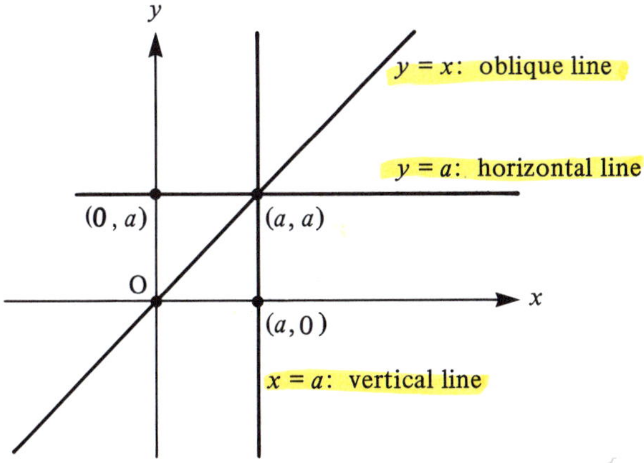

Figure 1.6

Every oblique line is represented by the set:

$$\{(x, y) \mid y = mx + b\}$$

where m and b are constants. The number m is called the **slope** of the line and the number b is the **y-intercept**. In Figure 1.6 the line $y = x$ has slope 1 and y-intercept 0. We define the slope of a line through the points $P_1(x_1, y_1)$ and $P_2(x_2, y_2)$ as

$$m = \frac{y_2 - y_1}{x_2 - x_1} = \frac{\text{change in } y}{\text{change in } x}.$$

This can be written in the form

$$y_2 - y_1 = m(x_2 - x_1).$$

In general,

$$y - y_1 = m(x - x_1) \tag{1}$$

is the **point-slope form** of an equation of a straight line. (Recall that an **equation** is a statement of equality between two expressions.) The slope of a horizontal line is zero, whereas the slope of a vertical line is undefined. Two lines are **parallel** if their slopes are equal.

EXAMPLES **(a)** The line through (2, 3) and parallel to the x-axis is the horizontal line $y = 3$. We verify this by use of Eq. (1), thus:

$$y - 3 = 0(x - 2) = 0$$
$$y = 3.$$

(b) The line through (2, 3) and parallel to the y-axis is the vertical line $x = 2$. The slope of $x = 2$ is undefined so that use of Eq. (1) is not feasible.

(c) The line through (2, 3) with slope -4 is

$$y - 3 = -4(x - 2) = -4x + 8$$

or

$$y = -4x + 11.$$

Note: The form $y = mx + b$ will be referred to as either a line or an equation, interchangeably.

(d) The lines $y = x + 1$ and $y = -x + 7$ intersect at a point that lies on both lines. We equate the two expressions which are equal to y and obtain

$$x + 1 = -x + 7.$$

Solving for x, we get

$$2x = 6 \quad \text{or} \quad x = 3.$$

Substituting $x = 3$ in either of the given equations we have $y = 4$. Therefore, the point (3, 4) is the point of intersection. This method of finding where two lines meet is called a **simultaneous solution** of the two equations. It is wise to check answers. When $x = 3$ and $y = 4$, we have

in $y = x + 1$: $\quad 4 = 3 + 1 = 4$

in $y = -x + 7$: $\quad 4 = -3 + 7 = 4.$

Thus, $x = 3$ and $y = 4$ are the correct coordinates of the point of intersection (Figure 1.7).

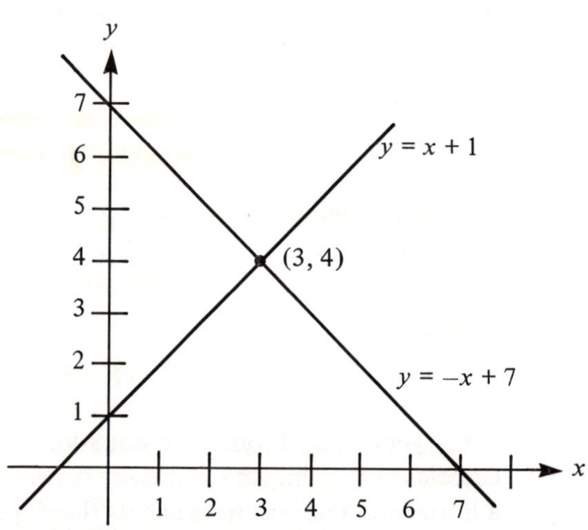

Figure 1.7

(e) An equation of the line through (2, 1) and (3, 4) can be found by use of the form

$$y - y_1 = m(x - x_1).$$

First we find the slope:

$$m = \frac{4 - 1}{3 - 2} = 3.$$

Then with (2, 1) as (x_1, y_1), we have

$$\begin{aligned} y - 1 &= 3(x - 2) \\ &= 3x - 6 \\ y &= 3x - 5, \quad \text{the desired equation.} \end{aligned}$$

The point (3, 4) could also be used for (x_1, y_1) to obtain the same equation.

PROBLEM SET 1.1

1. Plot the indicated numbers on the number line and find their absolute values:

(a) $-\sqrt{2}$ **(b)** $[3 - (-4)]$ **(c)** $\dfrac{-1}{-2}$

(d) $(-2)^3$ **(e)** $[-2(-3)^2]$ **(f)** $\dfrac{-3 + 1}{-1 + 3}$

2. Write in simplest form, without absolute value symbols:

(a) $|-4 - (-2)|$ **(b)** $|x - 1|$ **(c)** $|2 - x|$

3. Write in symbolic form:

(a) b is greater than c
(b) c is less than d
(c) a is less than or equal to b
(d) c is greater than or equal to a and less than b
(e) t is between x_i and x_{i-1} (between means the end points are not included)
(f) The interval of x includes both a and b
(g) The absolute value of x is less than 2

4. Write each set as an interval:

(a) $\{x \mid x > 1\}$ **(b)** $\{y \mid y \le 3\}$

(c) $\{t \mid 0 \le t < 4\}$ **(d)** $\left\{z \,\middle|\, \frac{1}{4} < z < \frac{3}{4}\right\}$

5. Find the values of all numbers K for which:

(a) $|K + 1| = 2$ **(b)** $|K - 2| < 4$

(c) $|K + 3| \le 1$ **(d)** $\dfrac{|K|}{3} > 3$

6. Write in symbols:

(a) x plus two equals four minus x implies that x equals 1.
(b) The quotient of b by c equals d if and only if b equals c times d.

7. Plot and label the given points on one cartesian coordinate (cc) system:

(a) (2, 5) **(b)** (3, 6) **(c)** (5, 7)
(d) (−2, 4) **(e)** (6, −3) **(f)** (10, 5000)

Note: choose convenient units in (f).

8. If $P_0 = (0, 0)$, $P_1 = (2, 4)$, $P_2 = (6, -2)$, and $P_3 = (10, 18)$, compute the indicated distances:

(a) P_0P_1 **(b)** P_1P_2 **(c)** P_2P_3 **(d)** P_3P_0

9. Find the slope of the line joining each given pair of points:

(a) (0, 0) and (3, 4) **(b)** (2, 3) and (4, 6)
(c) (4, −2) and (2, −1) **(d)** (1, 1) and $(1, \frac{1}{2})$
(e) $(2, \frac{1}{3})$ and $(3, \frac{1}{4})$ **(f)** (0.2, 2) and (3, 0.3)

10. Write an equation of the line through each given pair of points in problem 9.

11. Find an equation of the line parallel to the line in problem

(a) 9(b) and through (1, 2)
(b) 9(f) with y-intercept 3
(c) 9(a) and through $(\frac{1}{2}, \frac{3}{4})$

12. Find the points of intersection of the given lines:

(a) $y = 2x$ and $y = x + 2$ **(b)** $y = x + 4$ and $y = 2x + 3$
(c) $y = x - 4$ and $y = x/3$ **(d)** $y = -x$ and $y = x + 20$

1.2 THE FUNCTION CONCEPT

The basic ideas underlying the concept of function provide us with the necessary background for understanding the calculus.

Consider the relationship where with each production unit x we associate a particular number $P(x) = 2x - 10$, representing dollars of profit. This type of association is called a function, i.e., P is said to be a function of x. If $x = 15$, then $P = 2(15) - 10 = 30 - 10 = 20$ dollars of profit. Thus, P acts as a rule that associates with the number 15 one and only one number, 20.

DEFINITION *Let X and Y be sets and assume that $x \in X$ and $y \in Y$. A **function** f is a rule that associates with each element x in X a unique (exactly one) element y in Y. The set of elements in X constitute the **domain** of f and the set of elements in Y comprise the **range** of f.*

Functions are usually represented by the letters f, g, h, F, G, H, u, v, w, y, z, α, β, etc. The notation $y = f(x)$ means y is a function of x. Furthermore, $f(x)$ is read "f of x" or "f at x." In the profit function above, the value of P at $x = 15$ is indicated by $P(15)$. In other words, P is the rule that associates with the element 15 in the domain of P the unique element $P(15)$ in the range of P. Since the value of a function f of x is known only if the value of x is known, then if $y = f(x)$, we say that y depends on x. Accordingly, y is called the **dependent variable** and x is called the **independent variable**. If a relationship between x and y is not a function, it is simply called a relation. We present some illustrations.

EXAMPLES

(a) The *identity function* associates with each number x the number itself. With $x = 2$, the function f associates the unique number $f(2) = 2$; with $x = 3$, it associates $f(3) = 3$, etc. This function is written as $f(x) = x$.

(b) The *doubling function* associates with each number x twice that number, or $2x$. With $x = 2$ the rule associates $2(2)$ or 4; with $x = 3$, it associates $2(3)$ or 6, etc. This function is written as $f(x) = 2x$.

(c) The *squaring function* associates with each number x the square of x. If g is the rule, then $g(x) = x^2$; $g(2) = 2^2 = 4$, $g(3) = 3^2 = 9$, etc.

(d) The *square root function* associates with each nonnegative number t the square root of t. If h is the rule, then $h(t) = \sqrt{t}$; $h(0) = \sqrt{0} = 0$; $h(4) = \sqrt{4} = 2$; $h(5) = \sqrt{5}$, etc. The domain of h is the set of nonnegative numbers.

(e) The *absolute value function* associates with each number x the absolute value of x, or $|x|$. If F is the rule, then $F(x) = |x|$. This function associates with each number its distance from 0. Thus, $F(3) = |3| = 3$, $F(-4) = |-4| = 4$, etc. The domain of F is **R** (the set of real numbers) and its range is the set of nonnegative real numbers.

(f) The *reciprocal function* associates with each number $x\,(\neq 0)$ a unique number $1/x$. If G is the rule, then $G(x) = 1/x$; $G(2) = \frac{1}{2}$, $G(\frac{1}{2}) = 1/(1/2) = 2$, etc. The domain of G is the set of all reals except 0.

We usually describe a function f in terms of the expression for f. Thus, if $f(x) = 1/x$ we simply say "the function one over x"; if $f(x) = x^3$, we say "the function x cubed," and so on.

(g) A function can be represented in **tabular form**. Let us assume that a rule associates with advertising expense x the amount of sales revenue $S(x)$ as follows:

Advertising expense x	\$300	600	900	1200
Sales revenue $S(x)$	900	1800	2700	3600

When $x = 300$, $S(300) = 900$; when $x = 600$, $S(600) = 1800$; etc. This leads us to guess that $S(x) = 3x$.

1.3 GRAPHING FUNCTIONS

The locus (or position) of all points defined by

$$\{(x, y) \mid y = f(x),\ x \in \text{domain of } f\}$$

is called the graph of $f(x)$. Let us investigate the graphing of a variety of functions.

(*a*) If $f(x) = K$ where K is constant on $[0, 10]$, then the graph of f is a horizontal line segment. For example, if $f(x) = 20$, then f is always 20 units above the x-axis for the domain $[0, 10]$ (see Figure 1.8).

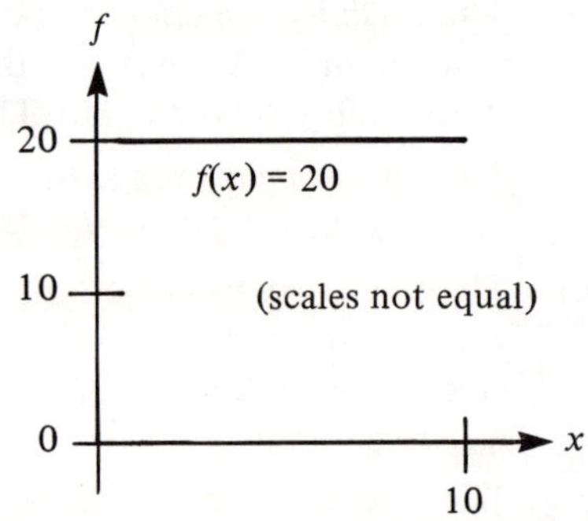

Figure 1.8

(*b*) Consider $f(x) = 3x - 2$, $-\infty < x < \infty$, where the notation $-\infty < x < \infty$ indicates the domain of f, i.e., all real numbers. When $x = 0$, $f(0) = -2$; when $f(x) = 0$, then $0 = 3x - 2$, $3x = 2$, $x = 2/3$. Thus $(0, -2)$ and $(\frac{2}{3}, 0)$ are points on the graph of f. (These are called intercept points.) We draw a line through these points. This line is the graph of $f(x) = 3x - 2$. Actually, the line extends to infinity in the positive and negative y directions but, for convenience, we exhibit only a small portion of the entire line. We could also have graphed f by use of other points such as (1, 1), (2, 4), etc. This function is called a **linear function** because its graph is a straight line. Any function of the form $y = mx + b = f(x)$ is a linear function.

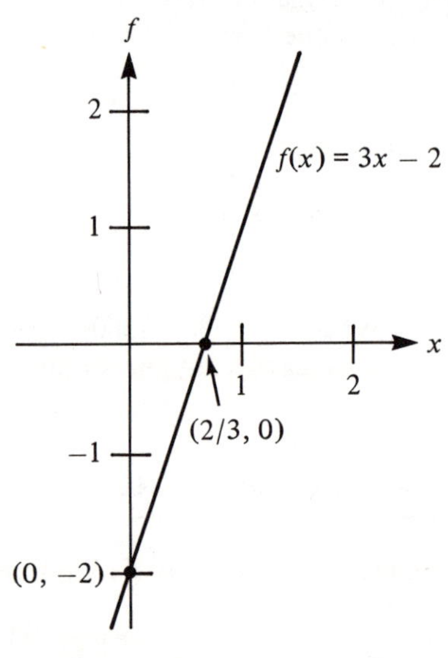

Figure 1.9

(*c*) Supply and demand are two economic forces whose interplay affects the state of an economy. Consider price p as a function of quantity q.

We say that market equilibrium exists when supply equals demand, that is, there exists an equilibrium price and quantity (see Figure 1.10). When this situation prevails the market place is said to enjoy stability. The supply and demand functions intersect at (q^*, p^*), called the point of equilibrium. This point is found by solving the system of equations S and D in two "unknowns," q and p. In other words, we solve S and D simultaneously. For example, consider the system

$$\left.\begin{aligned} D&: \quad 3q + 2p = 12 \\ S&: \quad q - p = -1 \end{aligned}\right\}$$

From S we get

$$q = p - 1.$$

Substituting in D, we have

$$\begin{aligned} 3(p - 1) + 2p &= 12 \\ 3p - 3 + 2p &= 12 \\ 5p &= 15 \\ p &= 3 \end{aligned}$$

When $p = 3$, $q = 3 - 1 = 2$, so that $(q^*, p^*) = (2, 3)$. Figure 1.11 illustrates this solution.

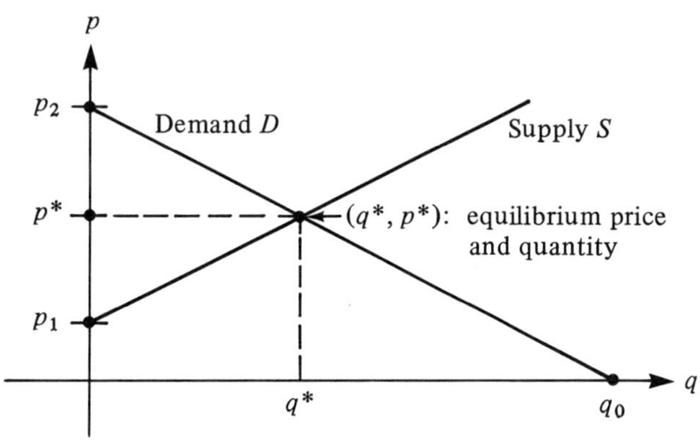

Figure 1.10

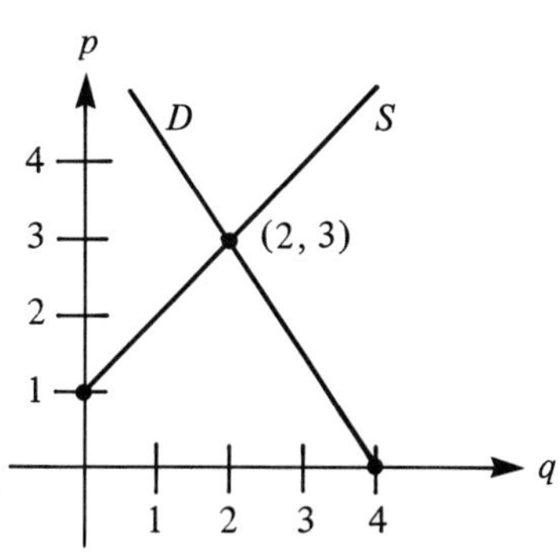

Figure 1.11

An alternate method for solving the system is to multiply S by 2, thus making the coefficients of p numerically the same for S and D. Then

$$\left.\begin{aligned} 3q + 2p &= 12 \\ 2q - 2p &= -2 \end{aligned}\right\}$$

Adding: $5q = 10$
Solving: $q = 2$
Substituting in S: $2 - p = -1$
$p = 3,$ as before.

(*d*) Let $C(q) = \frac{1}{10}q^2 + 2$, $0 \le q \le 8$, where q denotes the quantity of items produced and C is the cost of production. We find some points for graphing purposes:

$$C(0) = 2$$

$$C(2) = \frac{1}{10} \cdot 2^2 + 2 = 2.4$$

$$C(4) = \frac{1}{10} \cdot 4^2 + 2 = 3.6$$

$$C(6) = \frac{1}{10} \cdot 6^2 + 2 = 5.6$$

$$C(8) = \frac{1}{10} \cdot 8^2 + 2 = 8.4.$$

See Figure 1.12. The solid portion of the graph of C reflects values of C in [0, 8]. The dotted portion of the graph is shown to remind the reader that the graph of $(q^2/10) + 2$ is a **parabola** if the domain is all real numbers. The function $C(q) = (q^2/10) + 2$ is a special form of the general equation $f(q) = aq^2 + bq + c$, called a **quadratic function**. The lowest (or highest) point of a parabola which opens upward (or downward) is called a **vertex**. The vertex of the graph of C in Figure 1.12 is the point (0, 2). Some examples of quadratic functions are:

$$10x^2, \qquad t^2 + 5, \qquad 2y^2 + 3y, \qquad 3x^2 + \frac{3x}{4} - 1.$$

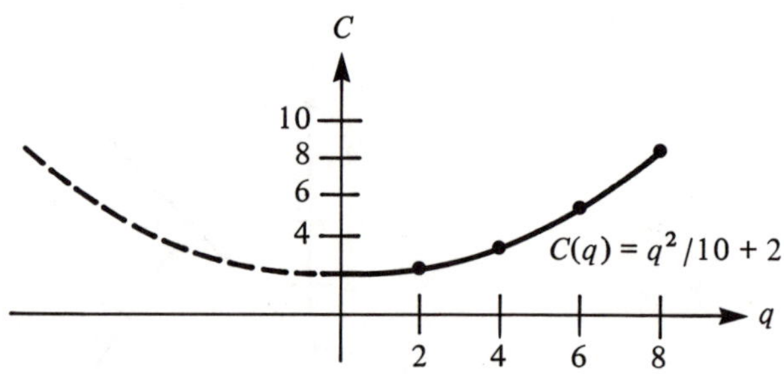

Figure 1.12

(*e*) A function and its domain are given by $y = f(x) = 1/x$, $x \neq 0$. This is equivalent to $xy = 1$ and $x = 1/y$. We see that x and y cannot

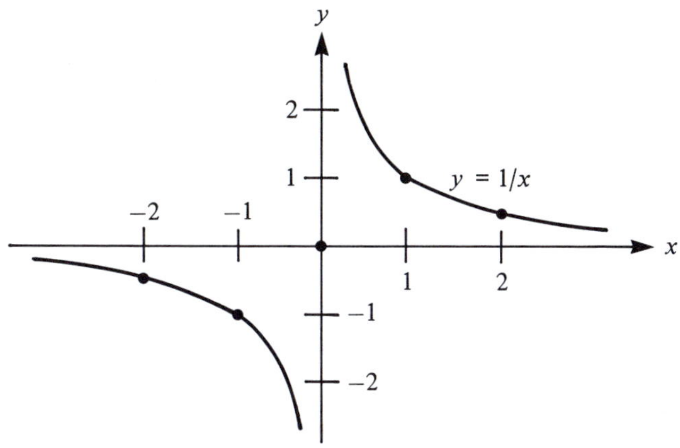

Figure 1.13

be zero since division by 0 is not permitted. We find a few points: $f(1) = 1$, $f(2) = 1/2$, $f(-1) = -1$, $f(-2) = -1/2$. As x approaches 0 from the right, through $x = 2$, $x = 1$, $x = \frac{1}{2}$, $x = \frac{1}{4}$, $x = \frac{1}{8}$, $x = \frac{1}{100}$, etc., then y becomes infinitely large in the positive direction. As x approaches 0 from the left, through the negative numbers, y becomes infinitely large in the negative direction. Refer to Figure 1.13. We say that $x = 0$, the y-axis, is a **vertical asymptote** and that $y = 0$, the x-axis, is a **horizontal asymptote**. The graph of $y = 1/x$ is called a **hyperbola**.

DEFINITION *If the limiting value of a function f is the number b as x increases indefinitely in the positive or negative direction, then the line y = b is called a* ***horizontal asymptote*** *of the graph of f. Similarly, if f becomes infinitely large, positively or negatively, as x approaches c from the left or right, then the line x = c is called a* ***vertical asymptote*** *of the graph of f.*

As illustrations, if $y = 1/(x - 1)$ then $x = 1$ is a vertical asymptote. If $y = 2x/(x + 2)$, then $xy + 2y = 2x$, $2x - xy = 2y$, $(2 - y)x = 2y$, $x = 2y/(2 - y)$. From this form we see that $y = 2$ is a horizontal asymptote.

(*f*) The square root function $f(x) = \sqrt{x}$ is defined on the interval $[0, \infty)$ and has the following points on its graph: (0, 0), (1, 1), $(2, \sqrt{2})$, (4, 2). The range of f is all nonnegative real numbers (Figure 1.14).

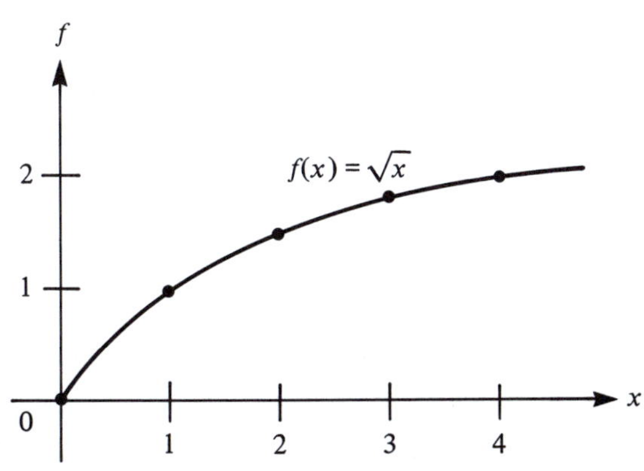

Figure 1.14

(*g*) The graph of the absolute value function $f(x) = |x|$, $-\infty < x < \infty$, consists of two line segments. One bisects quadrant I, the other bisects quadrant II, and they meet at the origin (see Figure 1.15). The range of f is the interval $[0, \infty)$.

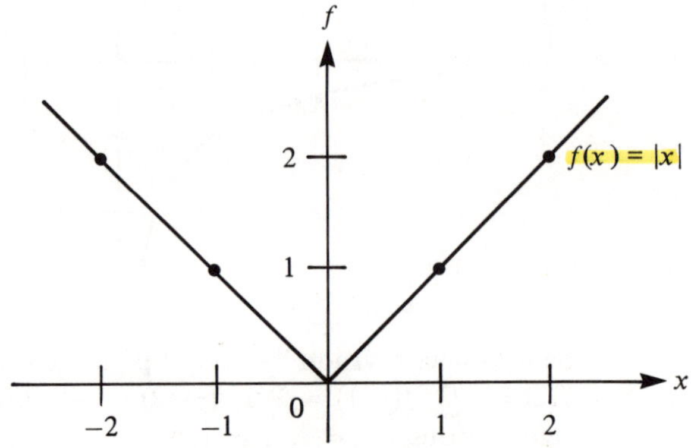

Figure 1.15

A knowledge of **symmetry** can be an aid in graphing. If $y = x^2$ then x can be replaced by $-x$ without altering the form or meaning of the function. This means the graph of $y = x^2$ is symmetric about the y-axis. In general, a parabola is symmetric about the vertical line which passes through its high (or low) point, i.e., the vertex. This vertical line is called the **axis of symmetry**. If we could fold the plane, like a sheet of paper, along the axis of symmetry, then the right and left halves of the parabola would coincide (see Figure 1.16). Thus, if $f(x) = x^2$, then $f(-x) = f(x)$. When this property applies, a function is said to be an **even function**. Examples of other even functions are: $1 - x^2$, $x^4 + 3x^2$, $1/x^2$.

If $f(x) = x^3$, then $f(-x) = (-x)^3 = -x^3 = -f(x)$. Because of this property, f is called an **odd function**. In general, if $f(-x) = -f(x)$, then f is an odd function. When this property applies we say the function has **symmetry about the origin** (see Figure 1.17). This means that each

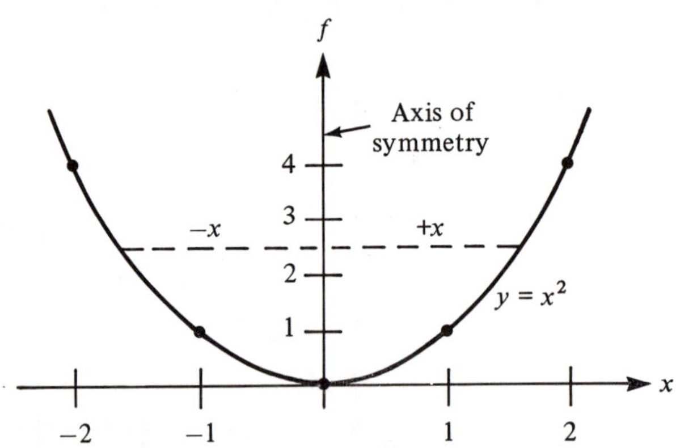

Figure 1.16 Reflection in the y-axis.

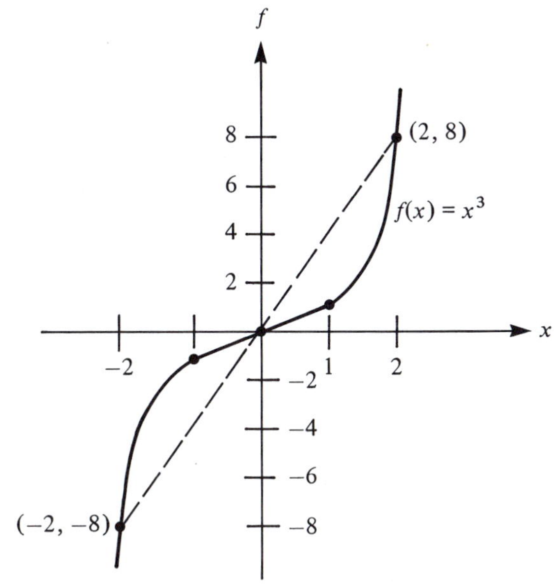

Figure 1.17 Reflection in the origin.

point on the curve of $y = x^3$ in quadrant I has its image in quadrant III. For example, $f(-1) = -1 = -f(1)$; $f(-2) = -8 = -f(2)$, etc. Some other odd functions are: $1/x$, $x - x^3$, $x/(x^2 - 1)$.

Let us investigate other graphing aids. Consider $f(x) = x^n$, $n > 0$. As n increases in value, the graph of f gets closer to the x-axis in the interval $[0, 1]$, and we say the graph becomes more **concave** on $[0, 1]$. This behavior is exhibited in Figure 1.18 for the functions x, x^2, x^3, and x^4.

As n decreases in value, the graph of $f(x) = x^n$, $n > 0$, gets farther from the x-axis in $[0, 1]$. That is, the graph becomes more **convex** in $[0, 1]$ as n gets smaller. This is shown in Figure 1.19 for the functions x, $x^{2/3}$, $x^{1/2}$, and $x^{1/3}$. We explore the concept of concavity a bit further.

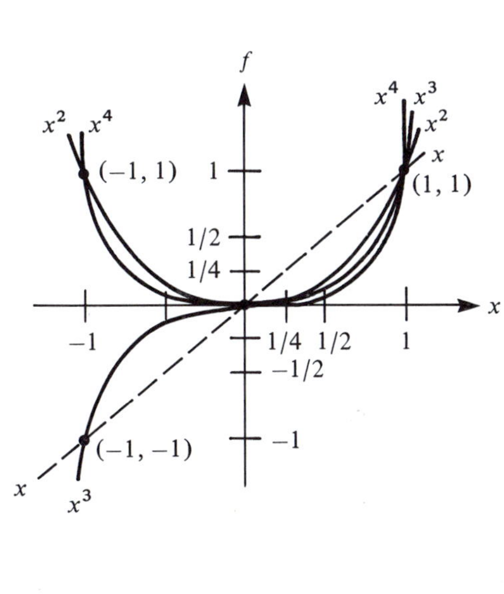

Figure 1.18

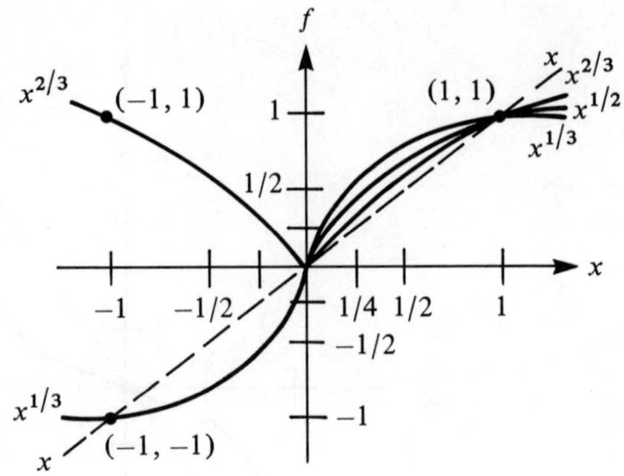

Figure 1.19

EXAMPLES **(a)** The graph of $y = 2x^2$ is a parabola opening up; it is concave upward.

(b) The graph of $y = -2x^2$ is a parabola opening down; it is concave downward.

(c) The graph of $y = -x^3$ is concave upward for $x < 0$ and concave downward for $x > 0$.

The graphs in Examples (a), (b), and (c) are shown in Figure 1.20.

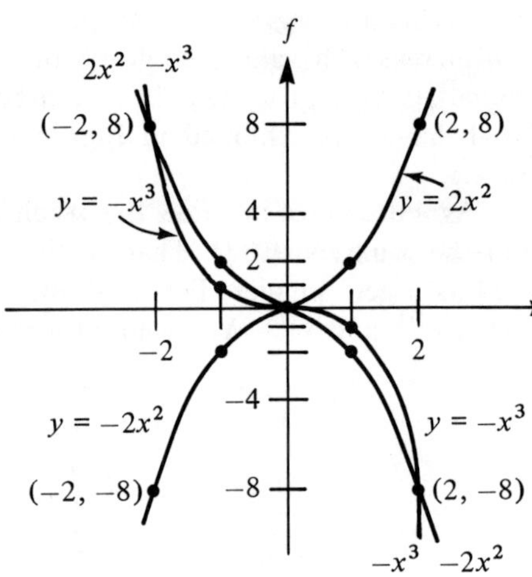

Figure 1.20

PROBLEM SET 1.2, 1.3

1. Describe the graphs of the given sets of points:

(a) $\{(x, y)\|x = 3/2\}$	**(b)** $\{(x, y)\|y = 6\}$
(c) $\{(x, y)\|y = 5x\}$	**(d)** $\{(x, y)\|y = 3x + 7\}$
(e) $\{(x, y)\|y = x^2 + 6\}$	**(f)** $\{(x, y)\|y = 1 - x^2\}$
(g) $\{(t, u)\|u = 2\sqrt{t}\}$	**(h)** $\{(q, p)\|p = 1/2q\}$

2. If f is the squaring function, $f(x) = x^2$, compute:

(a) $f(-3)$ **(b)** $f(\frac{2}{3})$ **(c)** $f(1.2)$

(d) $f(1.1) - f(1)$ **(e)** $\dfrac{f(1.1) - f(1)}{1.1 - 1}$

3. If g is the square root function, $g(t) = \sqrt{t}$, compute:

(a) $g(9)$ **(b)** $g(\frac{16}{25})$ **(c)** $g(0.04)$

(d) $g(1.21) - g(1)$ **(e)** $\dfrac{g(1.21) - g(1)}{1.21 - 1}$

4. Let $h(x) = 3x - 4$. Show that $\dfrac{h(b) - h(a)}{b - a} = 3$.

5. Assuming $F(u) = 3u^2$, verify that $\dfrac{F(t_2) - F(t_1)}{t_2 - t_1} = 6.3$ when $t_2 = 1.1$, $t_1 = 1$.

6. If $G(x) = 1/x$, show that $\dfrac{G(x + h) - G(x)}{h} = -\dfrac{1}{(x + h)x}$.

7. Assuming $f(x) = 1/(x + 1)$, find:

(a) the domain and range of f; **(b)** $f(a - 1)$ **(c)** $f\left(\dfrac{1}{x + 1}\right)$.

8. If $g(t) = 3t - 2$, complete the table:

t	-2.5	-1.5	0	1.5	2.5
$g(t)$					

9. Sketch graphs of:

(a) $y = 2x + 5$ **(b)** $y = 4 - 3x$ **(c)** $f(t) = 3t^2$
(d) $g(u) = 3u^3$ **(e)** $h(x) = 2 + \sqrt{x}$ **(f)** $F(u) = 2/u$

10. Describe the symmetry of the graphs of:

(a) $f(x) = 4 + x^2$ **(b)** $g(t) = 4t^3$ **(c)** $h(u) = u^4$
(d) $F(z) = 2z^2 + z^4$ **(e)** $G(x) = x^{2/3}$ **(f)** $H(t) = t^{1/3}$

11. Sketch the graph of each function, state the domain and range, and name the asymptotes.

(a) $y = -3/x$ **(b)** $y = -3x^2 + 1$ **(c)** $y = -2x^3 + 1$
(d) $f(x) = 1/(x - 3)$ **(e)** $g(t) = |t - 1|$ **(f)** $h(u) = \sqrt{u - 1}$

12. Solve the given systems:

(a) $\left.\begin{aligned} x + y &= 4 \\ 2x - y &= -1 \end{aligned}\right\}$ **(b)** $\left.\begin{aligned} 2x + y &= 5 \\ 3x - 2y &= 4 \end{aligned}\right\}$

(c) $\left.\begin{aligned} u + 3v &= 12 \\ -2u + 6v &= 0 \end{aligned}\right\}$ **(d)** $\left.\begin{aligned} \frac{2}{3}z + \frac{1}{3}w &= 7 \\ \frac{z}{6} + \frac{w}{9} &= 2 \end{aligned}\right\}$

13. Find the point of equilibrium (q^*, p^*) for the following supply-demand systems:

(a) $\left.\begin{aligned} D\text{:}\quad 5q + 8p &= 80 \\ S\text{:}\quad -q + 4p &= 12 \end{aligned}\right\}$ **(b)** $\left.\begin{aligned} D\text{:}\quad 3q + 4p &= 24 \\ S\text{:}\quad q - 3p &= -5 \end{aligned}\right\}$

(c) $\left.\begin{aligned} D\text{:}\quad 4q + 3p &= 26 \\ S\text{:}\quad 8q - 8p &= 24 \end{aligned}\right\}$ **(d)** $\left.\begin{aligned} D\text{:}\quad 6q + 5p &= 66 \\ S\text{:}\quad 4q - p &= 18 \end{aligned}\right\}$

1.4 FORMING NEW FUNCTIONS

In this section we will consider various processes for determining a single function from two (or more) other functions. A certain operation may construct a "new" function from "old" functions. Another process may build a "chain" of functions from given functions. Still a third action may transform a function into another which is the image of the original function when "reflected" in some line. In any case, let us contemplate the forming of new functions.

(*a*) ***Addition*** If f and g are functions having the same domain D, then their sum $f(x) + g(x)$ is also a function defined on D, and is written

$$[f + g](x) = f(x) + g(x).$$

EXAMPLE If $f(x) = x + 3$ and $g(x) = x^2 - 2x + 3$, then

$$\begin{aligned}[f + g](x) &= (x + 3) + (x^2 - 2x + 3)\\ &= x^2 - x + 6.\end{aligned}$$

If the domain D_f of f and the domain D_g of g are not the same but D_f and D_g have points in common, then $[f + g](x)$ is defined on the common points.

EXAMPLE If $f(x) = x - 2$ and $g(x) = \sqrt{x}$ then

$$\begin{aligned}f - g = f + (-g) &= [f + (-g)](x)\\ &= x - 2 - \sqrt{x}.\end{aligned}$$

Recall that subtraction is defined in terms of addition. The domain of f is all real numbers whereas the domain of g is all nonnegative real numbers. The domain of $f - g$ is the set of points common to both domains. Therefore, the function $f - g$ is defined for $x \geq 0$ only.

(*b*) ***Multiplication*** Assume that $f(x)$ and $g(x)$ are defined on the same domain D. Then their product fg is also defined on D where

$$\boxed{[fg](x) = f(x) \cdot g(x)}.$$

EXAMPLE Let $f(x) = x - 2$ and $g(x) = x + 2$. Then

$$[fg](x) = (x - 2)(x + 2) = x^2 - 4.$$

EXAMPLE If $f(x) = x$ and $g(x) = \sqrt{x}$, then

$$[fg](x) = x \cdot x^{1/2} = x^{3/2}.$$

The domain of g is the set of nonnegative real numbers. This set of points is common to the domains of both f and g. So, the domain of $fg = x^{3/2}$ is $[0, \infty)$. This is verified by the graph of fg in Figure 1.21.

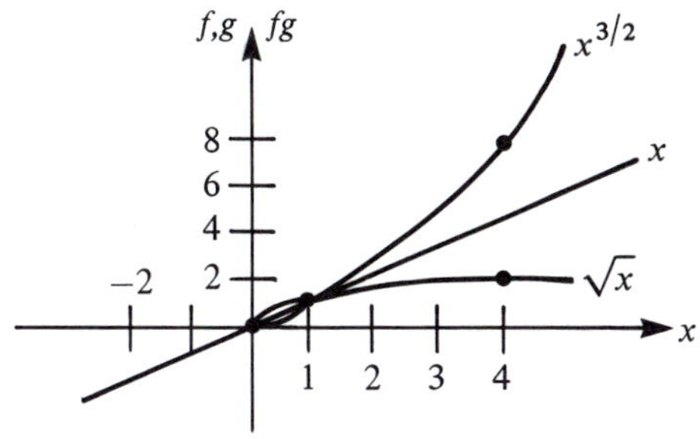

Figure 1.21

(c) ***Composition*** If f and g are functions such that the range of g lies in the domain of f, then the composition f of g is given by

$$\boxed{[f \circ g](x) = f[g(x)]}\,.$$

Notice that the process of composition effects a "chain," linking two functions. The domain of $f \circ g$ is the domain of g.

EXAMPLE Given $f(x) = x^2 - 1$ and $g(x) = \sqrt{1 - x^2}$. Since f of any number equals that number squared minus 1, then the composition

$$\begin{aligned} f[g(x)] &= [g(x)]^2 - 1 \\ &= (\sqrt{1 - x^2})^2 - 1 \\ &= 1 - x^2 - 1 \\ &= -x^2. \end{aligned}$$

The graph of g is a semicircle (the upper half) with radius 1. It follows that the domain of g is $[-1, 1]$. This interval also defines the domain of $f[g(x)]$ (see Figure 1.22).

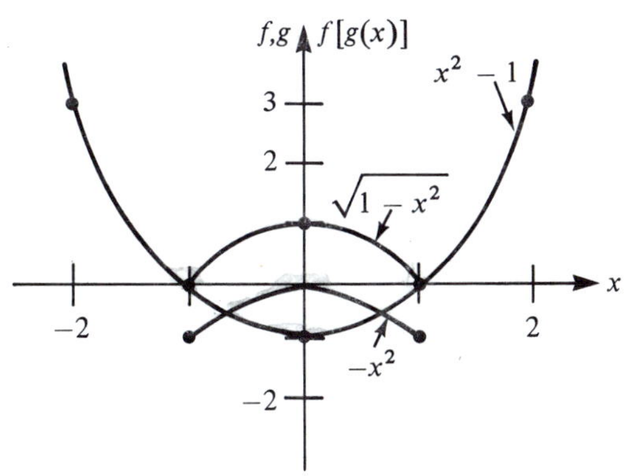

Figure 1.22

EXAMPLE If $f(x) = \sqrt{x}$ and $g(x) = -2$ then

$$f[g(x)] = \sqrt{g(x)} = \sqrt{-2},$$

which is not defined in the real number system. Since $f[g(x)]$ does not exist, it makes little sense to discuss its domain.

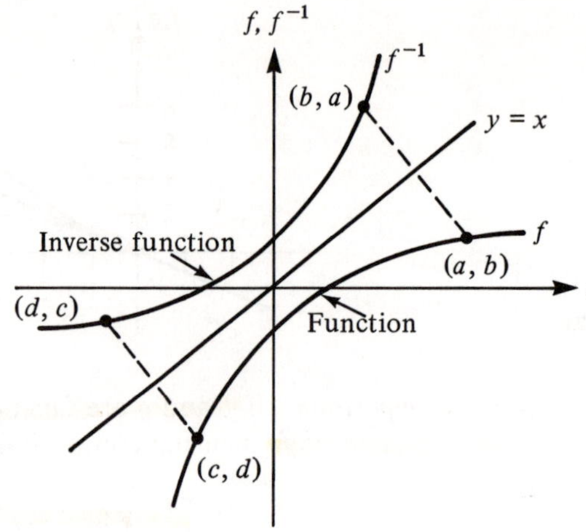

Figure 1.23

(d) ***Inverse Functions*** When we reflect the graph of a function $y = f(x)$ in the line $y = x$, we say the **image** of f is the graph of $y = f^{-1}(x)$. The symbol f^{-1} denotes the inverse function of f and does *not* at all mean the reciprocal $1/f$. This reflective property means that the points $(x, y) \in f$ and $(y, x) \in f^{-1}$ are symmetric about the line $y = x$. Thus, if (a, b) is on the graph of f then (b, a) is on the graph of f^{-1}, the inverse function of f (see Figure 1.23).

EXAMPLE If $y = f(x) = 2x - 1$, we can find the inverse function by first interchanging x and y to obtain

$$x = 2y - 1.$$

Solving for y, we have

$$y = \frac{x + 1}{2} = f^{-1}(x),$$

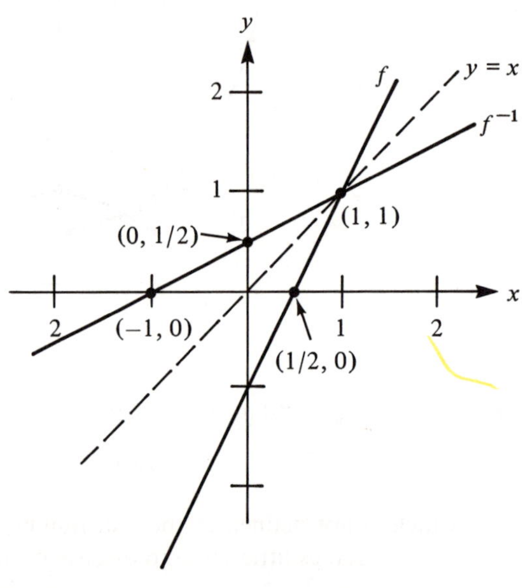

Figure 1.24

which represents a function of x, specifically, the inverse function $f^{-1}(x)$. The graphs of f and f^{-1} are shown on one cc system in Figure 1.24. The domain of f is the range of f^{-1}, as can be seen from the figure. For instance, the points $(0, -1)$, $(\frac{1}{2}, 0)$, $(1, 1)$ are on the graph of f while the points $(-1, 0)$, $(0, \frac{1}{2})$, $(1, 1)$ are on the graph of f^{-1}.

EXAMPLE To find the inverse function of $y = f(x) = x^2 + 1$, $x \geq 0$, we first interchange x and y:

$$x = y^2 + 1.$$

Solving for y, we get

$$y = \pm\sqrt{x - 1}.$$

But $y = -\sqrt{x - 1}$ is not relevant here. Its graph does not represent a reflection of f in the line $y = x$ because f is defined only for $x \geq 0$. Hence,

$$y = \sqrt{x - 1} = f^{-1}(x).$$

The graphs of f and f^{-1} are drawn in Figure 1.25. This illustration introduces an important point: If a function f does not have an inverse function, we can restrict the domain of f so that the restricted form will have an inverse that is a function.

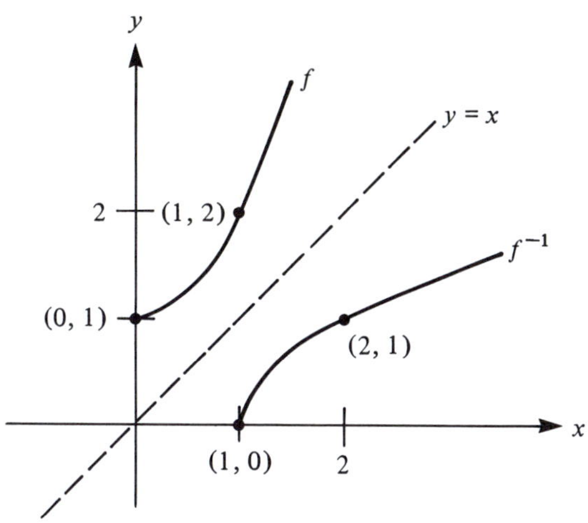

Figure 1.25

An interesting property of inverse functions is that

$$f[f^{-1}(x)] = f^{-1}[f(x)] = x.$$

This property provides us with a tool for checking. Applying this to $f(x) = 2x - 1$ and $f^{-1}(x) = (x + 1)/2$ from a previous example, we obtain

$$\begin{aligned} f[f^{-1}(x)] &= 2f^{-1}(x) - 1 \\ &= 2\left(\frac{x+1}{2}\right) - 1 \\ &= x + 1 - 1 = x \end{aligned}$$

and
$$f^{-1}[f(x)] = \frac{f(x) + 1}{2} = \frac{2x - 1 + 1}{2} = x.$$

For $f(x) = x^2 + 1, x \geq 0$, and $f^{-1}(x) = \sqrt{x - 1}$, we get

$$\begin{aligned} f[f^{-1}(x)] &= (f^{-1}(x))^2 + 1 \\ &= (\sqrt{x - 1})^2 + 1 \\ &= x - 1 + 1 = x \end{aligned}$$

and
$$\begin{aligned} f^{-1}[f(x)] &= \sqrt{f(x) - 1} \\ &= \sqrt{x^2 + 1 - 1} \\ &= \sqrt{x^2} = |x| = x \quad \text{for } x \geq 0. \end{aligned}$$

Other illustrations of algebraic functions and their inverses are as follows.

Function f	*Inverse function f^{-1} of f*
$3x + 4$	$\frac{x - 4}{3}$
$2x^2 + 3, \quad x \geq 0$	$\sqrt{\frac{x - 3}{2}}$
x^3	$x^{1/3}$
$\frac{1}{x}, \quad x \neq 0, \quad y \neq 0$	$\frac{1}{x}, \quad x \neq 0, \quad y \neq 0$

(e) Rational Functions. Miscellaneous For the most part we have discussed simple algebraic functions which basically fit the formula

$$\boxed{P(x) = a_0 + a_1x + a_2x^2 + \cdots + a_nx^n}.$$

This expression is called a **polynomial function** P; n is an integer, positive or zero, and the coefficients $a_0, a_1, a_2, \ldots, a_n$ are real numbers (for our purposes). The domain of P is the set of real numbers **R** and its range is a subset of **R**. If $a_n \neq 0$ we say that P is of degree n. If $n = 0$, $P(x) = a_0$; if $n = 1$, $P(x) = a_0 + a_1x$; if $n = 2$, $P(x) = a_0 + a_1x + a_2x^2$, and so on. Examples of polynomial functions are:

(a) $P(x) = 2x^2 + 3x - \frac{1}{2}$; degree 2; $a_2 = 2$; $a_1 = 3$; $a_0 = -\frac{1}{2}$

(b) $P(x) = 4$; degree 0; $a_0 = 4$

(c) $P(x) = x^7$; degree 7; $a_7 = 1$; $a_0 = a_1 = \cdots = a_6 = 0$

Examples which are not polynomials are:

(a) $f(x) = x^{1/3}$, because n is not an integer.

(b) $g(x) = 1/x^2$, because n is not a nonnegative integer.

We say that

$f(x) = 5$ is a constant polynomial of zero degree;

$f(x) = 2 + 3x$ is a linear polynomial of first degree;

$f(x) = 1 - 2x + x^2$ is a quadratic polynomial of second degree;

$f(x) = 2 + 3x - 4x^2 + 5x^3$ is a cubic polynomial of third degree;

and so on. Of course, we may have a cubic polynomial, for example, where all terms are missing except the term of highest degree, 3. More precisely, we say that P is a cubic polynomial provided $n = 3$ and $a_n \neq 0$. The sum, product, or composition of any two functions, $f(x)$ and $g(x)$, is in turn a polynomial. But the quotient of f and g may or may not be a polynomial. This brings us to the subject of forming rational functions.

The quotient (or ratio) of two polynomials, such as

$$f(x) = \frac{x - 2}{x - 1}, \qquad x \neq 1,$$

is called a **rational function**. The domain and range of f are all the reals except the number 1. We see that $x = 1$ is a vertical asymptote. There is another asymptote that is not so apparent but is easy to find. Let $y = f(x) = (x - 2)/(x - 1)$; then

$$\begin{aligned} xy - y &= x - 2 \\ xy - x &= y - 2 \\ (y - 1)x &= y - 2 \\ x &= \frac{y - 2}{y - 1}. \end{aligned}$$

In this form the function exhibits $y = 1$ as the horizontal asymptote. Figure 1.26 illustrates this function.

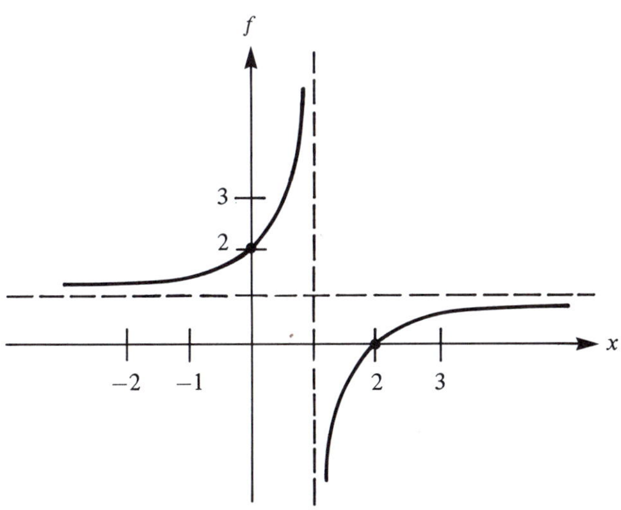

Figure 1.26

In general, if a function Q is defined by

$$y = Q(x) = \frac{P_1(x)}{P_2(x)}, \qquad P_2 \neq 0,$$

where P_1 and P_2 are polynomials, then we call $Q(x)$ a rational function. We assume that P_1 and P_2 have no common factors. The domain of $Q(x)$ is the set of all reals except those which are real **zeros** of P_2. The values of x for which a polynomial $P(x)$ equals 0 are called the zeros of P. For instance, if

$$P(x) = x^2 - 1,$$

then $x = 1$ and $x = -1$ are zeros of P, since $P(\pm 1) = 0$. Furthermore, the zeros of the polynomial P are the **roots** of the equation $P(x) = 0$. Thus, if

$$P(x) = 2x - x^2 = 0,$$

then $x = 0$ and $x = 2$ are the roots of the given equation. This can be verified as follows:

$$P(x) = 2x - x^2 = 0$$
$$x(2 - x) = 0$$
$$\left.\begin{array}{r} x = 0 \\ 2 - x = 0 \end{array}\right\}$$

[*Theorem*: If $ab = 0$ then $a = 0$ or $b = 0$]

Solving for x: $\quad x = 2.$

It can be shown that every polynomial equation $P(x) = 0$ of degree $n \geq 1$ has exactly n roots (real roots for our purposes). There is another property of roots that is useful to know:

Property *If $x = r$ is a root of $P(x) = 0$, then $(x - r)$ is a **factor** of the polynomial P.*

To illustrate, since $x = 3$ and $x = -3$ are roots of

$$P(x) = x^2 - 9 = 0,$$

it follows that $(x - 3)$ and $(x + 3)$ are factors of P. The graph of $x^2 - 9$ is a parabola, concave upward (Figure 1.27).

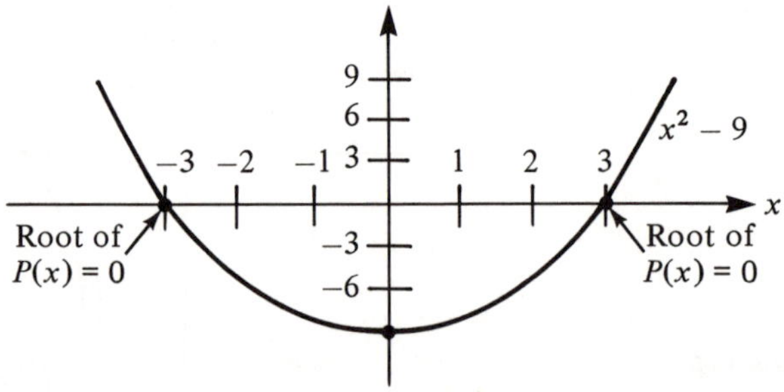

Figure 1.27

Let us consider the zeros of the quadratic polynomial $f(x) = ax^2 + bx + c, a > 0$. Figure 1.28 illustrates the three important cases. The roots of a quadratic equation $f(x) = 0$ are given by the quadratic formula:

$$x = \frac{-b \pm \sqrt{b^2 - 4ac}}{2a}.$$

The sign of $b^2 - 4ac$ determines the nature of the roots of $f(x) = 0$ (or the zeros of the polynomial $f(x)$). Let $D = b^2 - 4ac$. Then

Case (a) If $D > 0$, f has exactly two real zeros.

Case (b) If $D = 0$, f has one real zero.

Case (c) If $D < 0$, f has no real zeros.

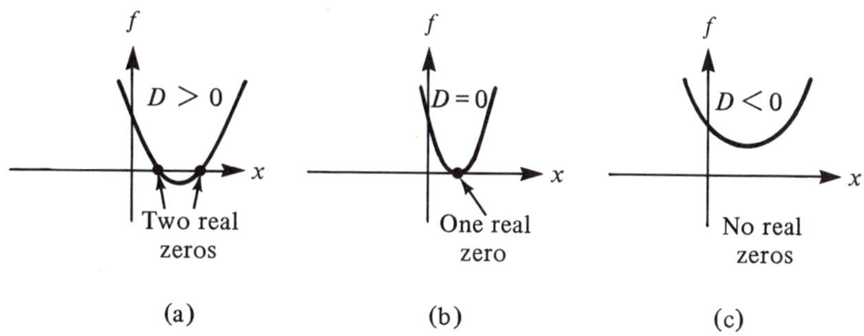

Figure 1.28 The zeros of $f(x) = ax^2 + bx + c, a > 0$.

The zeros of rational functions can also be found. If

$$f(x) = \frac{x - x^2}{x + 1}, \qquad x \neq -1,$$

then $f(x) = 0$ when $x - x^2 = 0$, $x(1 - x) = 0$, $x = 0$, $x = 1$. Thus, the zeros of f are $x = 0$ and $x = 1$.

It may be useful to know how to divide one polynomial by another. Consider the rational function

$$g(x) = \frac{x^2}{x - 2}, \qquad x \neq 2.$$

The division of x^2 by $x - 2$ involves the following procedure:

$$\begin{array}{lr|ll} & & x \;\; + 2: & \text{Quotient} \\ \text{Divisor:} & x - 2 & x^2 \qquad : & \text{Dividend} \\ & & \underline{x^2 - 2x} & \\ & & 2x & \\ & & \underline{2x - 4} & \\ & & 4: & \text{Remainder} \end{array}$$

Therefore, $$g(x) = \frac{x^2}{x - 2} = x + 2 + \frac{4}{x - 2}.$$

This process is sometimes called the **decomposition** of a rational function into a sum (or difference) of "simpler" functions. It is a handy tool for graphing and for making integration (in calculus) easier. Returning to function g above, as x approaches infinity the fraction $4/(x-2)$ approaches 0. This means that $g(x)$ approaches $x+2$ as x becomes infinitely large. Therefore, the graph of g approaches the line $x+2$, which acts as an **oblique asymptote** (refer to Figure 1.29).

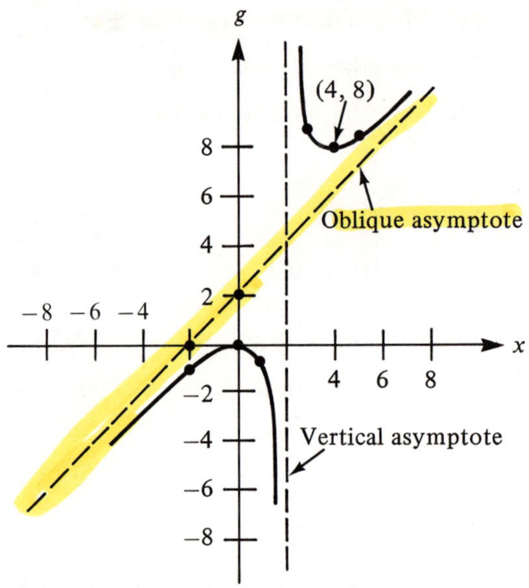

Figure 1.29

PROBLEM SET 1.4

1. If $f(x) = 4x + 1$ and $g(x) = 1 - x^2$, find:
   (a) $[f + g](x)$ **(b)** $[fg](x)$ **(c)** $f[g(x)]$
2. If $f(x) = x^2 + x$ and $g(x) = x - 1$, find
   (a) $[f + g](x)$ **(b)** $[fg](x)$ **(c)** $f[g(x)]$
3. Let $g(t) = 2t^2$ and $h(t) = 4 - t^2$. Find:
   (a) $[h + g](t)$ **(b)** $[g - h](t)$ **(c)** $g[h(t)]$
   (d) the domain in each case
4. Determine $f[g(x)]$ if **(a)** $f(x) = x$ and $g(x)$ is any function; **(b)** $g(x) = x$ and $f(x)$ is any function.
5. If $f(x) = 2 - x$ and $g(x) = 2/x$ find:
   (a) $f[f(x)]$ **(b)** $g[g(x)]$
6. What can you say about $f \circ g$ if $f(x) = \sqrt{x}$ and $g(x) = -1 - x^2$? About $g \circ f$? About $g \circ g$?
7. Show that if $f(u) = 3u$, then $f(a + b) = f(a) + f(b)$.
8. Verify: If $g(t) = t^2$, then $g(a - b) \neq g(a) - g(b)$.

9. **(a)** Find the inverse function f^{-1} of f; **(b)** sketch the graphs of f^{-1} and f; **(c)** state the domain and range of $f^{-1}(x)$:

(I) $f(x) = 3x - 2$ **(II)** $f(x) = \sqrt{x}$ **(III)** $f(x) = x^3, x \geq 0$
(IV) $f(x) = x^2 + 2, x \geq 0$ **(V)** $f(x) = 1 - x^2, x \leq 0.$

10. Which of the given functions have inverses that are functions?

(a) $1 - x$ **(b)** $1 - x^2$ **(c)** $1 - x^3$

11. Find the inverse function g^{-1} of g:

(a) $g(x) = 2x - \frac{1}{2}$ **(b)** $g(x) = \sqrt{x - 2}$

(c) $g(x) = \dfrac{1}{x}, x > 0$ **(d)** $g(x) = \dfrac{1}{x - 2}, x > 2$

12. Find all the asymptotes and sketch graphs of:

(a) $f(x) = \dfrac{1}{x - 3}$ **(b)** $g(x) = \dfrac{2}{x - 1}$ **(c)** $h(x) = \dfrac{x}{x - 1}$

13. Sketch graphs of:

(a) $f(x) = \dfrac{1}{2x - 1}$ **(b)** $g(x) = \dfrac{2}{2 - x}$ **(c)** $h(x) = \dfrac{x}{x - 2}$

(d) $F(x) = \dfrac{x - 1}{x - 3}$ **(e)** $G(x) = \dfrac{x^2}{x - 1}$

14. Find the zeros of the following polynomials:

(a) $3x - 6$ **(b)** $2x - \frac{3}{2}$ **(c)** $2(x - 1) + 3(x + 2)$

15. Find all zeros of the given polynomials:

(a) $x^2 - 16$ **(b)** $2x^2 - 50$ **(c)** $(x - 2)(x - 3)$

16. Determine all zeros of:

(a) $(x - 1)(2x + 3)$ **(b)** $x^3 - 8$ **(c)** $x(x^2 - 4)$

17. Find the roots of:

(a) $2x - \frac{1}{2} = 0$ **(b)** $3x^2 - 27 = 0$ **(c)** $x^3 - 64 = 0$

(d) $\dfrac{2}{x} = 3$ **(e)** $\sqrt{x - 1} = 2$ **(f)** $\dfrac{1}{x^2} = \dfrac{1}{9}$

18. Decompose each rational function if possible and sketch its graph:

(a) $f(x) = \dfrac{x - 1}{x - 2}$ **(b)** $f(x) = \dfrac{x + 2}{x - 3}$ **(c)** $f(x) = \dfrac{x + 3}{x - 1}$

(d) $g(x) = \dfrac{x^2}{x - 1}$ **(e)** $h(x) = \dfrac{x^2}{x - 3}$ **(f)** $F(x) = \dfrac{2x^2}{x - 2}$

1.5 REVIEW PROBLEMS

1. Express each statement as an inequality:

(a) Annual costs of a college education are more than $2000.
(b) I will pay no more than $30,000 for a house.
(c) The GNP (Gross National Product) is between 1.2 and 1.4 trillion dollars.

2. Define: **(a)** function **(b)** domain **(c)** range **(d)** absolute value **(e)** zero of a function.

3. Solve for x:

(a) $|x| = 2$ **(b)** $|x| < \frac{1}{2}$ **(c)** $|2x| > 3$

4. Find the domain and range of each function:

(a) $f(x) = 4x - 5$ **(b)** $y = 1 - \dfrac{x}{2}$

(c) $g(t) = t^2 - 4$ **(d)** $S = \{(1, 2), (2, 3), (3, 3)\}$

(e) $h(u) = 1 + \sqrt{u}$ **(f)** $y = |x - 2|$

(g) $F(x) = \dfrac{1}{2x - 3}$ **(h)** $u = 16 - x^2$

5. Let $f(t) = 4t^2 + 2t - 1$. Find:

(a) $f(2)$ **(b)** $f(-3)$

(c) $f(x + h)$ **(d)** $\dfrac{f(x + h) - f(x)}{h}$

6. If $f(x) = 1 - x^2$ and $g(x) = x - 1$, find:

(a) $f[g(x)]$ **(b)** $[fg](x)$ **(c)** $\dfrac{f(x)}{g(x)}$

7. Let $f(t) = 1 - 2t$ and $g(t) = \sqrt{1 - t}$. Find:

(a) $g[f(t)]$ **(b)** $g[f(\frac{1}{2})]$ **(c)** $f[g(1)]$

8. Find the distance between the given pairs of points:

(a) $(1, 2)$ and $(3, -1)$ **(b)** $(-2, 4)$ and $(3, 3)$

9. Find the slope of the line joining:

(a) $(3, -1)$ and $(-4, 5)$ **(b)** $(10, 20)$ and $(30, 10)$

10. Obtain an equation of the line through:

(a) $(4, 5)$ and $(-1, 6)$
(b) $(1, \frac{1}{2})$ and $(\frac{1}{3}, 1)$
(c) $(4, 8)$ and parallel to $2x - y = 3$
(d) $(6, 6)$ and parallel to $3x + 2y = 1$

11. Sketch graphs of:

(a) $y = 3x + 8$ **(b)** $y = (x/2) + 4$ **(c)** $y = 4x^2$
(d) $y = 2x^3$ **(e)** $y = 3x^4/2$

12. Determine the zeros of:

(a) $x^2 - \frac{4}{9}$ **(b)** $2x^2 - \frac{1}{8}$ **(c)** $1 - \sqrt{x}$
(d) $(x - 2)^2$ **(e)** $x^3 - \frac{8}{27}$

13. Sketch graphs of:

(a) $2x^2 - 18 = y$ **(b)** $12 - 3x^2 = y$
(c) $f(t) = (t - 1)^2$ **(d)** $g(z) = z^2 - 6z + 9$

14. Find all roots of the equations:

(a) $x - \frac{3}{2} = 2x$ **(b)** $4t^2 - 9 = 0$
(c) $x^2 - x - 2 = 0$ **(d)** $2x^3 - \frac{1}{4} = 0$

15. Sketch graphs of:

(a) $y = \sqrt{x - 4}$ **(b)** $y = 3 + \sqrt{x}$ **(c)** $f(x) = \sqrt[3]{x}$
(d) $g(t) = 1 + t^{2/3}$ **(e)** $h(u) = u^{3/2}$

16. Determine all roots of the equations:
(a) $2x^2 - 3x = 0$ (b) $x^2 - x^3 = 0$ (c) $x^2 - x^4 = 0$
(d) $(2x - 1)(x + 3) = 0$ (e) $(2x + 4)(3x - 6) = 0$

17. Sketch graphs of:
(a) $f(x) = \dfrac{1}{2x}$ (b) $g(t) = \dfrac{2}{3t}$ (c) $h(u) = \dfrac{1}{1 + u}$
(d) $F(x) = \dfrac{2}{1 - x}$ (e) $G(r) = \dfrac{r}{r - 1}$

18. Find all zeros of:
(a) $f(x) = \dfrac{2x - 4}{x - 1}$ (b) $g(x) = \dfrac{x^2}{x + 1}$ (c) $h(t) = \dfrac{t^2 - 4}{t + 3}$
(d) $F(x) = \dfrac{1 - x}{x + 2}$ (e) $G(t) = \dfrac{x^3 - 27}{2x - 1}$

19. If $f(x) = 1 + \sqrt{x}$, find the inverse function f^{-1} and sketch graphs of f and f^{-1}.

20. Find the inverse f^{-1} of $f(x) = x^2/2$, $x \geq 0$; sketch graphs of f and f^{-1}.

21. Decompose the following rational functions:
(a) $f(x) = \dfrac{x - 2}{x - 3}$ (b) $g(x) = \dfrac{x^2 - 4}{x - 1}$
(c) $h(x) = \dfrac{2x^2}{x + 1}$ (d) $F(x) = \dfrac{3x^2}{x - 4}$

22. Sketch graphs of:
(a) $f(x) = \dfrac{x - 1}{2x - 4}$ (b) $g(x) = \dfrac{4x^2}{2x - 6}$

1.6 ENRICHMENT TOPICS AND PROBLEMS

1. If $A = \{0, 1, 2, 3\}$, $B = \{2, 3, 4, 5\}$, and $U = \{0, 1, 2, 3, 4, 5, 6\}$, find:
(a) $A \cup B$ (b) $A \cap B$ (c) $A' \cap B'$

2. (a) Add: $x^2 - 2x$, $2 + 3x$, $1 - 3x^2$
(b) Subtract $a^2 - 2ab + 4$ from $2a^2 + 3ab - 6$

3. Multiply:
(a) $x(x - 1)(x + 2)$ (b) $(2a - b)(a + 3b)$

4. Factor:
(a) $5a^2 - 20b^2$ (b) $c^2 - 2c - 8$
(c) $y^2 - 5y + 6$ (d) $x(a - b) + y(a - b)$
(e) $4a^2 - (x - y)^2$ (f) $a + ab - c - bc$

5. Divide:
(a) $18x^2y^3$ by $3xy$ (b) $4a^2 - 9b^2$ by $2a - 3b$
(c) $2x^3 - 3x^2 + 1$ by $2x + 1$ (d) $x^4 + 4x^2 - 1$ by $x^2 - 2$

6. Reduce to simplest form:
(a) $\dfrac{84}{16}$ (b) $\dfrac{45x^2}{9x}$

(c) $\frac{5}{4} - 3 + \frac{5}{6}$ (d) $\frac{1}{x} + \frac{1}{x-1} + \frac{1}{x(x-1)}$

(e) $\frac{2}{x^2 - 4} - \frac{3}{x-2}$ (f) $\frac{2x+3}{4} - \frac{3x-4}{12}$

7. Perform the indicated operations:

(a) $44 \div 2\frac{3}{4}$ (b) $\frac{2x}{3y} \times \frac{9y}{4} \times \frac{6y}{3xy}$

(c) $\left(1 - \frac{d}{c-d}\right)\left(c + \frac{d^2}{c-2d}\right)$ (d) $\left(1 - \frac{2}{x+2}\right) \div \left(\frac{x}{4-x^2}\right)$

8. Simplify: $\left(1 + \frac{1}{2} - \frac{1}{3}\right) \div \left(1 + \frac{2}{3} - \frac{1}{4}\right)$

9. Solve for x:

(a) $\frac{x}{x-1} = \frac{2x}{x+1}$ (b) $\frac{2x}{x-1} = 2 - \frac{1}{x}$

10. A radio sold for \$100. The profit realized was $\frac{3}{7}$ of its cost. What was the cost of the radio?

11. The sum of \$2430 is invested in securities E and F. Security E pays annual dividends at the rate of 5% and F pays 6%. If total annual dividends are \$134, find the amount invested in each security.

12. Of a man's income, $\frac{1}{12}$ is spent on expense A, $\frac{2}{15}$ for expense B, and $\frac{3}{20}$ for expense C. If \$570 is left, find the man's income.

13. Beth and Linda sold candy bars at 5¢ and 10¢ per bar. In five days they sold 990 bars for a total amount of \$70.75. How many bars of each kind were sold?

14. How many pounds each of two kinds of coffee, priced at \$0.90 and \$1.20 per pound, must be used to get 30 pounds of a mixture priced at \$1 per pound?

15. If cost C is defined by $|C - \$50| < \$10$, find the range of values of C.

16. Solve the inequalities:

(a) $2(t+1) < 3(t-1)$ (b) $\frac{2t+3}{3} > \frac{3t-1}{2}$

(c) $|x| > |x-2|$

17. A closed box is to be made with square base and volume V. Let x be the length of a side of the base and y be the height of the box. Express the total area A as a function of x. [*Hint*: $A = 2x^2 + 4xy$; $V = x^2y$.]

18. If profit P depends on total sales revenue r according to the rule $P(r) = r^2 - r$ and r depends on the input of man-hours h according to $r(h) = h + 1$, find the composite function $P[r(h)]$.

19. In problem 18, find $P[r(h)]$ if $P(r) = \sqrt{2r+1}$ and $r(h) = h^2 - 1$.

20. An open box is to be made from a square piece of lumber, 4 ft. on a side, by cutting out equal squares from each corner and turning up the sides. Express the volume V as a function of the length x of an edge of a cut-out square.

21. Work problem 20 when the given piece of lumber is a square 12 in. on a side.

22. For the general quadratic function $y = f(x) = ax^2 + bx + c$, the form $x_v = -b/2a$ gives the x value at the vertex V and $f(x_v)$ gives the y value at V. Thus, for $f(x) = x^2 - 2x - 1$, we have $x_v = -(-2/2) = 1$ and $f(1) = -2$. Therefore, the vertex V is at $(1, -2)$. Find the vertex of the given parabolas:

(a) $f(x) = x^2 - 4x + 1$ (b) $f(x) = 2x^2 + 4x - 3$

23. Use the ideas in problem 22 as an aid in graphing the following parabolas:

(a) $f(x) = x^2 - 2x + 3$ **(b)** $g(t) = 4t^2 - 4t + 3$

24. Factor:

(a) $x^2 - 3x - 10$ **(b)** $2x^2 + 7x - 4$

25. Sketch graphs of:

(a) $y = x^{4/3}$ **(b)** $y = \dfrac{1}{\sqrt{x}}$ **(c)** $y = x^{-1/3}$

(d) $y = x^{-3/2}$ **(e)** $y = x^{-1}$

26. Factor:

(a) $8x^3 - 27$ **(b)** $3x^2 + 2x - 8$

27. Find all roots of:

(a) $2x^2 - 3x - 9 = 0$ **(b)** $x^3 - 64 = 0$

28. Solve each given system:

(a) $\left.\begin{aligned} x + y &= 10 \\ 2x - y &= 8 \end{aligned}\right\}$ **(b)** $\left.\begin{aligned} 5x + 6y &= 85 \\ 8x - 4y &= 0 \end{aligned}\right\}$

29. A demand function is defined by $D(x) = 9 - x^2$ and a supply function by $S(x) = x^2 + 1$. Find the point of intersection of D and S.

30. You give 12 coins in dimes and quarters to pay \$1.95 for a pen. Find the number of dimes and the number of quarters.

31. Find the total interest on \$1000 for 5 years at $6\frac{1}{2}\%$ simple interest; at $7\frac{1}{4}\%$ simple interest.

32. A store's profit for 1974 was \$5600, which was $12\frac{1}{2}\%$ less than its profit in 1973. Find the profit in 1973.

33. Simplify:

(a) $\sqrt{32} - \sqrt{18}$ **(b)** $2\sqrt[3]{24} - \sqrt[3]{81}$

34. If $A = \{1, 2\}$ and $B = \{3, 4\}$ then the cartesian product of A and B is

$$A \times B = \{(1, 3), (1, 4), (2, 3), (2, 4)\}.$$

Find $P \times Q$ if:

(a) $P = \{0, 2\}$ and $Q = \{3, 5\}$ **(b)** $P = \{2, 4, 6\}$ and $Q = \{1, 3, 5\}$

35. Find maximum profit if cost C and revenue R are defined as follows:

(a) $C(x) = 1 + 2x$ and $R(x) = 6x - x^2$
(b) $C(x) = 2 + 5x$ and $R(x) = 11x - x^2$

36. A binomial expansion is given by

$$(a + b)^n = a^n + na^{n-1}b + \frac{n(n-1)}{2!}a^{n-2}b^2 + \cdots + b^n$$

where a, b, n are real numbers and $n! = n(n-1)(n-2)\cdots 1$. Find:

(a) $(x + 3)^3$ **(b)** $(2x - 1)^4$ **(c)** $\left(2 - \dfrac{x}{2}\right)^5$

37. The compound amount A of the principal \$1000 for 5 years at 8% compounded quarterly is given by

$$A = 1000\left(1 + \frac{0.08}{4}\right)^{5(4)} = 1000(1 + 0.02)^{20} = \cdots \cong \$1485.90,$$

by use of the binomial expansion shown in problem 36. Apply the binomial expansion to find the compound amount of:

(a) \$2000 for 4 years at 8% compounded quarterly.
(b) \$5000 for 5 years at 12% compounded monthly.

38. Perform the indicated operations:

(a) $(x^{1/2}y^{-1/3})^6$ **(b)** $a^2b^3(a^{-1}b^{-1})$ **(c)** $\left(\frac{9}{4}\right)^{3/2} \div \left(\frac{8}{27}\right)^{2/3}$

39. Given $\log xy = \log x + \log y$; $\log \frac{x}{y} = \log x - \log y$; and $\log x^p = p \log x$. If $\log 2 = 0.3010$ and $\log 3 = 0.4771$, find:

(a) $\log \frac{\sqrt{8}}{3}$ **(b)** $\log \frac{2}{(9)^{1/3}}$ **(c)** $\log 288$

40. Describe the region bounded by the graphs of:

(a) $y = x$, $4x + 3y = 21$, and $y = 0$
(b) $y = 1 + x^2$, $y = 9 - x^2$, and $x = 0$

41. In finding the sum $\frac{1}{2} + \frac{3}{4} + \frac{5}{8} + \frac{7}{16}$ we first recognize 16 as the least common denominator (LCD). Expressing each fraction with this LCD, we have

$$\frac{8}{16} + \frac{12}{16} + \frac{10}{16} + \frac{7}{16} = \frac{37}{16}.$$

Use these notions to find the sums:

(a) $\frac{1}{3} + \frac{5}{6} + \frac{7}{9}$ **(b)** $\frac{2}{5} + \frac{3}{10} + \frac{7}{15} + \frac{9}{20}$

42. To reduce the expression $\frac{4}{x} + \frac{2}{x-2}$ to simplest form, we apply the theorem

$$\frac{a}{b} \pm \frac{c}{d} = \frac{ad \pm bc}{bd}, \qquad b, d \neq 0.$$

Accordingly,

$$\frac{4}{x} + \frac{2}{x-2} = \frac{4(x-2) + 2(x)}{x(x-2)} = \frac{4x - 8 + 2x}{x(x-2)} = \frac{6x-8}{x(x-2)},$$

the required form. Make use of these ideas to reduce to simplest form:

(a) $\frac{3}{x} - \frac{1}{x+1}$ **(b)** $\frac{2}{x} + \frac{3}{x-1} - \frac{4}{x+2}$

43. Solve for the unknown:

(a) $2(x - 3) = 4(1 - 2x)$ **(b)** $x^2 - x - 2 = 0$
(c) $x^2 - x - 6 = 0$ **(d)** $x^2 + 2x - 24 = 0$

44. Solve:

(a) $x^3 - 64x = 0$ **(b)** $x^4 - 16x^2 = 0$
(c) $t^3 - 2t + 1 = 0$ [*Hint*: Try $t = 1$]
(d) $6t^2 - t - 2 = 0$ [*Hint*: One factor is $(3t - 2)$]

45. To simplify the expression $(x^2 + 2x)/x$ we use the rule

$$\frac{a+b}{c} = \frac{a}{c} + \frac{b}{c}, \qquad c \neq 0.$$

Then

$$\frac{x^2 + 2x}{x} = \frac{x^2}{x} + \frac{2x}{x} = x + 2, \qquad x \neq 0.$$

Apply these concepts to simplify:

(a) $\dfrac{x^2 + 4x}{2x}$ **(b)** $\dfrac{2x^3 - x^2}{x}$ **(c)** $\dfrac{3t^2 - 6t + 9}{3t}$

46. Reduce to lowest terms:

(a) $\dfrac{x^2 - x - 12}{x - 4}$ **(b)** $\dfrac{t^2 + 3t - 18}{t - 3}$

47. An open box having a square bottom is to be made of material costing 10¢ per sq ft for the four sides and 15¢ per sq ft for the bottom. If the volume V of the box is to be 10 cu ft, express the total cost C as a function of x.

48. An **ellipse** has the general form

$$\frac{x^2}{a^2} + \frac{y^2}{b^2} = 1, \qquad a \neq b.$$

Thus, $(x^2/4) + (y^2/9) = 1$ is an ellipse whose x-intercepts are $a = \pm 2$ and y-intercepts are $b = \pm 3$ (see Figure 1.30).

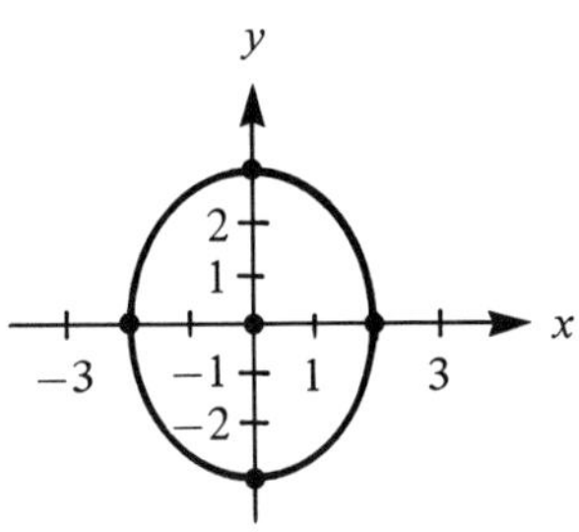

Figure 1.30

Graph the following ellipses:

(a) $\dfrac{x^2}{9} + \dfrac{y^2}{16} = 1$ **(b)** $\dfrac{x^2}{2} + \dfrac{y^2}{3} = 1$

A **hyperbola** has the general form

$$\frac{x^2}{a^2} - \frac{y^2}{b^2} = 1 \quad \text{or} \quad \frac{y^2}{a^2} - \frac{x^2}{b^2} = 1.$$

In the first form, $x = \pm a$ are the x-axis intercepts; in the second, $y = \pm a$ are the y-axis intercepts. Accordingly, the hyperbola $(x^2/9) - (y^2/4) = 1$ has $x = \pm 3$ as x-axis intercepts and has its graph as shown in Figure 1.31.

When $y = \pm 2,$
then $x = \pm\sqrt{18}$
$\cong \pm 4.2$

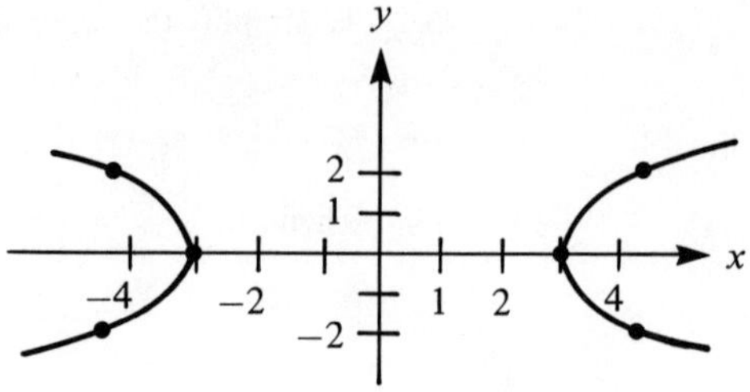

Figure 1.31

Likewise, the hyperbola $(y^2/4) - x^2 = 1$ has y-intercepts $y = \pm 2$; its graph is shown in Figure 1.32.

when $x = \pm 2$
then $y = \pm\sqrt{20}$
$\cong \pm 4.4$

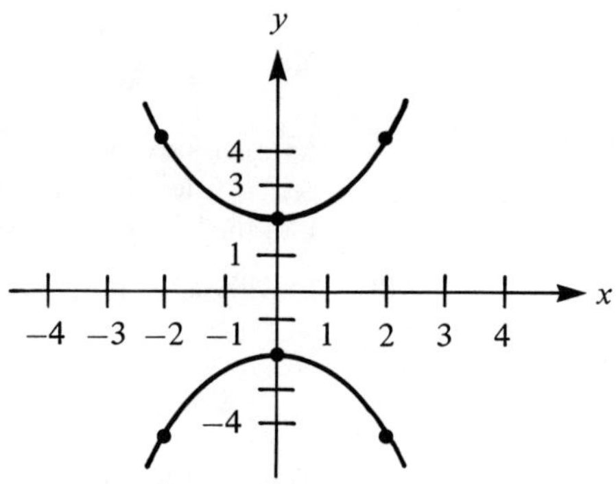

Figure 1.32

Sketch graphs of the following hyperbolas:

(c) $\dfrac{x^2}{16} - \dfrac{y^2}{9} = 1$ **(d)** $\dfrac{y^2}{8} - \dfrac{x^2}{25} = 1$

49. Expand:

(a) $(x + 2)^3$ **(b)** $(2x - 5)^2$ **(c)** $(1 - x)^4$

50. To rationalize the numerator of the fraction $\dfrac{\sqrt{4 + h} + 2}{h}$ we make use of the fact that the conjugate of any binomial of the form $a + b$ is $a - b$. Then the conjugate of $\sqrt{4 + h} + 2$ is $\sqrt{4 + h} - 2$. Multiplying the numerator and denominator by the conjugate $\sqrt{4 - h} - 2$, we obtain

$$\frac{\sqrt{4 + h} + 2}{h} \cdot \frac{\sqrt{4 + h} - 2}{\sqrt{4 + h} - 2} = \frac{4 + h - 4}{h(\sqrt{4 + h} - 2)} \qquad [(a + b)(a - b) = a^2 - b^2]$$

$$= \frac{h}{h(\sqrt{4 + h} - 2)}$$

$$= \frac{1}{\sqrt{4 + h} - 2}$$

Competency in rationalizing numerators or denominators is useful in certain areas of the differential calculus. Apply the above ideas in the following problems.

Rationalize the numerator:

(a) $\dfrac{1 - \sqrt{2}}{1 + \sqrt{2}}$ **(b)** $\dfrac{\sqrt{1 + h} - 1}{h}$ **(c)** $\dfrac{\sqrt{3 + h} - \sqrt{3}}{h}$

Rationalize the denominator:

(d) $\dfrac{\sqrt{7} + \sqrt{5}}{\sqrt{7} - \sqrt{5}}$ **(e)** $\dfrac{1 - \sqrt{x}}{1 + \sqrt{x}}$ **(f)** $\dfrac{h}{\sqrt{h + 9} - 3}$

2 DIFFERENTIAL CALCULUS

The Role of Derivatives in Decision-Making

IMPORTANT EQUATIONS

$$\lim_{x \to b} f(x) = L$$

$$f'(x) = \lim_{\Delta x \to 0} \frac{f(x + \Delta x) - f(x)}{\Delta x}$$

$$\frac{dy}{dt} = \frac{dy}{dx} \cdot \frac{dx}{dt}$$

$$dy = f'(x)\, dx$$

Differential calculus is essentially the mathematics of rates of change. Many decision problems in the modern business world are analyzed by considering the rate of change of a particular quantity (the dependent variable) with respect to the change in some other quantity (the independent variable). Management often has to make decisions as to what rate input should assume in order to obtain a desired optimum rate of output. Such analysis, based on considerations of rates of change, is often referred to as **Marginal Analysis**. Many of the tools developed in this chapter are aimed toward better decision-making in marginal analysis. Before discussing and applying these tools it is necessary to establish the meanings of two concepts, limits and continuity, which are basic to the study of differential calculus.

2.1 LIMITS OF FUNCTIONS

Consider the function f defined by

$$f(x) = \frac{(x - 2)(3x + 1)}{x - 2}, \qquad x \neq 2,$$

where f is defined for all values of x except $x = 2$. If $x \neq 2$ we may divide the numerator and denominator of f by $x - 2$ and obtain

$$f(x) = 3x + 1.$$

Let us illustrate what happens to $f(x)$ when values of x get closer and closer to 2. In Table 2.1 we choose values of x so that x approaches 2 through values that are less than 2. We symbolize this action by

$$x \to 2^-$$

and we say that "x approaches 2 from the left."

Table 2.1

$x\ (\neq 2)$	1.5	1.8	1.9	1.99	1.9999
$f(x) = 3x + 1$	5.5	6.4	6.7	6.97	6.9997

In Table 2.2 the values of x move closer and closer to 2 through values greater than 2. This is denoted by

$$x \to 2^+$$

which is read "x approaches 2 from the right."

Table 2.2

$x\ (\neq 2)$	2.4	2.3	2.1	2.01	2.0001
$f(x) = 3x + 1$	8.2	7.9	7.3	7.03	7.0003

From the tables it is not difficult to see that we can make the value of $f(x)$ as close to 7 as we please by taking values of x sufficiently close to 2. Stated differently: We can make the absolute value of the difference between $f(x)$ and 7 as small as we please by making the absolute value of the difference between x and 2 sufficiently small. That is,

$$|f(x) - 7|$$

becomes and remains less than any preassigned value whenever

$$0 < |x - 2|$$

is sufficiently small.

Note: We write $0 < |x - 2|$ since $x \neq 2$.

It can be seen from the tables that

$$|f(x) - 7| = 0.03 \qquad \text{whenever} \qquad 0 < |x - 2| = 0.01.$$

The last statement contains the implication: Given the positive number 0.03, there exists a positive number 0.01 such that the functional values $f(x)$ satisfy the inequality

$$|f(x) - 7| < 0.03$$

whenever the independent variable x satisfies the inequality

$$0 < |x - 2| < 0.01.$$

We could continue to generate other inequalities by preassigning values smaller than 0.03 and then finding other values of x sufficiently small so that

$$|f(x) - 7|$$

will be *arbitrarily* small whenever

$$0 < |x - 2|$$

is *sufficiently* small.

An algebraic investigation of the inequalities may help clarify the meaning of the important concept of a limit. From earlier discussions of inequalities we know that

$$|f(x) - 7| < 0.0003$$

implies that

$$-0.0003 < f(x) - 7 < 0.0003$$

or

$$7 - 0.0003 < f(x) < 7 + 0.0003$$

or

$$6.9997 < f(x) < 7.0003.$$

Similarly,

$$|x - 2| < 0.0001$$

implies

$$-0.0001 < x - 2 < 0.0001$$

or

$$1.999 < x < 2.0001.$$

That is, f lies between the values 6.9997 and 7.0003 when x lies between 1.999 and 2.0001. In Figure 2.1 we exhibit the limit concept geometrically for a preassigned value of 0.03.

Figure 2.1 serves to motivate and clarify the statement: Values of $f(x)$ on the vertical axis may be restricted to lie between $7 - 0.03$ and $7 + 0.03$ whenever values of x on the horizontal axis lie between $2 - 0.01$ and $2 + 0.01$. This limiting process is formally written as

$$\lim_{x \to 2} f(x) = 7.$$

For a limiting process in general we write

$$\boxed{\lim_{x \to b} f(x) = L}\,.$$

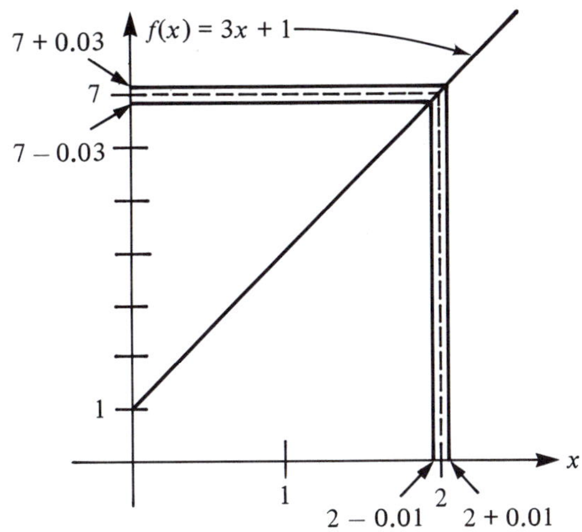

Figure 2.1

Such attributes of functions f may be stated as:

$$\left.\begin{array}{ll} & f(x) \text{ tends to a limit } L \\ \text{or} & f(x) \text{ approaches } L \\ \text{or} & f(x) \text{ has limit } L \end{array}\right\} \text{ as } x \text{ approaches } b.$$

A formal definition of the limit of a function follows.

DEFINITION 1 *The function values $f(x)$ approach a* **limit** *L as x approaches a number b if the absolute value of the difference between $f(x)$ and L can be made as small as we please by taking values of x sufficiently near b but not necessarily equal to b.*

From this definition we see that $f(x)$ need not be defined at $x = b$ in order for $\lim_{x\to b} f(x)$ to exist. Rather than presenting the purely theoretical aspects of Definition 1, we choose to embrace an intuitive approach, and to this end we will enlist the aid of "limit theorems." We state these theorems without proof.

Limit Theorems

(1) $\lim_{x\to c} x = c$

(2) *If k is a constant, then $\lim_{x\to c} k = k$ for any value of c.*

(3) *If a and b are any constants, then $\lim_{x\to c}(ax + b) = ac + b$.*

(4) *If $\lim_{x\to c} f(x) = A$ and $\lim_{x\to c} g(x) = B$, then $\lim_{x\to c}[f(x) \pm g(x)] = A \pm B$.*

(5) *If $\lim_{x\to c} f(x) = A$ and $\lim_{x\to c} g(x) = B$, then $\lim_{x\to c}[f(x) \cdot g(x)] = A \cdot B$.*

(6) *If $\lim_{x\to c} f(x) = A$ and $\lim_{x\to c} g(x) = B$, then $\lim_{x\to c}[f(x)/g(x)] = A/B, B \neq 0$.*

(7) *If $\lim_{x\to c} f(x) = A$, then $\lim_{x\to c} \sqrt[n]{f(x)} = \sqrt[n]{A}$ provided:*

(i) *$A > 0$ and n is a positive even integer*

or **(ii)** *n is a positive odd integer.*

A few examples of "rigorous" applications of these theorems now follow.

EXAMPLE 1 Find $\lim_{x\to 1} (2x^2 - x + 3)$.

Solution

$$\begin{aligned}
&\lim_{x\to 1} (2x^2 - x + 3) \\
&= \lim_{x\to 1} 2x^2 - \lim_{x\to 1} x + \lim_{x\to 1} 3 && \text{[by extension of Theorem 4]} \\
&= \lim_{x\to 1} 2 \cdot \lim_{x\to 1} x \cdot \lim_{x\to 1} x - \lim_{x-1} x + \lim_{x\to 1} 3 && \text{[by Theorem 5]} \\
&= 2 \cdot 1 \cdot 1 - 1 + 3 && \text{[by Theorems 1 and 2]} \\
&= 4.
\end{aligned}$$

EXAMPLE 2 Evaluate $\lim_{x\to 2} \sqrt{\dfrac{x^2+1}{5}}$.

Solution

$$\begin{aligned}
&\lim_{x\to 2} \sqrt{\frac{x^2+1}{5}} \\
&= \sqrt{\lim_{x\to 2} \frac{x^2+1}{5}} && \text{[by Theorem 7]} \\
&= \sqrt{\frac{\lim_{x\to 2} (x^2+1)}{\lim_{x\to 2} 5}} && \text{[by Theorem 6]} \\
&= \sqrt{\frac{\lim_{x\to 2} x^2 + \lim_{x\to 2} 1}{\lim_{x\to 2} 5}} && \text{[by Theorem 4]} \\
&= \sqrt{\frac{\lim_{x\to 2} x \cdot \lim_{x\to 2} x + \lim_{x\to 2} 1}{\lim_{x\to 2} 5}} && \text{[by Theorem 5]} \\
&= \sqrt{\frac{2 \cdot 2 + 1}{5}} && \text{[by Theorems 1 and 2]} \\
&= \sqrt{\frac{5}{5}} = 1.
\end{aligned}$$

The step-by-step procedure of finding limits by direct applications of the limit theorems, as shown in Examples 1 and 2, is cumbersome and tedious. In view of this we will forego this method of finding limits and will employ "short cuts" which are more practical. The next example illustrates this procedure.

EXAMPLE 3 Find $\lim_{x\to 2} C(x)$ and $\lim_{x\to 3} C(x)$, where C is a cost function defined by

$$C(x) = \frac{x^2 - x - 6}{x - 3}, \qquad x \neq 3,$$

and x denotes the number of items produced.

Solution

$$\lim_{x \to 2} C(x) = \lim_{x \to 2} \frac{x^2 - x - 6}{x - 3}$$
$$= \frac{2^2 - 2 - 6}{2 - 3}$$
$$= \frac{4 - 8}{-1} = \$4.$$

To find $\lim_{x \to 3} C(x)$ we proceed as follows:

$$\lim_{x \to 3} \frac{x^2 - x - 6}{x - 3} = \frac{3^2 - 3 - 6}{3 - 3}$$
$$= \frac{0}{0}, \quad \text{a meaningless form.}$$

We must be careful *not* to interpret 0/0 as equal to zero! This result is a meaningless one and forces us to try a new approach, one that will yield a meaningful result, hopefully. Upon investigation we see that the denominator, $x - 3$, causes the limiting process to be impractical. Therefore, we attempt to remove this "bottleneck" by factoring the numerator of the given function and simplifying. It then follows that

$$\lim_{x \to 3} C(x) = \lim_{x \to 3} \frac{(x - 3)(x + 2)}{x - 3}$$
$$= \lim_{x \to 3} (x + 2)$$
$$= 3 + 2 = \$5.$$

Figure 2.2 illustrates the above limiting process. Note the break in the graph at $x = 3$. (We denote this break by the small open circle at the point where $C(3)$ would normally occur.) This may mean that, for one reason or another, a production setup is not prepared to function properly when the number of items being produced reaches 3. The function C is said to be *discontinuous* at $x = 3$. (Continuity and discontinuity of functions will be discussed in the next section.)

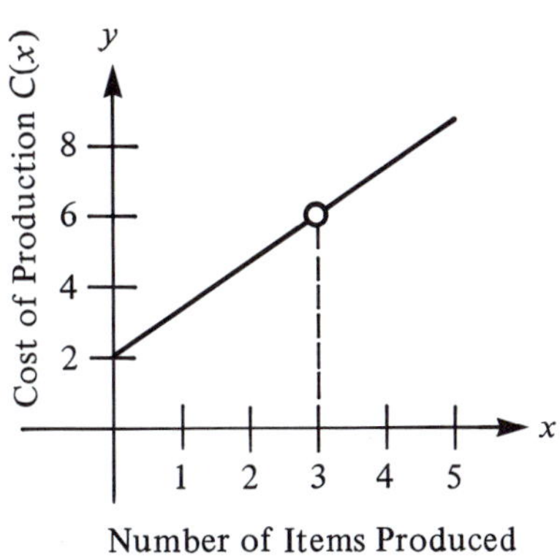

Figure 2.2

In the process of evaluating limits of the form

$$\lim_{x \to a} \frac{x^2 - a^2}{x - a}$$

it is possible to divide numerator and denominator by the quantity $(x - a)$ because we are considering values of x *close* to a but *not equal* to a.

You should remember that not all limits can be evaluated by simply substituting in the limiting value of the independent variable and performing the arithmetic. In many problems there is more to it than that: often one's diagnostic ability is severely challenged. We continue with a few more examples of evaluating limits of functions.

EXAMPLE 4 Find $\lim_{t \to 2} \dfrac{t^3 - 8}{t - 2}$.

Solution Substituting $t = 2$ into the given function yields 0/0, which is indeterminate. We must circumvent this temporary obstacle by *doing something algebraically*. After some investigation we note that the numerator $(t^3 - 8)$ is a difference of two cubes which can be factored. Accordingly, we have

$$\begin{aligned} \lim_{t \to 2} \frac{t^3 - 8}{t - 2} &= \lim_{t \to 2} \frac{(t - 2)(t^2 + 2t + 4)}{t - 2}, \qquad t \neq 2 \\ &= \lim_{t \to 2} (t^2 + 2t + 4) \qquad \left[\text{since } \frac{t - 2}{t - 2} = 1\right] \\ &= 2^2 + 2(2) + 4 = 12. \end{aligned}$$

EXAMPLE 5 Find $\lim_{h \to 0} \dfrac{\sqrt{9 + h} - 3}{h}$.

Solution Taking the limit directly gives 0/0, which is meaningless. After some examination we see that the numerator may be rationalized, thus eliminating the meaningless form and producing a form whose limit exists. In line with this procedure we have

$$\begin{aligned} \lim_{h \to 0} \frac{\sqrt{9 + h} - 3}{h} &= \lim_{h \to 0} \frac{\sqrt{9 + h} - 3}{h} \cdot \frac{\sqrt{9 + h} + 3}{\sqrt{9 + h} + 3} \\ &= \lim_{h \to 0} \frac{9 + h - 9}{h(\sqrt{9 + h} + 3)} \\ &= \lim_{h \to 0} \frac{h}{h} \cdot \frac{1}{\sqrt{9 + h} + 3} \\ &= \frac{1}{\sqrt{9 + 0} + 3} = \frac{1}{6}. \end{aligned}$$

In Example 4 the indeterminate form was circumvented by using the algebraic operations of factoring a difference of two cubes and simplifying. In Example 5 the bottleneck was avoided by employing the algebraic operation of rationalizing the numerator. Other problems may call for other types of algebraic operations that will effect a change from the 0/0 form to one whose limit can be evaluated. Remember, the only two acceptable answers to a limit problem (at this stage) are: a finite number, or undefined, i.e., ∞.

PROBLEM SET 2.1

1. Evaluate the following limits, showing each step (as in Examples 1 and 2) and stating the limit theorem which applies:

(a) $\lim_{x\to 1} (x^2 - 2x + 1)$ **(b)** $\lim_{x\to 3} \sqrt{\dfrac{x}{x^2 + 1}}$

Find each of the following limits:

2. $\lim_{x\to 1} \dfrac{2x + 1}{3}$

3. $\lim_{x\to 2} \dfrac{3x - 4}{x + 2}$

4. $\lim_{x\to 2} \dfrac{3x - 6}{x - 2}$

5. $\lim_{x\to 4} \sqrt{2x + 1}$

6. $\lim_{q\to 25} \dfrac{200 - q}{q}$

7. $\lim_{x\to 10} \left(x^2 - \dfrac{1}{x}\right)$

8. $\lim_{x\to 1} \dfrac{x^2 - 1}{x - 1}$

9. $\lim_{x\to 5} \dfrac{x^2 - 25}{x - 5}$

10. $\lim_{x\to 3} \dfrac{x^2 - 3x}{x - 3}$

11. $\lim_{r\to -1} \dfrac{1 - r}{1 - r^3}$

12. $\lim_{t\to 3} \dfrac{27 - t^3}{3 - t}$

13. $\lim_{x\to 2} \dfrac{x^2 - 16}{x^2 - 4}$

14. $\lim_{h\to 0} \dfrac{2 - \sqrt{4 - h}}{h}$

15. $\lim_{h\to 0} \dfrac{\sqrt{h + 2} - \sqrt{2}}{h}$

16. $\lim_{h\to 0} \dfrac{\sqrt{8 + h} - \sqrt{8}}{h}$

17. $\lim_{x\to 1} \dfrac{1 + x}{1 - \sqrt{x}}$

18. $\lim_{t\to 2} \dfrac{t^4 - 16}{t - 2}$

19. $\lim_{h\to 0} \dfrac{(2 + h)^3 - 8}{h}$

20. $\lim_{\Delta x\to 0} \dfrac{(1 + \Delta x)^3 - 1}{x}$

21. $\lim_{t\to 4} \dfrac{t^2 - 7t + 12}{t^2 - t - 12}$

22. Given the function

$$C(x) = \begin{cases} 20x + 100, & 0 \le x \le 10 \\ 10x + 200, & x > 10 \end{cases}$$

where x represents units of items produced and $C(x)$ is the cost of production in dollars.

(a) Graph that portion of C which is empirically meaningful.

(b) Discuss $\lim_{x\to 10^-} C(x)$ and $\lim_{x\to 10^+} C(x)$.

23. A certain cost function is defined by $C(q) = (q^2 + 17q - 60)/(q - 3), q \neq 3$, where q denotes the quantity produced (in hundreds) and C is the production cost (in thousands of dollars).

(a) Sketch a graph of C.
(b) Find $\lim_{q \to 3} C(q)$.

24. Assume that there exists a big demand for a new product, making sales revenue climb sharply at first. As competitors enter the market the amount of sales revenue changes "abruptly." These facts are formulated as follows:

$$S(t) = \begin{cases} t^2/20 & \text{for } 0 < t \leq 20 \\ 10 + (t/2) & \text{for } t > 20 \end{cases}$$

where $S(t)$ is sales revenue (in thousands of dollars) and t denotes time (in weeks).

(a) Sketch a graph of S.
(b) Discuss $\lim_{t \to 20} S(t)$.

25. The price P of the common stock of the ABC Co. enjoyed an increase in the first 4 months of the company's operations. In the second 4 months the price was approximately constant. For the last 4 months there was a decline in price. The following is a mathematical model of these data:

$$P(t) = \begin{cases} 10 + t & \text{for } 0 < t \leq 4 \\ 13 & \text{for } 4 < t \leq 8 \\ 96/t & \text{for } 8 < t \leq 12 \end{cases}$$

(a) Sketch a graph of P.
(b) Discuss $\lim_{t \to 4} P(t)$.

2.2 CONTINUITY OF FUNCTIONS

Continuous (and discontinuous) functions have important applications in business problems relating sales, cost, profit, and other areas. The concepts of continuity and limit (discussed in Section 2.1) constitute the basic ideas underlying the derivative and differentiation, which is the technique of finding a derivative.

In the study of $\lim_{x \to b} f(x) = L$ we were not concerned with the value of f at $x = b$. Our interest focused mainly on the limiting value of f as x approached b. However, in the study of continuity in this section we will be concerned with the value of f at $x = b$. Let us proceed with concrete examples before introducing the abstract definition of continuity.

EXAMPLE 1 Consider the total cost function $C(x) = (x + 500)$ dollars, where $0 \leq x < 5$. The function is the sum of the variable cost \$$x$ (x units at \$1 per unit) and the fixed cost \$500 (overhead and production setup costs). Since the domain of C is given by $0 \leq x < 5$ it follows that C is not defined for $x = 5$, which implies that C is discontinuous at $x = 5$. We may find the limit of C as x approaches 5 thus:

$$\lim_{x\to 5^-} C(x) = \lim_{x\to 5^-} (x + 500)$$
$$= 5 + 500 = \$505.$$

Figure 2.3 lends a geometric interpretation.

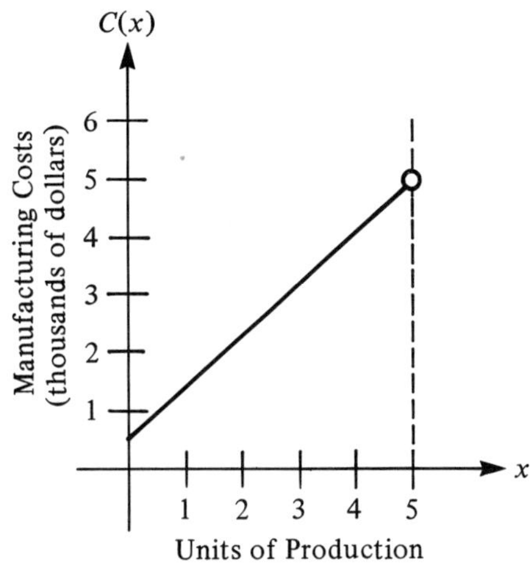

Figure 2.3

In many cases companies order large quantities of goods in order to benefit from significant discounts. The suppliers, in turn, are willing to give discounts so that they may get more large orders and thus effect economies in their cost setup. The next example illustrates these interacting forces of the marketplace.

EXAMPLE 2 Assume that the total cost C of purchasing x tons of material varies directly as x within each order interval. That is, assume that

$$C(x) = \begin{cases} 10x & \text{for} \quad 0 \le x \le 100 \\ 8x & \text{for} \quad 100 < x \le 200 \\ 6.5x & \text{for} \quad x > 200. \end{cases}$$

This price function has "gaps" or "jumps," as shown in Figure 2.4. In the closed interval $[0, 100]$ the total cost ranges from \$0 to \$1000. In the half-closed interval $(100, 200]$ the cost ranges from \$800 to \$1600. And in the open interval $(200, \infty)$ the cost is greater than \$1300. (Recall that the symbol ∞ means infinity.)

Let us review intervals briefly:

The closed interval $[a, b]$ contains values of x such that $a \le x \le b$.

The half-closed, half-open intervals $(a, b]$ and $[a, b)$ contain values of x such that $a < x \le b$ and $a \le x < b$, respectively.

The open interval (a, b) contains values of x such that $a < x < b$.

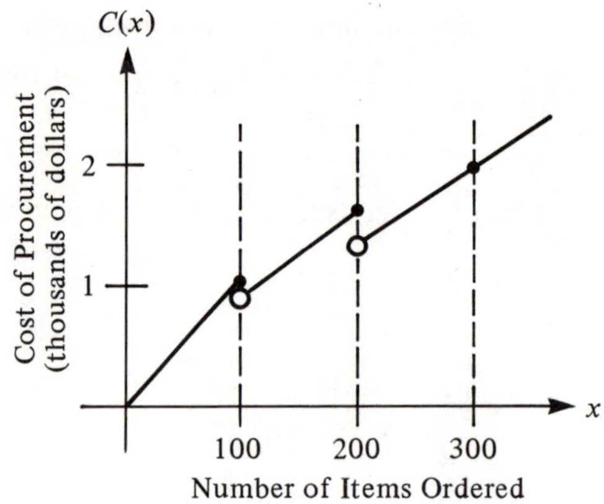

Figure 2.4

The function C of Example 2 is discontinuous at $x = 100$ and $x = 200$. (The mark at the end of a line such as A ○———• B indicates a discontinuity for the function at point A.) Let us examine the limits of the function C above at $x = 100$ and $x = 200$:

$$\lim_{x \to 100} C(x) = \begin{cases} \lim_{x \to 100^-} 10x = 10(100) = \$1000 \\ \lim_{x \to 100^+} 8x = 8(100) = \$800. \end{cases}$$

Thus, ordering "slightly more" than 100 tons results in a savings of \$200 in purchasing costs. Similarly,

$$\lim_{x \to 200} C(x) = \begin{cases} \lim_{x \to 200^-} 8x = 8(200) = \$1600 \\ \lim_{x \to 200^+} 6.5x = 6.5(200) = \$1300. \end{cases}$$

The savings, in this case, are \$300.

It now seems appropriate to define a continuous function.

DEFINITION 1 *A function $f(x)$, defined on an interval containing the number b, is said to be **continuous** at $x = b$ provided the following three conditions are met*:

(1) *f is defined at $x = b$*;

(2) *$\lim_{x \to b} f(x)$ is defined and is unique*;

(3) *$\lim_{x \to b} f(x) = f(b)$*.

*If any one (or more) of these three conditions is not satisfied, then the function f is said to be **discontinuous** at $x = b$.*

To clarify condition (2) above, when finding the limits "to the left" and "to the right" the two results must be the same number, i.e., a unique number, in order for the limit to exist (or to be defined).

EXAMPLE 3 Let f be defined by $f(x) = 1/(x - 1)$, $x > 1$. The graph of f is shown in Figure 2.5. From the graph it is clear that $x = 1$ is a vertical asymptote; the function is not defined at $x = 1$. A vertical line $x = k$ is called a vertical asymptote of the graph of f if $|f| \to \infty$ as $x \to k$. Thus, $f(1)$ does not exist and so condition (1) for continuity is violated. Hence, function f is discontinuous at $x = 1$.

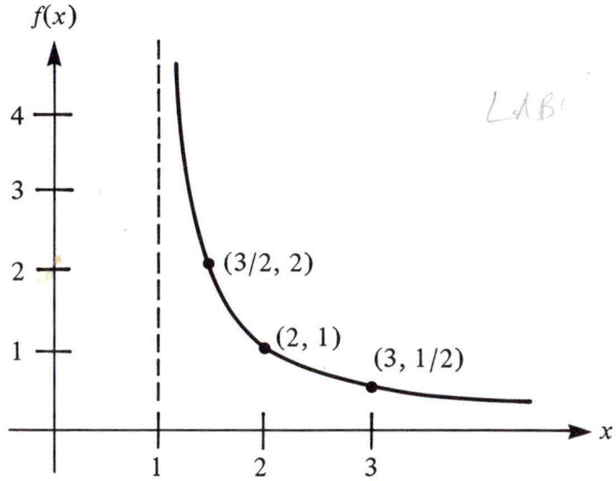

Figure 2.5

EXAMPLE 4 Given

$$g(x) = \begin{cases} \dfrac{(3x - 1)(x - 1)}{x - 1} & \text{if} \quad x \neq 1 \\ 2 & \text{if} \quad x = 1. \end{cases}$$

Discuss continuity for g.

Solution Applying the three conditions of the definition of continuity we have:

(1) $g(1)$ is defined and equals 2, by hypothesis.

(2)
$$\begin{aligned} \lim_{x \to 1} g(x) &= \lim_{x \to 1} \frac{(3x - 1)(x - 1)}{x - 1} \\ &= \lim_{x \to 1} (3x - 1) \\ &= 3(1) - 1 = 2 \end{aligned}$$

(3) $\lim_{x \to 1} g(x) = g(1) = 2.$

Therefore, $g(x)$ is continuous by definition since all three conditions for continuity are satisfied. Observe that we removed the discontinuity of g at $x = 1$ by arbitrarily defining $g(1)$ to be equal to 2. We say that examples of this type have a "removable discontinuity."

EXAMPLE 5 Let h be defined by

$$h(x) = \begin{cases} \sqrt{x} & \text{if} \quad 0 \le x \le 4 \\ 2 - \dfrac{x}{8} & \text{if} \quad x > 4. \end{cases}$$

The graph of h is shown in Figure 2.6. The break in the graph at $x = 4$ prompts us to investigate the three conditions for continuity at $x = 4$:

(1) $h(4) = \sqrt{4} = 2$

(2) $\lim_{x \to 4} h(x) = \begin{cases} \lim_{x \to 4^-} \sqrt{x} = \sqrt{4} = 2 \\ \lim_{x \to 4^+} \left(2 - \frac{x}{8}\right) = 2 - \frac{4}{8} = \frac{3}{2} \end{cases}$

(3) $\lim_{x \to 4} h(x) \neq h(4)$.

In condition (2) above the limit of h does not exist at $x = 4$ since the left-hand limit does not equal the right-hand limit; that is, $2 \neq 3/2$. Since the limit in (2) is not defined we say that $\lim_{x \to 4} h(x)$ is not defined. This fact justifies writing (3) above. Since condition (2) is not satisfied, function h is not continuous (discontinuous) at $x = 4$.

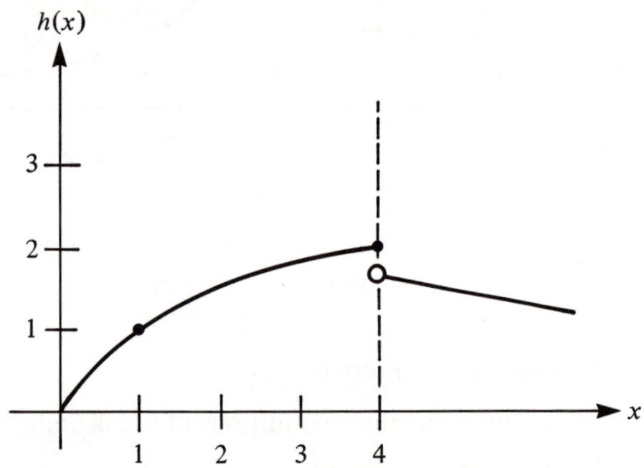

Figure 2.6

Remark: $\lim_{x \to b^-} f(x)$ must equal $\lim_{x \to b^+} f(x)$ in order for $\lim_{x \to b} f(x)$ to exist (be defined).

PROBLEM SET 2.2

Explain why each of the given functions is continuous or discontinuous at the point(s) specified or implied:

1. $f(x) = 500/x$ if $x = 0$; if $x = 50$.

2. $g(x) = 2x^2 + 3x - 5$ if $x = 2$.

3. $h(x) = x^2/20 - 20x + 1000$ if $x = 10$.

4. $F(x) = \sqrt[3]{9 - x}$ at $x = 1$.

5. $G(x) = \dfrac{x^2 - x - 6}{x - 3}$.

6. $H(x) = \begin{cases} \dfrac{x^2 - 1}{x - 1} & \text{if } x \neq 1 \\ 0 & \text{if } x = 1. \end{cases}$

7. $P(q) = \begin{cases} \dfrac{(q-2)(2q+5)}{q-2} & \text{for } q \neq 2 \\ 9 & \text{for } q = 2. \end{cases}$

8. $f(x) = 2 + \sqrt{x-5}.$

9. $g(t) = \dfrac{t^2 - 4}{t^4 - 16}.$

10. $h(x) = \begin{cases} \dfrac{x^2 - 1}{x - 1} & \text{if } x \neq 1 \\ -2 & \text{if } x = 1 \end{cases}$

11. $f(x) = \begin{cases} x + 2 & \text{for } 0 < x \leq 1 \\ -\dfrac{1}{2}(x+1) & \text{for } 1 < x \leq 3 \\ \dfrac{x^2}{6} & \text{for } x > 3 \end{cases}$

12. $g(v) = \dfrac{1}{\sqrt{v-1}}.$

13. $h(t) = \begin{cases} \dfrac{t^2 + 2t - 3}{t - 1} & \text{if } t \neq 1 \\ 6 & \text{if } t = 1. \end{cases}$

14. The total cost of transporting a truck-load of goods over a 1000-mile route is given by

$$C(x) = \begin{cases} \$0.90x & \text{if } 0 \leq x \leq 200 \\ \$0.80x & \text{if } 200 < x \leq 600 \\ \$0.70x & \text{if } 600 < x \leq 1000 \end{cases}$$

where x represents distance in miles.

(a) Sketch a graph of $C(x)$.
(b) Discuss the continuity of C at $x = 200$ and at $x = 600$.

15. Use the graph and discussion of Example 2 as a guide to work the following example:

$$C(x) = \begin{cases} 12.5x & \text{for } 0 \leq x \leq 200 \\ 10x & \text{for } 200 < x \leq 300 \\ 8x & \text{for } x > 300 \end{cases}$$

16. Assume that a demand function is defined by $D(q) = 100/(q - 10)$ for $10 < q \leq 50$.

(a) Sketch a graph of D.
(b) Discuss the continuity of D.

17. It is found that a supply function may be defined by

$$S(q) = \begin{cases} \sqrt{q} & \text{if } 0 \leq q \leq 100 \\ \dfrac{q}{12} & \text{if } q > 100 \end{cases}$$

where q denotes quantities of items of production.

(a) Sketch a graph of S.
(b) Discuss the continuity of S.

2.3 DERIVATIVES

In many business problems it is of great value to be able to evaluate the rate of change of a particular function with respect to the independent variable. Often the result is of great significance when the rate of change equals zero. For example, a firm may be concerned with the effects on total profits when promotional efforts such as advertising are increased. The firm would continue to increase its advertising outlay as long as the rate of change of its profits remained positive. In certain cases the firm would know that its efforts had produced optimality when the rate of change of profits with respect to advertising outlay had reached zero.

The derivative of a function f is synonymous with its rate of change. The geometric counterpart of the derivative of f is the slope of the line tangent to f at the point of interest. To clarify, we now develop the concept of the rate of change of a productivity function P by means of a concrete example invoking geometric considerations.

EXAMPLE 1 Let $P(d) = 20d - 2d^2$ be a productivity function where P denotes the number of items installed per week and d represents the total dollars (in thousands) paid to the labor force for the week's work. To facilitate drawing the graph of P we first complete the square. Thus,

$$\begin{aligned} P &= -2(d^2 - 10d + 25 - 25) \\ &= -2(d - 5)^2 + 50 \\ P - 50 &= -2(d - 5)^2 \end{aligned}$$

From this form of the equation (and from discussion in Chapter 1) we see that the point $V(5, 50)$ denotes the vertex of a parabola which is concave downward. Figure 2.7 illustrates this equation.

We arbitrarily choose four points on the curve of P, namely, points T, D, E, and F, and we draw the three lines TD, TE, and TF. These lines

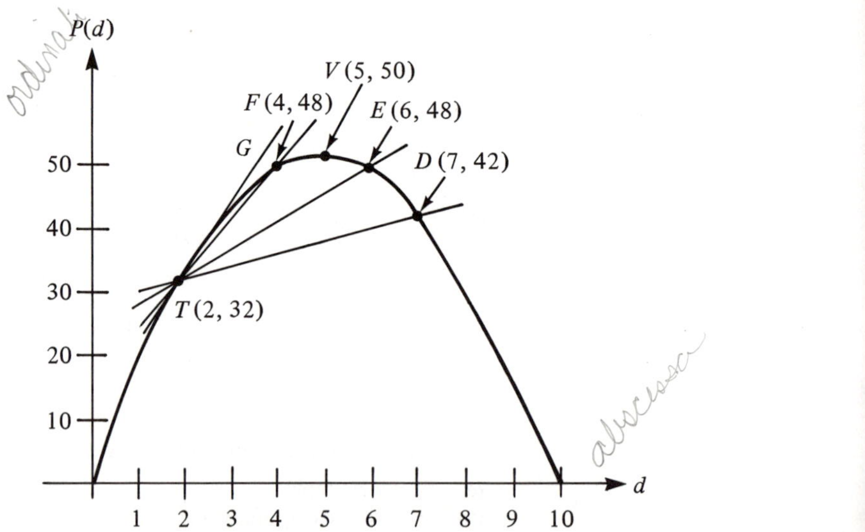

Figure 2.7

are called *secant lines* (or simply secants) and each line "cuts" the graph of P in only two points. The secant TD cuts off the part called the curve segment TD or simply *arc* TD. Secant TE cuts off a smaller segment of the curve or arc TE and secant TF cuts off still a smaller segment, arc TF. If we continued to draw more secant lines through point T in such a way that smaller and smaller arcs of the curve were cut off, we would reach the stage where a "limiting line" would cut off a "limiting portion" having a measure of zero. This limiting line, TG in Figure 2.7, has only the point T in common with curve P and is called the *tangent line* of P at point T.

Since the slope of a secant line is basic to a discussion and development of the rate of change of a function, we accordingly find the slopes of the secant lines TD, TE, and TF.

As a point moves along curve P from point T to point D, the change in ordinate P, call it ΔP, is

$$42 - 32 = 10 \qquad \text{(see Figure 2.7)}$$

while the change in abscissa d, call it Δd, is

$$7 - 2 = 5.$$

Since the slope of a straight line is defined to be the quotient of the change in the dependent variable by the change in the independent variable it follows that the slope of the secant TD is

$$\frac{\Delta P}{\Delta d} = \frac{10}{5} = 2.$$

The procedure is continued as the slopes of the other secants are found and tabulated in Table 2.3.

Table 2.3

Secant	Change in P: ΔP	Change in d: Δd	Slope of secant: $\Delta P/\Delta d$
TD	$42 - 32 = 10$	$7 - 2 = 5$	$10 \div 5 = 2$
TE	$48 - 32 = 16$	$6 - 2 = 4$	$16 \div 4 = 4$
TF	$48 - 32 = 16$	$4 - 2 = 2$	$16 \div 2 = 8$

An extension of the table to include additional secants would exhibit smaller and smaller values of Δd. Our main interest is in the limiting value of the ratio $\Delta P/\Delta d$ as Δd gets closer and closer to zero. This limiting value is called the **instantaneous rate of change** of P with respect to d at the point T, and is written:

$$P'(d) = \lim_{\Delta d \to 0} \frac{\Delta P}{\Delta d}.$$

This instantaneous rate of change is also defined as the **first derivative** of P with respect to d. The derivative exists if the limit exists.

We summarily conclude that finding the limiting line by considering a series of secant lines which cut off smaller and smaller arcs of the curve is an activity synonymous with that of calculating the slopes of these secants as the change in the independent variable approaches zero. Actually, for a given function f and a given point Q on f the *slope of the tangent* to f at Q is defined to be *numerically equal to the value of the derivative* of f at Q. Thus, the analytical meaning of a derivative has a supporting geometric interpretation.

Note: In the limiting process of the productivity function, $P'(d) = \lim_{\Delta d \to 0}(\Delta P/\Delta d)$, the change in P, that is, ΔP, also approaches zero as $\Delta d \to 0$. To avoid confusion, we recall the statement of the definition of a limit: Δd gets closer and closer to zero without actually reaching zero.

Let us now take leave of the productivity function, which has served its purpose, and continue with a discussion of derivatives in general.

At this stage questions probably arise in the reader's mind: Can we find the derivative of a function without the burden of a lengthy discussion as given above? Is there a convenient formula for doing so? If there is a formula, what is it and how is it applied? The answer is: Yes, there is a convenient formula for finding the derivative of a function. We proceed to develop the formula.

Let $y = f(x)$ be a function whose graph is shown in Figure 2.8. As we move along f from point P to point Q, the values of x undergo a change of Δx from x_0 to $x_0 + \Delta x$ and the y values undergo a change of Δy from $f(x_0)$ to $f(x_0 + \Delta x)$. When we let Δx approach zero it follows that Δy also approaches zero and the secant PQ assumes the limiting position defined by the tangent line T. Let us keep in mind the geometric interpretations exhibited in Figure 2.8 as we continue with more details of the theory underlying the derivative $f'(x)$.

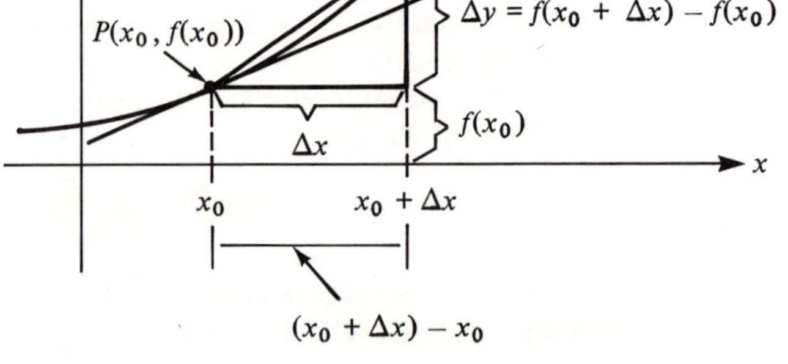

Figure 2.8

The fraction

$$\frac{f(x_0 + \Delta x) - f(x_0)}{\Delta x}$$

is commonly referred to as the **difference quotient**. If the limit of the difference quotient exists as $\Delta x \to 0$ then this limit is called the instantaneous rate of change of y with respect to x at $P(x_0, f(x_0))$. These notions are incorporated in a formal definition as follows.

DEFINITION 1 *The instantaneous rate of change of $y = f(x)$ with respect to x at a point $P(x_0, f(x_0))$ on f is defined to be the* ***derivative*** *of f with respect to x at point P, provided the limit exists.*

Note: In the limiting process of Definition 1 the point P is fixed.

The derivative is formulated as follows:

$$f'(x_0) = \lim_{\Delta x \to 0} \frac{\Delta y}{\Delta x} = \lim_{\Delta x \to 0} \frac{f(x_0 + \Delta x) - f(x_0)}{\Delta x} \tag{1}$$

where $f'(x_0)$ is read "f prime at x_0." Symbols other than $f'(x)$ may be used for the derivative. As examples, we have

$\dfrac{dy}{dx}$, read "dee y dee x,"

$D_x f(x)$, read "the derivative of $f(x)$ with respect to x,"

and y', read "y prime."

Equation (1) above indicates the derivative of $f(x)$ at $x = x_0$. For any x, Eq. (1) becomes

$$\boxed{f'(x) = \lim_{\Delta x \to 0} \frac{\Delta y}{\Delta x} = \lim_{\Delta x \to 0} \frac{f(x + \Delta x) - f(x)}{\Delta x}}. \tag{2}$$

Remarks: The derivative $f'(x)$ is a function, namely, the function derived from the function f. In Eq. (1) above, if x_0 is in the domain of f, then the limit (if it exists) acts as a rule which associates with each x_0 a number $f'(x_0)$. Thus, the derived function f' is the set of all ordered pairs resulting from the operation. The symbol $\dfrac{dy}{dx}$ denotes the derivative of y with respect to x. The $\dfrac{d}{dx}$ is an "operator" which operates on y to yield the derivative of y.

EXAMPLE 2 Given $f(x) = 2x^2 + 1$, find the average rate of change $\Delta y/\Delta x$ and the instantaneous rate of change $f'(x)$.

Solution

$$\begin{aligned} \Delta y &= f(x + \Delta x) - f(x) \\ &= 2(x + \Delta x)^2 + 1 - (2x^2 + 1) \\ &= 2x^2 + 4x\,\Delta x + 2(\Delta x)^2 + 1 - 2x^2 - 1 \\ &= 4x\,\Delta x + 2(\Delta x)^2. \end{aligned}$$

Dividing both sides of the equation by Δx, we obtain

$$\frac{\Delta y}{\Delta x} = \frac{4x\,\Delta x + 2(\Delta x)^2}{\Delta x}.$$

Simplification yields

$$\begin{aligned}\frac{\Delta y}{\Delta x} &= \frac{\Delta x}{\Delta x}(4x + 2\,\Delta x)\\ &= 4x + 2\,\Delta x. \qquad (3)\end{aligned}$$

This average rate of change represents the slope of the secant PQ in Figure 2.9.

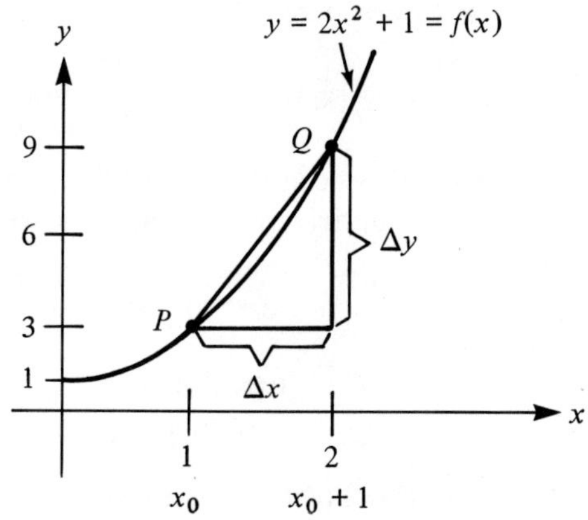

Figure 2.9

To find the slope of the secant PQ we let $x_0 = 1$ and $\Delta x = 2 - 1 = 1$ (see Figure 2.9). Substituting x_0 for x in Eq. (3) above we have

$$\frac{\Delta y}{\Delta x} = 4x_0 + 2\,\Delta x$$

which, evaluated at $x_0 = 1, \Delta x = 1$, becomes

$$\begin{aligned}\frac{\Delta y}{\Delta x} &= 4(1) + 2(1)\\ &= 4 + 2 = 6, \quad \text{the slope of secant } PQ.\end{aligned}$$

To find the instantaneous rate of change we find the derivative $f'(x)$ and evaluate it at $x_0 = 1$. Thus,

$$\begin{aligned}f'(x) &= \lim_{\Delta x \to 0} \frac{\Delta y}{\Delta x}\\ &= \lim_{\Delta x \to 0} (4x + 2\,\Delta x)\\ &= 4x.\end{aligned}$$

Then $f'(x_0) = 4(1) = 4$, the instantaneous rate of change of f at the point where $x = x_0 = 1$.

PROBLEM SET 2.3

Use methods of this section to find $\Delta y/\Delta x$ and $f'(x)$ where $y = f(x)$. Sketch graphs of the odd-numbered problems 1–13.

1. $f(x) = 3x + 4$

2. $f(x) = \dfrac{x}{2} - 3$

3. $f(x) = x^2$

4. $f(x) = 3x^2 - 5x - 8$

5. $f(x) = x^2 + x$

6. $f(x) = x^2 - x + 1$

7. $f(x) = x - 2x^2$

8. $f(x) = 1 + 3x - 4x^2$

9. $f(x) = x^3$

10. $f(x) = 2 - 3x^3$

$f(x) = \sqrt{x}$

12. $f(x) = \dfrac{1}{\sqrt{x}}$

13. $f(x) = \dfrac{1}{x}$

14. $f(x) = \dfrac{1}{x - 1}$

15. Discuss the ideas pertinent to Example 1 of this section when the productivity function is replaced by $P(d) = 30d - 3d^2$.

Apply Eq. (1) of this section to find $f'(x_0)$ for the indicated value of x_0:

16. $f(x) = x^2 + 10, \quad x_0 = 5$

17. $f(x) = 2 - 3x^2, \quad x_0 = 1$

18. $f(x) = 2x^2 + 4x - 5, \quad x_0 = 2$

19. $f(x) = 4x^2 - 3x + 10, \quad x_0 = 3$

20. $f(x) = x^3 + 1, \quad x_0 = 1$

21. $f(x) = \sqrt{x + 1}, \quad x_0 = 3.$

22. The research department of the ABC Corp. estimates total monthly costs (associated with the production of x transistors required to meet demand) as $C(x) = 10{,}000 + 5x + 0.02x^2$ dollars. Find the rate of change of total costs at the instant when 1000 transistors are produced, i.e., find $C'(1000)$.

23. Find an equation of the line tangent to each curve at the indicated point. Make use of the point-slope form of an equation of a straight line: $y - f(x_0) = f'(x_0)(x - x_0)$.

(a) $y = x^2 - 3x + 8$, point with ordinate 6
(b) $y = x^2 - x + 10$, point with abscissa 2
(c) $y - 2 = (x - 2)^2$, point where slope of the tangent is zero
(d) $y = x^4 - 8x^2$, point where slope of the tangent is zero

24. The Vietronics Co. has a revenue function given by $R(x) = 5x$ (R is gross revenue and x is quantity sold) and a production cost function given by $C(x) = 2 + (x^2/10)$. Find the rate of change of profits P with respect to the demand x when $x = 25$ units. [*Hint*: Profits P equal Revenues R minus Costs C.]

25. The monthly demand D (in thousands of cartons) of a certain vaccine is given by $D(q) = 10/(q + 1)$ where q denotes quantity of an item.

(a) Sketch a graph of D
(b) Find the rate of change of D when $q = 9$, i.e., find $D'(9)$.

2.4 DIFFERENTIATION

The process of finding a derivative is called differentiation. If $f(x)$ is a rather complicated function the method for finding $f'(x)$ as outlined in the previous section is a lengthy and tedious method. In view of this fact let us introduce more convenient and time-saving techniques for finding derivatives.

2.4.1 Techniques of Differentiation

THEOREM 1 *If $f(x) = c$, a constant for all values of x, then $f'(x) = 0$.*

Conclusion:

The derivative of a constant is zero.

EXAMPLE 1 Given $f(x) = 8$, find $f'(x)$.

Solution $f'(x) = 0$ by Theorem 1.

THEOREM 2 *If $f(x) = x^n$, n a positive integer, then $f'(x) = nx^{n-1}$.*

We choose to prove this important theorem.

Proof

$$f(x) = x^n \qquad \text{[by hypothesis]}$$

$$f(x + \Delta x) = (x + \Delta x)^n \qquad \text{[by substituting } x + \Delta x \text{ for } x\text{]}$$

$$\begin{aligned} f(x + \Delta x) - f(x) &= (x + \Delta x)^n - x^n \qquad \text{[by substitution]} \\ &= \left[x^n + nx^{n-1}\,\Delta x + \frac{n(n-1)}{2}x^{n-2}(\Delta x)^2 + \cdots + (\Delta x)^n\right] - x^n \qquad \text{[by the binomial theorem]} \\ &= nx^{n-1}\,\Delta x + \frac{n(n-1)}{2}x^{n-2}(\Delta x)^2 + \cdots + (\Delta x)^n \qquad \text{[by simplifying]} \end{aligned}$$

Dividing both sides of the equation by Δx and taking the limit as $\Delta x \to 0$, we obtain

$$\begin{aligned} \lim_{\Delta x \to 0} \frac{f(x + \Delta x) - f(x)}{\Delta x} &= \lim_{\Delta x \to 0}\left[\frac{nx^{n-1}\,\Delta x + \frac{n(n-1)}{2}x^{n-2}(\Delta x)^2 + \cdots + (\Delta x)^n}{\Delta x}\right] \\ &= \lim_{\Delta x \to 0}\left[nx^{n-1} + \frac{n(n-1)}{2}x^{n-2}\,\Delta x + \cdots + (\Delta x)^{n-1}\right] \\ &\qquad \text{[by dividing numerator and denominator by } \Delta x\text{]} \\ &= nx^{n-1} = f'(x) \end{aligned}$$

Conclusion:

The derivative of a variable raised to a positive integral power is equal to the power times the variable raised to the power less 1.

EXAMPLE 2 If $f(x) = x^3$ then $f'(x) = 3x^{3-1} = 3x^2$.

In General:

$D_x x^n = nx^{n-1}$ for any real number n. (It was proved above for positive integral values of n, but it will be assumed at this time for all real values of n.)

EXAMPLE 3 $D_x x^{2/3} = \frac{2}{3}x^{(2/3)-1} = \frac{2}{3}x^{-1/3} = \frac{2}{3x^{1/3}}$.

THEOREM 3 *$D_x cu(x) = c \cdot D_x u(x)$ if c is a constant.*

Conclusion:

The derivative of a constant times a function is equal to the constant times the derivative of the function.

EXAMPLE 4

$$D_t(2t^{-2}) = 2 \cdot D_t t^{-2}$$
$$= 2(-2t^{-3}) = -\frac{4}{t^3}.$$

THEOREM 4 *If $F(x) = g(x) \pm h(x)$, then $F'(x) = g'(x) \pm h'(x)$ provided $g'(x)$ and $h'(x)$ exist.*

Conclusion:

The derivative of the sum or difference of a finite number of functions is the sum or difference of the derivatives (if they exist) of the functions.

EXAMPLE 5 Let x be the number of items demanded of a manufactured product, p be the price per unit, and $p = a - bx$ be the demand relation, where a and b are constants. Then the total revenue function R becomes

$$R(x) = px = (a - bx)x = ax - bx^2.$$

Applying Theorems 2, 3, and 4 above we obtain

$$\begin{aligned} R'(x) &= a \cdot D_x x - b \cdot D_x x^2 \\ &= a \cdot 1 - b \cdot 2x \\ &= a - 2bx. \end{aligned}$$

The rate of change of R is R' which is called the **marginal revenue** function. Observe that the revenue function was formed by taking the product of the number x of units demanded by the price p per unit. Moreover, the marginal revenue function was derived by finding the rate of change of R with respect to x. Marginal revenue $R'(x)$ is defined to be the change in total revenue due to the sale of one additional unit of output x.

EXAMPLE 6 Assume that the cost of producing x tables is given by the function $C(x) = 15 + 2x + 50/x$ dollars, $x \geq 15$. Upon examination we see that the given cost function includes a constant term, 15, representing a fixed cost (for example, overhead) which does not depend on the number x of tables produced. Next, the function also includes the term $2x$, implying that cost increases linearly as the number of tables produced. Finally, the cost function depends on the term $50/x$, which decreases in value as x increases, implying a decrease in cost as more tables are produced. To find **marginal cost** we find

$$\begin{aligned} C'(x) &= D_x(15 + 2x + 50x^{-1}) \\ &= 0 + 2 + 50(-x^{-2}) \\ &= 2 - \frac{50}{x^2}. \end{aligned}$$

If we wish to know the marginal cost when production is 10 tables per certain period, we simply evaluate

$$\begin{aligned} C'(10) &= 2 - \frac{50}{(10)^2} \\ &= 2 - \frac{5}{10} \\ &= 2 - \frac{1}{2} = \$1.50. \end{aligned}$$

Thus, the rate of change of C is \$1.50 at the instant that $x = 10$. Keep in mind that marginal cost is the rate of change of cost for each additional unit of output x.

We conclude this section with the differentiation of a rather formidable function.

EXAMPLE 7 Find $F'(x)$ if $F(x) = 2x^{3/2} - 3x^{2/3} - x^{-1/3}$

Solution

$$\begin{aligned} F'(x) &= 2\left(\frac{3}{2}x^{1/2}\right) - 3\left(\frac{2}{3}x^{-1/3}\right) - \left(-\frac{1}{3}x^{-4/3}\right) \\ &= 3\sqrt{x} - \frac{2}{x^{1/3}} + \frac{1}{3x^{4/3}} \end{aligned}$$

PROBLEM SET 2.4.1

Use the theorems and examples of this section as guides to differentiating the following functions.

1. $y = 7$

2. $y = 3$

3. $y = \pi + 2\sqrt{5}$

4. $f(x) = 2x - 1$

5. $f(x) = 1 - 3x$

6. $g(x) = x^2 + 2$

7. $L(x) = x^2 + 2x - 3$

8. $f(x) = 1.5 - 0.5x - 3x^2$

9. $F(x) = \dfrac{3x^2}{2} - \dfrac{2x}{3} + 0.75$

10. $G(x) = x^3 - x$

11. $H(x) = 2x^3 - 3x^2 + 4x + 5$

12. $f(t) = t^6 - 6t^5 - 30t^4 - 120t^3 - 360t^2 - 720t$

13. $g(x) = \dfrac{3}{5x^4} + \dfrac{4}{3\sqrt{x}} + \dfrac{5}{2x^{1/3}} + 6$

14. $h(x) = 3x^{-1} + 8x^{-2} + 10x^{-3} + 2$

15. $y = (2x - 1)(x + 1)$

16. $y = (x + 3)^2$

17. $f(x) = (\sqrt{x} + 1)^2$

18. $g(x) = (x + 3)^3$

19. $h(t) = (t^2 - 1)(t^2 + 2)$

20. $F(x) = \dfrac{2x^3 - \sqrt{x}}{x}$

21. $G(x) = \dfrac{2x^4 - 3x^3 + 4x^2 - 5x + 6}{x^2}$

22. A firm's total revenue function is given by $R(x) = 100x - 5x^2$. Find the marginal revenue $R'(x)$; find $R'(10)$.

23. The cost of producing x radios is defined by $C(x) = 30 + 3x + (100/x)$, $x \geq 30$. Find the marginal cost for any x; then for $x = 20$.

24. If total revenue from sales of x cars at price p is $R(x) = px$, where p depends on x according to the rule $p = 50 - 0.02x$, find:
   (a) The marginal revenue function
   (b) Marginal revenue at $p = 49$.

25. The cost of production $C(x)$ is given by the function $C(x) = 200 + 7.68x - 0.01x^3$, x being the number of items produced. Find:
   (a) The marginal cost function
   (b) The number of items produced for which the marginal cost is zero.

26. A supply function is given by $S(x) = 1 + (x^{3/2}/6)$ where x denotes the number (in thousands) of items produced.
   (a) Find the marginal supply function
   (b) Determine the value of x for which $S'(x) = 3$.

2.4.2 Additional Techniques. The Chain Rule

THEOREM 5 Derivative of a Product

If $f(x) = u(x) \cdot v(x)$, then $f'(x) = u(x) \cdot v'(x) + v(x) \cdot u'(x)$, provided $u'(x)$ and $v'(x)$ exist. Briefly, if $f = uv$, then $f' = uv' + vu'$.

Conclusion:

The derivative of the product of two functions is equal to: The first function times the derivative of the second *plus* the second function times the derivative of the first.

To clarify this formula, we offer a nonrigorous proof.

Proof Let $y = uv$ where u and v are differentiable functions of x. Since $y = f(x)$, a change Δx results in corresponding changes Δu, Δv, and Δy. So

$$\begin{aligned} y + \Delta y &= (u + \Delta u)(v + \Delta v) \\ &= uv + u\,\Delta v + v\,\Delta u + \Delta u\,\Delta v. \end{aligned}$$

Subtracting $y = uv$, we obtain

$$\Delta y = u\,\Delta v + v\,\Delta u + \Delta u\,\Delta v.$$

After dividing both sides by Δx, we get

$$\frac{\Delta y}{\Delta x} = u\frac{\Delta v}{\Delta x} + v\frac{\Delta u}{\Delta x} + \Delta u\frac{\Delta v}{\Delta x}.$$

The function u is differentiable, so it is also continuous (this can be shown). Therefore,

$$\Delta u \to 0 \quad \text{as} \quad \Delta x \to 0$$

This means that the term

$$\Delta u\frac{\Delta v}{\Delta x} \to 0 \cdot \frac{\Delta v}{\Delta x} = 0 \quad \text{as} \quad \Delta x \to 0.$$

Accordingly, as $\Delta x \to 0$,

$$\frac{\Delta y}{\Delta x} \to u\frac{dv}{dx} + v\frac{du}{dx}$$

That is, if $y = uv$, then

$$\frac{dy}{dx} = u\frac{dv}{dx} + v\frac{du}{dx}$$

or

$$f'(x) = u(x) \cdot v'(x) + v(x) \cdot u'(x) \qquad (y = f(x))$$

as was to be shown. (We will occasionally give detailed proofs in order to clarify the underlying meanings of certain important formulas.)

EXAMPLE 8 If $f(x) = (2x - 1)(x^2 + 3x - 5)$, then

$$\begin{aligned} f'(x) &= (2x - 1) \cdot D_x(x^2 + 3x - 5) + (x^2 + 3x - 5) \cdot D_x(2x - 1) \\ &= (2x - 1)(2x + 3) + (x^2 + 3x - 5) \cdot 2 \\ &= 4x^2 + 4x - 3 + 2x^2 + 6x - 10 \\ &= 6x^2 + 10x - 13. \end{aligned}$$

THEOREM 6 Derivative of a Quotient

If $f(x) = \dfrac{u(x)}{v(x)}$, $v(x) \neq 0$, *then*

$$f'(x) = \frac{v(x) \cdot u'(x) - u(x) \cdot v'(x)}{v^2(x)}, \qquad \textit{provided } u'(x) \textit{ and } v'(x) \textit{ exist.}$$

Briefly, if $f = \dfrac{u}{v}$, *then* $f' = \dfrac{vu' - uv'}{v^2}$.

Conclusion:

The derivative of a quotient of two functions is: The denominator times the derivative of the numerator minus the numerator times the derivative of the denominator, all divided by the square of the denominator.

EXAMPLE 9 If $g(x) = \dfrac{x^3}{2x^2 + 3}$, then

$$g'(x) = \frac{(2x^2 + 3) \cdot D_x x^3 - x^3 \cdot D_x(2x^2 + 3)}{(2x^2 + 3)^2}$$
$$= \frac{(2x^2 + 3) \cdot 3x^2 - x^3 \cdot 4x}{(2x^2 + 3)^2} = \frac{2x^4 + 9x^2}{(2x^2 + 3)^2}.$$

EXAMPLE 10 **(a)** If $h(x) = 100/x^2$, then

$$h'(x) = 100 \cdot D_x x^{-2} = 100(-2x^{-3}) = -\frac{200}{x^3}.$$

(b) If $g(x) = x^2/100$, then

$$g'(x) = \frac{1}{100} \cdot D_x x^2 = \frac{1}{100}(2x) = \frac{x}{50}.$$

Conclusion:

If the numerator or the denominator of a quotient is a constant, then the quotient rule for finding the derivative need not be applied. Faster methods may be used, as exhibited in (a) and (b) of Example 10.

THEOREM 7 **The Chain Rule**

If $y = f(x)$ and $x = g(t)$, then $y = f[g(t)]$ and

$$D_t y = f'[g(t)] \cdot g'(t) \tag{1}$$

or equivalently,

$$\boxed{\frac{dy}{dt} = \frac{dy}{dx} \cdot \frac{dx}{dt}}. \tag{2}$$

Experience has shown that students remember the chain rule better when it is written in form (2) above rather than in form (1).

EXAMPLE 11 Use the chain rule to find dy/dt if $y = x^3 - 2x$ and $x = t^2$.

Solution By the chain rule,

$$\begin{aligned}\frac{dy}{dt} &= \frac{dy}{dx} \cdot \frac{dx}{dt} \\ &= \frac{d}{dx}(x^3 - 2x) \cdot \frac{d}{dt}(t^2) \\ &= (3x^2 - 2) \cdot 2t \\ &= [3(t^2)^2 - 2] \cdot 2t \\ &= (3t^4 - 2) \cdot 2t = 2t(3t^4 - 2).\end{aligned}$$

An important special case of the chain rule follows.

COROLLARY 1 *If $f(x) = [u(x)]^n$ and n is rational, then*

$$f'(x) = n[u(x)]^{n-1} \cdot D_x u(x).$$

In brief, if $f = u^n$ then $f' = nu^{n-1} \cdot D_x u$.

Conclusion:

The derivative of a function to a power is the power times the function to the power less 1, times the derivative of the function.

EXAMPLE 12 Given $f(x) = (x^2 - 2x)^4$, find **(a)** $f'(x)$; **(b)** $f'(2)$, i.e., the slope of the tangent to f at $x = 2$.

Solution **(a)** By Corollary 1 we have

$$\begin{aligned} f'(x) &= 4(x^2 - 2x)^3 \cdot D_x(x^2 - 2x) \\ &= 4(x^2 - 2x)^3(2x - 2) \\ &= 8(x^2 - 2x)^3(x - 1) \end{aligned}$$

(b)

$$\begin{aligned} f'(2) &= 8(4 - 4)^3(2 - 1) \\ &= 8(0)(1) = 0 \end{aligned}$$

Another special case, the chain rule for roots, is now presented.

COROLLARY 2 *If $y = [f(x)]^{p/q}$ where p and q are positive integers, then*

$$\frac{dy}{dx} = \frac{p}{q}[f(x)]^{(p/q)-1} \cdot \frac{d}{dx} f(x).$$

EXAMPLE 13 Given $y = (1 - 3x)^{2/3}$, find dy/dx.

Solution

$$\begin{aligned} \frac{dy}{dx} &= \frac{2}{3}(1 - 3x)^{(2/3)-1} \cdot D_x(1 - 3x) \\ &= \frac{2}{3}(1 - 3x)^{-1/3}(-3) \\ &= \frac{-2}{(1 - 3x)^{1/3}} \end{aligned}$$

EXAMPLE 14 The demand for a certain commodity is defined by $D(q) = \sqrt{25 - q^2}$ where q denotes quantity of items produced.

(a) Find the marginal demand function at $q = 3$.

(b) Find an equation of the line tangent to $D(q)$ at the point where $D = 3$.

Solution **(a)** Marginal demand equals

$$\begin{aligned} D'(q) &= \frac{d}{dq}(25 - q^2)^{1/2} \\ &= \frac{1}{2}(25 - q^2)^{-1/2}(-2q) \\ &= -\frac{q}{\sqrt{25 - q^2}}. \end{aligned}$$

Then $D'(3) = -\frac{3}{\sqrt{25 - 9}} = -\frac{3}{4}$.

(b) When $D = 3$ we have

$$3 = \sqrt{25 - q^2}$$

or

$$9 = 25 - q^2 \qquad \text{[squaring both sides]}$$

$$q^2 = 25 - 9 = 16$$

$$q = \pm 4$$

where the -4 is discarded because q (quantity of items) cannot be negative. Thus, we have the point (4, 3). The slope of $D(q)$ is

$$D'(q) = -\frac{q}{\sqrt{25 - q^2}}$$

as shown in (a). At the point (4, 3) this slope is

$$D'(4) = -\frac{4}{\sqrt{9}} = -\frac{4}{3}.$$

Using the point (4, 3) as (x_1, y_1) and $-4/3$ as m in the equation of a line,

$$y - y_1 = m(x - x_1) \qquad \text{[point-slope form]}$$

we obtain

$$y - 3 = -\frac{4}{3}(x - 4) \qquad [y_1 = D = 3, x_1 = q = 4]$$

or

$$3y - 9 = -4x + 16$$

which simplifies to

$$4x + 3y = 25,$$

the desired equation of the tangent line.

Note: Another important technique, called Implicit Differentiation, is introduced in Section 3.3, Enrichment Topics and Problems, as well as in Section 5.4.

PROBLEM SET 2.4.2

Find the derivative of each function in exercises 1–13; evaluate the derivative at the indicated point.

1. $f(x) = (2x - 1)(x^2 - x + 1); \quad x = 1$
2. $g(t) = (1 + t)(1 - t)^2; \quad t = 0$
3. $f(x) = x^2(2x - 3)^2; \quad x = 2$
4. $u(x) = (3x + 2)\sqrt{x}; \quad x = 1$
5. $h(u) = \sqrt{u} \cdot (u - 3)^3; \quad u = 4$
6. $F(x) = (x - 1)^3 \cdot (x + 2)^4; \quad x = 0$
7. $G(x) = (x^2 + 1)(2x - 1)^3; \quad x = 1$
8. $H(t) = t(1 - t)(1 + t)^2; \quad t = 2$

9. $y = \dfrac{10}{x+2} + 3; \quad x = 8$ **10.** $f(u) = \dfrac{u^2 - 2u}{5} - 6; \quad u = 2$

11. $g(t) = \dfrac{2t-1}{4t+1}; \quad t = 0$ **12.** $u(x) = \dfrac{1+2x}{1-2x}; \quad x = 0$

13. $h(z) = \dfrac{5 - z^2}{3 + z^2}; \quad z = 1$

14. The IPM Company has found its total cost function to be defined by $Q(x) = (2x + 1)^{2/3}$ where x is the number of units of a product sold.

(a) Sketch the graph of Q.
(b) At what rate is cost increasing when $x = 13$?

Find the indicated derivative by using the chain rule for each of the following exercises 15–22:

15. dy/dt if $y = 2x + 3$ and $x = 3t - 2$
16. dy/du if $y = x^2 + 1$ and $x = 1 - u^2$
17. dy/dx if $y = \sqrt{u}$ and $u = x^{1/3}$
18. dy/dt if $y = x^2 - x + 3$ and $x = t^3$
19. dy/dz if $y = x^3 + 3x$ and $x = 1 - z$
20. dz/du if $z = t^2 - t^3$ and $t = 2u - 1$
21. du/dx if $u = 1/t$ and $t = 1 - x$
22. dt/dq if $t = (u - 2)^2$ and $u = q^2$

23. If profit P depends on sales s according to $P(s) = 1 + 0.01s^3$ and s depends on advertising v according to $s(v) = 1 + v^{2/3}$, find the rate of change of profit with respect to advertising.

For each of exercises 24–28 find the derivative and use it to obtain an equation of the line tangent to the given curve at the indicated point.

24. $f(x) = (x^2 - 3x)^3; \quad (1, f(1))$.

25. $g(u) = 2\sqrt{u} - \dfrac{2}{\sqrt{u}}; \quad (1, g(1))$.

26. $h(t) = (t^2 - 3)^3; \quad (2, h(2))$.

27. $F(x) = \sqrt{x^2 + 2x + 4}; \quad (0, F(0))$.

28. $G(z) = \dfrac{z^2 + 1}{z^2 - 1}; \quad (0, G(0))$.

29. The total sales revenue (in millions of dollars) of the PTC Corp. is given by $R(x) = -1 + \sqrt{x + 1}$ where x (≤ 8) denotes the number of years PTC has been in operation.

(a) Find the rate at which revenue is increasing at the end of 8 years.
(b) Find an equation of the line tangent to R at the point where $x = 3$.

2.4.3 Derivatives of Higher Order

If $y = f(x)$ is a function having f' as its first derivative, then f' is also a function. If f' is differentiable then the derivative of f' is called the second derivative of f and is denoted by f'', read "f double prime." Other symbols used to represent the second derivative of f are: $D_x^2 f$, d^2y/dx^2, and

y''. The second derivative of f indicates the *rate of change of the slope* of the tangent to f at the point of tangency. No new rules are needed for finding the second derivative or derivatives of higher order since we simply "take the derivative of a derivative."

The derivative of the second derivative, if it exists, is called the third derivative and is written $f'''(x)$ or $D_x^3 f$ or d^3y/dx^3 or y'''. Derivatives of still higher order have analogous symbols, for example, $f^{(4)}(x), f^{(5)}(x), \ldots, f^{(n)}(x)$, etc.

EXAMPLE 15 If $f(x) = 2x^4$, then

$$\begin{aligned} f'(x) &= 8x^3 \\ f''(x) &= 24x^2 \\ f^{(3)}(x) &= 48x \\ f^{(4)}(x) &= 48 \\ f^{(5)}(x) &= 0 \end{aligned}$$

and

$$f^{(n)}(x) = 0, \quad n \geq 5.$$

Since velocity and acceleration are terms familiar to most people we will make use of them to motivate our study of derivatives.

An object moving in a straight line according to the equation of motion

$$s = f(t)$$

is said to have **velocity** $v(t)$ defined by

$$v(t) = s' = f'(t).$$

That is, the velocity of the object is the first derivative of its function of motion s with respect to time t.

The **acceleration** $a(t)$ is defined to be the rate of change of the velocity of the object with respect to time, i.e.,

$$a(t) = v'(t) = s'' = f''(t).$$

To summarize:

If $$s = f(t)$$

then velocity $$v(t) = \frac{ds}{dt}$$

and acceleration $$a(t) = \frac{d^2s}{dt^2}.$$

EXAMPLE 16 An object is moving along a straight line according to the equation of motion $s(t) = t^3 - 3t^2 + 5t + 1$, $t \geq 0$, where s is the distance (in meters) of the object from the origin after t seconds.

(a) Find time t_0 such that acceleration $a(t_0) = 0$.

(b) Find velocity $v(t_0)$ and position $s(t_0)$.

Solution **(a)**

$$\begin{aligned} v(t) &= \frac{ds}{dt} \\ &= \frac{d}{dt}(t^3 - 3t^2 + 5t + 1) \\ &= 3t^2 - 6t + 5 \\ a(t) &= 6t - 6 \\ &= 0 \quad \text{for} \\ 6t &= 6 \quad \text{or} \quad t = 1 \text{ (second)} = t_0. \end{aligned}$$

(b)

$$\begin{aligned} v(t_0) &= 3t_0^2 - 6t_0 + 5 \\ &= 3(1) - 6(1) + 5 \\ &= 2 \text{ meters per second.} \\ s(t_0) &= t_0^3 - 3t_0^2 + 5t_0 + 1 \\ &= 1 - 3 + 5 + 1 \\ &= 4 \text{ meters.} \end{aligned}$$

As another instance, recall the revenue function

$$\begin{aligned} R(x) &= px \\ &= (a - bx)x \qquad [\text{for } p = a - bx] \\ &= ax - bx^2, \quad a, b > 0. \end{aligned}$$

Then
$$\begin{aligned} R'(x) &= a - b \cdot 2x \\ &= a - 2bx \end{aligned}$$
and
$$R''(x) = -2b.$$

Thus, the rate of change of the marginal revenue is negative. In other words, it is decreasing for any output level x.

EXAMPLE 17 Assume that sales revenue (in hundreds of dollars) is given by $R(q) = (40 - 2q)^3/36$, $q < 20$, where q denotes quantity of goods sold. Find the rate of change of marginal revenue when $q = 17$ (that is, find $R''(17)$).

Solution

$$\begin{aligned} R'(q) &= \frac{3}{36}(40 - 2q)^2(-2) \\ &= -\frac{1}{6}(40 - 2q)^2 \end{aligned}$$

Then
$$\begin{aligned} R''(q) &= -\frac{2}{6}(40 - 2q)(-2) \\ &= \frac{2}{3}(40 - 2q), \end{aligned}$$

the rate of change of marginal revenue at any $q < 20$. At $q = 17$, we have

$$\begin{aligned} R''(17) &= \frac{2}{3}(40 - 34) \\ &= \frac{2}{3}(6) = 4, \quad \text{or } \$400, \end{aligned}$$

an increasing rate of change.

PROBLEM SET 2.4.3

1. If $f(x) = x^3 - 4x + 10$, find f', f'', and $f^{(3)}$.

2. Given $g(x) = 3x^4 - 7x^2 + 20x$, find $g', g'', \ldots, g^{(5)}$.

3. Find x such that $h''(x) = 0$ when $h(x) = \left(1 - \dfrac{x}{2}\right)^4$.

4. A demand function is defined by $y = \dfrac{1}{1 + x}$. Find $\dfrac{d^3y}{dx^3}$.

5. If $F(p) = (4p + 3)/2p$, find $F''(p)$.

6. Given $f(x) = 4x - 3x^3 - x^4$, find $f''(0)$.

7. Evaluate $g''(-2)$ and $g^{(5)}(0)$ if $g(x) = x^5 - 3x^3 + 10$.

8. For the following position functions find the instant $t_0 \geq 0$ when acceleration a is zero; then find velocity $v(t_0)$ and position $s(t_0)$.

(a) $s(t) = 120 - 6t + t^3$

(b) $s(t) = (2t - 3)^4 + 6t$

(c) $s(t) = \dfrac{1}{t} - \dfrac{t^3}{3}$.

9. Given the revenue function $R(x) = 15x + (3x^2/2) - (2x^3/3)$ where x denotes quantity of items sold, find:

(a) The marginal revenue
(b) The rate of change of marginal revenue
(c) $R''(4)$.

10. The cost of production of a firm is defined by $C(x) = (x^3/1000) + 10$ where x is the number of units produced. Find:

(a) The marginal cost
(b) The rate of change of marginal cost
(c) $C'(100)$ and $C''(100)$.

11. The total engine maintenance cost (dollars per kilometer) of a freight company is given by $M(x) = (1600/x) + 10^{-6}x$ where x represents the interval (in thousands of kilometers) between engine overhauls. Find:

(a) $M'(x)$
(b) The value of x for which $M'(x) = 0$
(c) $M''(10)$.

12. A demand function "behaves" according to the rule $D(q) = 100 - \frac{1}{2}(q + 1)^4$ where q indicates the amount of goods in thousands.

(a) At what rate is demand changing when $q = 2$?
(b) At what rate is marginal demand changing when $q = 3$?

13. Find the rate of change of the *slope* of the curve $f(x) = x^4 - x^3$ at the point where $x = \frac{1}{2}$.

14. At what rate is the slope of the tangent to $g(t) = \sqrt{5 + 4t}$ changing when $t = 1$?

15. Find acceleration a in m per sec per sec at time $t = 1$ sec if the position function is defined by $s(t) = (t^2 - t - 1)^2$.

2.5 DIFFERENTIALS

Let us define dx and dy as individual algebraic entities and call them "**differential** x" and "**differential** y," respectively. If $y = f(x)$ and f is differentiable with respect to the independent variable x, then we will let dx be equivalent to Δx. That is, $dx = \Delta x$. The first derivative of $y = f(x)$ is defined by

$$f'(x) = \lim_{\Delta x \to 0} \frac{\Delta y}{\Delta x} = \frac{dy}{dx}.$$

In Figure 2.10 we see that dy and Δy are approximately equal to each other. The approximation gets better as $\Delta x \to 0$. Thus,

$$\frac{\Delta y}{\Delta x} = f'(x) + \varepsilon$$

where $\varepsilon \to 0$ as $\Delta x \to 0$ (refer to Figure 2.10).

Let us consider the differential dx as an independent variable and let differential dy depend on both x and dx. This leads to the definition

$$\boxed{dy = f'(x)\, dx}\,.$$

From this definition and Figure 2.10 it is clear that dy is also dependent on the tangent line PT. The tangent line is determined by x and the function f.

One of the chief advantages of the differential notation comes from dx and dy being separate entities. This enables us to write

$$\frac{(dy)}{(dx)} = f'(x)$$

or

$$dy = f'(x)\, dx, \tag{1}$$

thus treating dy/dx as if it were a fraction.

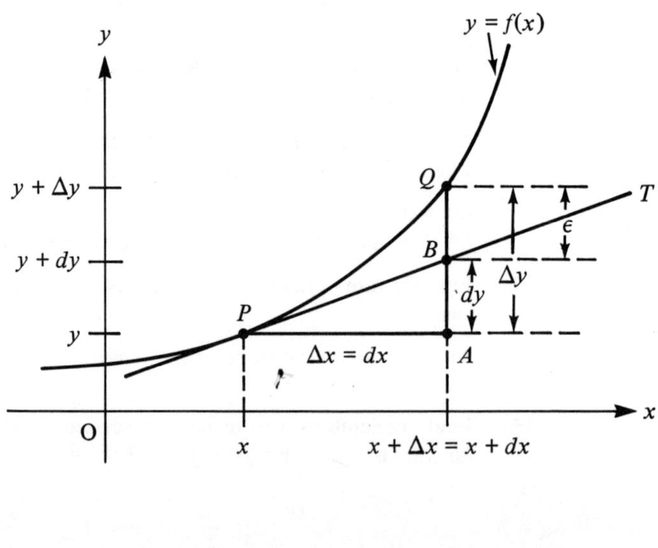

Figure 2.10

Formulas for derivatives have their counterparts in formulas for differentials. If u and v are differentiable functions of x, then

(1) $d(cu) = c\,du$

(2) $d(u^n) = nu^{n-1}\,du$

(3) $d(u \pm v) = du \pm dv$

(4) $d(uv) = u\,dv + v\,du$

(5) $d\left(\dfrac{u}{v}\right) = \dfrac{v\,du - u\,dv}{v^2}, \qquad v \neq 0.$

Several illustrations follow.

If $f(x) = 2x^2 + 1$, then $df = f'(x)\,dx = 4x\,dx$.

If $f(x) = (1 - x^2)^2$, then

$$df = f'(x)\,dx = 2(1 - x^2)(-2x\,dx) = -4x(1 - x^2)\,dx.$$

If $g(t) = (3t + 1)^{1/3}$, then

$$dg = g'(t)\,dt = \frac{1}{3}(3t + 1)^{-2/3}(3)\,dt = \frac{dt}{(3t + 1)^{2/3}}.$$

If $y = u/2 + 2/u$, then

$$dy = \frac{dy}{du}(du) = \left(\frac{1}{2} - \frac{2}{u^2}\right)du.$$

Thus, to find the differential of a function, say $f(x)$, we merely find the derivative of $f(x)$ and multiply by dx. Since $\Delta y \to dy$ as $\Delta x \to 0$ we make use of differentials to compute approximations. We emphasize that $\Delta x = dx$ must be "sufficiently" small if the approximations are to be "good." As an example of this, let us apply differentials to approximate $\sqrt[3]{26}$.

EXAMPLE 1 Let $y = \sqrt[3]{x} = f(x) = x^{1/3}$ and let $x = 27$, $dx = -1$. The desired answer will be represented by $y + dy \cong y + \Delta y$. Thus,

$$\begin{aligned} y + dy &= y + f'(x)\,dx \qquad [\text{since } dy = f'(x)\,dx] \\ &= \sqrt[3]{x} + \frac{1}{3}x^{-2/3}\,dx\,\bigg|_{\substack{x=27\\dx=-1}} \\ &= \sqrt[3]{27} + \frac{1}{3}\cdot\frac{1}{(27)^{2/3}}(-1) \\ &= 3 - \frac{1}{3}\cdot\frac{1}{9} \\ &\cong 3 - \frac{1}{3}(0.111) \\ &= 3 - 0.0370 \\ &= 2.963, \text{ to three decimal places.} \end{aligned}$$

A table of cube roots gives the cube root of 26 as 2.962. Our approximation by differentials differs from this answer by only 0.001.

As an alternate approach to this example, if $y = f(x)$ then $y + \Delta y = f(x + \Delta x)$, since some change Δx results in a corresponding change Δy. Then

$$\begin{aligned} f(x + \Delta x) &= (x + \Delta x)^{1/3} && [f(x) = x^{1/3}] \\ &= (27 - 1)^{1/3} && [\text{substituting}] \\ &= \sqrt[3]{26}. \end{aligned}$$

Since $$y + \Delta y \cong y + dy \qquad [\Delta y \cong dy]$$

then $$\sqrt[3]{26} \cong 2.963, \text{ to three decimal places}$$

as computed above.

EXAMPLE 2 Consider another application. Find the approximate error in the area A of a square, assuming a side x of length 80 m and a possible error of measurement of $\frac{1}{16}$ m.

Solution

$$A = x^2$$

$$\begin{aligned} dA &= 2x\,dx \bigg|_{\substack{x=80 \\ dx=1/16}} \\ &= 2(80) \cdot 1/16 \\ &= 10 \text{ m}^2. \end{aligned}$$

That is, the approximate error dA in the area for the given conditions is 10 square meters.

We define **relative error** in measuring a quantity u with an error of Δu as $\Delta u/u$. To find the **percentage error** we evaluate $(\Delta u/u) \cdot 100\%$.

EXAMPLE 3 If the percentage error in measuring the radius r of a spherical ball is 3%, find the percentage error in computing its volume V.

Solution

$$V = \frac{4}{3}\pi r^3$$

$$dV = 4\pi r^2\,dr.$$

The relative error is

$$\begin{aligned} \frac{dV}{V} &= \frac{4\pi r^2\,dr}{(4/3)\pi r^3} \\ &= 3 \cdot \frac{dr}{r} \end{aligned}$$

and the percentage error is

$$\begin{aligned} \frac{dV}{V}(100\%) &= 3 \cdot \frac{dr}{r}(100\%) \\ &= 3(0.03)(100\%) \\ &= 9\%. \end{aligned}$$

PROBLEM SET 2.5

Find the differential dy for the functions 1–15.

1. $y = 2x + 1$ **2.** $y = x^2 - 5x + 10$

3. $y = 3\sqrt{x}$ **4.** $y = \dfrac{2}{x} + 4$

5. $y = (2x - 3)^2$ **6.** $y = (x - 2)^3$

7. $y = \dfrac{4}{x - 2}$ **8.** $y = \dfrac{3x - 2}{2x + 1}$

9. $y = \sqrt{2x + 1}$ **10.** $y = \sqrt[3]{1 + x^3}$

11. $y = \dfrac{3}{t} - \dfrac{t}{3} + 3$ **12.** $y = u^2(u^2 - 1)^2$

13. $y = z(z + 1)(z - 1)$ **14.** $y = \dfrac{3x^3 - 4x^2 - 1}{x^2}$

15. $y = x^5 + 5x^4 + 20x^3 + 60x^2 + 120x$

Use differentials to approximate the numbers indicated in problems 16–23:

16. $\sqrt{8}$ **17.** $\sqrt{65}$ **18.** $\sqrt[3]{63}$ **19.** $\sqrt[3]{995}$

20. $\sqrt[4]{83}$ **21.** $\sqrt[4]{15}$ **22.** $(1.2)^{-3}$ **23.** $(99)^{-2}$

24. The diameter of a circular toy is found to be 8.28 ± 0.04 inches by measurement. To what degree of accuracy can the circumference C and area be found? [$C = 2\pi r$]

25. The formula $F = (9/5)C + 32$ defines Fahrenheit temperature in terms of centigrade temperature. If the Fahrenheit temperature can be read to the nearest 1°, to what degree of accuracy can centigrade temperature be found?

26. Use differentials to find the approximate change in profit P if $P = x^2 - 20x$ dollars and x changes from 30 to 33 units of production. [*Hint*: $x = 30$, $dx = 3$.]

27. A company contracts to dip-paint circular advertising signs, each of diameter 3 feet. Upon delivery each sign is found to be 1 inch in diameter oversize. About what percentage of additional paint will be needed?

28. The ATD Co. finds that the supply for x units of its product is determined by the price function $y = (8 + x^2)$ dollars. If x changes from 10 to 11 units, find the corresponding percentage change in price.

29. The profit P of a firm can be determined by $P(x) = 300 + 100x - x^2$ where x denotes quantity of production. Find the approximate percentage change in profit when production changes from 40 to 45 units. [*Hint*: $x = 40$, $dx = 5$.]

30. The total cost of manufacturing x units of product K_{22} is defined by $C(x) = 2 + x/2 + x^2/4$. Approximate the percentage change in cost C when x increases from 20 to 22 units.

2.6 THE LAW OF THE MEAN

This section concerns a basic and important property of the derivative, known as the law of the mean, or the mean value theorem for derivatives. Geometrically, the theorem asserts that if the graph of f, continuous on

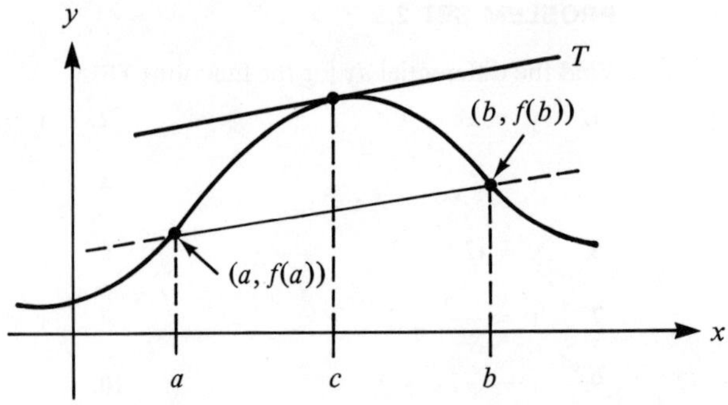

Figure 2.11

$[a, b]$, has a nonvertical tangent for each point in (a, b), then the graph has at least one tangent line T parallel to the line through $(a, f(a))$ and $(b, f(b))$ (see Figure 2.11). In other words, there exists at least one number c in the open interval (a, b) such that the *slope* of the chord with endpoints $(a, f(a))$ and $(b, f(b))$ equals the *slope* of the curve $f(x)$ at some $x = c$. Let us state these ideas as a theorem without proof.

THEOREM 1 The Mean Value Theorem for Derivatives

Let f be a function continuous on the closed interval $[a, b]$ and differentiable on the open interval (a, b). Then there is at least one number c in (a, b) such that

$$\frac{f(b) - f(a)}{b - a} = f'(c).$$

This theorem has important uses in establishing many other theorems in differential calculus. An interesting application is one in which the derivative is interpreted as velocity. For instance, if a car moves so that its positions are $f(t_2)$ at time t_2 and $f(t_1)$ at time t_1, then the "difference quotient"

$$\frac{f(t_2) - f(t_1)}{t_2 - t_1}$$

represents the ratio

$$\frac{\text{change in position}}{\text{change in time}}.$$

That is, the theorem states that at some time between t_1 and t_2 the velocity of the car equals its average velocity. In particular, if a car covers 200 kilometers in 4 hours, then at some time the car must have gone 50 kilometers per hour.

EXAMPLE 1 Find all values of c in the interval $(1, 3)$ that satisfy the conditions of the law of the mean for $f(x) = x^2 + 1$.

Solution

$$b = 3, \quad f(b) = f(3) = 3^2 + 1 = 10$$
$$a = 1, \quad f(a) = f(1) = 1^2 + 1 = 2$$
$$f'(x) = 2x, \qquad f'(c) = 2c$$

Then

$$f'(c) = \frac{f(b) - f(a)}{b - a}$$

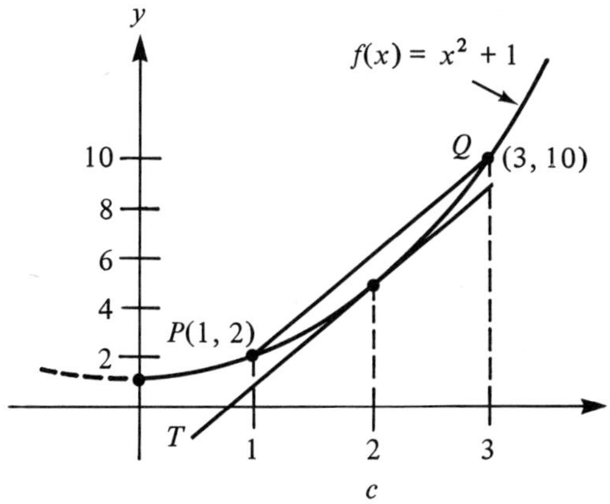

Figure 2.12

becomes

$$2c = \frac{10 - 2}{3 - 1} = 4$$
$$c = 2,$$

the required answer since $2 \in (1, 3)$ (see Figure 2.12). Geometrically, the slope of chord PQ equals the slope of the tangent line T at $c = 2$.

The equation for the law of the mean can be written as

$$f(b) - f(a) = f'(c)(b - a)$$

or

$$f(b) = f(a) + f'(c)(b - a).$$

When b is near a then c is near a, since c is between a and b. Therefore, $f'(c)$ is near $f'(a)$; that is,

$$f'(a) \cong f'(c).$$

This leads to the approximation

$$f(b) \cong f(a) + f'(a)(b - a),$$

which is useful in finding approximations of certain numbers. We illustrate.

EXAMPLE 2 Approximate $(26)^{1/3}$ to three decimal places.

Solution Let $f(x) = x^{1/3}$, $a = 27$, $b = 26$, $b - a = -1$. Then

$$\begin{aligned} f(b) = (26)^{1/3} &\cong f(a) + f'(a)(b - a) \\ &= 27^{1/3} + \frac{1}{3}(27)^{-2/3}(-1) \\ &= 3 - \frac{1}{3(9)} \\ &= 3 - \frac{0.1111}{3}\cdots \\ &\cong 3 - 0.03703 \\ &= 2.96397 \\ &= 2.964 \text{ to three decimal places.} \end{aligned}$$

PROBLEM SET 2.6

Find all values of c in the indicated interval for the given function so that the law of the mean is satisfied.

1. $f(x) = x^2 + 2$ in $(0, 3)$
2. $f(x) = 4x - x^2$ in $(0, 2)$
3. $f(x) = x^3$ in $(-2, 2)$
4. $f(x) = 1 - x^3$ in $(-1, 1)$
5. $f(x) = \sqrt{x}$ in $(0, 4)$
6. $f(x) = 1/x$ in $(1, 3)$
7. $f(x) = \sqrt{2x + 4}$ in $(0, 6)$
8. $f(x) = x^2 + 2x - 6$ in $(0, 2)$
9. $f(x) = x^2 - 3x + 5$ in $(2, 4)$

Use $f(b) \cong f(a) + f'(a)(b - a)$ to approximate the following numbers to three decimal places:

10. $\sqrt{99}$
11. $(7.9)^{1/3}$
12. $(4.01)^{3/2}$
13. $(8.1)^{2/3}$
14. $\left(\frac{1}{99}\right)^2$
15. $(2.02)^5$
16. $(8.01)^{1/3} - (8.01)^{-1/3}$

2.7 NEWTON'S METHOD

An effective procedure for determining a root of an equation is the one called Newton's Method. If $y = f(x)$ is a continuous function, then by the Intermediate Value Theorem (which we will not discuss) there is at least one root r of $f(x) = 0$ between any two numbers a and b for which $f(a)$ and $f(b)$ are opposite in sign (see Figure 2.13).

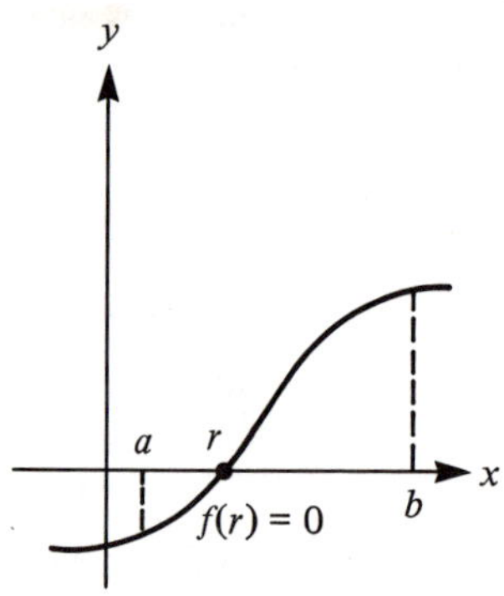

Figure 2.13

Let us assume that $f(a) < 0$ and $f(b) > 0$, that f is differentiable, and $f'(x) \neq 0$ in the closed interval $a \leq x \leq b$. The tangent to f at $P(x, y)$ [Figure 2.14] intercepts the x-axis at $x = x_1$. Let x_1 be a first approximation to the root r. The value of x_1 may be found by inspection or by the following procedure.

If, for example,

$$f(x) = x^2 - 2 = 0$$

then

$$f(0) = -2$$

$$\left.\begin{aligned} f(1) &= -1 \\ f(2) &= 2 \end{aligned}\right\} \quad \text{opposite signs}$$

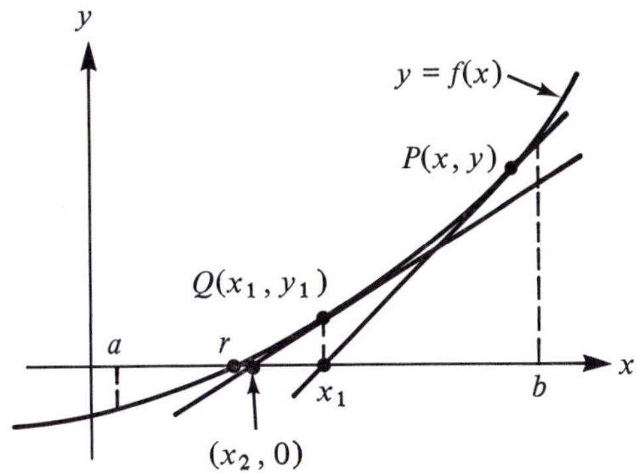

Figure 2.14

Since $f(1)$ and $f(2)$ have opposite signs, there exists a root between $x = 1$ and $x = 2$. We choose a value between 1 and 2, say 3/2, for which $f(3/2) = 1/4$. Since $f(1) < 0$ and $f(3/2) > 0$, there is a root between 1 and 3/2. We choose 4/3 as the value between 1 and 3/2. Then $f(4/3) = -2/9$. Since $f(3/2) > 0$ and $f(4/3) < 0$ there is a root between 4/3 and 3/2. We would let x_1, our first approximation to r, be some value between 4/3 and 3/2, perhaps 1.4. At this point Newton's Method is applied. We have eliminated much of the arithmetic in arriving at this point. This is convenient because Newton's Method involves tedious arithmetic. In cutting down on the number of times we must use the method we have eliminated much of the tedious computations. We proceed to develop the formula(s) for use in Newton's Method.

An equation of the tangent line through the points $Q(x_1, y_1)$ and $(x_2, 0)$ [the x-intercept of the tangent through Q] is

$$y_1 - 0 = f'(x_1)(x_1 - x_2) \qquad \text{[two-point form of a line]}$$

or $$x_1 - x_2 = \frac{y_1}{f'(x_1)}$$

or $$x_1 - x_2 = \frac{f(x_1)}{f'(x_1)} \qquad [y_1 = f(x_1)].$$

Solving for x_2, we obtain

$$x_2 = x_1 - \frac{f(x_1)}{f'(x_1)},$$

the second approximation to r.

For the function discussed above, where we arrived at $x_1 = 1.4$ as the first approximation to r, we would find the second approximation thus:

$$x_2 = 1.4 - \frac{f(1.4)}{f'(1.4)}.$$

Knowing the value of x_2, we may want the value of x_3, the third and still better approximation to r. This value can be found to be

$$x_3 = x_2 - \frac{f(x_2)}{f'(x_2)}.$$

This recursive process can be continued until we get as good an approximation as we wish. However, as stated above, our preliminary explorations would preclude using Newton's Method more than twice.

The general formula for this recursive method is

$$x_{n+1} = x_n - \frac{f(x_n)}{f'(x_n)}, \qquad f'(x_n) \neq 0.$$

EXAMPLE 1 Find to two decimal places the positive root of $x^2 + x - 1 = 0$.

Solution Let $f(x) = x^2 + x - 1$. Then $f(0) = -1$, $f(1) = 1$; the change in sign implies that a root is between $x = 0$ and $x = 1$. In this interval we choose $x = 0.5$; then $f(0.5) = -0.25$. Since $f(0.5) < 0$ and $f(1) > 0$, a root lies between 0.5 and 1. Choose $x_1 = 0.6$ since $f(0.5)$ is much closer to 0 than $f(1)$ is. By Newton's Method, we have

$$x_2 = x_1 - \frac{f(x_1)}{f'(x_1)} \qquad \text{and} \qquad f(x_1) = -0.04$$

where $x_1 = 0.6$ and

$$f'(x_1) = 2x_1 + 1 = 2(0.6) + 1 = 2.2.$$

Substituting, we get

$$\begin{aligned} x_2 &= 0.6 - \frac{-0.04}{2.2} \\ &= 0.6 + 0.018 = 0.618. \end{aligned}$$

At this point, we do some investigating.

$$\begin{aligned} f(0.6) &= -0.04 && [\text{from above}] \\ f(0.62) &= +0.0044 && [\text{after some computation}] \end{aligned}$$

The change in sign means that a root lies between 0.60 and 0.62, but closer to 0.62 since 0.0044 is closer to 0 than is -0.04. We may draw these conclusions because the interval of interest, $[0.60, 0.62]$ in this case, is small. Since $x_2 = 0.618$ is closer to 0.62 than to 0.60, we have sufficient reason to stop here and name $x_2 \cong 0.618$ or 0.62 to two decimal places as the desired approximation of the positive root of $x^2 + x - 1 = 0$.

The quadratic formula provides us with a check:

$$\begin{aligned} x &= \frac{-1 \pm \sqrt{1 + 4}}{2} \\ &= \frac{-1 + 2.236}{2} = 0.618, \end{aligned}$$

where we deleted the negative sign so as to get the positive root.

We chose an easy example so that Newton's Method would be easy to see and also that we might check the answer without much trouble. The example is illustrated graphically in Figure 2.15.

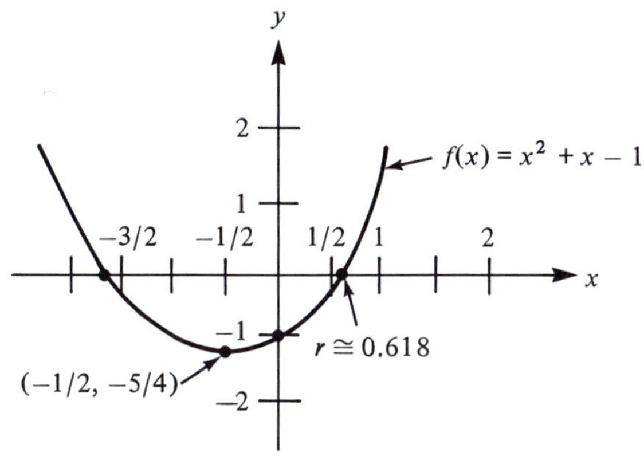

Figure 2.15

Let us further illustrate Newton's Method in another example.

EXAMPLE 2 Find the root to three decimal places of $f(x) = x^3 - 3x - 3 = 0$.

Solution $f(0) = -3$, $f(1) = -5$, $f(2) = -1$, $f(3) = 15$. A root exists between 2 and 3 since $f(2)$ and $f(3)$ have opposite signs. Since -1 is much closer to 0 than 15 is, we choose 2.11 as a value between 2 and 3. Then

$$f(2.11) = (2.11)^3 - 3(2.11) - 3 = 0.064$$

Since $f(2) = -1$ and $f(2.11) = 0.064$, a root exists between 2 and 2.11. $f(2.11)$ is much closer to 0 than is $f(2)$. Accordingly, we choose $x_1 = 2.11$ and find x_2:

$$x_2 = x_1 - \frac{f(x_1)}{f'(x_1)}$$

where $f(x_1) = f(2.11) = 0.064$ and

$$f'(x_1) = 3x_1^2 - 3 = 3(2.11)^2 - 3 = 10.356$$

(a calculator is helpful). Then

$$\begin{aligned} x_2 &= 2.11 - \frac{0.064}{10.356} \\ &= 2.11 - 0.00618 \\ &= 2.10382 \quad \text{or} \quad 2.104 \text{ to three decimal places.} \end{aligned}$$

Usually, when the fraction $f(x_1)/f'(x_1)$ or $f(x_2)/f'(x_2)$ is as small as it is here, we consider this good enough for our purposes. Moreover, $x_2 = 2.104$ is considered a sufficiently good approximation to the root because $f(2.104) = 0.0097$, which is sufficiently close to zero (for us).

Since the problem of finding a root of $f(x) = 0$ is equivalent to finding where the graph of f crosses the x-axis, it is appropriate to examine the graph of $f(x) = x^3 - 3x - 3$ in the previous example, shown in Figure 2.16. [The relative minimum $(1, -5)$ and relative maximum $(-1, -1)$ were found by methods of optimization to be discussed in Chapter 3.]

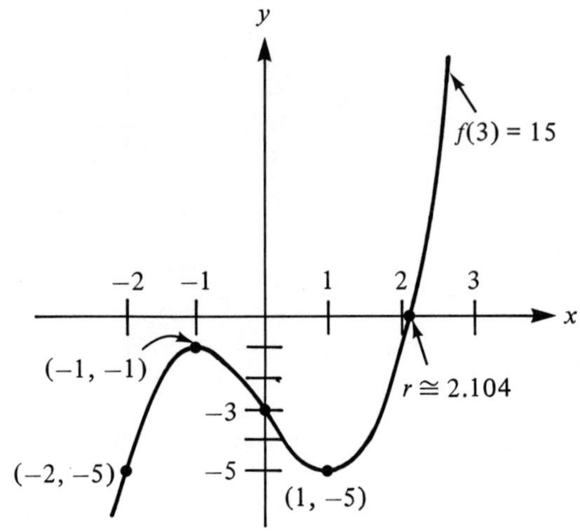

Figure 2.16

PROBLEM SET 2.7

A calculator will be useful in working the following problems.

Find the indicated root, to two decimal places, by using Newton's Method in problems 1–4.

1. $\sqrt[3]{7}$ [*Hint*: if x is the root, then $x^3 = 7$]
2. $\sqrt[3]{9}$
3. $\sqrt[4]{10}$
4. $\sqrt[4]{15}$

Use Newton's Method to find the indicated root, to two decimal places, of the given equations in problems 5–10:

5. $x^2 - 2x - 2 = 0$; positive root
6. $x^2 - x - 3 = 0$; positive root
7. $2x^2 + x - 1/2 = 0$; positive root
8. $2x^2 + 2x - 1/2 = 0$; positive root
9. $x^3 + 3x - 5 = 0$; root between 1 and 2
10. $x^3 - 3x - 1 = 0$; smallest positive root
11. A certain cost function is given by $C(x) = x^2 + 1$ and a sales revenue function by $R(x) = 2x + 3$. Use Newton's Method to find the value of $x > 0$ to three decimal places so that profit $P = 0$.

2.8 REVIEW PROBLEMS

Find the following limits:

1. $\lim_{x \to 1} (3x + 4x^2)$
2. $\lim_{x \to 2} (x^3 - 2x^2)$

3. $\lim_{x \to 2} \frac{3x^2 - 3x}{x - 1}$

4. $\lim_{x \to -1} (1 - |x|)$

5. $\lim_{x \to 0} \left(\frac{x^2}{x} + 1\right)$

6. $\lim_{x \to 2} \frac{x^2 - 4}{x - 2}$

7. $\lim_{x \to 1} \frac{x - 1}{\sqrt{x} - 1}$

8. $\lim_{x \to 4} \frac{2 - \sqrt{x}}{4 - x}$

9. $\lim_{h \to 0} \frac{(1 + h)^2 - 1}{h}$

10. $\lim_{h \to 0} \frac{(1 + h)^3 - 1}{h}$

11. $\lim_{x \to 0} \frac{x}{1 - \sqrt{1 + x}}$

12. $\lim_{t \to 3} \frac{t^2 - 2t - 3}{t - 3}$

Discuss continuity for the functions in exercises 13–16.

13. $f(x) = \frac{x}{x^2 - 4}$

14. $g(x) = \frac{1}{2x - 1}$

15. $y = \begin{cases} 2x & \text{if } 0 \le x \le 2 \\ x - 1 & \text{if } x > 2 \end{cases}$

16. $h(t) = \begin{cases} 2 + t & \text{if } 0 \le t \le 3 \\ 1 + \frac{2t}{3} & \text{if } 3 < t \le 6 \end{cases}$

17. Find an equation of the line tangent to the curve of $f(x) = 3x^2 - 5$ at the point (2, 7).

18. Determine an equation of the line that is tangent to:

(a) the parabola $y = x^2$ and is parallel to the line $y = 3x$

(b) the curve $y = x^3 - x + 1$ and is parallel to the line $x + 3y = 6$.

19. Use $\lim_{\Delta x \to 0}(\Delta y/\Delta x)$ to find the first derivative of the specified empirical relation:

(a) The supply function $y = S(x) = 2 + x^2$

(b) The demand function $y = D(x) = 8 - x^3$

(c) The cost function $y = C(x) = 4 + \sqrt{x}$

In exercises 20–25 find $D_x y$:

20. $y = (3x - 1)(x^2 + 10)$

21. $y = x^2(1 - 3x)^2$

22. $y = \frac{2}{x^2 + 3}$

23. $y = \frac{x^2}{x^2 + 1}$

24. $y = \frac{2x + 3}{2x - 3}$

25. $y = \frac{2x}{1 - 2x}$

26. Find an equation of the line tangent to each curve of exercises 20–25 at the point where $x = 1$.

27. Determine the velocity and acceleration at time t for each function:

(a) $s(t) = t^3 + t^2$ **(b)** $s(t) = 6t^{1/2} + t^{3/2}, \quad t > 0.$

Find the indicated derivative for each given relation in exercises 28–33:

28. $f(x) = x\sqrt{x^2 - 2}; f'(x)$

29. $f(x) = \frac{x^2}{x - 1}; D_x f$

30. $g(w) = (2 + 4w)^{1/2}; g'(w)$

31. $h(x) = \sqrt{x^2 + 2x - 1}; h''(x)$

32. $f(x) = (1 - 3x)^{1/3}; f^{(3)}(x)$

33. $u(r) = r^{5/3}; \dfrac{d^3u}{dr^3}$

Find the indicated derivative (by the chain rule) of each given composite function:

34. dy/dt if $y = 2x^2$ and $x = 2t$

35. dy/dz if $y = x^3$ and $x = 1 - 2z$

36. dy/du if $y = 2\sqrt{t}$ and $t = 1/u$

37. A new machine produces revenue according to $R(t) = 8(10)^4 - (10^6/t)$, where t is time in years. As t increases the machine undergoes a decrease in efficiency while costs of its maintenance increase by the rule $M(t) = 500 + t + 500t^2$.
   (a) Neglecting other machine costs, find the marginal revenue and marginal maintenance costs.
   (b) Assuming that profit is largest when marginal revenue equals marginal cost, how long should the machine be used?

38. If the total cost C of making g gallons of a cough syrup is given by $C(g) = 20 + 40g^{1/2}$ dollars, find the marginal cost when output $g = 64$ gallons.

39. Assume that the cost of making x units is C dollars where $C(x) = 10x - 0.3x^2 + 0.003x^3$. Find:
   (a) Marginal cost at $x = 100$ units
   (b) The outputs x for which $C'(x) = \$10$.

40. Find the points on the curve $y = 2x^3 - 3x^2 - 12x + 20$ where the tangent is parallel to the x-axis.

41. Assume that total revenue r (in hundred thousands of dollars) is a function of total sales s (in thousands) as follows:

$$r(s) = \begin{cases} 4s & \text{if } 0 \le s \le 1 \\ 10 - \dfrac{6}{s} & \text{if } s > 1 \end{cases}$$

   (a) Find marginal revenue when $s = 2$.
   (b) Sketch the graph of r.

42. Find d^2y/dx^2, the rate of change of the marginal function dy/dx, of each given function:
   (a) $y = 3x^3 + 4x^2 - 2x$
   (b) $y = 2x^4 - 6x^3 + 7$
   (c) $y = \dfrac{1}{2x + 1}$
   (d) $y = \dfrac{2x^3 - 3x^2}{2x}$
   (e) $y = (3x + 4)^{1/3}$
   (f) $y = \dfrac{(4x + 1)^{3/2}}{4}$

43. Let the cost of producing and selling q units be represented by $C(q) = 2 - 2q^2 + q^4$, where C is in millions of dollars and q is in thousands of units, $q < 2$. Find:
   (a) Marginal cost at $q = 1$
   (b) The rate of change of marginal cost
   (c) The output q for which $C''(q) = 0$.

44. A profit function is given by $P(x) = \frac{2}{3}x - \frac{4}{3}x^2$, where x denotes amount of production.
   (a) Find the value of x for which marginal profit is zero.
   (b) Determine the rate of change of marginal profit.

45. Given the sales revenue function $S(x) = 48x^{1/3} + 2x$ where x is the amount of sales units.

(a) Find marginal sales at $x = 64$.

(b) For what value of x does the rate of change of marginal sales equal $-1/96$?

46. Estimate the change in profit P if $P(x) = \sqrt{x}$ dollars and x increases from 81 to 83 units of production.

47. The total cost Q of manufacturing p units of a product is given by $Q(p) = 12 + 2p + p^2$. Estimate the percentage change in cost Q when p increases from 10 to 11 units.

48. Find differential dy for the following functions:

(a) $y = 4x^2 + 5x^4$ **(b)** $y = x(2x - 3)^3$

(c) $y = \dfrac{3x + 1}{1 - 4x}$ **(d)** $y = \sqrt{x}(1 - 2x)$

Use differentials to approximate the following numbers:

49. $\sqrt{82}$ **50.** $\sqrt[3]{124}$ **51.** $\sqrt[4]{82}$

52. $\sqrt{0.98}$ **53.** $\sqrt[3]{0.96}$ **54.** $(15)^{-1/4}$

55. $(1.99)^4 - (1.99)^3$ **56.** $(2.01)^6 + (2.01)^{3/2}$

Find all values of c such that the law of the mean holds for the given function in the indicated interval:

57. $f(x) = 9 - x^2$ in $(1, 2)$ **58.** $f(x) = x^2 - 2x$ in $(0, 1)$

59. $f(x) = 8 + x^3$ in $(0, 2)$ **60.** $f(x) = 1 + \sqrt{x}$ in $(1, 4)$

61. $f(x) = \dfrac{1}{x - 1}$ in $(2, 3)$

Make use of $f(b) \cong f(a) + f'(a)(b - a)$ to compute the following numbers to three decimal places:

62. $\sqrt{24}$ **63.** $\sqrt{120}$ **64.** $\sqrt[3]{26}$ **65.** $\sqrt[3]{126}$

66. $(80)^{-1/2}$ **67.** $(80)^{1/4}$ **68.** $(0.95)^{3/4}$ **69.** $(31)^{1/5}$

70. Use Newton's Method to find the positive root to three decimal places of $x^3 - 3x^2 - 4 = 0$.

2.9 ENRICHMENT TOPICS AND PROBLEMS

1. When a merchant prices an item at \$9.98 he hopes the consumer will think in terms of 9 and not 10. When he prices at \$7.95 he hopes the consumer "thinks" 7 and not 8, and so on. These ideas may be formalized mathematically by $y = [x]$, where x is price in dollars and $[x]$ denotes the *greatest integer* less than or equal to x. For example, $[\$0.95] = \$0$, $[\$1.55] = \$1$, etc. Sketch a graph of the so-called "step function," $y = [x]$ for the domain $\$0 \leq x \leq \$4$.

2. ***L'Hospital's Rule.*** There is a technique called L'Hospital's Rule which enables us to evaluate limits of indeterminate forms without the burden of initial algebraic manipulations. We state (without proof) only that aspect of the rule which deals with the indeterminate form 0/0. (L'Hospital is pronounced Lo'-pee-tall.)

THEOREM 1 **(L'Hospital's Rule)**

If

(1) *f and g are functions whose domains include the open intervals (b, c) and (c, a), and d is an element of the domains of f and g;*

(2) *$\lim_{x\to d} f(x) = 0$ and $\lim_{x\to d} g(x) = 0$;*

(3) *$f'(x)$ and $g'(x)$ exist;*

(4) *$g'(x) \neq 0$ for $b < x < c$ or $c < x < a$;*

(5) *$\lim_{x\to d} (f'(x)/g'(x))$ exists or is equal to $+\infty$ or $-\infty$;*

then it is true that

$$\lim_{x\to d} \frac{f(x)}{g(x)} = \lim_{x\to d} \frac{f'(x)}{g'(x)}.$$

Conclusion:

The limit as x approaches d of a quotient of two functions is equal to the limit as x approaches d of the quotient of the derivatives of the respective functions, provided the conditions of Theorem 1 are satisfied.

Let us apply L'Hospital's Rule.

EXAMPLE 1 Find $\lim_{x\to 1} \dfrac{1 - x^2}{2 - \sqrt{x^2 + 3}}$.

Solution Taking the limit directly yields $(1 - 1)/(2 - \sqrt{4})$ or $0/0$. Thus, the conditions of Theorem 1 are satisfied so we continue as follows:

$$\lim_{x\to 1} \frac{1 - x^2}{2 - \sqrt{x^2 + 3}} = \lim_{x\to 1} \frac{(-2x)}{-\frac{1}{2}(x^2 + 3)^{-1/2}(2x)}$$

$$= \lim_{x\to 1} 2\sqrt{x^2 + 3} = 2\sqrt{1 + 3} = 4.$$

Note: Do not apply L'Hospital's Rule unless both numerator and denominator have limit zero in the limiting process.

Apply L'Hospital's Rule to find the following limits:

(a) $\lim_{x\to 2} \dfrac{3x^2 - 10x + 8}{2x^2 - 5x + 2}$ **(b)** $\lim_{x\to 1} \dfrac{x^3 - x^2 - 4x + 4}{x^3 - 1}$

(c) $\lim_{x\to 2} \dfrac{4 - x^2}{1 - \sqrt{x - 1}}$ **(d)** $\lim_{x\to 0} \dfrac{(x + 1)^{10} - 1}{x}$

3. If $y = x^2 + x$, find $D_{x^2}y^2$, the rate of change of y^2 with respect to x^2. [*Hint*: Let $u = x^2$, then $dy^2/du = (dy^2/dx)/(du/dx) = \cdots$.

4. Given $f'(x) = x$ and $f(2) = 2$, find $f(x)$.

5. Find $\lim_{n\to\infty}(\sqrt{n^2 + 1} - n)$.

6. Determine $\lim_{n\to\infty}(\sqrt{n^2 + n} - n)$.

7. Obtain the rate of change of the function $g(t) = t(t^2 - 2)(t^2 + 1)$ at $t = 1$.

8. Find equations of the lines tangent to $f(x) = x^4 - 5x^2 - 18$ at the point where $f = 18$.

9. ***Time Rates.*** The LMN Company's annual sales revenue (in millions of dollars) is given by $R(t) = 10 + 2t + 0.25t^2$. Time t is in years and $t = 0$ corresponds to 1970, $t = 1$ to 1971, $t = 2$ to 1972, and so on. Marginal revenue is $R'(t) = 2 + 0.5t$.

$$\text{At } t = 1,\ R'(1) = 2 + 0.5 = 2.5 \quad \text{or} \quad \$2.5 \text{ million for 1971.}$$
$$\text{At } t = 2,\ R'(2) = 2 + 1 = 3 \quad \text{or} \quad \$3 \text{ million for 1972.}$$

A comparison of these two results is not meaningful without a knowledge of previous performance as a reference. A more meaningful index for comparison is the one given by $R'(t)/R(t)$, called the **relative time rate**. When this rate is multiplied by 100 we get

$$\frac{R'(t)}{R(t)}(100),$$

which is called the **percentage time rate**.

For 1971 the percentage time rate for the LMN Co. is

$$\frac{R'(1)}{R(1)}(100) = \frac{2.5(100)}{10 + 2(1) + 0.25(1)} = \frac{250}{12.25} \cong 16.95\%.$$

For 1972 the percentage time rate is

$$\frac{R'(2)}{R(2)}(100) = \frac{3(100)}{10 + 2(2) + 0.25(4)} = \frac{300}{15} \cong 20\%,$$

an increase of 3.05% over the previous year. If conditions remain the same, the percentage for 1974 can be found to be about 18.18%, a decrease of 1.82% from 1972. This would be an indication of a decline in growth and would, perhaps, call for a reappraisal of management's policies. With these procedures in mind, find the percentage time rates in 1971, 1972, and 1974 for the following situations:

(a) The ABC Company's annual earnings are defined by $E(t) = 40 + 2t + t^2/2$, t is time in years, E is in thousands of dollars.

(b) The profit function for the DMK Corp. is defined by $P(t) = 5 + t^3/5$, P is in thousands of dollars, t is time in years.

(c) A manufacturer's earnings from the sales of a new product are given by $E(t) = ((1 + t)^4/4) - 2t^3 + 10$, where t is time in years and E is in hundred thousands of dollars.

Elasticity of Demand. If p is the price of an item and D denotes the demand for the item, then the elasticity of demand is defined by

$$\eta = \frac{p}{D} \cdot \frac{dD}{dp},$$

where the symbol η is the Greek letter "eta." This formula represents the approximate percent change in demand which corresponds to a one percent increase in the price. For example, if

$$D(p) = 200 - p^2 \quad \text{and} \quad p = 10$$

then

$$D(10) = 200 - 10^2 = 100$$

and

$$\frac{dD}{dp} = -2p\Big|_{p=10} = -20.$$

From these data the elasticity of demand can be found:

$$\begin{aligned}\eta &= \frac{p}{D} \cdot \frac{dD}{dp}\\ &= \frac{10}{100}(-20) = -2.\end{aligned}$$

The result -2 indicates that an increase of one percent in the price will effect an approximate decrease of 2% in the demand. For instance, if a 3% price increase is planned, then a decrease of 3(2) or 6% in demand should be expected. With these ideas and methods in mind, work the following problems 10 and 11:

10. The demand (in hundreds of cartons) for a certain brand of detergent is given by $D(p) = 98 - p - (p^2/4)$, where p denotes the price (in dollars) per carton.

(a) Find the elasticity of demand when $p = \$4$.

(b) What should be the percent decrease in demand when a 3% price increase is implemented?

11. The demand for Michael's Luscious Pastry is defined by $D(x) = 40 - (x + 1)^3/2$, where x denotes price per pastry in dollars.

(a) Find the elasticity of demand when $x = 1$.

(b) If an increase in price of 12% is planned, what percent decrease in demand should be expected?

12. Use differentials to approximate the numbers:

(a) $(31)^{-2/5}$ **(b)** $(8.01)^3 - \dfrac{1020}{\sqrt[3]{8.01}}$

13. Find the error E in the answer to problem 12(a) if $|E| \le M(dx)^2$ where $|f''(x)| \le M$, assuming $f(x) = x^{-2/5}$, $x = 32$, $dx = -1$.

14. The total surface area of a right cylindrical tank is given by $S = 2\pi r^2 + 2\pi rh$ where r is the radius and h is the height of the tank. Let $h = 12$ feet and let r increase from 4 feet to 4 feet 3 inches. Use differentials to find the approximate change in S.

15. The total cost of manufacturing a product is given by $C(x) + 1 = \frac{1}{4}(2x - 10)^2$. Find the approximate percentage change in cost C when x increases from 50 to 52 units of production.

16. Assume that the profits P of the PLT Co. are defined by $P(q) = 96q - 2q^2$ where q denotes quantity of items sold. What is the approximate percentage change in profits when sales increase from 20 to 23 units?

17. Find a number $c \in (a, b)$ such that $\dfrac{f(b) - f(a)}{b - a} = f'(c)$ in the following examples:

(a) $f(x) = x^2 + 4x + 2, \quad [a, b] = [-1, 3]$

(b) $f(x) = x^3 - 3x^2 + 4, \quad [a, b] = [1, 3]$

18. Use $f(b) \cong f(a) + f'(a)(b - a)$ to estimate the following numbers:

(a) $(1.02)^{10}$; let $f(x) = x^{10}$, $a = 1$, $b = 1.02$

(b) $\dfrac{1}{100.1}$; let $f(x) = \dfrac{1}{x}$, $a = 100$, $b = 100.1$

19. Use differentials to approximate the following numbers to three decimal places:

(a) $\sqrt{143.5}$ **(b)** $(2.005)^5$ **(c)** $(8.002)^2 - \dfrac{126}{\sqrt[3]{8.002}}$

20. A box has a base that is square and a height that is one meter less than a side of the base. The volume of the box is 5 cubic meters. Find the length of a side of the base to two decimal places by Newton's Method.

3 OPTIMIZATION ANALYSIS

Decision-Making for Optimality

IMPORTANT EQUATIONS

$$f''(x_0) > 0 \rightarrow \min f \text{ at } [x_0, f(x_0)]$$

$$f''(x_0) < 0 \rightarrow \max f \text{ at } [x_0, f(x_0)]$$

$$q(x) = \frac{Q(x)}{x}$$

$$Q(x) = \frac{cx}{2} + \frac{sd}{x}$$

3.1 EXTREMA

Our intention in this section is to apply rates of change to the maximization or minimization of algebraic functions. In later sections we will extend and relate these applications in a definitive way to decision-making for optimum results. The approach used will not be analytically rigorous but will instead appeal to geometric intuition.

Consider Figure 3.1, which exhibits the graph of a continuous function $y = f(x)$. Points of interest $A, B, C, \ldots, G$ are shown on the graph. At the points A and E function f is increasing, thus defining f as an **increasing function** at points A and E. At these same points the slope f' of the tangent to the curve is positive. At point C, f is decreasing and so f is called a **decreasing function** at C. At this point the slope f' is negative. At points B, D, and F, the tangent to f is parallel to the x-axis, that is, the slope is zero at these three points. Point B marks the transition from positive to negative slopes while point D marks the transition from negative to positive slopes. Point C is a transition point with a special name: an **inflection point**, since it separates concavity downward (to the left of C) from concavity upward (to the right of C). Point E is also an inflection point since the graph of f is concave up to the left of E and concave down to the right of E.

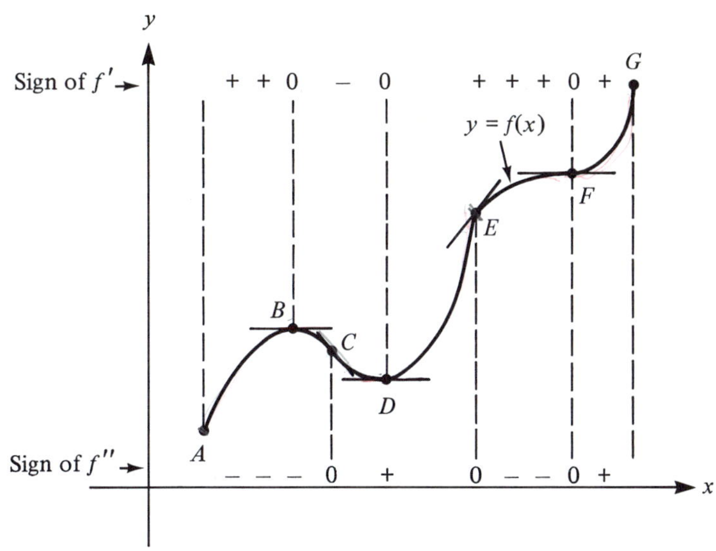

Figure 3.1

DEFINITION 1 *A **point of inflection** is defined to be a point which separates concavity upward from concavity downward or vice versa.*

As we move along the curve of f from point A to point G we see that point A is the lowest point of f and G is the highest point. In view of this we say that f assumes an absolute minimum value at A and an absolute maximum value at G. Point B is defined to be a **relative maximum** point and point D is a **relative minimum** point. At F there is neither a relative maximum nor a relative minimum, but instead a point of inflection as stated above.

From Figure 3.1 we see that the algebraic sign of the second derivative is positive in the open intervals C to E and F to G. In other words, the slope of f is changing positively in these two intervals. For these ranges where $f''(x) > 0$, the tangent "moves" counter-clockwise and the graph of f is concave up. Similarly, the algebraic sign of f'' is negative in the open intervals A to C and E to F. That is, the slope of f is changing negatively in these latter two intervals. Moreover, we write $f''(x) < 0$ and see that the tangent to f "moves" clockwise while the curve is concave down. It would be wise for you now to refer again to Figure 3.1 in order to establish a firm understanding of the "behaviors" of f', f'', and points of relative maximum, relative minimum, and inflection.

Let us now become acquainted with the "**first derivative test**" for finding **extrema**, i.e., relative maximum and relative minimum points. We do so by a concrete example.

EXAMPLE 1 Examine $f(x) = x^3 - 3x^2 + 1$ for extrema.

Solution Given

$$f(x) = x^3 - 3x^2 + 1,$$

then

$$\begin{aligned} f'(x) &= 3x^2 - 6x \\ &= 3x(x-2) \\ &= 0 \qquad \text{when } x = 0 \quad \text{or} \quad x = 2. \end{aligned}$$

These values of x for which $f'(x) = 0$ will be referred to as *critical values*, or briefly cv. Applying the first derivative test, we first evaluate the first derivative f' a "little" to the left of the critical value $x = 0$ and then evaluate f' a "little" to the right of this cv. Accordingly, we formally have

$$\left.\begin{aligned} f'(-1) &= 3(-1)(-1-2) > 0 \\ f'(+1) &= 3(+1)(+1-2) < 0 \end{aligned}\right\} \quad \wedge \rightarrow \text{a relative maximum}$$

at $(0, f(0))$ or $(0, 1)$ (see Figure 3.2).

Remark 1: To clarify the phrases "a little to the left" and "a little to the right" of a cv, consider $f'(x) = (x-1)(x-2)(x-3)$. A little to the left of $x = 2$ implies not as far as $x = 1$ since $f'(1) = 0$. A little to the right of $x = 2$ implies not as far as $x = 3$ since $f'(3) = 0$.

Remark 2: The tangent lines $/$ and $\backslash$ have positive and negative slopes, respectively. The device $\wedge$ is used to indicate pictorially that the slope changes from positive to negative and to serve as a "framework" showing concavity downward.

To continue the test, we evaluate f' a "little" to the left of the other cv, $x = 2$, and then evaluate f' a "little" to the right of $x = 2$. Thus, we have

$$\left.\begin{aligned} f'(1) &= 3(1)(1-2) < 0 \\ f'(3) &= 3(3)(3-3) > 0 \end{aligned}\right\} \quad \vee \rightarrow \text{a relative minimum}$$

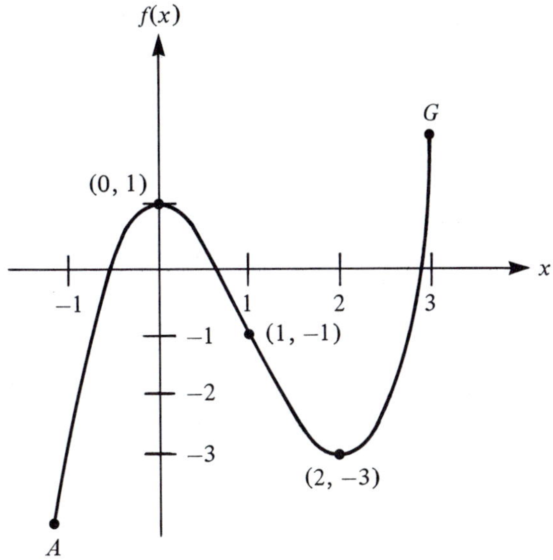

Figure 3.2

at $(2, f(2))$ or $(2, -3)$. The device $\vee$ in this case implies a change of slope from negative to positive, indicating concavity up. To facilitate the graphing of f a few more points are found: $f(-1) = -3$ and $f(3) = 1$. Points A and G in Figure 3.2 are included merely to illustrate the absolute minimum and absolute maximum points, respectively.

In summary, to find extrema of f by the first derivative test:

(1) Find f' and set f' equal to zero.

(2) Find a value or values of the independent variable for which $f' = 0$; these are critical values, cv.

(3) Test $f'(\text{cv}^-)$ and $f'(\text{cv}^+)$ in that order where cv^- and cv^+ signify to the left of cv and to the right of cv, respectively.

(4) If $\left\{\begin{matrix} f'(\text{cv}^-) < 0 \\ f'(\text{cv}^+) > 0 \end{matrix}\right\}$ then $\vee \rightarrow$ a relative minimum for f at the point $(\text{cv}, f(\text{cv}))$.

(5) If $\left\{\begin{matrix} f'(\text{cv}^-) > 0 \\ f'(\text{cv}^+) < 0 \end{matrix}\right\}$ then $\wedge \rightarrow$ a relative maximum for f at the point $(\text{cv}, f(\text{cv}))$.

In many situations a more convenient test for finding extrema is the "**second derivative test**." We introduce this test by applying it to the previous example. Thus,

$$f(x) = x^3 - 3x^2 + 1.$$

Then $\quad f'(x) = 3x^2 - 6x = 0 \quad$ for $x = 0$ or $x = 2$: cv

and $\quad f''(x) = 6x - 6.$

We evaluate $f''(x)$ at the cv $x = 0$, and get

$$f''(0) = 6(0) - 6 < 0 \rightarrow \text{a relative maximum}$$

at $(0, f(0))$ or $(0, 1)$ which checks the answer found by the first derivative test. Next, we evaluate $f''(x)$ at the other cv $x = 2$, and obtain

$$f''(2) = 6(2) - 6 > 0 \rightarrow \text{a relative minimum}$$

at $(2, f(2))$ or $(2, -3)$, which again checks.

Since the second derivative of f is the rate of change of the first derivative of f (that is, the rate of change of the slope of f), it is understandable that a positive $f''(\text{cv})$ indicates a relative minimum for f at $x = \text{cv}$. By similar reasoning it should be clear that a negative $f''(\text{cv})$ indicates a relative maximum for f at $x = \text{cv}$.

If $f''(\text{cv}) = 0$, then we say the second derivative test fails and we revert to the first derivative test. If the latter test fails, we assume that no extrema exist for f at $x = \text{cv}$.

EXAMPLE 2 Find extrema for $f(x) = x^4$.

Solution

$$\begin{aligned} f'(x) &= 4x^3 \\ &= 0 \quad \text{for } x = 0: \quad \text{cv} \\ f''(x) &= 12x^2 \end{aligned}$$

and

$$\begin{aligned} f''(\text{cv}) &= f''(0) \\ &= 0 \rightarrow \text{no test.} \end{aligned}$$

We are forced to apply the first derivative test:

$$\left.\begin{aligned} f'(-1) &= 4(-1)^3 < 0 \\ f'(+1) &= 4(+1)^3 > 0 \end{aligned}\right\} \quad \vee \rightarrow \text{relative minimum}$$

at $(0, f(0))$ or $(0, 0)$ (see Figure 3.3).

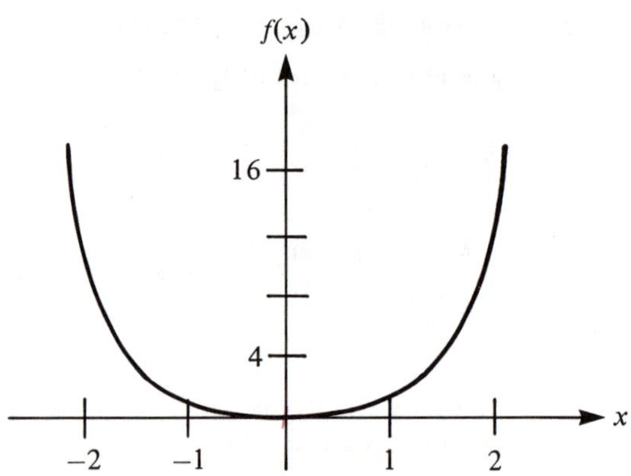

Figure 3.3

Let us summarize the mechanics of the second derivative test:

(1) If $f'(x) = 0$ at $x = \text{cv}$ and $f''(\text{cv}) > 0$, then $f(x)$ attains a relative minimum at $(\text{cv}, f(\text{cv}))$.

(2) If $f'(x) = 0$ at $x = \text{cv}$ and $f''(\text{cv}) < 0$, then $f(x)$ attains a relative maximum at $(\text{cv}, f(\text{cv}))$.

$$\begin{array}{|ll|} f''(x_0) > 0 \rightarrow \min f & \text{at } [x_0, f(x_0)] \\ f''(x_0) < 0 \rightarrow \max f & \text{at } [x_0, f(x_0)] \end{array}$$

Note: The first and second derivative tests for extrema are valid only if the function f is continuous and f' and f'' exist and are continuous functions.

To find an inflection point (the point separating concavity up from concavity down or vice versa) we proceed as follows. From Example 1 we have

$$f(x) = x^3 - 3x^2 + 1.$$

Then
$$\begin{aligned} f'(x) &= 3x^2 - 6x \\ &= 3x(x-2) \\ &= 0 \quad \text{for } x = 0 \text{ or } x = 2\text{:} \quad \text{cv for extrema.} \end{aligned}$$

Also,
$$\begin{aligned} f''(x) &= 6x - 6 \\ &= 6(x-1) \\ &= 0 \quad \text{for } x = 1\text{:} \quad \text{cv for inflection point.} \end{aligned}$$

Thus, $x = 1$ is a cv for testing for an inflection point on f. Next, we test $f''(x)$ a "little" to the left of $x = 1$, then a "little" to the right. The results are:

$$\left.\begin{aligned} f''(0) &= 0 - 6 < 0 \\ f''(2) &= 12 - 6 > 0 \end{aligned}\right\} \quad \cap \text{ to } \cup \rightarrow \text{an inflection point}$$

at $(1, f(1))$ or $(1, -1)$ (see Figure 3.2). In this case the point of inflection occurs at a place where the tangent to f has a negative slope. In the next example the tangent to f has slope zero at the point of inflection. But let's first summarize the test for an inflection point on $f(x)$:

(1) Find $f''(x)$ and set it equal to zero.

(2) Solve for x such that $f''(x) = 0$; this or these values of x are critical values, cv, for testing for inflection points.

(3) If $\left\{\begin{aligned} f''(\text{cv}^-) &< 0 \\ f''(\text{cv}^+) &> 0 \end{aligned}\right\}$ or $\left\{\begin{aligned} &> 0 \\ &< 0 \end{aligned}\right\}$ then $\cap$ to $\cup$ or $\cup$ to $\cap \rightarrow$ an inflection point at $(\text{cv}, f(\text{cv}))$.

The devices "$\cap$ to $\cup$" or "$\cup$ to $\cap$" are used simply to indicate pictorially the change from concavity down to concavity up or vice versa.

Observe that finding

(1) $f'(x)$ and solving $f'(x) = 0$ are necessary steps for determining extrema (relative maximum and minimum points); and that finding

(2) $f''(x)$ and solving $f''(x) = 0$ are necessary steps for isolating points of inflection.

We now return to an example where the tangent to f has slope zero at the point of inflection.

EXAMPLE 3 Find extrema and inflection point(s) for $f(x) = 2 - (x - 2)^3$.

Solution

$$\begin{aligned} f'(x) &= -3(x-2)^2 \\ &= 0 \quad \text{for } x = 2: \quad \text{cv for extrema.} \\ f''(x) &= -6(x-2) \\ &= 0 \quad \text{for } x = 2: \quad \text{cv for inflection point.} \end{aligned}$$

Since $f''(2) = 0$: no test, we revert to the first derivative test and obtain

$$\left.\begin{aligned} f'(1) &= -3(1-2)^2 < 0 \\ f''(3) &= -3(3-2)^2 < 0 \end{aligned}\right\} \text{ no change in sign of the slope}$$

implies no extrema for f at $x = 2$. Testing for an inflection point we have

$$\left.\begin{aligned} f''(1) &= -6(1-2) > 0 \\ f''(3) &= -6(3-2) < 0 \end{aligned}\right\} \cup \text{ to } \cap \rightarrow \text{an inflection point}$$

at $(2, f(2))$ or $(2, 2)$ (see Figure 3.4).

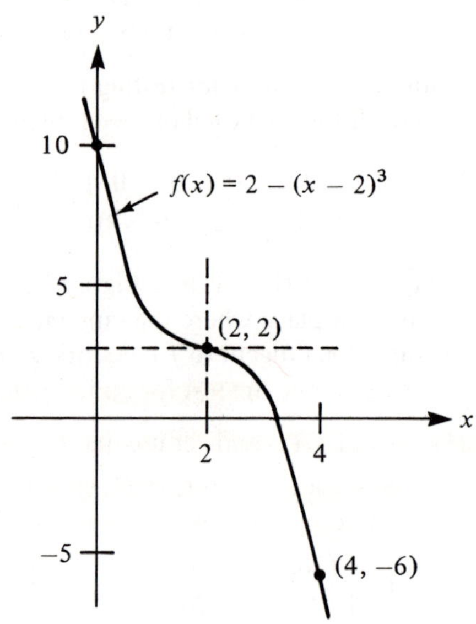

Figure 3.4

There is another convenient way of finding in inflection point. The next theorem illustrates this method.

THEOREM 1 The Third Derivative Test for Inflection Points

If

(1) $f''(x_0) = 0$,

(2) $f^{(3)}(x)$ *is continuous on an open interval that contains* x_0,

(3) $f^{(3)}(x_0) \neq 0$,

then

(4) f *has a point of inflection at* $(x_0, f(x_0))$.

EXAMPLE 4 Find points of inflection on the curve $f(x) = x^4 - 4x^3 + 10$.

Solution

$$\begin{aligned} f'(x) &= 4x^3 - 12x^2 \\ f''(x) &= 12x^2 - 24x = 12x(x-2) \\ &= 0 \quad \text{for } x = 0,\ x = 2. \end{aligned}$$

Thus, points of inflection can occur only where $x = 0$ and $x = 2$. Continuing:

$$\begin{aligned} f^{(3)}(x) &= 24x - 24 \\ f^{(3)}(0) &= -24 \neq 0. \end{aligned}$$

According to Theorem 1, f has a point of inflection at $(0, f(0))$ or $(0, 10)$. Also,

$$f^{(3)}(2) = 48 - 24 = 24 \neq 0.$$

Thus, another inflection point occurs at $(2, f(2))$ or $(2, -6)$.

PROBLEM SET 3.1

For each of the following functions, use either the first or the second derivative test to find extrema and points of inflection. Sketch graphs of the odd-numbered problems.

1. $f(x) = 2x^2 - 8x$

2. $g(x) = 3x^2 - 18x$

3. $h(x) = 20 + 8x - x^2$

4. $F(x) = 9 + 16x - 4x^2$

5. $G(x) = x^3 - 6x^2$

6. $H(x) = 2x^3 - 24x$

7. $y = 3x^3 - 81x + 80$

8. $y = 1 + 6x^2 - x^3$

9. $f(x) = 15x^2 - 2x^3$

10. $g(x) = 8x - (x^4/4)$

11. $h(x) = 108x - x^4$

12. $F(t) = t^4 - 8t^2$

13. $G(t) = t + (1/t)$

14. $H(t) = t + (9/t)$

15. $f(x) = \dfrac{x}{x-1}$

16. $h(x) = \dfrac{x+1}{x-1}$

17. $g(x) = \dfrac{1}{x^2+1}$

18. $F(t) = (2t-3)^2$

19. $h(x) = \sqrt{9 - x^2}$

20. $u(z) = \sqrt{4 - z^2}$

3.2 APPLICATIONS

We will now apply the ideas of the preceding section to real-world situations, involving optimization problems. These will include cost and revenue functions together with their attendant functions, average cost and average revenue. Other types of empirical problems will be investigated in succeeding sections as we continue to put the power of the calculus to concrete use.

Quadratic Cost Functions

In the study of economic price theory there are occasions when the relation between marginal cost and average cost is of interest and value.

Let $Q(x)$ be a total cost function and $q(x)$ be the **average cost** function (x denotes quantity of items produced) where $q(x)$ is defined by

$$\boxed{q(x) = \frac{Q(x)}{x}, \qquad x \neq 0}\,.$$

Note that $Q(x)/x$ is simply cost per unit. In order for the empirical situation to be meaningful, (1) infinitely large values of x are not permitted and (2) Q must increase as x increases.

EXAMPLE 1 Consider the total cost function defined by $Q(x) = 2x^2 - 12x + 50$.

(a) Find the level of output at which total cost Q and average cost q are a minimum.

(b) Verify that marginal cost Q' is equal to average cost q at the low point of q.

Solution **(a)**

$$\begin{aligned} Q'(x) &= 4x - 12 \\ &= 0 \quad \text{for } x = 3\text{:} \quad \text{cv} \\ Q''(x) &= 4 > 0 \rightarrow \text{minimum } Q \text{ at } (3, Q(3)) \end{aligned}$$

or (3, 32). Thus the level of output for which Q is a minimum is $x = 3$ units. Average cost is

$$\begin{aligned} q(x) &= \frac{Q(x)}{x} \\ &= \frac{2x^2 - 12x + 50}{x} \\ &= 2x - 12 + 50x^{-1}. \end{aligned}$$

Then

$$\begin{aligned} q'(x) &= 2 + 50(-x^{-2}) \\ &= 2 - 50x^{-2} \\ &= 2 - \frac{50}{x^2} \\ &= 0 \quad \text{when } 2 = \frac{50}{x^2} \end{aligned}$$

or

$$\begin{aligned} 2x^2 &= 50 \\ x^2 &= 25 \\ x &= 5\text{:} \quad \text{cv} \end{aligned}$$

($x = -5$ also, but this negative value is not relevant since x denotes units of production). Then

$$q''(x) = \frac{100}{x^3}\bigg|_{x=5} > 0 \rightarrow \text{minimum } q$$

at $(5, q(5))$ or $(5, 8)$, the low point of q. The notation $f(x)|_{x=a} > 0$ means $f(x)$ evaluated at $x = a$ is positive.

(b) It remains to show that $Q'(5) = 8$. $Q'(5) = 4(5) - 12 = 8$. This is verification that marginal cost equals average cost at the low point of average cost. We refer to Figure 3.5 and see that the graph of Q' passes through the point $(5, 8)$ at which point $Q'(x) = q(x)$.

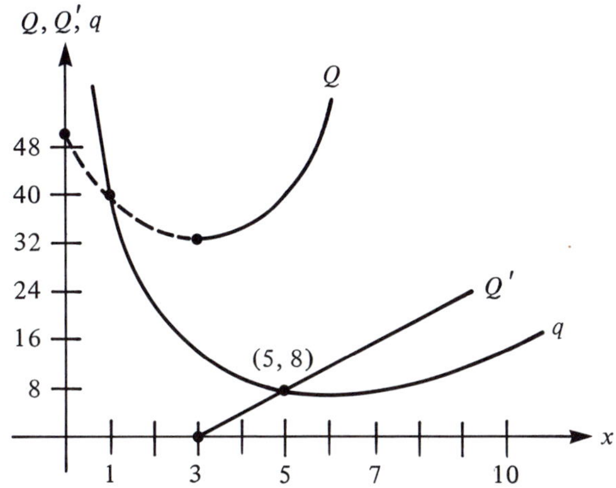

Figure 3.5

The behavior of Q in the interval $0 \leq x \leq 3$ does not satisfy the condition stated in (2) above, namely, that Q must increase as x increases (the dotted portion of Q in Figure 3.5). According to some economists' interpretation, $Q(0) = 50$ is not significant and does not represent "cost of overhead" in a function of this type. Therefore, Q is meaningful as a cost function only in the domain $x \geq 3$.

Note: We say that $q'(5) = 0$ and $q''(5) > 0$ are necessary and sufficient conditions for average cost q to be a minimum at the point $(5, 8)$. From Figure 3.5 it is clear that marginal values (values of Q') are less than average values (values of q) for $x < 5$. However, for $x > 5$ marginal values are greater than average values. A knowledge of such relationships can be useful. For instance, if management knows from experience how average values are running, it can reasonably approximate marginal values.

Cubic Cost Functions

In this discussion of cubic cost functions we will confine ourselves to those functions whose first derivative does not vanish, i.e., $Q'(x) \neq 0$.

EXAMPLE 2 A total cost function is defined by $Q(x) = x^3 + 12x + 16$. Show that marginal cost $Q'(x)$ equals average cost $q(x)$ at the minimum point of q and sketch graphs.

Solution

$$Q'(x) = 3x^2 + 12$$
$$= 0 \quad \text{for no real value of } x.$$

Thus, the slope is never zero; that is, there is no horizontal tangent to the curve of $Q(x)$.

$$Q''(x) = 6x$$
$$= 0 \quad \text{for } x = 0: \quad \text{cv}$$

Testing for an inflection point, we have

$$\left.\begin{array}{l} Q''(-1) < 0 \\ Q''(+1) > 0 \end{array}\right\} \quad \cap \text{ to } \cup \rightarrow \text{an inflection point}$$

at $(0, Q(0)) = (0, 16)$. As graphing aids, we find $Q(1) = 29$, $Q(2) = 48$, $Q(3) = 79$ (refer to Figure 3.6).

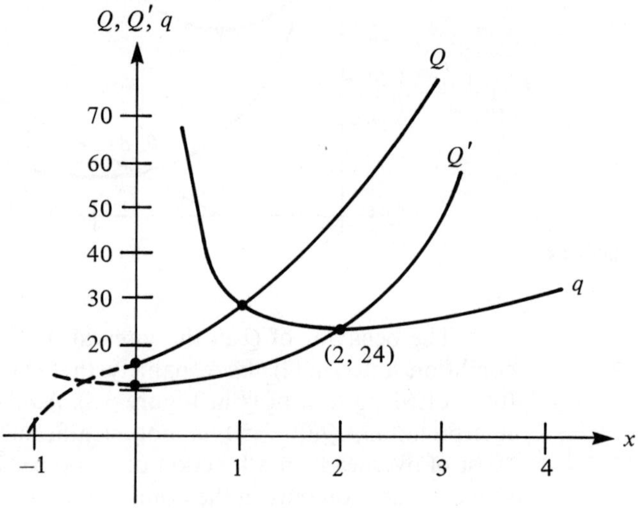

Figure 3.6

By definition, average cost

$$q(x) = \frac{Q(x)}{x}$$
$$= \frac{x^3 + 12x + 16}{x}$$
$$= x^2 + 12 + \frac{16}{x}.$$

Then

$$q'(x) = 2x - \frac{16}{x^2}$$
$$= \frac{2x^3 - 16}{x^2}$$
$$= 0 \quad \text{for } 2x^3 = 16$$
$$x^3 = 8$$
$$x = 8^{1/3} = 2: \quad \text{cv.}$$

$$\text{Furthermore,} \qquad q''(x) = 2 + \frac{32}{x^3}$$

$$q''(2) = 2 + \frac{32}{8} > 0 \rightarrow \text{minimum } q$$

at $(2, q(2))$ or $(2, 24)$.

To see if $Q'(x) = q(x)$ at $x = 2$, we have

$$Q'(2) = 3(4) + 12 = 24 = \text{minimum } q,$$

as was to be shown.

Some interesting sidelights to this example are worthy of note. Since

$$q(x) = \frac{Q(x)}{x}, \qquad x \neq 0,$$

then $$Q(x) = xq(x)$$

and $$Q'(x) = xq'(x) + q(x) \qquad \text{[differentiating by the product rule].} \tag{1}$$

A necessary condition for finding minimum q is that

$$q'(x) = 0.$$

Substituting $q'(x) = 0$ in Eq. (1) above yields

$$Q'(x) = q(x)$$

or $$Q'(x) = \frac{Q(x)}{x}.$$

In particular, $Q'(x_0) = Q(x_0)/x_0$ where x_0 is the critical value for which q is a minimum. The quotient

$$\frac{Q(x_0)}{x_0}$$

represents the slope of the line OT, which is the tangent to Q at the point $(x_0, Q(x_0))$ (see Figure 3.7). Therefore, in cost problems of this type, the critical value $x = x_0$ may be approximated by simply drawing the graph

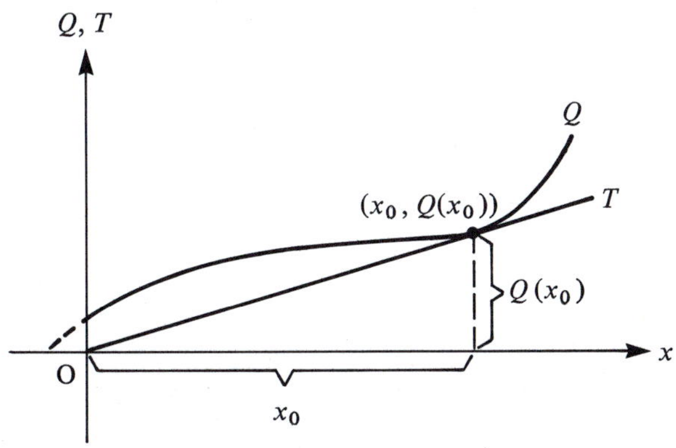

Figure 3.7

of the tangent to Q from the origin O and observing the value of the abscissa at the point of tangency. This important value of x can then be used as the abscissa of the low point of average cost q. In such cost problems, $Q'(x_0) = Q(x_0)/x_0$ and $Q''(x_0) > 0$ are considered to be necessary and sufficient conditions for minimum q.

Revenue Functions

Let a total revenue function be represented by

$$R(x) = xp$$

where x is the quantity of units demanded and p is the price per unit. We have already learned that

$$\text{marginal revenue} = R'(x).$$

We now define the **average revenue** function to be

$$r(x) = \frac{R(x)}{x}, \qquad x \neq 0.$$

It can be proved that marginal revenue $R'(x)$ equals average revenue $r(x)$ at the high point of r. That is, the output level $x = x_0$ (the critical value) that produces maximum average revenue is the same value of x at which average revenue equals marginal revenue. In introductory price theory the graphs R, R', r, and P (the demand curve) are often shown on one cc-system (coordinate system) so that comparisons can be conveniently made.

EXAMPLE 3 Let the demand function be $P(x) = 20 - 4x$ and the total revenue function be $R(x) = 12x - x^3$. Show that average revenue $r(x)$ equals marginal revenue $R'(x)$ at the high point of r. Sketch graphs of R, R', r, and P on one cc-system.

Solution

$$R'(x) = 12 - 3x^2 = 0$$

when

$$3x^2 = 12$$

or

$$x^2 = 4$$

$$x = \pm 2: \quad \text{cv};$$

but $x = -2$ is not applicable (x is assumed to be nonnegative).

Then

$$R''(x) = -6x$$

$$R''(2) < 0 \rightarrow \max R \text{ at } (2, R(2)) \text{ or } (2, 16).$$

The average revenue function is given by

$$r(x) = \frac{R(x)}{x} = \frac{12x - x^3}{x} = 12 - x^2$$

$$\begin{aligned}
&\text{Continuing: } r'(x) = -2x = 0\\
&\text{when} \qquad x = 0: \quad \text{cv}\\
&\text{Then} \qquad r''(x) = -2 < 0 \rightarrow \max r \text{ at } (0, r(0)) \text{ or } (0, 12).
\end{aligned}$$

We now show that $r = R'$ at (0, 12), the high point of r.

$$\begin{aligned}
R'(0) &= 12 - 3(0)^2\\
&= 12\\
&= \max r, \text{ as desired.}
\end{aligned}$$

Figure 3.8 illustrates this example.

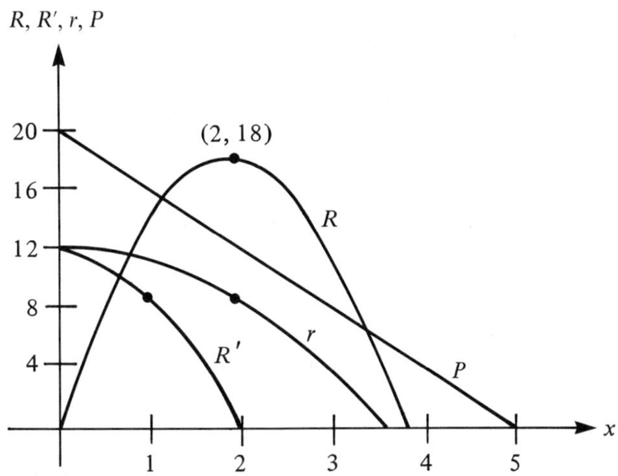

Figure 3.8

We have investigated quadratic and cubic cost functions and a cubic revenue function. There are many other types of cost and revenue functions for which optimization methods give useful results.

PROBLEM SET 3.2

1. Given the total cost function $Q(x) = x^2 - 2x + 8$.
   - **(a)** For what values of x is Q a valid cost function?
   - **(b)** Does $Q(0) = 8$ represent the overhead cost?
   - **(c)** Find the output level x for which each of Q and q attains a minimum.
   - **(d)** Show that average cost equals marginal cost at the low point of q.
   - **(e)** Sketch graphs on one cc-system.
2. The total cost of producing an item is defined by $Q(x) = x^2 + 4x + 16$.
   - **(a)** Find the level of output at which Q is a minimum.
   - **(b)** Find $q(x)$, the average cost, then determine the level of the output for which q is a minimum.
   - **(c)** Verify that marginal cost equals average cost at the low point of q.
   - **(d)** Sketch graphs on one cc-system.
3. Given the total cost function $Q(z) = (z^2/4) + 9$ where z denotes quantity of items manufactured. Verify that marginal cost equals average cost at the low point of the curve of average cost and sketch the graphs on one cc-system.

4. A total cost function is given by $Q(x) = x^3 + 9x + 16$. Show that marginal cost Q' equals average cost q at the low point of q. Sketch the graphs on one cc-system.
5. A total revenue function is defined by $R(x) = 27x - x^3$ and the demand function by $P(x) = 40 - 2x$. Verify that average revenue r equals marginal revenue R' at the high point of r. Sketch the graphs on a cc-system.
6. Given the total revenue function $R(x) = 6x^2 - x^3$ and the demand function $P(x) = 100 - 4x$. Show that marginal revenue equals average revenue at the point where average revenue is a maximum. Sketch the graphs on a cc-system.
7. A total cost function is defined by $Q(x) = 4 + x^{3/2}$ where x denotes the number (in hundreds) of items produced. For what value of x is average cost a minimum?
8. An average cost function is given by $q(x) = \frac{3}{x}(x^3 - 10x^2)$. Find the value of x such that average cost is a minimum.
9. If total revenue is defined by $R(q) = 12q^2 - q^4$ where q denotes quantity of items sold, find the value of q for which average revenue $r(q)$ is a maximum.
10. An average revenue function is given by $r(x) = 3x - 3x^{-1} - 2x^3$, x denoting thousands of units of production. For what value of x will average revenue be a maximum? [*Hint*: $r'(x) = 0$ for $0 < x < 3$.]

3.3 OPTIMUM LOT SIZE

Let us now consider simple inventory problems where the price of a commodity x is uniform and fixed for a given time period. Our chief concern will be to find the optimum lot size (of units of x) which minimizes the costs associated with inventory.

Note: The optimum lot size is often referred to as the economic order quantity, EOQ.

Procedures for the optimization of inventory costs involve two main types of costs, namely, setup and carrying costs. If a firm produces its own lot of items, then *setup costs* represent the expenses associated with the production of these items. But if the firm is supplied with these items by an outsider, then setup costs are called *reorder costs.*

The other type, *carrying costs,* involve expenses pertinent to the carrying of inventory items until they are sold. Such expenses include charges for storage, maintenance, depreciation, insurance, taxes, and incidentals.

As a general rule, setup costs decrease and carrying costs increase where "large" inventories are concerned; the converse is true for "small" inventories.

From the assumption that demand for an item is uniform throughout a given time period, it follows that inventory decreases at a constant rate. For a given lot size of x units, it is reasonable to use $x/2$ as the average inventory for a given period between production runs. This stems from adding the initial inventory to the final inventory and dividing by 2. That is, $(x + 0)/2 = x/2$ = average inventory. This implies that the purchasing of inventory stock is timed precisely so as to have shipments on hand just as the level of inventory units reaches zero.

Carrying costs vary directly as the number of units in the average inventory. Let us now formulate a typical problem.

Let x = lot size,
d = demand (uniform and fixed) per period,
c = carrying cost per item per period,
s = setup cost per run.

Then total inventory cost Q for the period may be expressed as

$$\boxed{Q(x) = \frac{x}{2}(c) + \frac{d}{x}(s)}$$

where $cx/2$ denotes the total cost of *carrying* inventory, d/x is the number of runs, and ds/x is the total *setup* cost per period. Applying the optimization techniques learned in previous sections we obtain

$$\begin{aligned} Q'(x) &= \frac{c}{2} - \frac{ds}{x^2} \\ &= \frac{cx^2 - 2ds}{2x^2} \\ &= 0 \quad \text{for} \\ cx^2 &= 2ds \\ x^2 &= \frac{2ds}{c} \\ x &= \sqrt{\frac{2ds}{c}}: \quad \text{cv.} \end{aligned}$$

Then
$$\begin{aligned} Q''(x) &= -ds(-2x^{-3}) \\ &= \frac{2ds}{x^3} \\ Q''(\text{cv}) &= \left.\frac{2ds}{x^3}\right|_{\text{cv}} > 0 \rightarrow \text{minimum } Q \end{aligned}$$

for $x = \sqrt{(2ds/c)}$, the optimum lot size. Observe that this most economic lot size varies jointly as d and s and inversely as c.

To illustrate the ideas and procedures associated with the concept of optimum lot size, we offer a concrete example.

EXAMPLE 1 In the general problem above, let

annual demand $d = 10^6$ units of production,
annual carrying cost c = \$0.80 per unit
and setup cost per run r = \$1000.

The total inventory cost Q per year equals carrying cost plus setup cost or

$$\begin{aligned} Q(x) &= \frac{x}{2}(0.80) + \frac{10^6}{x}(1000) \\ &= 0.4x + 10^9 x^{-1}. \end{aligned}$$

Differentiating and simplifying, we obtain

$$\begin{aligned} Q'(x) &= 0.4 + 10^9(-x^{-2}) \\ &= 0.4 - \frac{10^9}{x^2} \\ &= \frac{0.4x^2 - 10^9}{x^2} \\ &= 0 \quad \text{when} \\ 0.4x^2 - 10^9 &= 0 \end{aligned}$$

or

$$\begin{aligned} 0.4x^2 &= 10^9 \\ x^2 &= \frac{10^9}{4(10^{-1})} = \frac{10^{10}}{4} \end{aligned}$$

Therefore

$$\begin{aligned} x &= \sqrt{\frac{10^{10}}{4}} \\ &= \frac{10^5}{2} \quad \text{or} \quad 50{,}000\text{:} \quad \text{cv.} \end{aligned}$$

Employing the second derivative test, we get

$$\begin{aligned} Q''(x) &= -10^9(-2x^{-3}) \\ &= \left.\frac{2(10)^9}{x^3}\right|_{\text{cv}} > 0 \rightarrow \text{minimum } Q \end{aligned}$$

for $x = 50{,}000$ units, the optimum lot size.

There are many variations on this "theme" and there may be costs other than those designated in the simple model above. Consider the case where more costs enter the picture. For example, if we let

$$\begin{aligned} w &= \text{warehousing cost per unit,} \\ v &= \text{variable setup cost per unit,} \\ c, d, s &= \text{same as above,} \end{aligned}$$

then total cost per period may be written as

$$\begin{aligned} Q(x) &= \frac{x}{2}(c) + \frac{d}{x}(s) + wx + vx \\ &= \frac{c}{2}(x) + dsx^{-1} + (w + v)x. \end{aligned} \tag{1}$$

The next example illustrates an application of this cost formula.

EXAMPLE 2 Assume that a manufacturer's reordering costs for $d = 8000$ units per month are: $s = \$300$, $w = \$0.10$ per unit, $c = \$1$, and $v = \$0.20$ per unit, where c, d, s, w, and v are as in Eq. (1) above. Find the optimum lot size.

Solution Substituting the given data in Eq. (1) we obtain

$$\begin{aligned} Q(x) &= \frac{x}{2}(1) + 8000(300)x^{-1} + (0.10 + 0.20)x \\ &= \frac{x}{2} + 24(10)^5x^{-1} + 0.3x. \end{aligned}$$

Next, we differentiate and simplify:

$$\begin{aligned} Q'(x) &= 0.5 + 24(10)^5(-x^{-2}) + 0.3 \\ &= 0.8 - \frac{24(10)^5}{x^2} \\ &= \frac{0.8x^2 - 24(10)^5}{x^2} \\ &= 0 \quad \text{when} \\ 0.8x^2 - 24(10)^5 &= 0 \\ \text{or} \qquad 0.8x^2 &= 24(10)^5 \\ x^2 &= \frac{24(10)^5}{8(10)^{-1}} = 3(10)^6 \\ x &= \sqrt{3(10)^6} \\ &= (10)^3\sqrt{3} \cong 1732: \quad \text{cv.} \\ \text{Then} \qquad Q''(x) &= -24(10)^5(-2x^{-3}) \\ &= \left.\frac{48(10)^5}{x^3}\right|_{\text{cv}} > 0 \rightarrow \text{minimum } Q \text{ or } C_{\min} \end{aligned}$$

at $x \cong 1732$ units, the optimum lot size which minimizes cost. If the dollar amount of $C_{\min}$ is desired, then $Q[(10)^3\sqrt{3}]$ may be evaluated.

PROBLEM SET 3.3

1. In Example 1 find optimum lot size and minimum cost if $d = 10^4$ units, $c = \$1$ per unit, and $s = \$800$ per run.
2. Work Example 2 when $d = 10^4$ units, $c = \$1.30$ per unit, $s = \$400$ per run, $w = \$0.15$, and $v = \$0.20$.
3. A firm uses 100 "contraptions" each of 240 working days per year for the purpose of assembling machines, one contraption per machine. It buys these contraptions for \$80 per unit and raises the money for this expense by a short-term loan at 8% simple interest per annum. Other inventory charges include \$1.60 per contraption per year and a fixed charge of \$150 per shipment. Approximate the optimum lot size. [*Hint*: Use $d = 100(240)$, $c = 80(0.08) + 1.60$, $s = \$150$.]

4. Work problem 3 above for the case where the firm uses 120 contraptions; purchasing costs are \$100 per unit; the short-term loan is at 10% annual simple interest; other inventory charges of \$1.80 per contraption; and a fixed charge per shipment of \$160.

5. Given a total inventory cost function $Q(x) = \frac{x}{2}(1.20) + \frac{10^4(120)}{x} + 0.60x$; find the optimum lot size.

6. A total cost function is defined by $Q(x) = T + (w/v) + Dx^{1/2}$ where

Q = dollar cost per pound of food distributed,
x = area (in square miles) served by the main warehouse,
T = dollar amount of taxes per pound of food distributed,
w = warehouse storage costs (in dollars) per pound per month,
v = volume in pounds of frozen food handled by the warehouse per month,
D = amount of delivery costs (gas, repairs, etc.) per month.

Find the optimum area, i.e., the region to be served, such that costs are minimized. [*Hint*: First substitute $v = kx$ in the given equation (here $k = v/x$ is a "constant" which represents the volume of business in pounds per square mile per month).]

7. Determine the optimum lot size for the inventory cost function

$$Q(z) = \frac{z}{2}\left(\frac{3}{2}\right) + \frac{15(10^4)}{z} + 50(1 + 0.06z).$$

8. If $Q(x) = (x/16) + (1000/\sqrt{x})$ defines a total inventory cost function, find minimum cost.

9. Find the optimum lot size for the inventory cost function $Q(x) = 10x + 10^7/(x - 10)$.

10. An inventory cost function is given by $Q(x) = \frac{36x}{10^3} + \frac{18(10^3)}{x^2}$. What is the optimum lot size?

11. Determine the optimum lot size for the inventory cost function

$$Q(x) = \frac{x}{100} + \frac{10^8}{4x - 100}.$$

3.4 OTHER APPLICATIONS

We continue our optimization analysis by solving other types of practical problems arising in various business operations. No additional mathematical tools are needed for their solution. However, diagnostic abilities will need sharpening in order to correctly interpret the wording of problems and to follow through with proper mathematical models.

EXAMPLE 1 A rancher wants to enclose a rectangular field with fencing. Three of the sides need fencing and a fourth side, bordering a river, requires no fence. (Assume the river side to be a straight line.) Find dimensions of the field such that maximum area is enclosed.

Solution We draw a pertinent figure as shown and see that the perimeter

$$P = x + 2y$$

and the area

$$A = xy.$$

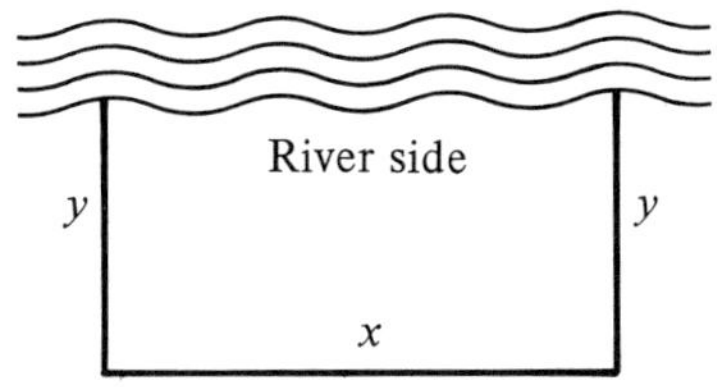

Since A is a function of two variables and we have not as yet discussed derivatives of such functions, we proceed to change A to a function of one variable. Accordingly, we solve P for x, obtaining

$$x = P - 2y$$

and substitute this value of x in A. Then

$$\begin{aligned} A &= (P - 2y)y \\ &= Py - 2y^2 = A(y). \end{aligned}$$

Furthermore,

$$\begin{aligned} A'(y) &= P - 4y \\ &= 0 \quad \text{for } P = 4y \end{aligned}$$

or

$$y = \frac{P}{4}: \quad \text{cv.}$$

Then

$$A''(y) = -4 < 0 \rightarrow \text{maximum } A$$

when $y = P/4$ and $x = P - 2(P/4) = P/2$, the desired dimensions.

In a real-world situation where, for example, 600 yards of fencing were available, then A_{max} occurs when $y = 600/4$ or 150 yards and $x = 600/2$ or 300 yards.

In the next case a different analytical approach is required in order to find an optimal solution.

EXAMPLE 2 If 25 lemon trees are planted per acre in a certain region of favorable soil type, the yield will be about 370 lemons per tree. For each additional tree planted per acre, the yield per tree will be reduced by about 10 lemons. What is the optimum number of trees to plant per acre?

Solution Let x be the number of *additional* trees to plant for optimum results. Then $(25 + x)$ is the adjusted number of trees and $(370 - 10x)$ is the corresponding adjusted number of lemons per tree. We may now write

$$f(x) = (25 + x)(370 - 10x)$$

where f represents total yield. We proceed with the second derivative test.

$$\begin{aligned} f'(x) &= (25 + x)(-10) + (370 - 10x) \cdot 1 \\ &= -250 - 10x + 370 - 10x \\ &= 120 - 20x \\ &= 0 \quad \text{for } x = 6\text{:} \quad \text{cv} \\ f''(x) &= -20 < 0 \rightarrow f_{\max} \quad \text{at } x = 6 \end{aligned}$$

additional trees. Therefore, the optimum number of trees to plant per acre is (25 + 6) or 31.

In the next two cases optimization involves the minimizing of functions.

EXAMPLE 3 A manufacturer wishes to minimize the cost of making a container from thin sheet metal. The container is to have a square base, vertical sides, no lid, and a volume of 100 m^3. If the bottom costs 80¢ per m^2 and the sides 50¢ per m^2, find the dimensions of the container so that its cost is minimized.

Solution Using the dimensions shown in the figure, we can write the formula for volume as

$$V = x^2y = 100.$$

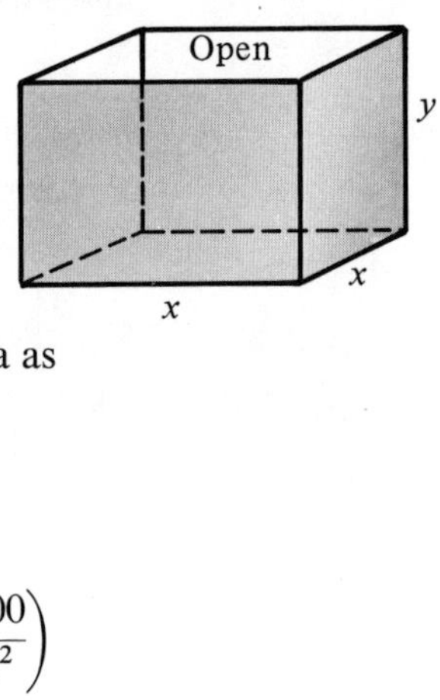

From this equation we see that

$$y = \frac{100}{x^2}.$$

Furthermore, we can write the formula for area as

$$A = x^2 + 4xy.$$

Then cost may be written as

$$\begin{aligned} C &= 80x^2 + 50(4x)\left(\frac{100}{x^2}\right) \\ &= 80x^2 + \frac{20{,}000}{x}. \end{aligned}$$

We now apply the mechanics of the second derivative test.

$$\begin{aligned} C'(x) &= 160x + 20{,}000(-x^{-2}) \\ &= 160x - \frac{20{,}000}{x^2} \\ &= \frac{160x^3 - 20{,}000}{x^2} \\ &= 0 \quad \text{when} \\ 160x^3 &= 20{,}000 \end{aligned}$$

or $$x^3 = \frac{20{,}000}{160} = 125$$

Therefore $$x = \sqrt[3]{125} = 5: \quad \text{cv.}$$

Then $$C''(x) = 160 - 20{,}000(-2x^{-3})$$
$$= 160 + \frac{40{,}000}{x^3}\bigg|_{\text{cv}} > 0 \to C_{\text{min}}$$

at $x = 5$, $y = 100/5^2 = 4$. Therefore, the dimensions for minimum cost are 5 m × 5 m × 4 m.

The following example deals with the important concept of maximizing profit.

EXAMPLE 4 Let a demand function be defined by $p = 99 - (q/2)$ where p is the price per unit and q is the quantity demanded. Let $c = (q/2) + 1$ denote cost per unit of demand. Then total sales revenue may be expressed as $R = pq$ and total cost as $Q = cq = (q/2 + 1)q$. Since profit P equals total sales revenue minus total costs, we may write

$$\begin{aligned} P &= R - Q \\ &= pq - cq \\ &= \left(99 - \frac{q}{2}\right)q - \left(\frac{q}{2} + 1\right)q \\ &= 99q - \frac{q^2}{2} - \frac{q^2}{2} - q \\ &= 98q - q^2. \end{aligned}$$

Then $$\begin{aligned} P' &= 98 - 2q \\ &= 0 \quad \text{for } q = 49: \quad \text{cv} \\ P'' &= -2 < 0 \to \text{maximum } P \text{ (or } P_{\text{max}}) \end{aligned}$$

at $q = 49$ units. If P_{max} is desired we evaluate $P(49) = 98(49) - (49)^2 = 49(98 - 49) = 49(49) = \$2401$.

This result, i.e., $q = 49$ units which determine P_{max}, can be checked by another approach:

$$\text{Revenue } R = 99q - \frac{q^2}{2}$$
$$\text{Marginal revenue } R' = 99 - q$$
$$\text{Cost } Q = \frac{q^2}{2} + q$$
$$\text{Marginal cost } Q' = q + 1.$$

Equating the R' to the Q' yields

$$99 - q = q + 1$$

or $$98 = 2q,$$
$$q = 49 \text{ units, as above.}$$

The inherent implication is that maximum profit occurs when

$$\text{Marginal revenue} = \text{Marginal cost}.$$

This is not surprising in view of the following analysis:

$$\begin{aligned} \text{Write}\quad P &= R - Q \\ \text{Then}\quad P' &= R' - Q' \qquad \text{[by the definition of differentiation]} \\ &= 0 \quad \text{when} \\ R' &= Q'. \end{aligned}$$

But R' is marginal revenue and Q' is marginal cost. Furthermore, $P'(q) = 0$ when $q = 49$ and $P''(49) < 0$. We have, essentially, established necessary and sufficient conditions for maximum profit.

The last example of this section is a cost-minimizing problem.

EXAMPLE 5 A small storage tank, in the form of a right circular cylinder with no top, is to be constructed so as to hold 324 m^3 of water. Material costs include $\$3/\pi$ per m^3 for the bottom and \$2 per m^3 for the lateral surface (the number $3/\pi$ is used to facilitate arithmetic computations). What is the cost of the least expensive tank that can be manufactured to contain 324 m^3?

Solution We draw a sketch of the tank and from the dimensions used we can write the volume as

$$V = \pi x^2 y = 324.$$

This formula is valid since the volume of a right circular cylinder can be proved to be:

$$\text{area of base} \times \text{height}.$$

The problem asks for minimum cost, so we need a cost equation. This, in turn, means we need an equation for the area because the cost figures are quoted in terms of m^3.

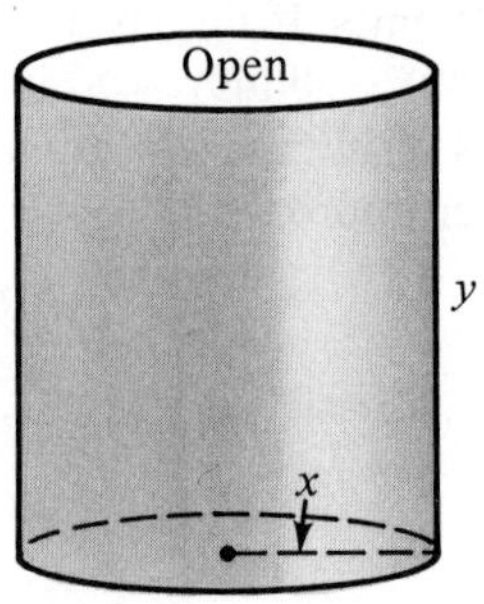

$$\text{Total area } A = \text{area of bottom} + \text{area of lateral surface}$$

or

$$A = \pi x^2 + 2\pi xy.$$

Then

$$\text{cost } C = \frac{\$3}{\pi}(\pi x^2) + \$2(2\pi xy)$$
$$= (3x^2 + 4\pi xy) \text{ dollars}.$$

Since $$y = \frac{324}{\pi x^2} \quad \text{[from the formula for } V\text{]},$$

then cost becomes a function of x as follows:

$$\begin{aligned} C(x) &= 3x^2 + 4\pi x\left(\frac{324}{\pi x^2}\right) \\ &= 3x^2 + 1296x^{-1}. \end{aligned}$$

Continuing with the familiar optimization analysis, we get

$$\begin{aligned} C'(x) &= 6x - \frac{1296}{x^2} \\ &= \frac{6x^3 - 1296}{x^2} \\ &= 0 \quad \text{when} \\ 6x^3 &= 1296 \end{aligned}$$

or $$\begin{aligned} x^3 &= 216 \\ x &= \sqrt[3]{216} = 6: \quad \text{cv.} \end{aligned}$$

Then $$\begin{aligned} C''(x) &= 6 - 1296(-2x^{-3}) \\ &= 6 + \frac{2592}{x^3}\bigg|_{\text{cv}} > 0 \rightarrow C_{\min} \end{aligned}$$

at $x = 6$ m. To find $C_{\min}$ we simply evaluate

$$\begin{aligned} C(6) &= 3(6)^2 + \frac{1296}{6} \\ &= 108 + 216 = \$324. \end{aligned}$$

In many situations it is necessary to set up two functions. One is an auxiliary function and the other is the objective function to be optimized. The auxiliary function furnishes us with the variable whose value is substituted in the objective function f. This substitution changes f from a function of two variables to a function of one variable. This latter form is then analyzed by techniques of optimization outlined earlier. In a later section optimal solutions for functions of two or more variables will be introduced and discussed.

Note: Another important application, called "related rates" and involving rates of increase and decrease, is introduced in Section 3.6, Enrichment Topics and Problems.

PROBLEM SET 3.4

1. A rancher wishes to enclose a rectangular field for pasturing. He has 800 meters of fencing material available for use. What are the dimensions of the field for maximum area?

2. Work problem 1 above if 1000 meters of fencing are available and one side of the field borders a river, thus needing no fence.

3. A production manager knows from experience that each of his 30 employees can produce very close to 150 units of work per day. For more than 30 employees, the production per employee drops (for various reasons such as overcrowding, behavioral problems, etc.) by three units for each additional employee hired in excess of 30. How many employees should the manager use to enjoy maximum production?

4. A telephone company consents to build a new exchange for 120 subscribers (or less) at a uniform charge of \$60 each. To attract more customers, the company is willing to lower this uniform charge by thirty cents for each subscriber over the 120. For example, if 140 persons subscribe, then each pays \$60 − 0.30(20) or \$54. Find the total number of subscribers for which company income will be a maximum.

5. A box with square base, vertical sides, and a volume of 64 m^3 is to be made from thin sheet metal. Find the dimensions for minimum cost if the bottom costs \$0.80 per m^2, the sides cost \$1 per m^2, and the lid costs \$1.20 per m^2.

6. Work problem 5 if there is no lid, volume is 10 m^3, the bottom costs \$1 per m^3, and the sides cost \$0.50 per m^3.

7. A closed rectangular container with a square base is to be constructed from 54 m^2 of lumber. Find the dimensions for maximum volume.

8. For a certain company the cost (overhead and wholesale price) for handling x clothes dryers is $Q(x) = (150 + 72x)$ dollars. The x dryers are sold at \$$p$ per unit per month where $x = 60 - p/3$, the demand function. What retail price per unit is required for profit P to be a maximum? [*Hint*: From $x = 60 - p/3$ we have $p = 180 - 3x$, the demand price. Revenue $R = xp = x(180 - 3x)$. Then $P = R - Q$, etc.]

9. The demand price function for a certain firm is estimated to be $p = 60 - q/2$, where p is amount of price per unit and q is quantity of demand. The cost function is found to be $Q = 2q$. Show that $P_{\max}$ occurs when marginal revenue equals marginal cost.

10. A cylindrical tank is to be constructed to hold 27 m^3 of water and to have an open top. Find the height and base radius of the tank when its surface area is a minimum.

11. A rectangular field with 2000 m^2 of area is to be fenced off and divided into two equal regions by a fence parallel to one side of the rectangle. The middle fence costs \$1 per linear meter and the rest of the fence costs \$2 per linear meter. Minimize the cost of all the fencing.

12. A pan is to be made from a piece of tin 12 cm square by cutting out equal squares from the four corners and turning up the sides. Find dimensions of the largest pan which can be so made.

13. A lot is in the shape of a right triangle of sides 50, 120, and 130 meters. Find the dimensions of the rectangular warehouse of maximum floor area which can be built on the lot in the position shown in the figure on the next page. *Hint*: By similar triangles we have the equation:

$$\frac{y}{50} = \frac{120 - x}{120}.$$

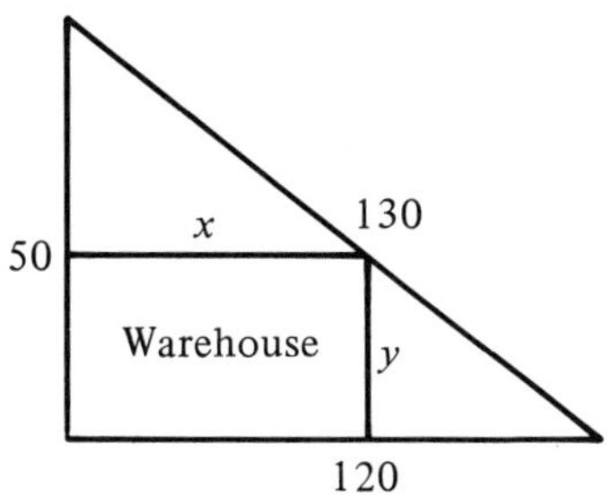

14. A manufacturer can sell x units of his product per day at p dollars per unit where $p = 410 - 4x$. The cost Q of producing the x units per day is $Q(x) = (100 + 10x + x^2)$ dollars. Find:

(a) The number of units of production per day.
(b) The price to be charged per unit so that profit will be as large as possible.

15. A specialty store offers all of its merchandise for sale at \$10 per unit. The total cost of making x units of the product is found to be $Q(x) = \$(40 + 2x + 0.02x^2)$.

(a) Write the profit function.
(b) How many units must be made to maximize profit?
(c) Show that marginal revenue equals marginal cost when profit is a maximum.

16. On each page of a book under production the printed matter is to have 96 in^2 of area. Plans call for a 1-inch margin at the top and bottom of each page and a $1\frac{1}{2}$-inch margin on each side. Find the most economical overall dimensions of each printed page.

3.5 REVIEW PROBLEMS

For each of the following problems 1–10 find extrema and points of inflection. Sketch graphs of the odd-numbered problems.

1. $f(x) = \dfrac{x^3}{3} - x^2 - 3x$

2. $g(x) = 27x - x^3$

3. $h(x) = \dfrac{3x^4}{2} - 12x^2$

4. $F(x) = 32x - x^4$

5. $G(x) = \dfrac{4}{1 + x^2}$

6. $H(x) = \dfrac{x - 2}{x - 3}$

7. $f(x) = \dfrac{x^2 + 1}{x}$

8. $u(x) = \dfrac{x^2 + 4}{x}$

9. $g(x) = \sqrt{16 - x^2}$

10. $h(x) = \sqrt{25 - x^2}$

11. A box is to have a square base, vertical sides, a lid, and a volume of 30 m^3. If the lid and bottom each cost 15¢ per m^2 and the sides cost 5¢ per m^2, what should be the dimensions for minimum cost?

12. Work problem 11 for the case where there is no lid and volume is to be 96 m^3.

13. The OK Airline Co. will make a special ski run if 100 people buy tickets at \$10 each. For each additional person over the 100 the company agrees to decrease the fare by 5¢. For example, if 120 persons take the trip, the fare for each would be \$9. For what total number of passengers will the company's revenue be maximized?

14. A company wishes to make open cardboard boxes having optimum capacity. Each box is to be made from a piece of cardboard 4 m by 4 m in size by cutting out equal squares from each of the four corners and folding up the sides. What are the dimensions of the cut-out squares if the volume is to be a maximum?

15. If the total cost function $Q(x) = 3x^2 - 6x + 27$, find the level of output x at which Q and the average cost $q(x)$ are a minimum. Verify that marginal cost Q' equals average cost at the low point of the curve of q. Sketch the graphs.

16. Given the total revenue function $R(x) = 24x + 2x^2 - x^3$ and the demand function $p = 150 - 25x$. Show that average revenue r equals marginal revenue R' at the high point of the graph of r. Sketch graphs of R, R', r, and p.

17. A total cost function is defined by $Q(x) = \frac{Bx}{2} + \frac{CD}{x} + (E + F)x$ where x denotes lot size, B, D, E, and F are constants representing various costs associated with inventory, and C is total number of units. Find the optimum lot size.

18. Work problem 17 when $B = \$1$, $C = 10{,}000$ units, $D = \$9$ per unit, $E = \$0.20$, $F = \$0.30$.

19. The World Wide Containers Corp. wishes to manufacture a closed cylindrical can having a volume of $16\pi\ \text{cm}^3$. For what values of the radius and height will the surface area of the can be a minimum?

20. The manager of a small chain of stores finds that the cost of buying and distributing x items of product K-11 is represented by $Q(x) = \$(300 + 4x)$ per month. If $\$p$ is charged per item then $x = (6 - (p/2))$ items are sold per month. Find the number of items of K-11 which yield maximum profit P and find $P_{\max}$.

3.6 ENRICHMENT TOPICS AND PROBLEMS

Find extrema of the following problems 1–6 and sketch graphs of the even-numbered problems.

1. $f(x) = \frac{x^5}{20} - x^2 + 3$

2. $g(t) = \frac{1}{t} + \frac{t}{16}$

3. $h(u) = \frac{u^3}{3} - \frac{u^2}{2} - 6u$

4. $F(z) = 3z^{1/3} - z$

5. $G(t) = t + \frac{9}{t}$

6. $H(x) = 2\sqrt{x} - x$

7. Determine the maximum and minimum values of $f(x) = x^4 + \frac{4x^3}{3} - 48x + 5$ on the interval $[-3, 3]$.

8. Find the inflection point for $g(t) = t^{3/2} - \frac{9t}{2} + 6\sqrt{t}$.

9. Which point on the parabola $f(x) = x^2/4$ is closest to the point (1, 2)?

10. Find the dimensions of the rectangle with area 100 units^2 such that its perimeter is a minimum.

11. Town C is 12 km east of A and town D is 3 km north of C. A road is to be built from A to D (see figure). The cost from A to B is \$100,000 per km; any other path costs \$200,000 per km. Minimize the cost of construction. [*Hint*: Cost $C = x + 2d$ (in hundred thousands of dollars); $x + y = 12$; $d^2 = y^2 + 9$; therefore $C = (12 - y) + 2\sqrt{y^2 + 9}$.]

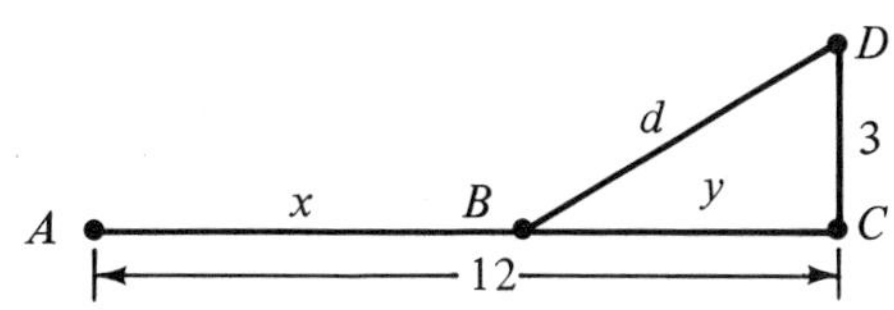

12. ***Implicit Differentiation.*** Assume that y is a function of x and that y satisfies the equation $2xy - 3y = 6x - 1$. We could solve for y and differentiate the explicit form. But we can find dy/dx without first solving for y. The given expression is in implicit form. By a technique called **implicit differentiation**, we differentiate both sides with respect to x, keeping in mind that y is a function of x. Thus,

$$\frac{d}{dx}(2xy) + \frac{d}{dx}(-3y) = \frac{d}{dx}(6x) + \frac{d}{dx}(-1)$$

$$2x\frac{dy}{dx} + y \cdot 2\frac{dx}{dx} - 3\frac{dy}{dx} = 6\frac{dx}{dx} + 0$$

$$(2x - 3)\frac{dy}{dx} = 6 - 2y$$

$$\frac{dy}{dx} = \frac{6 - 2y}{2x - 3}$$

The method of differentiating implicitly is most useful when the given implicit form does not lend itself conveniently to solving explicitly for y.

EXAMPLE Find dy/dx if $x^2y^3 - y^4 = x - y + 2$.

Solution

$$x^2 \cdot 3y^2\frac{dy}{dx} + y^3 \cdot 2x - 4y^3\frac{dy}{dx} = 1 - \frac{dy}{dx}$$

$$(3x^2y^2 - 4y^3 + 1)\frac{dy}{dx} = 1 - 2xy^3$$

$$\frac{dy}{dx} = \frac{1 - 2xy^3}{3x^2y^2 - 4y^3 + 1}.$$

Find dy/dx by implicit differentiation:

(a) $x + 3xy + y = 1$ **(b)** $x^3 + y^3 - 3xy = 0$
(c) $x^3 - 3xy + 6y^2 = 4$ **(d)** $\sqrt{x} + \sqrt{y} = 1$

13. ***Related Rates.*** If x and y are known to be functions of another variable, for example t, and they are differentiable with respect to t, then an implicit form such as $x^2 + y^2 = 1$ can be differentiated implicitly as follows:

$$2x\frac{dx}{dt} + 2y\frac{dy}{dt} = 0$$

or

$$x\frac{dx}{dt} + y\frac{dy}{dt} = 0.$$

Important applications to *problems of motion* occur when related variables are dependent on another variable such as time t.

EXAMPLE Trucks A and B leave point O simultaneously, A going due east at 40 km/hr and B going due north at 30 km/hr. How fast are they separating after t hours?

Solution After t hours A has traveled $40t$ km and B has gone $30t$ km. From the figure we see that the variables are related by $z^2 = x^2 + y^2$. Differentiating implicitly we obtain

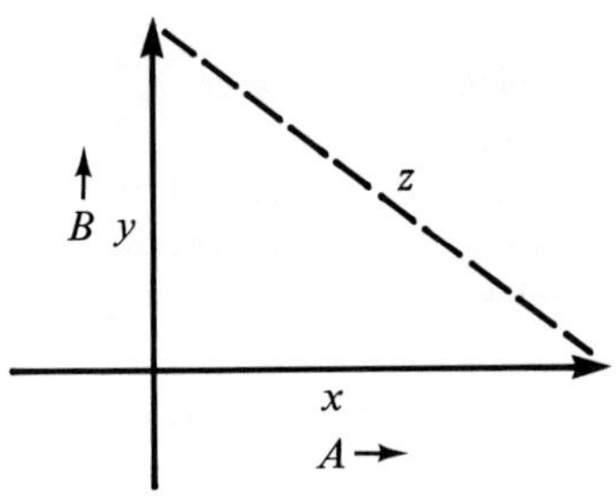

$$2z\frac{dz}{dt} = 2x\frac{dx}{dt} + 2y\frac{dy}{dt}$$

or

$$z\frac{dz}{dt} = x\frac{dx}{dt} + y\frac{dy}{dt}$$

After t hours, $z^2 = (40t)^2 + (30t)^2 = 2500t^2$

or $z = 50t$

$(-50$ is discarded since z is a distance, which is positive). Substituting, we get

$$50t\frac{dz}{dt} = 40t \cdot 40 + 30t \cdot 30$$

$$\frac{dz}{dt} = \frac{2500t}{50t} = 50 \text{ km/hr},$$

the rate at which trucks A and B are separating after t hours.

(a) The length of a rectangle is increasing at the rate of 3 cm/sec and the width is increasing at the rate of 2 cm/sec. At what rate is the area increasing when the length is 7 cm and the width is 4 cm?

(b) A spherical balloon expands because of solar radiation. If its radius r increases at the rate of 2 cm per minute, how fast is the volume V changing at the instant when $r = 4$ cm? Use $V = \frac{4}{3}\pi r^3$.

(c) A conical tank of radius $r = 12$ m and height $h = 8$ m is leaking water at the rate of 3 m^3/hr. At what rate is h decreasing when the water is 4 m deep? Use $V = \frac{1}{3}\pi r^2 h$ and $r/h = 12/8$.

4 EXPONENTIAL AND LOGARITHMIC FUNCTIONS

Applications of e^u and $\ln|u|$

IMPORTANT EQUATIONS

$$f(x) = ae^{kx}$$

$$x = \log_b y$$

$$D_x \ln|u| = \frac{1}{u} \cdot D_x u$$

$$D_x e^u = e^u D_x u$$

4.1 EXPONENTIAL FUNCTIONS

We are familiar with power functions of the form $f(x) = x^n$ which exhibit the exponent n as a constant and the base x as the independent variable. If we make the switch so that the base is the constant b and the exponent is the independent variable, we then have a so-called **exponential function** defined by

$$y = f(x) = b^x$$

where $b > 0$ but $b \neq 1$. The function f is defined and positive for all real values of x.

For $b > 1$, the graph of f increases as x increases. For $0 < b < 1$, f decreases as x increases. Figure 4.1 illustrates this function. In either case,

(1) there is no x-intercept since there is no value of x for which b^x is zero;

(2) the y-intercept is the point (0, 1);

(3) the graph appears entirely above the x-axis;

(4) the x-axis is an asymptote.

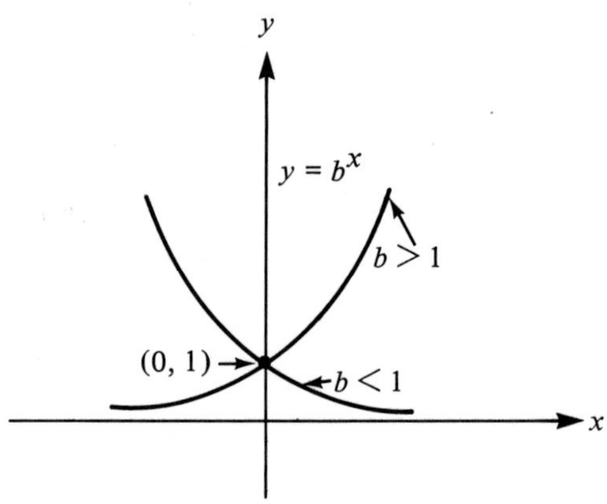

Figure 4.1

Consider the case where $b = 2$. Then the exponential function $y = b^x$ becomes $y = 2^x$. The following tabular form shows some values in the domain of 2^x and the corresponding values in its range.

x	0	1	2	3	4	5
2^x	1	2	4	8	16	32

The graph of 2^x is shown in Figure 4.2.

We say that the elements in the domain of $f(x) = b^x$, $b > 1$, increase (decrease) linearly while the associated elements in the range increase (decrease) geometrically. That is, the elements $\{\ldots -2, -1, 0, 1, 2, \ldots\}$

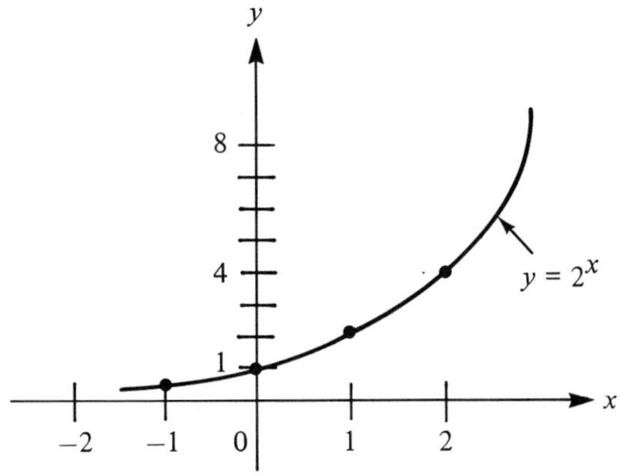

Figure 4.2

increase arithmetically when $f(x) = 2^x$. The geometric progression of values of f constitutes an exponential trend of *growth* or of *decay* (erosion). These trends are of particular significance when the base b is the constant e (defined below). The general forms which represent these exponential trends are given as follows:

$$f(x) = ae^{kx} \quad \text{(growth)}$$

and

$$f(x) = ae^{-kx} \quad \text{(decay)}$$

where a and k are positive constants.

The number e is an irrational number approximately equal to 2.71828. One method of computing the approximate value of e to five decimal places is to evaluate the first ten terms of the series

$$1 + 1 + \frac{1}{2!} + \frac{1}{3!} + \frac{1}{4!} + \cdots + \frac{1}{9!} + \cdots \tag{1}$$

Recall the meaning of factorial: $2! = 2 \cdot 1 = 2$, $3! = 3 \cdot 2 \cdot 1 = 6$, $4! = 4 \cdot 3 \cdot 2 \cdot 1 = 24$, and so on. In general, we obtain the formula $n! = n(n-1)(n-2)(n-3)\cdots 1$. The sum of the series in Eq. (1) is the number e, by definition. This sum is exhibited below:

$$\begin{aligned}
1 &= 1.000000 \\
1 &= 1.000000 \\
1/2! &= 0.500000 \\
1/3! &= 0.166667 \\
1/4! &= 0.041667 \\
1/5! &= 0.008333 \\
1/6! &= 0.001389 \\
1/7! &= 0.000198 \\
1/8! &= 0.000025 \\
1/9! &= 0.000003 \\
\cdots & \\
\hline
e &= 2.718282
\end{aligned}$$

The series can be obtained by expanding $(1 + h)^{1/h}$ by the binomial expansion, simplifying, and letting h approach zero. In this process the limit of $(1 + h)^{1/h}$ as h approaches zero is the number e.

We will discuss in a later paragraph the growth characteristic of e as applied to the compound-interest process. The process will be idealized in that interest will be compounded n times per year as we let n increase without bound.

Because of its fundamental importance in many facets of life, the function e^x is commonly called *the* exponential function. Its graph, along with that of e^{-x}, is presented in Figure 4.3.

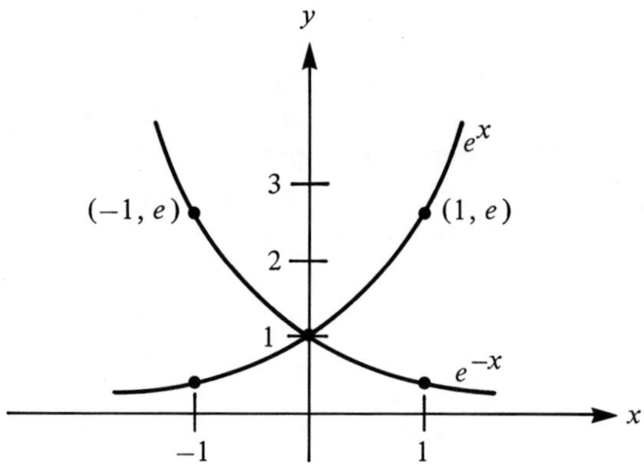

Figure 4.3

We will not go into a detailed account of the interesting theory underlying the number e, but will rather concern ourselves with uses and pertinent applications of this fascinating number.

Some basic properties of exponential functions are as follows:

Property 1 $e^{-x} = \dfrac{1}{e^x}$

Property 2 $e^{x+y} = e^x \cdot e^y$

Property 3 $e^{x-y} = e^x \div e^y$

Property 4 $e^{bx} = (e^x)^b$ *if b is rational*

Property 5 $e^x \to +\infty$ *as* $x \to +\infty$

Property 6 $e^x \to 0$ *as* $x \to -\infty$

Property 7 *Domain of* e^x: *all real numbers*
Range of e^x: *all positive numbers*

(Properties 5–7: see Figure 4.3)

A few application of these properties are shown next.

(a)
$$\frac{1}{3e^{-x}} + e^{-x} + e^{-2x} = \frac{e^x}{3} + \frac{1}{e^x} + \frac{1}{e^{2x}}$$
$$= \frac{e^x \cdot e^{2x} + 3e^x + 3}{3e^{2x}}$$
$$= \frac{e^{3x} + 3e^x + 3}{3e^{2x}} \quad \text{as a single fraction.}$$

(b) $e^{2t} + 3e^t - 10 = (e^t + 5)(e^t - 2)$ in factored form.

(c)
$$\lim_{x\to 0} \frac{e^{3x} - 1}{e^x - 1} = \frac{e^0 - 1}{e^0 - 1} \quad \left[= \frac{0}{0} \text{ or meaningless}\right]$$
$$= \lim_{x\to 0} \frac{(e^x - 1)(e^{2x} + e^x + 1)}{e^x - 1}$$
$$= \lim_{x\to 0} (e^{2x} + e^x + 1) = e^0 + e^0 + 1 = 3$$

Exponential trends abound in the business environment. Examples include the computing of present values and compound amounts at continuous interest rates, studies of short-term trends in earnings (either growth or decay), forecasts and projections of sales, consumer demand indices, various cost relationships, and many others.

If interest is at $i\%$ compounded f times per year, then the compound amount A of a principal P may be written as

$$A = P\left(1 + \frac{i}{f}\right)^{nf}$$

where n denotes the number of years. When f, the frequency of compounding interest, is large and an approximation is acceptable, another formula, involving e, may be used. This latter formula is convenient in its application when exponential tables for e^x are available. Let us develop this formula.

The factor $(1 + i/f)^{nf}$ can be written in the equivalent form

$$\left[\left(1 + \frac{i}{f}\right)^{f/i}\right]^{in}.$$

But the limiting value of the expression inside the square brackets can be proven to be the number e, when f becomes infinitely large. That is,

$$\lim_{f\to\infty} \left(1 + \frac{i}{f}\right)^{f/i} = e.$$

In view of this, it follows that

$$\left[\left(1 + \frac{i}{f}\right)^{f/i}\right]^{in} \to e^{in}$$

for increasingly large values of f. Accordingly, the compound amount

$$A = P\left(1 + \frac{i}{f}\right)^{nf} \qquad \text{[annual compounding]}$$

may be approximated by the amount

$$A = Pe^{in} \qquad \text{[continuous compounding].}$$

We illustrate how these ideas may be applied.

EXAMPLE 1 Given $P = \$100$, $i = 6\%$, $f = 12$, and $n = 8$. Find the compound amount (a) by annual compounding and (b) by continuous compounding.

Solution (a) By annual compounding we have

$$\begin{aligned} A &= 100\left(1 + \frac{0.06}{12}\right)^{8(12)} \\ &= 100(1 + 0.005)^{96} \\ &= 100(1.61414271) \qquad \text{[from compound interest tables]} \\ &\cong \$161.41. \end{aligned}$$

(b) By the continuous compounding method we get

$$\begin{aligned} A &= 100e^{0.06(8)} \\ &= 100e^{0.48} \\ &\cong 100(1.618) \qquad \text{[from the exponential table at the back of the text, using interpolation]} \\ &= \$161.80. \end{aligned}$$

The answers in (a) and (b) differ by $0.39 in favor of the continuous compounding of interest. The trend in today's financial circles is toward more frequent compounding of interest. It is no longer unusual to see advertisements soliciting the consumer's dollar by using as their main attraction the daily compounding of interest.

The formula $A = P(1 + i)^n$ enables us to find the compound amount A that results when a principal P is invested for n periods at $i\%$ per period. If we solve this formula for P we obtain the **present value** of the amount A due in n periods at $i\%$ per period. That is, if P is invested at the indicated rate, it would accumulate to the amount A after n periods. Thus,

$$P = \frac{A}{(1 + i)^n} = A(1 + i)^{-n}$$

is the formula for finding present value.

Let us now consider how the number e plays an important role in the calculation of present value. Solving the equation $A = Pe^{in}$ for P (the principal or present value) we obtain

$$P = Ae^{-in}. \qquad (1)$$

This formula for present value is applied to a practical situation in the next example.

EXAMPLE 2 Find the present value of \$5000 compounded continuously at 8% for 10 years.

Solution In Eq. (1) above, $A = 5000$, $i = 8\%$, and $n = 10$. Then

$$\begin{aligned} P &= 5000e^{-0.08(10)} \\ &= 5000e^{-0.8} \\ &\cong 5000(0.449) \qquad \text{[from the exponential table]} \\ &= \$2245. \end{aligned}$$

This means that you would have to invest about \$2245 *now* if you expect to have \$5000 in 10 years, money being worth 8% compounded continuously.

Of course, the concept of continuous compounding of interest is an idealization and is not practiced in the business world. Its use, however, should point out the advantages of frequent compounding of interest. Additionally, it drives home the fact that there is a limit to the amount of growth of assets by the compound-interest process.

Business problems may involve functions whose graphs rise steeply at the start then flatten out and eventually approach a horizontal line without ever reaching it. A case in point is the "modified" exponential function

$$y = a - ae^{-kx}$$

whose graph is commonly called a *learning curve*. This curve reflects a rapid increase in learning at first, as unforeseen weaknesses are uncovered and corrected. However, as efficiency increases with time, the learning level flattens out and tends to an upper limit. This upper bound represents the optimum performance that is possible under the conditions of the particular situation. In the next example we relate a business problem to the notions underlying the learning curve.

EXAMPLE 3 The number of units assembled per day, x days after the start of a production run, is given by the function $y = 50 - 50e^{-0.2x}$. What optimum performance can be expected and what percent of this optimum can be achieved after 15 days?

Solution We find the upper bound of performance by a limiting process:

$$\begin{aligned} \lim_{x \to \infty} y &= \lim_{x \to \infty} (50 - 50e^{-0.2x}) \\ &= 50 - 50 \lim_{x \to \infty} e^{-0.2x} \\ &= 50 - 50 \lim_{x \to \infty} \frac{1}{e^{0.2x}} \\ &= 50 - 50(0) = 50 \text{ units,} \end{aligned}$$

the optimum performance to be expected. In the limiting process, as x increases indefinitely the expression $e^{0.2x}$ also increases indefinitely, so that the value of $1/e^{0.2x}$ approaches zero.

To answer the second part of the example's question, we substitute $x = 15$ into the original function and obtain

$$\begin{aligned} y &= 50 - 50e^{-0.2(15)} \\ &= 50(1 - e^{-3}) \\ &= 50(1 - 0.05) \qquad \text{[from the exponential table]} \\ &= 50(0.95) = 47.5 \text{ units,} \end{aligned}$$

the daily output after 15 days. Then

$$\frac{47.5}{50} \quad \text{or} \quad 95\%$$

represents the maximum performance attainable after 15 days (see Figure 4.4).

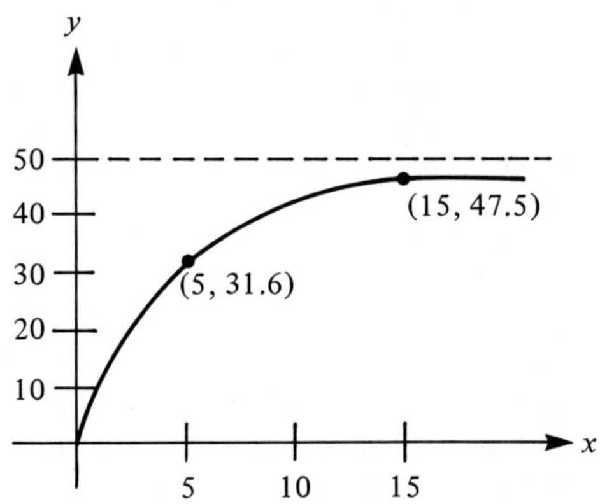

Figure 4.4

PROBLEM SET 4.1

1. Sketch the graphs of:
   (a) $y = 3^x$ (b) $y = (\frac{1}{2})^x$ (c) $y = 10^x$
   (d) $y = (0.9)^x$ (e) $y = e^{-2x}$ (f) $y = e^{-x/3}$
2. If \$500 is invested at 5% compounded continuously, to what amount will it grow after 16 years?
3. Find the compound amount of \$2000 after 10 years if money is worth $5\frac{1}{2}\%$ compounded continuously.
4. What amount should be invested now at 6% per year compounded continuously in order to have \$1000 in 15 years?
5. Determine the present value of \$8000 if money is invested at $6\frac{1}{4}\%$ compounded continuously for 8 years.
6. Given $D(t) = D_0 e^{\phi t}$, find t when $D(t) = 10{,}000$, $D_0 = 2466$, and $\phi = 0.5$ [D_0 is the initial value of D].
7. The number of items manufactured per day x days after the start of a production run is defined by $y = 200(1 - e^{-0.1x})$. What optimum performance

can be achieved and what percent of this optimum will be attained in 20 days? Sketch a graph.

8. The weekly output of a finished product x weeks after the start of a production run is given by $y = 400 - 400e^{-0.25x}$. What is the upper bound for units of the finished product? What percent of this upper limit can be achieved in 10 weeks?

9. The product sales function of the ATS Co. is given by $S(x) = 2000(1 - e^{-0.01x})$ where x denotes advertising promotional expenses in thousand of dollars. Find $S(10)$ and $S(20)$; sketch a graph.

10. The Best Belted Tire Co. picked a random sample of 120 tires (its own) which it subjected to a rigorous driving test. It was found that the number of tires still in usable condition after having been driven for m miles was given by the function $u(m) = 10^4e^{-0.00015m}$. How many tires were in usable condition after 50,000 miles of test driving? after 60,000 miles?

11. Assume that an initial investment of \$5000 in an established growth stock increases in value at a continuous rate of 15% per year because of stock splits and appreciation. How much will the investor have at the end of 10 years?

12. Express as a single fraction:

(a) $\frac{e^x}{2} - \frac{e^x}{3} + \frac{4}{e^x}$ **(b)** $1 + e^{-x} - 2e^{-x}$

(c) $e^{-t} + \frac{1}{e^{2t}} - \frac{1}{e^{3t}}$ **(d)** $\frac{1}{1 - e^u} - \frac{1}{1 + e^u}$

13. Factor:

(a) $2e^x - 4e^{2x}$ **(b)** $6x^2e^{-2x} + 9xe^{-3x}$

(c) $e^{2x} - e^x - 12$ **(d)** $e^{2t} + 2e^t - 15$

14. Let R denote sales revenue, d be the rate of decline (decay constant), t be the time in months, and R_0 be the initial sales revenue level. Then sales revenue at any time t is given by $R(t) = R_0e^{-dt}$.

(a) Find sales revenue when $t = 2$, $d = 0.10$, and $R_0 = \$1000$.

(b) How long will it take for sales revenue to amount to \$100,000 if $R_0 =$ \$250 when $d = 0.2$?

4.2 LOGARITHMIC FUNCTIONS

We return to the function $y = b^x$, which defines y as a function of x. This equation also determines x as a function of y. To this end we need to solve the equation for x and write the result as a function of y. The accepted form is written as

$$x = \log_b y$$

which is read "x equals the logarithm of y to the base b." Thus, the *exponent x is a logarithm.* Formally, we have the definition:

DEFINITION 1 *If $b^x = N$ and $b > 0$, $b \neq 1$, then the exponent x is called the* ***logarithm*** *of N to the base b.*

The exponential form $b^x = N$ may be written in the equivalent logarithmic form $x = \log_b N$. Some examples of exponential forms written in the equivalent logarithmic forms are:

$$\begin{aligned} 10^2 &= 100 \rightarrow \quad 2 = \log_{10} 100 \\ 8^{2/3} &= 4 \quad \rightarrow 2/3 = \log_8 4 \\ 2^{-3} &= 1/8 \rightarrow -3 = \log_2(1/8). \end{aligned}$$

(Recall that $\rightarrow$ means "implies.") Some examples of logarithmic equations expressed in equivalent exponential equations are:

$$\begin{aligned} \log_{10} 1000 &= 3 \quad \rightarrow 10^3 = 1000 \\ \log_2 32 &= 5 \quad \rightarrow \ 2^5 = 32 \\ \log_3 \frac{1}{9} &= -2 \rightarrow 3^{-2} = \frac{1}{9}. \end{aligned}$$

If we are confronted with the exponential form $x = e^y$, it is natural to want to solve the expression for y. (We have worked before with explicit forms exhibiting y as the dependent variable and x as the independent variable.) The solution for y is

$$\begin{aligned} y &= \log_e x \\ &= \ln x \end{aligned}$$

which is read "the **natural logarithm** of x." It is quite common to express

$$\log_e N \quad \text{as} \quad \ln N \tag{1}$$

and

$$\log_{10} N \quad \text{as} \quad \log N. \tag{2}$$

In Equations (1) and (2) the natural logarithm and the **common logarithm** are indicated, respectively. An understanding of this nomenclature will facilitate the discussion and applications of logarithms.

How are $x = e^y$ and $y = e^x$ related? Notice that a simple interchange of the variables has been effected. However, this interchange is most significant in that it introduces the concept of **inverse function**. In this case we say that the function $x = e^y$ is the inverse function of $y = e^x$ and vice versa. This leads to the following definition:

DEFINITION 2 *If the point (a, b) is on the graph of $f(x)$, then the point (b, a) is on the graph of $f^{-1}(x)$, the **inverse function** of $f(x)$.*

Caution: $f^{-1}(x)$ is not to be confused with $1/f(x)$.

Implementing this definition with respect to the two exponential functions, we have

$$\begin{aligned} y &= e^x = f(x) & & \\ x &= e^y & & \text{[interchanging variables]} \\ y &= \ln x = f^{-1}(x) & & \text{[solving for } y\text{].} \end{aligned}$$

Figure 4.5 shows how f and f^{-1} are related graphically. Observe that each graph is the reflection of the other across the line $y = x$ (dotted line). One curve is said to be the *mirror image* of the other when reflected in the line $y = x$. We see how Definition 2 comes into play: for $(0, 1) \in f$ there is $(1, 0) \in f^{-1}$; for $(1, e) \in f$ there is $(e, 1) \in f^{-1}$, and so on. Observe that the domain of f^{-1} is the range of f and the range of f^{-1} is the domain of f.

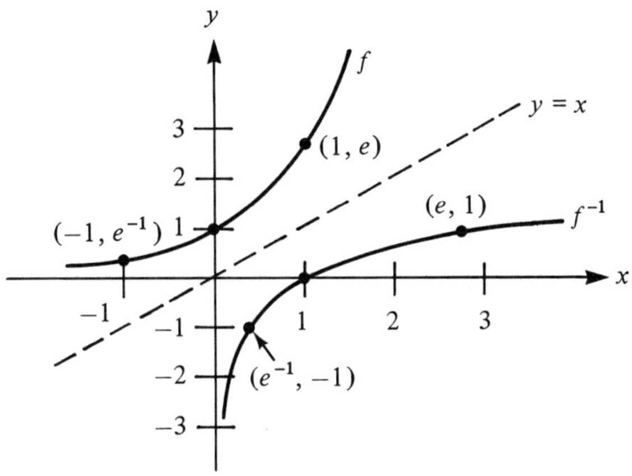

Figure 4.5

Certain basic properties of ln x are evident from Figure 4.5:

Property 1 $\ln x$ *is positive for* $x > 1$

Property 2 $\ln x$ *is zero for* $x = 1$

Property 3 $\ln x$ *is negative for* $0 < x < 1$

Property 4 $\ln x$ *approaches positive infinity (but slowly) as* x *approaches positive infinity, i.e.,* $\ln x \to +\infty$ *as* $x \to +\infty$

Property 5 $\ln x$ *approaches negative infinity as* $x \to 0^+$

Property 6 $\ln x$ *is undefined for* $x = 0$ *and for* $x < 0$.

There are three algebraic properties of logarithms whose use is advantageous for computational purposes. They are the following:

Property (a) $\log_b xy = \log_b x + \log_b y$

Property (b) $\log_b \frac{x}{y} = \log_b x - \log_b y$

Property (c) $\log_b x^a = a \log_b x$.

In words,

(a) The logarithm of a product of factors is equal to the sum of the logarithms of the factors.

(b) The logarithm of a quotient is equal to the logarithm of the numerator minus the logarithm of the denominator.

(c) The logarithm of a base to a power is equal to the power times the logarithm of the base.

Let us develop a proof for (a).

Proof Let $x = b^m$ and $y = b^n$. Then

$$\begin{aligned} m &= \log_b x \quad \text{and} \quad n = \log_b y && \text{[definition of logarithm]} \\ xy &= b^m \cdot b^n = b^{m+n} && \text{[law of exponents]} \\ \log_b xy &= m + n && \text{[definition of logarithm]} \\ &= \log_b x + \log_b y && \text{[substitution]}, \end{aligned}$$

which was to be shown.

The proofs of (b) and (c) can be developed similarly.
It can be shown that

$$\log_a N = \frac{\log_b N}{\log_b a}, \qquad b > 1. \tag{3}$$

That is, the logarithm to a base of a number is equal to the logarithm of the number divided by the logarithm of the base, all to a different base. For example,

$$\begin{aligned} \log 2 &= \frac{\ln 2}{\ln 10} && \text{[different base is } e\text{]} \\ \log_2 3 &= \frac{\log 3}{\log 2} && \text{[different base is 10]} \\ \ln 10 &= \frac{\log_3 10}{\log_3 e} && \text{[different base is 3].} \end{aligned}$$

In Eq. (3) above, if $N = b$ then

$$\log_a b = \frac{\log_b b}{\log_b a} = \frac{1}{\log_b a} \qquad [\text{since } \log_b b = 1]. \tag{4}$$

Equation (4) is useful in certain situations. For instance, if we wished to calculate $\log_2 10$ and tables for base 2 were not available, we could write, using Eq. (4),

$$\begin{aligned} \log_2 10 &= \frac{1}{\log 2} \\ &= \frac{1}{0.3010} \cong 3.32 && \text{[from base 10 logarithm tables].} \end{aligned}$$

Base 10 tables are referred to as common log tables. Base e tables are referred to as natural log tables.

Let us now compute the logarithms of numbers by applications of the properties of logarithms.

EXAMPLE 1 Compute log 72 without using tables if $\log 2 = 0.3010$ and $\log 3 = 0.4771$.

Solution

$$\begin{aligned} \log 72 &= \log(8)(9) \\ &= \log(2^3)(3^2) \\ &= \log 2^3 + \log 3^2 && \text{[by Property (a)]} \\ &= 3\log 2 + 2\log 3 && \text{[by Property (c)]} \\ &= 3(0.3010) + 2(0.4771) \\ &= 1.8572. \end{aligned}$$

EXAMPLE 2 Given $\log 2 = 0.3010$ and $\log 5 = 0.6990$, evaluate $\log(50/\sqrt{10})$.

Solution

$$\begin{aligned} \log \frac{50}{\sqrt{10}} &= \log 50 - \log \sqrt{10} && \text{[Property (b)]} \\ &= \log(2)(25) - \log 10^{1/2} \\ &= \log 2 + \log 5^2 - \log 10^{1/2} && \text{[Property (a)]} \\ &= \log 2 + 2\log 5 - \tfrac{1}{2}\log 10 && \text{[Property (c)]} \\ &= 0.3010 + 2(0.6990) - \tfrac{1}{2}(1) \\ &= 0.3010 + 1.3980 - 0.5000 \\ &= 1.1990 \end{aligned}$$

We continue the application of properties of logarithms, this time in the solution of logarithmic equations.

EXAMPLE 3 Solve the equation: $\ln x + \ln(x - 1) = \ln 6$.

Solution Given $\ln x + \ln(x - 1) = \ln 6$. Then

$$\ln[x(x - 1)] = \ln 6 \qquad \text{[since } \ln a + \ln b = \ln(ab) \text{ by definition]}$$

Therefore

$$\begin{aligned} x(x - 1) &= 6 && \text{[since } \ln a = \ln b \to a = b\text{]} \\ x^2 - x - 6 &= 0 \\ (x - 3)(x + 2) &= 0 \\ x - 3 &= 0, \quad x = 3 \\ x + 2 &= 0, \quad x = -2. \end{aligned}$$

Check $x = 3$:

$$\ln 3 + \ln(3 - 1) = \ln 3 + \ln 2 = \ln(3)(2) = \ln 6.$$

Thus, $x = 3$ checks out and therefore is a solution. Check $x = -2$:

$$\ln(-2) + \cdots.$$

We needn't continue the check since logarithms of negative numbers are undefined. So $x = -2$ is not a solution.

EXAMPLE 4 Solve:

(a) $5^x = 1$
(b) $3^x = 81$
(c) $e^{1/x} = 4^3$
(d) $3^x = 9^{1-x}$.
(e) How long will it take a sum of money to double itself if invested at 6%, compounded annually?

Solution **(a)** Taking the common log of both sides we get

$$\log 5^x = \log 1.$$

Then

$$x \log 5 = 0$$

$$x = \frac{0}{\log 5} = 0.$$

(b) Taking the common log of both sides we have

$$\log 3^x = \log 81 = \log 3^4.$$

$$x \log 3 = 4 \log 3$$

$$x = \frac{4 \log 3}{\log 3} = 4.$$

(c) Taking the natural log of both sides we obtain

$$\frac{1}{x} \ln e = 3 \ln 4$$

or

$$\frac{1}{x}(1) = 3 \ln 2^2.$$

Therefore,

$$\frac{1}{x} = 6 \ln 2$$

and

$$x = \frac{1}{6 \ln 2}.$$

(d) Taking the common log of both sides gives us

$$\begin{aligned} x \log 3 &= (1 - x) \log 9 \\ &= \log 3^2 - x \log 3^2 \\ &= 2 \log 3 - 2x \log 3 \end{aligned}$$

Adding $2x \log 3$ to both sides yields

$$3x \log 3 = 2 \log 3$$

Solving for x:

$$x = \frac{2 \log 3}{3 \log 3} = \frac{2}{3}$$

An alternate way is to proceed thus:

$$3^x = (3^2)^{1-x} = 3^{2-2x}$$

Therefore,

$$x = 2 - 2x$$
$$3x = 2$$
$$x = \frac{2}{3}.$$

(e) Applying the compound-interest formula, we have

$$A = P(1 + i)^n,$$
$$2 = 1 \cdot (1 + 0.06)^n$$

or

$$2 = (1.06)^n.$$

We take the common log of both sides and get

$$\log 2 = n \log 1.06$$

or

$$n = \frac{\log 2}{\log 1.06}$$
$$= \frac{0.3010}{0.0253} \quad \text{[from common log tables]}$$
$$\cong 12 \text{ years.}$$

We now introduce a short table of logarithms as an aid in computing common logarithms. The use of natural log tables (in back of this text) is similar.

N		0	1	2	3	4	5	6	7	8	9
100	00	000	043	087	130	173	217	260	303	346	389
101		432	475	518	561	604	647	689	732	775	817
		---	---	---	---	---	---	---	---		
		---	---	---	---	---	---	---	---		
		---	---	---	---	---	---	---	---		
159	20	140	167	194	222	249	276	303	330	358	385
160		412	439	466	493	520	548	575	602	629	656
161		683	710	737	763	790	817	844	871	898	925

EXAMPLE 5 Use the short table above to evaluate $(1.01)^3(15.9)^2/161$.

Solution Let $N = (1.01)^3(15.9)^2/161$. Then

$$\begin{aligned}\log N &= \log(1.01)^3 + \log(15.9)^2 - \log 161\\ &= 3 \log 1.01 + 2 \log 1.59(10) - \log 1.61(10)^2\\ &= 3 \log 1.01 + 2(\log 1.59 + 1) - (\log 1.61 + 2)\\ &= 3(0.00432) + 2(1.20140) - 2.20683\\ &= 0.20893.\end{aligned}$$

Therefore,

$$N = \text{antilog } 0.20893 \cong 1.617815.$$

Thus, the value of the given fraction is found to be 1.617815 by use of logs. The following operations were involved in finding the antilog.

$$\left\{\begin{array}{l} \left.\begin{array}{l} 0.20871 - 1.617 \\ 0.20893 - ? \end{array}\right\} x \\ 0.20898 - 1.618 \end{array}\right\}$$

$$\frac{x}{0.001} = \frac{0.00022}{0.00027} = \frac{22}{27}$$

$$x = 0.001\left(\frac{22}{27}\right)$$

$$= 0.001(0.815) = 0.000815$$

Adding this to 1.617 we obtain

$$1.617 + 0.000815 = 1.617815, \quad \text{the antilog of } 0.20893.$$

For additional information on the above procedures, consult the explanations in the Appendix titled "Some Examples of the Use of Logarithms." Tables for common logs and natural logs are given in back of this text.

We have seen how the use of logarithms acts as a labor-saving device in many numerical computations. However, it has not been our main purpose to introduce logarithms for their computational value, but rather to acquaint ourselves with them and to extend their use in differential and integral calculus.

PROBLEM SET 4.2

1. Write the following exponential equations in the equivalent logarithmic form:

   (a) $y = 10^x$ **(b)** $16^{1/2} = 4$ **(c)** $2^{-2} = 1/4$
   (d) $64^{2/3} = 16$ **(e)** $\sqrt{81} = 9$ **(f)** $10^{-3} = 0.001$.

2. Express each of the given logarithmic equations in the equivalent exponential form:

   (a) $\log_2 64 = 6$ **(b)** $\log_5 1 = 0$ **(c)** $\ln a = b$
   (d) $\log_b d = e$ **(e)** $\log 0.1 = -1$ **(f)** $\log_9 \frac{1}{3} = -\frac{1}{2}$.

3. Find the inverse functions of each of the following:

   (a) $y = e^{2x}$ **(b)** $x = \ln 2y$ **(c)** $x = 2^y$
   (d) $2y = e^{-x}$ **(e)** $x - 1 = \ln y$.

4. Sketch the graphs of the functions in problem 3 together with their respective inverses.

5. If common log tables were available, how would you compute:

   (a) $\log_3 2$ **(b)** $\log_4 5$ **(c)** $\log_2 27$ **(d)** $\log_5 57$

6. Given $\log 2 = 0.3010$ and $\log 3 = 0.4771$, compute without using tables:

   (a) $\log 36$ **(b)** $\log \frac{27}{8}$ **(c)** $\log \frac{3\sqrt{6}}{32}$ **(d)** $\log \sqrt[3]{\frac{2}{81}}$

7. Given $\log 3 = 0.4771$ and $\log 5 = 0.6990$ compute without using tables:

   (a) $\log 135$ (b) $\log \frac{25}{27}$ (c) $\log \sqrt{\frac{3}{125}}$ (d) $\log \frac{5(3)^{2/3}}{9}$

8. Solve the equation: $\log 2 + \log x = \log(x + 1)$.

9. Solve: $\ln(x + 2) - \ln 3 = \ln 2x$.

10. Solve: $\log t + \log(t + 1) = \log 2$.

11. Solve: $\ln(x + 1) + \ln(x + 2) = \ln(x + 10)$.

12. Use common logs (and tables) to evaluate:

    (a) $\sqrt{898}$ (b) $(2.02)^4(3.03)^3$ (c) $\frac{(1.98)^4(9.95)^3}{795}$

13. Evaluate by use of natural logs (and tables):

    (a) $(6.3)^{1/3}$ (b) $(e^5)^{1/2}$ (c) $\frac{1.984 - e^{1/4}}{e}$

14. How long will it take a sum of money to double itself if invested at 5%, compounded annually? Use $\log 2 = 0.3010$ and $\log 1.05 = 0.0212$.

15. Work problem 14 when the sum of money is \$1000 and the rate of interest is 10%. Use $\log 2 = 0.3010$ and $\log 1.1 = 0.0414$.

16. Work problem 14 when the amount invested is \$5000 and the rate of return is 15%, due to appreciation, stock splits, etc.

17. Work problem 14 when the amount invested is \$10,000 and the rate of return is 20% (assume this is an established growth stock enjoying great public and institutional favor).

18. Solve:

    (a) $7^x = 1$ (b) $2^x = 128$ (c) $e^{-x} = \frac{1}{e^2}$ (d) $2^x = 16^{1-x}$.

19. Given the cost function $C(x) = e^x + e^{-x}$, solve for x when $C = 2$.

20. If $\ln(2x + 1) = 2$, determine the value of x.

21. Find the value of x such that:

    (a) $\log_4 x = 3$
    (b) $\log_x 1/4 = 2$
    (c) $\log_5(1/125) = x$
    (d) $\log_2 \frac{1}{\sqrt[3]{16}} = x$.

22. How long will it take for a house valued at \$20,000 to depreciate to a value of \$10,000 if the depreciation rate per year is 5%? [*Hint*: $V_t = V_0(1 - r)^t \rightarrow 10{,}000 = 20{,}000(1 - 0.05)^t$.]

23. A small computer costs \$50,000 and depreciates at the rate of 10% annually. How long will it take to depreciate to a "scrap" value of \$10,000? See problem 22 for the depreciation function to use.

4.3 THE DIFFERENTIATION OF ln $|u(x)|$

We first develop the proof for the derivative of $\ln u(x)$ where we use $u(x) = x$ for convenience.

THEOREM 1 *If* $f(x) = \ln x$, $x > 0$, *then* $f'(x) = \dfrac{1}{x}$.

Proof

$$
\begin{aligned}
f'(x) &= \lim_{\Delta x \to 0} \frac{f(x+\Delta x) - f(x)}{\Delta x} && \text{[by definition]} \\
&= \lim_{\Delta x \to 0} \frac{\ln(x+\Delta x) - \ln x}{\Delta x} && \text{[since } f(x) = \ln x\text{]} \\
&= \lim_{\Delta x \to 0} \frac{1}{\Delta x} \ln\left(\frac{x+\Delta x}{x}\right) && \text{[logarithm property (b)]} \\
&= \lim_{\Delta x \to 0} \frac{1}{x} \cdot \frac{x}{\Delta x} \ln\left(1 + \frac{\Delta x}{x}\right) && \left[\text{multiplying by } \frac{x}{x},\ x \neq 0\right] \\
&= \lim_{\Delta x \to 0} \frac{1}{x} \ln\left(1 + \frac{\Delta x}{x}\right)^{x/\Delta x} && \text{[logarithm property (c)]} \\
&= \frac{1}{x} \lim_{\Delta x \to 0} \ln\left(1 + \frac{\Delta x}{x}\right)^{x/\Delta x} && \text{[property of limits]} \\
&= \frac{1}{x} \lim_{n \to \infty} \ln\left(1 + \frac{1}{n}\right)^{n} && \left[\text{letting } n = \frac{x}{\Delta x}\right] \\
&= \frac{1}{x} \ln \lim_{n \to \infty} \left(1 + \frac{1}{n}\right)^{n} && \text{[property of limits]} \\
&= \frac{1}{x} \ln e && \text{[definition of } e\text{]} \\
&= \frac{1}{x} && [\ln e = 1].
\end{aligned}
$$

Thus, if $f(x) = \ln x$, then $f'(x) = 1/x$, $x \neq 0$.

In General:

If $y = f(u(x)) = \ln|u(x)|$, then by the chain rule,

$$D_x \ln|u(x)| = \frac{1}{u(x)} \cdot D_x u(x)$$

provided $u(x) \neq 0$ and $D_x u(x)$ exists. More compactly we write

$$\boxed{D_x \ln|u| = \frac{1}{u} \cdot D_x u}$$

or

$$\boxed{\frac{d}{dx}(y) = \frac{1}{u} \cdot \frac{du}{dx}}.$$

EXAMPLE 1 If $y = \ln|x^2 - 3|$, then $\dfrac{dy}{dx} = \dfrac{1}{x^2 - 3} \cdot 2x$.

EXAMPLE 2 If $f(t) = \ln(5/\sqrt{t})$ then

$$\begin{aligned} f'(t) &= \frac{\sqrt{t}}{5} \cdot D_t 5t^{-1/2} \\ &= \frac{\sqrt{t}}{5} \cdot 5\left(-\frac{1}{2}t^{-3/2}\right) \\ &= \frac{-t^{1/2}}{2t^{3/2}} \\ &= \frac{-1}{2t}. \end{aligned}$$

An alternative method is to first apply the properties of logs, which state that

$$\begin{aligned} f(t) = \ln \frac{5}{t^{1/2}} &= \ln 5 - \ln t^{1/2} \\ &= \ln 5 - \frac{1}{2}\ln t. \end{aligned}$$

Then $f'(t) = 0 - \frac{1}{2} \cdot \frac{1}{t} \cdot 1 = -\frac{1}{2t}$, as before.

In certain situations where the derivative of a rather complicated log function is desired, the initial application of properties of logs lends convenience to the procedure. The following examples illustrate.

EXAMPLE 3 Find $f'(x)$ if $f(x) = \ln\left[\frac{x(1-3x)^{1/3}}{(2x+3)^2}\right]$.

Solution First we write

$$\begin{aligned} f(x) &= \ln x(1-3x)^{1/3} - \ln(2x+3)^2 \\ &= \ln x + \frac{1}{3}\ln(1-3x) - 2\ln(2x+3). \end{aligned}$$

Then we differentiate and simplify:

$$\begin{aligned} f'(x) &= \frac{1}{x} + \frac{1}{3} \cdot \frac{1}{1-3x}(-3) - 2 \cdot \frac{1}{2x+3} \cdot 2 \\ &= \frac{1}{x} - \frac{1}{1-3x} - \frac{4}{2x+3}. \end{aligned}$$

Observe the convenience of this method over that of applying $f'(u) = (1/u) \cdot D_x u$ directly.

The use of logarithms also simplifies manipulations in another type of problem, as follows.

EXAMPLE 4 Find $\frac{dy}{dx}$ if $y = \frac{(2x - 1)^3(2 - x)^4}{x^3}$.

Solution Taking the natural log of both sides we have

$$\ln y = \ln(2x - 1)^3(2 - x)^4 - \ln x^3$$

or

$$\ln y = 3 \ln(2x - 1) + 4 \ln(2 - x) - 3 \ln x$$

Differentiating, we get

$$\frac{1}{y}\frac{dy}{dx} = 3 \cdot \frac{1 \cdot 2}{2x - 1} + 4 \cdot \frac{1(-1)}{2 - x} - 3 \cdot \frac{1}{x}$$

$$= \frac{6}{2x - 1} - \frac{4}{2 - x} - \frac{3}{x}$$

Solving for dy/dx, we obtain

$$\frac{dy}{dx} = y\left[\frac{6}{2x - 1} - \frac{4}{2 - x} - \frac{3}{x}\right].$$

We may substitute for y its original given value. In this example we have found dy/dx by a method called **logarithmic differentiation**.

PROBLEM SET 4.3

Differentiate the following functions:

1. $y = \ln(2x^2)$
2. $f(x) = \ln|x^3|$
3. $g(u) = \ln(u^4 + u^2)$
4. $h(w) = w \ln|4w|$
5. $F(t) = \ln\sqrt{1 + t^2}$
6. $G(u) = \frac{\ln u}{u}$
7. $H(x) = \frac{x}{3} \ln \frac{3}{x}$
8. $y = (2 - \ln x)^2$
9. $f(t) = \frac{t^2\sqrt{1 - 2t}}{(3t - 1)^3}$ [use logarithmic differentiation]
10. $g(u) = \ln\left[\frac{x^2}{1 + x^2}\right]$
11. $h(x) = \frac{e^{2x}(1 - e^x)^2}{(1 + e^x)^2}$ [use logarithmic differentiation]
12. $f(x) = \frac{x^2(2x - 1)^3}{\sqrt{2x - 1}}$ [use logarithmic differentiation]
13. If a certain cost function is given by $C(x) = \ln(x^2 + 1)$, $x \geq 0$, find minimum cost and sketch the graph.
14. An index of inflation behaves according to the rule $I(c) = c - \ln c$, where c defines an index of consumer prices. For what value of c is the inflation function a minimum? Sketch the graph.
15. Let a profit function be defined by $P(x) = (\ln x)/x$ where x is the quantity of items produced (in thousands) and profit is in millions of dollars. For what value of x is profit a maximum? Sketch the graph.

16. Find an equation of the line tangent to each given curve at the indicated point:

(a) $f(x) = x \ln x$; $(1, 0)$

(b) $g(x) = \ln \frac{1}{x}$; $(e^{-1}, 1)$

(c) $h(t) = \frac{\ln t}{t}$; $\left(2, \frac{\ln 2}{2}\right)$

(d) $F(t) = \ln(t^2 + 1)$; $(1, \ln 2)$.

4.4 THE DIFFERENTIATION OF $e^{u(x)}$

In the last section we learned that the chain rule plays a role in the operation

$$D_x \ln y = \frac{1}{y}\frac{dy}{dx}.$$

Similarly, if

$$z = \ln y \quad \text{and} \quad y = f(x)$$

then

$$\frac{dz}{dx} = \frac{dz}{dy} \cdot \frac{dy}{dx} \qquad \text{[by the chain rule]}$$

$$= \frac{1}{y} \cdot \frac{dy}{dx}$$

This result is used in the following development of the derivative of e^x.

Let $y = a^x$. Taking the natural log of both sides we obtain

$$\ln y = x \ln a = (\ln a) \cdot x.$$

Differentiating, we get

$$\frac{1}{y} \cdot \frac{dy}{dx} = \ln a$$

or

$$\frac{dy}{dx} = y \cdot \ln a$$

$$= a^x \cdot \ln a \qquad [\text{since } y = a^x].$$

If we let $a = e$, then

$$y = e^x$$

and

$$\frac{dy}{dx} = e^x \cdot \ln e$$

$$= e^x \cdot 1 \qquad [\text{since } \ln e = 1]$$

$$= e^x.$$

It is interesting that the derivative of e^x is e^x itself!

In General:

If u is a function of x and $D_x u(x)$ exists, then

$$\boxed{D_x e^{u(x)} = e^{u(x)} \cdot D_x u(x)}$$

or briefly,

$$\boxed{\frac{d}{dx} e^u = e^u \cdot \frac{du}{dx}}.$$

EXAMPLE 1 If $y = 2e^{2x}$, find dy/dx.

Solution

$$\begin{aligned}\frac{dy}{dx} &= \frac{d}{dx}(2e^{2x}) \\ &= 2\frac{d}{dx}e^{2x} \\ &= 2e^{2x} \cdot \frac{d}{dx}(2x) \\ &= 2e^{2x} \cdot 2 = 4e^{2x}.\end{aligned}$$

EXAMPLE 2 Find $f'(t)$ if $f(t) = t^3e^{-t^2}$.

Solution

$$\begin{aligned}f'(t) &= t^3 \cdot D_te^{-t^2} + e^{-t^2} \cdot D_tt^3 \\ &= t^3 \cdot e^{-t^2} \cdot D_t(-t^2) + e^{-t^2} \cdot 3t^2 \\ &= t^3e^{-t^2}(-2t) + 3t^2e^{-t^2} \\ &= -2t^4e^{-t^2} + 3t^2e^{-t^2} \\ &\text{or} \quad t^2e^{-t^2}(3 - 2t^2).\end{aligned}$$

EXAMPLE 3 Given $y = (e^x + e^{-x})/2$, show that $y'' = y$.

Solution

$$\begin{aligned}y' &= \frac{1}{2} \cdot D_x(e^x + e^{-x}) \\ &= \frac{1}{2}[e^x + e^{-x}(-1)] \\ &= \frac{1}{2}(e^x - e^{-x}).\end{aligned}$$

Continuing, we have

$$\begin{aligned}y'' &= \frac{1}{2} \cdot D_x(e^x - e^{-x}) \\ &= \frac{1}{2}[e^x - e^{-x}(-1)] \\ &= \frac{1}{2}(e^x + e^{-x}) = y.\end{aligned}$$

EXAMPLE 4 A certain cost function is defined by $C(x) = e^{x/2} - x$, $x \geq 2 \ln 2$, where x (in thousands) denotes quantity of production and cost is in tens of thousands of dollars. Find $C_{\min}$.

Solution

$$\begin{aligned}C'(x) &= e^{x/2} \cdot \frac{1}{2} - 1 \\ &= \frac{e^{x/2}}{2} - 1 = 0\end{aligned}$$

when $$\frac{e^{x/2}}{2} = 1$$

or $\qquad e^{x/2} = 2$

or $\qquad x/2 = \ln 2 \qquad$ [by definition of log].

Then $\qquad x = 2 \ln 2$: cv.

We apply the second derivative test and obtain

$$C''(x) = \frac{1}{2} e^{x/2} \cdot \frac{1}{2}$$

$$= \frac{1}{4} e^{x/2}.$$

Then $$C''(2 \ln 2) = \frac{1}{4} e^{(2 \ln 2)/2}$$

$$= \frac{1}{4} e^{\ln 2} > 0 \rightarrow C_{\min}$$

at $x = 2 \ln 2 = \ln 2^2 = \ln 4$. The log form $x = \ln 4$ has as its equivalent exponential form $e^x = 4$. Consulting the exponential tables we find that $x \cong 1.39$. To determine $C_{\min}$ we evaluate

$$\begin{aligned} C(2 \ln 2) &\cong C(1.39) \\ &= e^{1.39/2} - 1.39 \\ &= e^{0.695} - 1.39 \\ &\cong 2 - 1.39 \qquad \text{[from exponential tables]} \\ &= 0.61 \text{ tens of thousands of dollars} \\ &= \$6100. \end{aligned}$$

Thus, for $x \cong 1.39$ or 1390 items of production, the minimum cost is \$6100. Figure 4.6 illustrates this equation, where cost C is defined by the solid curve. To facilitate the graphing, several points were determined:

$C(0) = e^0 - 0 = 1$; $\qquad C(1) = e^{1/2} - 1 \cong 0.65$;

$C(2) = e^{2/2} - 2 = e - 2 \cong 0.7$; $\qquad C(3) = e^{3/2} - 3 \cong 1.5$.

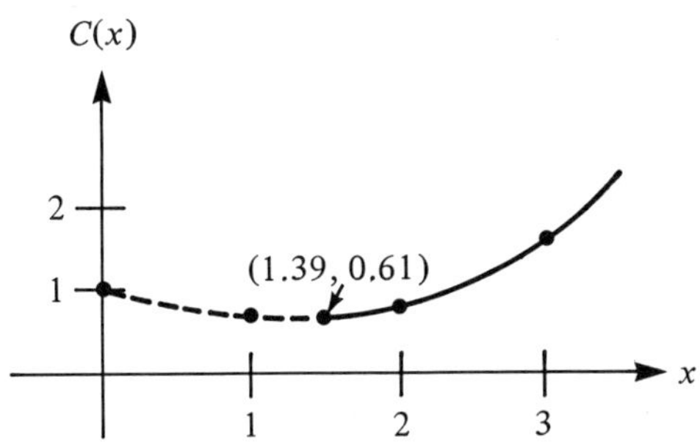

Figure 4.6

PROBLEM SET 4.4

Find the first derivative of the following exponential functions:

1. $y = 3e^{3x}$ $\qquad$ **2.** $f(x) = \dfrac{4}{e^{2x}}$ $\qquad$ **3.** $g(x) = e^{1-2x}$

4. $h(x) = e^{x^2}$
5. $F(t) = 2e^{2-t}$
6. $G(u) = ue^{u-1}$
7. $H(w) = \dfrac{e^w}{1 - e^w}$
8. $y = e^{2 \ln x}$
9. $f(t) = \ln e^{3t}$
10. $g(x) = e^x \ln x$
11. $h(u) = (e^u - 1)^3$
12. $F(x) = \dfrac{2}{(1 - 2e^x)^2}$
13. $G(t) = \dfrac{e^{4t} - e^t + e}{e^t}$
14. $H(x) = e^{-x}(1 - e^x - e^{2x})$

Find the second derivative of:

15. $y = xe^x$
16. $y = x \ln x$
17. $f(x) = \dfrac{\ln x}{e^x}$
18. $g(u) = \dfrac{e^{u^2}}{4}$
19. $h(v) = \dfrac{e^{2v} - 1}{e^{-v}}$
20. $F(z) = \sqrt{2 - e^{2z}}$.
21. Let p denote price per unit, and sales S be $S(p) = 1000e^{-0.2p}$. Find:
    (a) The total revenue $R = pS$
    (b) The marginal revenue $R'(p)$
    (c) The rate of change of marginal revenue at $p = 10$.
22. Let total costs $C = 0.3S$. With R and S as defined in problem 21 find:
    (a) Net profits $P = R - C$ at $p = \$2$
    (b) Marginal profits $P'(p)$ at $p = \$5$
    (c) $P''(10)$.
23. Find points of extrema for each of the following:
    (a) $f(x) = xe^x$ **(b)** $g(x) = x - \ln x$ **(c)** $h(x) = e^{-x^2}$
24. The manager of a certain firm ascertains that the function $P(x) = 4xe^{-x}$ defines closely his firm's profits in millions of dollars. The x (>0) denotes an index that measures the inflationary trend of the nation's economy.
    (a) For what index value is profit a maximum?
    (b) Find the maximum profit.
    (c) Determine the point of inflection and discuss its significance.
    (d) Sketch a graph of P.
25. Given the cost function $C(x) = xe^{\ln x} + 2$ where x ($\geq e^{-1}$) denotes quantity produced in millions and C is in thousands of dollars.
    (a) Sketch a graph of C.
    (b) How many units must be produced for cost to be a minimum?
26. A total revenue function is given by $R(q) = 10qe^{-0.002q}$ where q is the quantity of a commodity. How many items of q will produce maximum revenue?
27. Work problem 26 when $R(q) = 50qe^{-0.005q}$.

4.5 REVIEW PROBLEMS

1. Sketch the graphs of:
    (a) $y = 4^x$ **(b)** $y = \dfrac{e^{x/3}}{2}$ **(c)** $y = 2e^{-x/2}$
    (d) $y = 1 + e^x$ **(e)** $y = \ln 3x$ **(f)** $x = 1 - \ln(y/2)$.

2. How long will it take \$3000 to double if interest is at 6% compounded continuously?

3. Work problem 2 if the interest rate is $5\frac{1}{2}\%$.

4. Evaluate:

(a) $\lim_{x \to 0} \left(1 + \frac{x}{y}\right)^{y/x}$ (b) $\lim_{h \to \infty} \left(1 + \frac{1}{h}\right)^{h}$.

5. Find the present value of \$4000 invested at 7% compounded continuously for 10 years.

6. How much should be invested now if one wishes to have \$5000 in 5 years and money is worth $6\frac{1}{2}\%$ compounded continuously?

7. The number of items manufactured per day, x days after the start of a production run, is given by $y = 300(1 - e^{-0.2x})$. What is the optimum performance possible and what percent of the optimum can be attained in 15 days?

8. Because of stock splits, price appreciation, and investors' favor, a growth stock enjoys appreciation of value at a continuous rate of 20% per year. To what amount will an investment of \$10,000 grow in 5 years?

9. Express as a single fraction:

(a) $\frac{2e^{-x}}{3} + \frac{3e^{2x}}{4} - \frac{e^{-2x}}{5}$ (b) $\frac{1 + x}{e^x} + \frac{2 - x}{e^{2x}} - \frac{x^2}{e^{2x}}$.

10. Factor:

(a) $16e^{2t} - 81e^{4t}$ (b) $e^{2x} + 4x - 32$.

11. Solve for x:

(a) $y = e^{x+1}$ (b) $2 = e^x - e^{-x}$ (c) $e^{2x} - e^2 = 0$

(d) $\ln \frac{x}{2} = 2$ (e) $\ln(x + 1) = 3$ (f) $e^{x^2 - x} = 1$.

12. Write in logarithmic form:

(a) $2^6 = 64$ (b) $3^4 = 81$ (c) $2^{-1} = 0.5$

(d) $e^{\ln x} = x$. (e) $y = 10^{2x}$.

13. Change to exponential form:

(a) $\log 0.01 = x$ (b) $\log 2 = 0.3010$ (c) $\ln 1.03 = 0.03$

(d) $\log_5 \sqrt[3]{25} = \frac{2}{3}$ (e) $\log_2 x = -4$.

14. Given $\log 3 = 0.4771$ and $\log 5 = 0.6990$ find:

(a) $\log 225$ (b) $\log \frac{125}{81}$ (c) $\log 25\sqrt{27}$

(d) $\log \left(\frac{625}{243}\right)^{1/3}$ (e) $\log \left(\frac{5}{2} + \frac{7}{3} - \frac{19}{6}\right)$.

15. Use logs to evaluate:

(a) $(9.99)\sqrt{99.5}$ (b) $(2.015)^3(1.989)^4$ (c) $\frac{(10.05)^2(9.955)^3}{(99.89)^2}$

16. Find the inverse function of:

(a) $y = e^{-2x}$ (b) $x = \ln(y - 2)$ (c) $2x = e^{3y}$

(d) $2y = \ln x^4$ (e) $\frac{y}{2} = e^{x-1}$.

17. Solve the following equations:

(a) $\ln x + \ln(2x + 1) = \ln 21$ (b) $\log(x - 1) - \log(x + 3) = \log \frac{x}{10}$

(c) $\ln(x - 2) + \ln x = 1$ (d) $\log(x^2 + 2x + 2) = 1$

18. How long will it take for a sum of money to double itself if invested at 6% compounded annually?

19. Work problem 18 when the rate is 8%.

20. How long will it take for an investment to triple itself if money is worth $5\frac{1}{2}\%$ compounded annually?

21. A building is bought for \$80,000. How many years will it take to depreciate to a value of \$20,000 if the depreciation rate is 6% per year? Use $V_t = V_0(1 - r)^t \rightarrow 20{,}000 = 80{,}000(1 - 0.06)^t$.

22. Solve:

(a) $8^{2x} = 16^{x+1}$ (b) $e^{2x} = 32e^{-2x}$.

Differentiate the following functions:

23. $f(x) = \ln(x^2 + x)$

24. $f(x) = \ln \sqrt{x - 1}$

25. $g(t) = \ln\left(\frac{1}{1 - t^2}\right)$

26. $h(u) = \frac{1}{2} \ln\left(\frac{1 + u}{1 - u}\right)$

27. $F(x) = e^{3x} \cdot \ln 4x$

28. $G(t) = \frac{e^{2t} - e^{-2t}}{4}$

29. $H(u) = \frac{e^u - e^{-u}}{e^u + e^{-u}}$

30. $y = (e^x + e^{-x})^3$

31. $f(x) = e^{1/x} \cdot \ln \frac{1}{x}$

32. $g(x) = \frac{e^{\ln x}}{x^2}$

33. $h(u) = \frac{1 + e^u}{1 - e^u}$

34. $F(x) = x^2 \ln(1 - x^2)$

35. $G(v) = 2v \ln \sqrt{v}$

36. $H(w) = \left(1 + \frac{1}{w}\right) \ln \frac{1}{w}$

37. $y = \ln\left[\frac{e^x\sqrt{x^2 - 1}}{x^2 + 1}\right]$

38. $y = \frac{x^3(2x + 3)^4}{\sqrt{3x - 4}}$

39. $y = \frac{(2x - 1)^3(1 - 3x)^4}{\sqrt{x}}$

40. $y = \ln\left[\frac{2z + 1}{2z - 1}\right]^{3/2}$

41. Find $f''(x)$ if $f(x)$ is:

(a) $e^{2x-1} \cdot \ln(x^{-2})$ (b) $\frac{2}{\ln e^{2x}}$

(c) $\ln e^{\sqrt{x}}$ (d) $e^{\ln \sqrt{x-1}}$

42. A supply function is defined by $S(x) = e^{-x} + x$ where x denotes quantity of goods in thousands.

(a) For what value of x does $S_{\min}$ exist?
(b) Why is there no inflection point?
(c) Find the rate of change of marginal supply at $x = 1$.
(d) Sketch a graph of S.

4.6 ENRICHMENT TOPICS AND PROBLEMS

1. A profit function is given by $P(x) = (\ln x)/2x$ (in millions of dollars) where x is units of produce in thousands.
   (a) Find x such that profit is maximized.
   (b) Sketch a graph of P for $x \geq 1$.
2. Find an equation of the line tangent to
   (a) $f(x) = \dfrac{\ln x}{e^x}$ at the point where $x = 1$.
   (b) $g(x) = e^{\ln x} + \ln x$ at the point where $x = 2$.
   (c) $h(x) = \sqrt{e^x} \cdot \ln x$ at the point where $x = 4$.
3. If inflation reduces the value of a dollar at an average annual rate of 4% compounded continuously, after how many years will the dollar be worth only 50¢?
4. Work problem 3 if the rate is 5%.
5. For what rate of inflation, compounded continuously, will the purchasing power of the dollar be only 67¢ after 10 years?
6. A sum of $1000 is borrowed at the interest rate of 15% compounded continuously. How much will the borrower repay if he makes two annual payments, i.e., one payment after each of 2 years?
7. Sketch graphs of:
   (a) $y = 3^{|x|}$ (b) $y = \left(\dfrac{1}{2}\right)^{|x|}$ (c) $y = e^{-x^2}$
   (d) $y = \ln \dfrac{1}{x}$ (e) $y = 10 - 2e^{-x}$.
8. If $2000 is deposited in an account which earns 6% compounded continuously, compute the balance after 5 years.
9. Work problem 8 if the annual rate of interest is 8%.
10. Mr. Jay wishes to have $10,000 on hand at the end of 15 years for his son's college education. How much should he deposit now in his account earning 6% compounded continuously if he is to have the amount desired?
11. Work problem 10 if Mr. Jay is to attain his goal in 16 years with money earning $6\frac{1}{4}$% compounded continuously.
12. How long will it take an investment to double in value if it earns 8% interest compounded continuously?
13. An investment earns 10% interest compounded continuously. How long will it take to triple in value?
14. Sketch graphs after finding extrema and inflection points for:
   (a) $f(x) = e^{-x} - e^x$ (b) $f(x) = 2x - 2 \ln x$.
15. Let consumers' total weekly outlay for a product be given by $C(p) = 10^6 pe^{-0.005p}$, where p denotes the market price in dollars per unit. Find the value of p for which C is a maximum.
16. Differentiate:
   (a) $f(x) = xe^{-2x^2}$ (b) $g(x) = x \ln \sqrt{x}$
   (c) $h(t) = \ln(t + \sqrt{1 + t^2})$ (d) $F(u) = e^{u + 2 \ln u}$

5 MULTIVARIABLE DIFFERENTIATION

Applications of Multivariable Functions

IMPORTANT EQUATIONS

$$\frac{\partial P}{\partial x} = P_x = P_1(x, y)$$

$$\frac{\partial P}{\partial y} = P_y = P_2(x, y)$$

$$\frac{\partial}{\partial x}\left(\frac{\partial f}{\partial x}\right) = \frac{\partial^2 f}{\partial x^2} = f_{xx}$$

$$\frac{\partial}{\partial y}\left(\frac{\partial f}{\partial y}\right) = \frac{\partial^2 f}{\partial y^2} = f_{yy}$$

$$\frac{\partial}{\partial x}\left(\frac{\partial f}{\partial y}\right) = \frac{\partial^2 f}{\partial x\,\partial y} = f_{yx}$$

$$\frac{\partial}{\partial y}\left(\frac{\partial f}{\partial x}\right) = \frac{\partial^2 f}{\partial y\,\partial x} = f_{xy}$$

$$\frac{dy}{dx} = \frac{\partial y}{\partial x}\frac{dx}{dx} + \frac{\partial y}{\partial u}\cdot\frac{du}{dx}$$

$$\frac{dy}{dx} = -\frac{H_x}{H_y}$$

In this chapter we extend our study of the differential calculus to consider functions of several variables. We will discuss rates of change, extrema, Lagrange multipliers, least squares, and other pertinent applications, mainly in connection with functions of two and three variables. Such functions occur frequently in the business environment. In fact, it is usually the case in real-world situations that the dependent variable is a function of more than one independent variable. As an instance, $P = f(q, s)$ may represent profits as a function of output q and the amount of sales promotion s. Other examples are:

(a) Interest rates may depend on the degree of inflation, availability of money, the international monetary situation, and so on.

(b) The annual sales of the IGM Corp. might be defined by the multivariable function $S = R^{\alpha}F^{\beta}M^{\gamma}$, where $0 < \alpha, \beta, \gamma < 1$ and R, F, M represent the amounts allocated to research, plant facilities, and marketing, respectively.

These are only a few of the many examples which exist throughout the business and industrial world.

Actually, we have already encountered functions of two variables in a previous section on extrema. Two instances come to mind: the open box with square base whose area $A = x^2 + 4xy$ and volume $V = x^2y$; and the open tank whose volume $V = \pi x^2 y$ and surface area $A = \pi x^2 + 2\pi xy$. In each case area and volume are functions of two variables, which necessitated solving one equation for y, say, substituting into the other equation, and optimizing this latter equation. The optimization was possible because of rules and procedures for finding extrema of a function of one variable.

Consider the equation $z = f(x, y)$. Three variables are involved here, one of which, z, is the dependent variable and the other two, x and y, are the independent variables. How can we develop a "feel" for understanding the basic ideas? One feasible way is to think of the ordered pairs (x, y) as the domain of the function and $f(x, y)$ as the values in the range. Analogous to the statement for functions of one variable, we may assign to each ordered pair of values (x, y) in the domain a unique and corresponding value in the range $f(x, y)$ [see Figure 5.1].

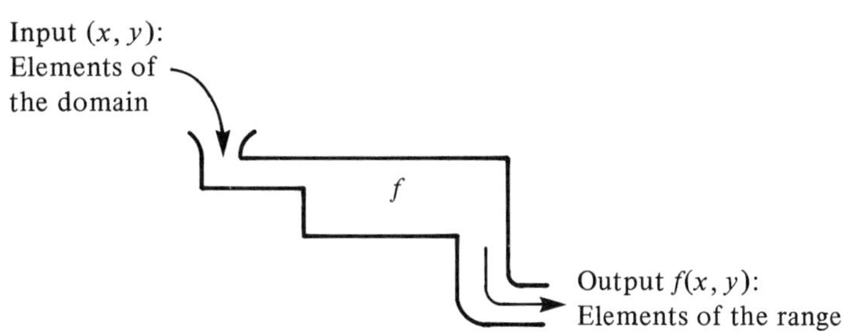

Figure 5.1

5.1 GRAPHS

We graphed functions of one variable in E^2, that is, two-dimensional space. We intuitively ask: Why not go to E^3, three-dimensional space, for

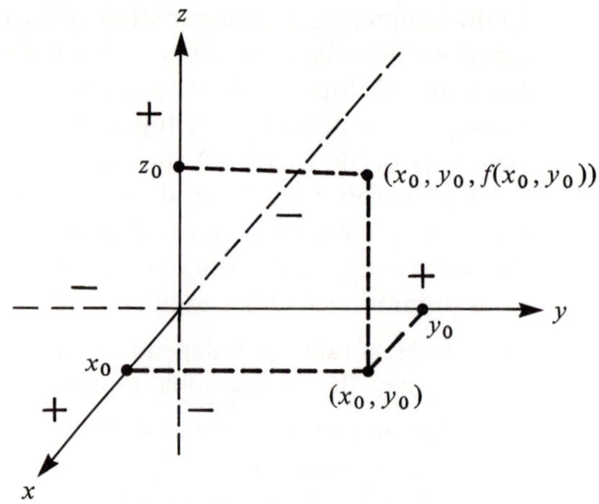

Figure 5.2

graphing functions of two variables? Accordingly, we draw a system of three axes mutually perpendicular in the "number space" (see Figure 5.2). In the xy-plane, which is horizontal, we locate members of the domain of f, i.e., we locate points of the form (x, y). One of these members is shown in Figure 5.2 as (x_0, y_0). Next we locate the corresponding member of the range on a vertical line through (x_0, y_0) and parallel to the z-axis. This corresponding point is an "ordered triple" and is indicated by $(x_0, y_0, f(x_0, y_0))$ as shown in the figure. We say that the set of ordered triples of the form $(x, y, f(x, y))$ which correspond to ordered pairs (x, y) in the domain of f constitutes a subset of E^3. The totality of such points comprise the "graph" of f. This graph will usually (for our purposes) be a plane or curved surface. Let us draw and discuss some of these graphs.

EXAMPLE 1 Discuss the graph of $z = f(x, y) = 4 - x^2$.

Solution The given function does not depend on y but does depend on x. Let us first graph the curve of $z = 4 - x^2$ in the zx-plane (Figure 5.3). This curve, call it C, is known as the **directrix** of the surface S whose graph we wish to draw. A line L not in the zx-plane but through a point of C and

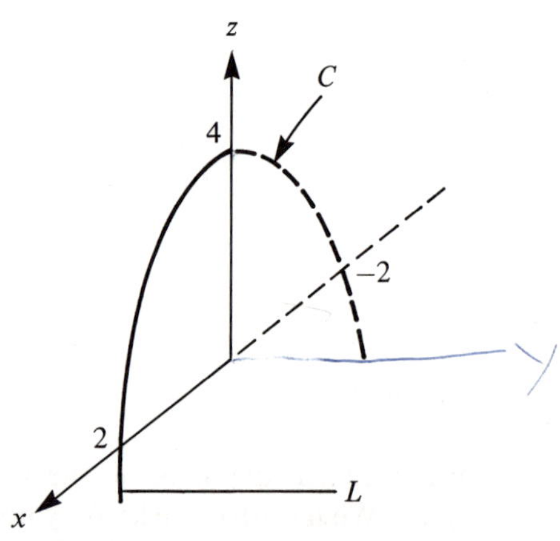

Figure 5.3

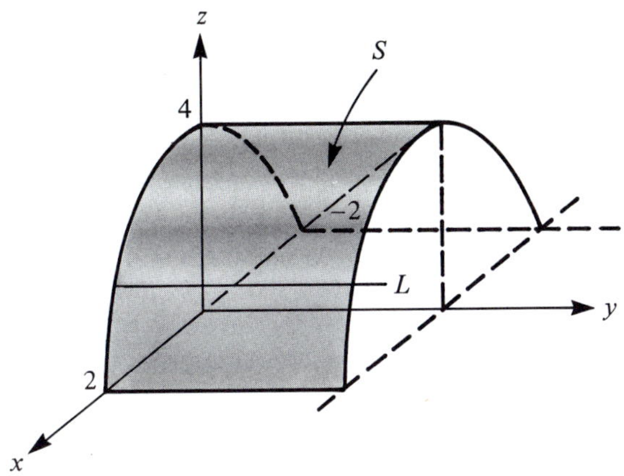

Figure 5.4

perpendicular to the plane of C is called a generating line, or simply a **generator**. We may think of the required surface S as being "generated" by moving the generator L parallel to itself along directrix C (Figure 5.4). The desired graph of $z = 4 - x^2$ is a parabolic cylinder having generators parallel to the y-axis. Figure 5.4 exhibits only a portion of the cylinder. The parabolic curve C is referred to as a **trace** of the cylinder in the zx-plane. A number of positions of L are shown as it moves parallel to itself along the parabolic curve C.

The graph, a parabolic cylinder, gets its name from the name of the trace, in this case a parabola. This method of naming cylinders also holds true for elliptic cylinders and hyperbolic cylinders. (See problem 48 of Section 1.6 for a discussion of ellipses and hyperbolas.)

EXAMPLE 2 Discuss the graph of the cylinder formed when the generator L, a vertical line through (0, 1, 0), is moved parallel to itself along the directrix C defined by $x^2 + y^2 = 1$.

Solution Since the trace of the cylinder is given by $x^2 + y^2 = 1$ we say that the cylinder is a right circular cylinder (a special case of an elliptic cylinder). This is so because its directrix is a circle (with radius 1 and center at the origin) and its generator is a line moving along the trace perpendicular to the xy-plane. Figure 5.5 shows the trace, $x^2 + y^2 = 1$, with four

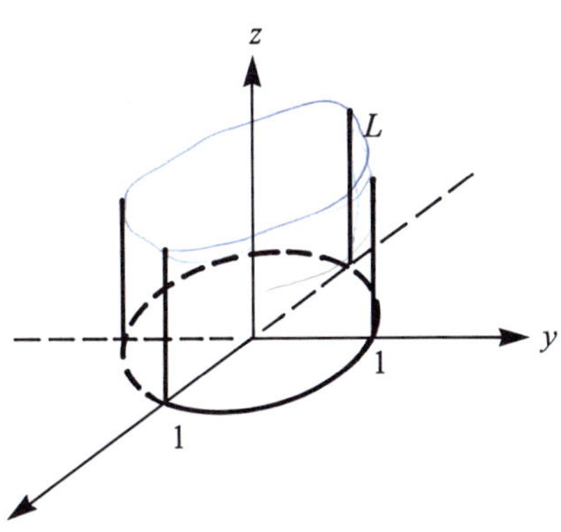

Figure 5.5

positions of the generator L. Figure 5.6 exhibits that part of the right circular cylinder located in the first octant, where x, y, $z > 0$. There are eight octants in three-dimensional space, four above the xy-plane and four below.

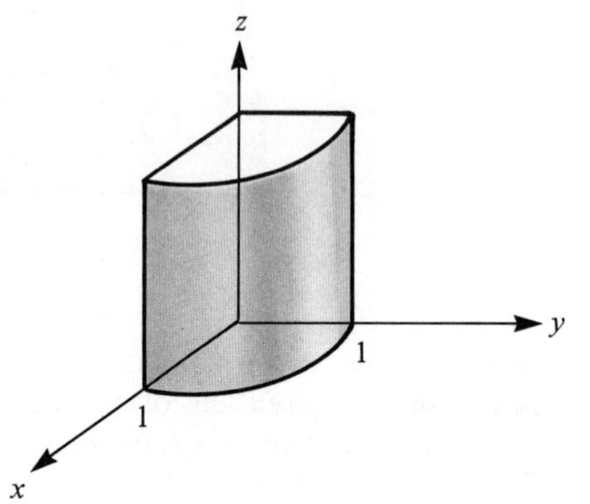

Figure 5.6

EXAMPLE 3 Discuss the graph of $z = 2 - x - 2y$.

Solution If $z = 0$, then $x + 2y = 2$ is the trace in the xy-plane, indicated by the line C_1 in Figure 5.7. Line C_1 is obtained by finding the x and y intercepts and joining them with a line. When $x = 0$, then $z = 2 - 2y$ is the trace in the zy-plane, labeled C_2 in the figure. If $y = 0$, then $z = 2 - x$ is the trace in the zx-plane, labeled C_3 in the figure. We visualize the graph of $z = 2 - x - 2y$ as a plane "slashing" through the three axes in the oblique manner shown. Figure 5.7 exhibits only that part of the plane situated in the first octant.

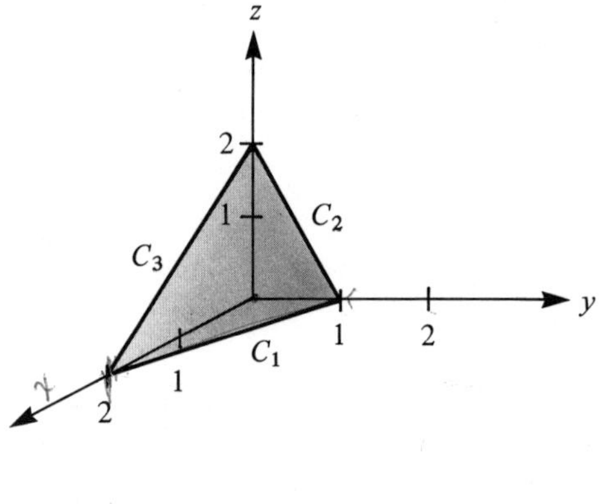

Figure 5.7

We will confine our graphing to functions whose "surfaces" are fairly easy to see. Many graphs of functions of more than one variable are very difficult to visualize and thus almost impossible to sketch.

5.2 PARTIAL DERIVATIVES

How do we view rates of change for functions of two variables?

Consider the production function $P(x, y) = xy$, where x denotes the number of labor man-hours of the LMN Co. and y is the amount of outlay for plant facilities. If we desire the rate of change of P with respect to x, then we must pretend that y is fixed. We cannot cope with two or more variables changing simultaneously. Finding the derivative of P with respect to x is a known procedure from previous discussion. The new matter is one of notation: we want a symbol indicating the first derivative of P with respect to x when y is fixed. Such a symbol is

$$\frac{\partial P}{\partial x}$$

which is read "the partial of P with respect to x." It is a **partial derivative** of P, but is alluded to simply as a **partial** of P, or the first partial of P. With the introduction of partial derivatives we need a contrasting name for the first derivative previously discussed. Let's agree to call dy/dx an *ordinary* derivative and $\partial y/\partial x$ a partial derivative. Note the difference in the two "operators" d/dx and $\partial/\partial x$.

Returning to the example $P = xy$, we have

$$\frac{\partial P}{\partial x} = y, \quad y \text{ constant when differentiating,}$$

and

$$\frac{\partial P}{\partial y} = x, \quad x \text{ constant when differentiating.}$$

Other symbols equivalent to $\partial P/\partial x$ are P_x and $P_1(x, y)$; for $\partial P/\partial y$ we have P_y and $P_2(x, y)$. The sub-index 1 refers to the first variable in (x, y); similarly, the sub-index 2 refers to the second variable in (x, y). P_x is read "P sub x" and P_y, "P sub y." When $P = f(x, y)$ then P_x means the first derivative of P with respect to x evaluated at (x, y). If it is explicitly a requirement to find P_x at (a, b), then we write

$$P_x\Big|_{(a,\,b)} \quad \text{or} \quad \frac{\partial P}{\partial x}\Big|_{(a,\,b)}$$

and similarly for P_y or $\partial P/\partial y$.

The limit concept is the basis of a partial derivative, as it is for the ordinary derivative. Thus, if $z = f(x, y)$, then

$$z_x = \lim_{\Delta x \to 0} \frac{\Delta z}{\Delta x} = \lim_{\Delta x \to 0} \frac{f(x + \Delta x, y) - f(x, y)}{\Delta x} \tag{1}$$

provided the limit exists. In Eq. (1) the numerator denotes the increment (or change) in f due to a change from (x, y) to $(x + \Delta x, y)$ in the xy-plane. This is a change in the x-direction. We may refer to z_x as the instantaneous rate of change of z in the x-direction. We can define and describe z_y

similarly:

$$z_y = \lim_{\Delta y \to 0} \frac{\Delta z}{\Delta y} = \lim_{\Delta y \to 0} \frac{f(x, y + \Delta y) - f(x, y)}{\Delta y} \tag{2}$$

provided the limit exists.

A geometric interpretation of Eq. (2) evaluated at the point (x_0, y_0) is offered in Figure 5.8. A similar geometric illustration can be drawn for Eq. (1). Figure 5.8 implies that curve C results from the intersection of the plane $x = x_0$ with the surface $z = f(x, y)$. The line tangent to z at (x_0, y_0, z_0) has slope defined by z_y.

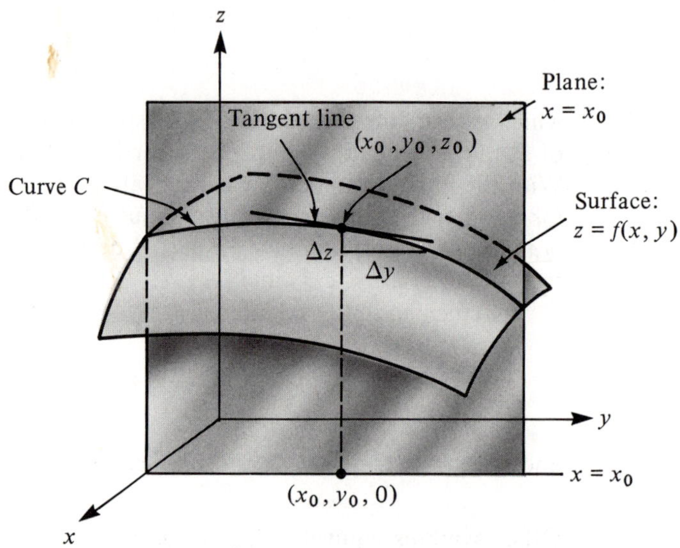

Figure 5.8

These notions may be extended to functions of three or more variables. For instance, we could find F_t holding r, s, and u fixed in $F(r, s, t, u)$; or F_u might be found holding r, s, and t constant; and so on. Let us now perform some partial differentiations.

EXAMPLE 1 Find the two first partial derivatives of $f(x, y) = 2x + 3y - 4xy$.

Solution

$$f_x = 2(1) + 0 - 4y(1) = 2 - 4y,$$

where we kept in mind that x varies and y is constant. Also,

$$f_y = 0 + 3(1) - 4x(1) = 3 - 4x$$

where we concentrated on the fact that x is constant while y is changing.

EXAMPLE 2 If $g(x, y) = x^3 + 2x^2y + y^2$, find g_x and g_y.

Solution

$$g_x = 3x^2 + 2y(2x) + 0 = 3x^2 + 4xy.$$
$$g_y = 0 + 2x^2(1) + 2y = 2x^2 + 2y.$$

EXAMPLE 3 If $H(x, y) = e^{xy} \cdot \ln(xy)$, find H_x and H_y at the point $(x, y) = (1, 1)$.

Solution

$$H_x = e^{xy} \cdot \frac{\partial}{\partial x} \ln(xy) + \ln(xy) \cdot \frac{\partial}{\partial x} e^{xy}$$

$$= e^{xy} \cdot \frac{1}{xy} \cdot y + \ln(xy) \cdot e^{xy} \cdot y$$

$$= \frac{e^{xy}}{x} + ye^{xy} \cdot \ln(xy).$$

Then $$H_x(1, 1) = e + e \cdot \ln 1 = e + e(0) = e.$$

$$H_y = e^{xy} \cdot \frac{\partial}{\partial y} \ln(xy) + \ln(xy) \cdot \frac{\partial}{\partial y} e^{xy}$$

$$= e^{xy} \cdot \frac{1}{xy} \cdot x + \ln(xy) \cdot e^{xy} \cdot x$$

$$= \frac{e^{xy}}{y} + xe^{xy} \cdot \ln(xy).$$

Then $$H_y(1, 1) = e + e \ln 1 = e.$$

PROBLEM SET 5.1, 5.2

1. Sketch graphs of the following surfaces:

(a) $z = f(x, y) = x^2$ **(b)** $z = f(x, y) = x^2 + 2$
(c) $z = 9 - y^2$.

2. Sketch and identify the following surfaces:

(a) $z = f(x, y) = 2x^2 + 2y^2$ **(b)** $z = 4x^2 + 4y^2$
(c) $z = x^2 + 2y^2$.

3. Graph the functions and identify each graph:

(a) $z = f(x, y) = 1 - 2x - 3y$ **(b)** $z = 3y + 3$
(c) $x + 2y + 3z = 6$.

4. Graph the surfaces defined by:

(a) $x^2 + y^2 = 16$ **(b)** $yz = 6$ **(c)** $x^2 + (y - 1)^2 = 1$.

Find the first partial derivatives of the following functions:

5. $f(x, y) = 2x^2y + 3xy^2$

6. $u = e^{2x - 3y}$

7. $w = \ln(xy + 1)$

8. $z = \sqrt{x^2 + y^2}$

9. $F(x, y) = \dfrac{x}{y} + \dfrac{y}{x}$

10. $g(x, y) = \sqrt{9 - x^2 - y^2}$

11. $h(x, y) = y^2 e^{x^2}$

12. $u = \ln(x^2 + y^2)^{1/2}$

13. $V = \dfrac{x + e^y}{y + e^x}$

14. $w = x^2y + yz^2 + zy^2$

15. $u = e^{xz} - e^{yz}$

16. $w = \dfrac{1 + xyz}{1 - xyz}$.

17. Sam's Service Station has annual profits P determined by the function $P(x, y, z) = xy + 2xz + 3yz - 3x^2 - 2y^2 - z^2$, where x is the number of employees; y is the number of gasoline pumps; and z is the value of other supplies: oil, grease, tires, and miscellaneous items. Find P_x, P_y, and P_z and interpret the results. [*Hint*: $P_x = y + 2z - 6x$ which indicates the rate of change of profits with respect to the number of persons employed when the number of pumps and the value of other supplies are held fixed. Now find P_y and P_z and interpret each similarly.]

18. Work problem 17 when $P(x, y, z) = 2xy + 3xz/2 + 4yz - 2x^2 - 3y^2 - z^2/3$.

5.3 PARTIAL DERIVATIVES OF HIGHER ORDERS. THE TOTAL DERIVATIVE

A **second partial derivative** is the first partial derivative of a first partial derivative. That is, if two successive partial differentiations take place, the result is a partial derivative of the second order, or simply a second partial. Given the function $f(x, y)$, then the second partial of f with respect to x (twice) is

$$\boxed{\frac{\partial}{\partial x}\left(\frac{\partial f}{\partial x}\right) = \frac{\partial^2 f}{\partial x^2} = f_{xx}}\,. \tag{1}$$

Similarly,

$$\boxed{\frac{\partial}{\partial y}\left(\frac{\partial f}{\partial y}\right) = \frac{\partial^2 f}{\partial y^2} = f_{yy}}\,. \tag{2}$$

There are four partials of a function of two variables. Two have been presented in Eqs. (1) and (2) above; the remaining two partials are:

$$\boxed{\frac{\partial}{\partial x}\left(\frac{\partial f}{\partial y}\right) = \frac{\partial^2 f}{\partial x\,\partial y} = f_{yx}} \tag{3}$$

and

$$\boxed{\frac{\partial}{\partial y}\left(\frac{\partial f}{\partial x}\right) = \frac{\partial^2 f}{\partial y\,\partial x} = f_{xy}}\,. \tag{4}$$

In Eqs. (3) and (4) the results can be proven to be identical. In other words, the order of the partial differentiations is immaterial. The proof of this is valid provided all the derivatives involved are continuous functions. Moreover, the validity holds for partial derivatives of all orders. The second partials f_{xx} and f_{yy} are often referred to as pure second partials, whereas f_{xy} and f_{yx} are called mixed second partials.

The above ideas may be extended to functions of three or more variables. For instance, if

$$u = F(x, y, z)$$

then

$$u_{xxy} = \frac{\partial}{\partial y}\left(\frac{\partial^2 u}{\partial x^2}\right) = \frac{\partial^3 u}{\partial y\,\partial x^2},$$

$$u_{yxz} = \frac{\partial}{\partial z}\left(\frac{\partial^2 u}{\partial x\,\partial y}\right) = \frac{\partial^3 u}{\partial z\,\partial x\,\partial y}, \qquad \text{and so on.}$$

In particular, u_{zyy} is read "the third partial of u with respect to z once and y twice."

EXAMPLE 1 Find all second partials of $f(x, y) = x^2 - 3xy + y^2 - 2x + 4y - 5$.

Solution

$$\begin{aligned} f_x &= 2x - 3y - 2; & f_{xx} &= 2; & f_{xy} &= -3. \\ f_y &= -3x + 2y + 4; & f_{yy} &= 2; & f_{yx} &= -3 = f_{xy}. \end{aligned}$$

EXAMPLE 2 Find the second partials of $F(x, y) = e^{xy} + \ln(xy)$.

Solution

$$F_x = e^{xy}(y) + \frac{1}{xy}(y) = ye^{xy} + \frac{1}{x} = \frac{\partial F}{\partial x}$$

$$\begin{aligned} F_{xx} &= \frac{\partial}{\partial x}\left(\frac{\partial F}{\partial x}\right) \\ &= ye^{xy}(y) - \frac{1}{x^2} = y^2e^{xy} - \frac{1}{x^2} \end{aligned}$$

$$\begin{aligned} F_{xy} &= \frac{\partial}{\partial y}\left(\frac{\partial F}{\partial x}\right) \\ &= ye^{xy}(x) + e^{xy}(1) + 0 = xye^{xy} + e^{xy} \end{aligned}$$

$$F_y = e^{xy}(x) + \frac{1}{xy}(x) = xe^{xy} + \frac{1}{y} = \frac{\partial F}{\partial y}$$

$$\begin{aligned} F_{yy} &= \frac{\partial}{\partial y}\left(\frac{\partial F}{\partial y}\right) \\ &= xe^{xy}(x) - \frac{1}{y^2} = x^2e^{xy} - \frac{1}{y^2} \end{aligned}$$

$$\begin{aligned} F_{yx} &= \frac{\partial}{\partial x}\left(\frac{\partial F}{\partial y}\right) \\ &= xe^{xy}(y) + e^{xy}(1) + 0 = xye^{xy} + e^{xy} = F_{xy}. \end{aligned}$$

If a production function is defined by $P(l, k)$, then the marginal productivity for l (the amount of labor input) is written as P_l. In a similar manner, the marginal productivity for k (the amount of capital input) is given by P_k. Moreover, the rate of marginal productivity for l twice is P_{ll} and for k twice, P_{kk}. The mixed second partials are given by P_{lk} (for l first and k second) and by P_{kl} (for k first and l second).

EXAMPLE 3 A production function is represented by $P(l, k) = 8l - l^2 + 3lk + 50k - k^2$, where l and k denote the amount of labor and capital inputs, respectively.

(a) Find the marginal productivities for l and k if $l = 2$, $k = 5$.
(b) Find the rates of change of the marginal productivities.

Solution **(a)**

$P_l = 8 - 2l + 3k$, the marginal productivity for l,
$P_k = 3l + 50 - 2k$, the marginal productivity for k.

If $l = 2$ (hundreds of man-hours) and $k = 5$ (thousands of dollars) then

$$P_l \Big|_{\substack{l=2\\k=5}} = 8 - 2(2) + 3(5) = 19$$

and

$$P_k \Big|_{\substack{l=2\\k=5}} = 3(2) + 50 - 2(5) = 46.$$

(b) $P_{ll} = -2$, i.e., P_l is changing at a decreasing rate for all l.
$P_{kk} = -2$ or a change similar to that of P_{ll}.
$P_{lk} = 3$ and $P_{kl} = 3$, i.e., the mixed second partials are equal, each one indicating that the marginal productivities are increasing.

The Total Derivative

In connection with composite functions of a single variable we have already learned that if $y = f(u)$ and $u = g(x)$, then by the chain rule we have

$$\frac{dy}{dx} = \frac{dy}{du} \cdot \frac{du}{dx}$$

provided the derivatives exist. Analogous to this is the chain rule for functions of two variables. We state the analog and omit the proof.

THEOREM 1 Chain Rule for Functions of Two Variables

If $y = f(x, u)$ and $u = g(x)$, then the total derivative of y with respect to x is

$$\boxed{\frac{dy}{dx} = \frac{\partial y}{\partial x}\frac{dx}{dx} + \frac{\partial y}{\partial u} \cdot \frac{du}{dx}} \tag{5}$$

$$= f_x(1) + f_u\frac{du}{dx} \quad \text{or} \quad f_x + f_u\frac{du}{dx},$$

provided the derivatives exist.

In Eq. (5) the derivatives $\partial y/\partial x$ and $\partial y/\partial u$ are partials because y is directly a function of two variables. As a result, one of these variables is held fixed while the rate of change is being found in the direction of the other variable.

EXAMPLE 4 Find the total derivative dy/dx if $y = u^2 - 2ux + 3x^2 = g(u, x)$ and $u = 2x^2 - 1 = f(x)$.

Solution

$$\begin{aligned}
\frac{dy}{dx} &= \frac{\partial y}{\partial u}\frac{du}{dx} + \frac{\partial y}{\partial x}\frac{dx}{dx} \\
&= (2u - 2x)(4x) + (-2u + 6x)(1) \\
&= 8ux - 8x^2 - 2u + 6x \\
&= 8(2x^2 - 1)(x) - 8x^2 - 2(2x^2 - 1) + 6x \\
&= 16x^3 - 8x - 8x^2 - 4x^2 + 2 + 6x \\
&= 16x^3 - 12x^2 - 2x + 2
\end{aligned}$$

where the answer is a function of x alone.

If we had been seeking du/dt, given $u = f(t, r)$ and $r = g(t)$, then we would write

$$\frac{du}{dt} = \frac{\partial u}{\partial t}\frac{dt}{dt} + \frac{\partial u}{\partial r}\frac{dr}{dt} = u_t + u_r\frac{dr}{dt}. \tag{6}$$

For a particular situation we would find the derivatives in Eq. (6) and simplify to an answer in t alone. We do this only when the substitutions are "nice." In complicated situations we leave the answer in t and r.

An extension can be made of these notions for functions of three or more variables. For instance, if

$$u = F(x, y, z)$$

where x, y, z are functions of t alone, then the total derivative of u with respect to t is

$$\begin{aligned}\frac{du}{dt} &= \frac{\partial F}{\partial x}\frac{dx}{dt} + \frac{\partial F}{\partial y}\frac{dy}{dt} + \frac{\partial F}{\partial z}\frac{dz}{dt}\\ &= F_x\frac{dx}{dt} + F_y\frac{dy}{dt} + F_z\frac{dz}{dt}.\end{aligned}$$

There are many other relations between u, x, y, z, and t which are interesting to explore, but which we will forego here.

EXAMPLE 5 Given $u = F(x, y, z) = x^2 + y^2 + z^2$ and $x = e^{2t}$, $y = \ln 2t$, and $z = 2t$, find the total derivative du/dt.

Solution

$$\begin{aligned}\frac{du}{dt} &= \frac{\partial F}{\partial x}\frac{dx}{dt} + \frac{\partial F}{\partial y}\frac{dy}{dt} + \frac{\partial F}{\partial z}\frac{dz}{dt}\\ &= 2xe^{2t}(2) + 2y\left(\frac{2}{2t}\right) + 2z(2)\\ &= 4xe^{2t} + \frac{2y}{t} + 4z\Big|_{\text{at } x=e^{2t},\ y=\ln 2t,\ z=2t}\\ &= 4e^{2t}(e^{2t}) + \frac{2}{t}(\ln 2t) + 4(2t)\\ &= 4e^{4t} + \frac{2\ln 2t}{t} + 8t.\end{aligned}$$

The total derivative plays a vital role in finding the slope of a function whose variables are related *implicitly*. In the next section we will see how this is done.

PROBLEM SET 5.3

Find the second partials for each of the functions in problems 1–14:

1. $f(x, y) = x + xy + 1$

2. $g(x, y) = x^2 + xy^2$

3. $u = 2x^3y^2 - 3x^2y$

4. $z = \dfrac{x}{y} + \dfrac{y}{x}$

5. $w = \sqrt{x + y}$

6. $F(x, y) = xe^y$

7. $G(x, y) = e^x \ln y$

8. $H(x, y) = \dfrac{\ln x}{2y}$

9. $u = (x - 1)e^y$

10. $w = \dfrac{1 + e^x}{1 + e^y}$

11. $z = e^{2x-3y}$

12. $u = \ln \sqrt{x + y}$

13. $w = \dfrac{1}{x} \ln \dfrac{1}{y}$

14. $z = \dfrac{x^3y^2 - x^2y^3}{xy}$

Find the third partials, z_{xyy} and z_{yxx}, for the following functions:

15. $z = 2xy^2 + 3x^2y$

16. $z = \sqrt{2x + 4y}$

17. $z = e^{xy}$

18. $z = \ln(2x + 3y)$

19. A production function is defined by $P(l, k) = 10l + 40k - l^2 + 5lk - k^2$, where l and k represent the amount of labor and capital inputs, respectively.
   (a) Find the marginal productivities for l and for k when $l = 30$ and $k = 4$.
   (b) Determine the rates of change of the marginal productivities when $l = 6$ and $k = 2$.

20. Work problem 19 if $P(l, k) = 3l^3 + k^3 - l^2k^{1/2}$ and $l = 4 = k$.

Find the total derivative of u with respect to x using the chain rule in exercises 21–25:

21. $u = x^2 + xy - y^2$ when $y = 3x + 4$.

22. $u = 2xy^2 + 3x^2y$ when $y = e^x$.

23. $u = e^{x^2y}$ when $y = \ln \sqrt{x}$.

24. $u = xy + yz + xz$ when $y = (2x - 1)^2$ and $z = 1/x$.

25. $u = e^{x+y}/e^{z-y}$ when $y = 1 - x^2$ and $z = \ln(x/2)$.

26. Let profit P be a function of sales revenue r and earnings n, where r and n depend on advertising a. Specifically, let $P(r, n) = \sqrt{r^2 + n^2}$ when $r = a^2$ and $n = -5a$. Find the total derivative dP/da by the chain rule at the instant when $r = 3$, $n = 4$, and $a = 5$.

27. Work problem 26 when $P = \sqrt{r^2 - n^2}$, $r = 5a^{1/3}$, $n = 4a^{1/2}$ at the instant when $r = 5$, $n = 4$, and $a = 1$.

5.4 IMPLICIT DIFFERENTIATION

When we see a particular function of x such as $y = 2x^2 - 3x + 4$ we should realize that y is **explicitly** defined in terms of x. In fact, we can write $y = f(x) = 2x^2 - 3x + 4$. In many instances, however, functions are not defined explicitly. Consider, for example, the function

$$x^2 + y^5 = xy \tag{1}$$

which may be expressed as

$$H(x, y) = x^2 + y^5 - xy = 0.$$

It is very difficult to solve Eq. (1) for y explicitly in terms of x. The equation $H(x, y) = 0$ may have a root $y = f(x)$, which satisfies the equation so that

$$x^2 + [f(x)]^5 - x[f(x)] = 0$$

for all values of x in the domain of f. In such cases we say that y is an implicit function of x, or that the equation relates x and y **implicitly**. If an equation defines y as one or more differentiable functions of x, we can differentiate implicity in order to find the derivative of y with respect to x.

We return to $H(x, y) = 0$ and substitute $y = f(x)$ so that

$$H[x, f(x)] = 0.$$

Let us differentiate with respect to x using the chain rule:

$$H_x \frac{dx}{dx} + H_f \frac{df}{dx} = 0$$

or

$$H_x + H_f \frac{df}{dx} = 0.$$

Since $y = f(x)$ we may write this as

$$H_x + H_y \frac{dy}{dx} = 0.$$

Solving for dy/dx we have

$$\boxed{\frac{dy}{dx} = -\frac{H_x}{H_y}}, \tag{2}$$

which is valid at each point (x, y) for which $H_y \neq 0$ and $H(x, y) = 0$.

Remark: An aid to remembering Eq. (2) is to "think diagonally." As cases in point:

$$\text{If} \quad G(t, z) = 0 \quad \text{then} \frac{dz}{dt} = -\frac{G_t}{G_z}.$$

$$\text{If} \quad F(u, v) = 0 \quad \text{then} \frac{dv}{du} = -\frac{F_u}{F_v}.$$

The reader should realize that in Eq. (2) we are finding the slope of a function (defined implicitly) without solving the function for y in terms of x.

EXAMPLE 1 Find dy/dx at the point $(1, 2)$ if $x^2 + y^2 + 2x = 3y + 1$ and $y = f(x)$.

Solution Let $H(x, y) = x^2 + y^2 + 2x - 3y - 1 = 0$. Then by formula (2) above we have

$$\begin{aligned} \frac{dy}{dx} &= -\frac{H_x}{H_y}\bigg|_{(1,2)} \\ &= -\frac{2x + 2}{2y - 3}\bigg|_{(1,2)} \\ &= -\frac{2(1) + 2}{2(2) - 3} \\ &= -4 \end{aligned}$$

EXAMPLE 2 Given $e^{2xy-2} + \ln(xy) = 2x - y$; find dy/dx at $(1, 1)$.

Solution Let $H(x, y) = e^{2xy-2} + \ln(xy) - 2x + y = 0$. Then

$$\frac{dy}{dx} = -\frac{H_x}{H_y}$$

$$= -\left.\frac{2ye^{2xy-2} + \dfrac{1}{x} - 2}{2xe^{2xy-2} + \dfrac{1}{y} + 1}\right|_{(1,\,1)}$$

$$= -\frac{2 + 1 - 2}{2 + 1 + 1}$$

$$= -\frac{1}{4}$$

EXAMPLE 3 Given that $rs^2 = r^2s + rs - 1$. Find the rate of change of earnings r with respect to sales s by (a) the method of Example 1 and (b) the method of problem 12, Section 3.6 (Enrichment Topics and Problems), p. 117.

Solution (a) Let $H(s, r) = rs^2 - r^2s - rs + 1 = 0$. Then

$$\frac{dr}{ds} = -\frac{H_s}{H_r}$$

$$= -\frac{r(2s) - r^2(1) - r(1)}{s^2(1) - s(2r) - s(1)}$$

$$= -\frac{2rs - r^2 - r}{s^2 - 2rs - s}.$$

(b) Employing term by term differentiation with respect to s, as in problem 12, Section 3.6, we have:

$$rs^2 = r^2s + rs - 1$$

$$r(2s)\frac{ds}{ds} + s^2\frac{dr}{ds} = r^2\frac{ds}{ds} + s(2r)\frac{dr}{ds} + r\frac{ds}{ds} + s\left(\frac{dr}{ds}\right).$$

Simplifying and solving for dr/ds yields:

$$2rs + s^2\frac{dr}{ds} = r^2 + 2sr\frac{dr}{ds} + r + s\frac{dr}{ds}$$

$$s^2\frac{dr}{ds} - 2rs\frac{dr}{ds} - s\frac{dr}{ds} = r^2 + r - 2rs$$

$$(s^2 - 2rs - s)\frac{dr}{ds} = r^2 + r - 2rs$$

$$\frac{dr}{ds} = \frac{r^2 + r - 2rs}{s^2 - 2rs - s}$$

$$= -\frac{2rs - r^2 - r}{s^2 - 2rs - s},$$

which checks the answer in (a) above.

PROBLEM SET 5.4

Find dy/dx by implicit differentiation for each implicit function in examples 1–8. Use the two methods of this section.

1. $x^2 + y^2 + 2x - 3y = 6$

2. $2x^2 + 3x^2y - 4xy^2 = 0$

3. $x^2 + (y + 1)^2 = x^2y^2 - 3$

4. $xy = (x + y)^2 - 12$

5. $e^x + \ln y = e^x \ln y + e$

6. $xe^y + y \ln x = \ln 2$

7. $(x + y)^2 = x - y + 1$

8. $(x - y)^2 = x^2 + y - 2x.$

9. In exercises 1, 2, 3, and 4 above, evaluate dy/dx at (2, 2).

10. Evaluate dy/dx at (0, 0) in exercises 7 and 8.

11. The management of the ICN Corp. is interested in the rate of change of earnings r with respect to sales s. It learns from its management scientist that r and s are related implicitly by the equation $r^3 + r^2s^2 + s^3 = 4s + 5$. Find the desired rate $\dfrac{dr}{ds}$ at $(r, s) = (1, 2)$ by the method of example 1.

12. Assume that the amount of sales s and the amount of cost c are related implicitly by the expression $10\sqrt{c} - cs + 6\sqrt{s} = 2$. Find ds/dc at $(c, s) = (4, 9)$.

13. Find an equation of the line tangent to the curve defined by $(x - y)^2 = x + y$ at (1, 0).

14. Work exercise 13 if the curve is defined by $e^y - \ln x - e^y \ln x = 1$.

15. A certain supply function is defined implicitly by $y^2 - x^2 - 2y - 2x - 4 = 0$ where x (in thousands) is the quantity of goods supplied by the producers. Find dy/dx by implicit differentiation and evaluate this rate of change at the point where $x = 1$.

5.5 EXTREMA FOR FUNCTIONS OF TWO VARIABLES

We studied the theory for finding extrema of functions of one variable where the first-derivative and second-derivative tests were used to find relative maximum and relative minimum points. We now extend the theory to functions of two variables. In this extension we find that the underlying theory is basically similar to that of functions of one variable.

Consider the function $z = f(x, y)$ defined for certain points (x, y) in the domain of f. If z assumes an extreme value at $(x, y) = (a, b)$, then z may have a relative minimum at (a, b) and

$$\left.\frac{\partial z}{\partial x}\right|_{(a,\,b)} = \left.\frac{\partial z}{\partial y}\right|_{(a,\,b)} = 0.$$

See Figure 5.9. Similarly, z may have a relative maximum at (a, b), so that

$$\left.\frac{\partial z}{\partial x}\right|_{(a,\,b)} = \left.\frac{\partial z}{\partial y}\right|_{(a,\,b)} = 0.$$

See Figure 5.10. In other words, both partials vanish in both cases.

In Figure 5.11 we see an example of the two first partials of z vanishing, but where z has neither a relative maximum nor a relative minimum at $[a, b, f(a, b)]$. In this case, z is said to have a "saddle point" at $[a, b, f(a, b)]$.

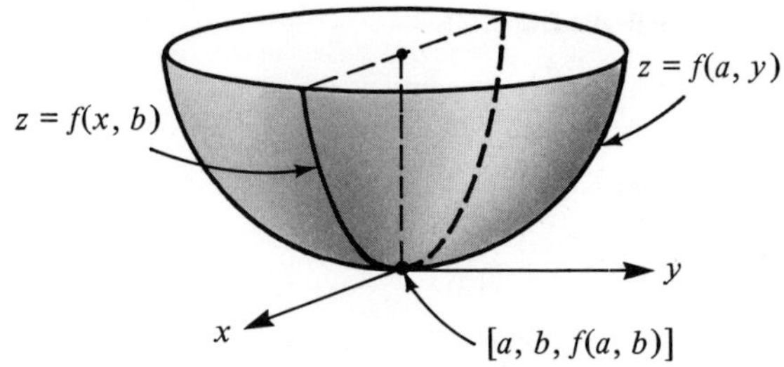

Figure 5.9

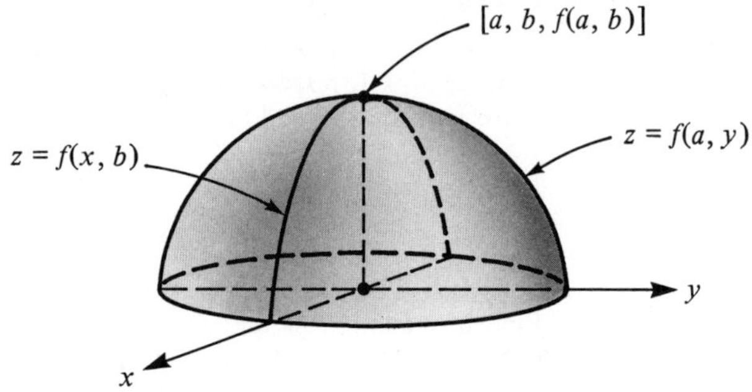

Figure 5.10

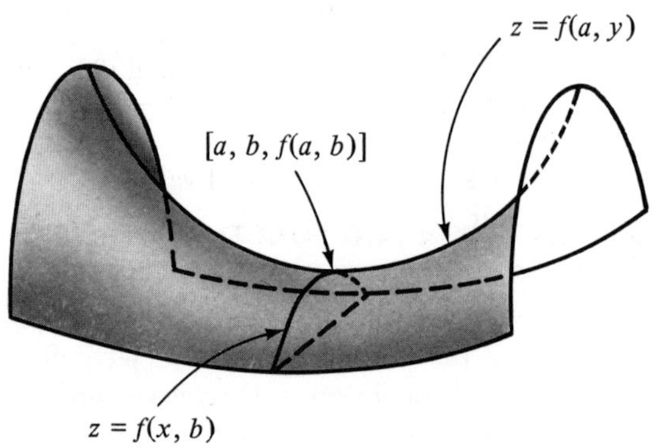

Figure 5.11

The graph of z increases in certain neighborhoods of the saddle point and decreases in other neighborhoods of the point.

In each of the above three cases, when y is a constant, $y = b$, then z is a function of x only and $z = f(x, b)$. Likewise, when x is a constant, $x = a$, then z is a function of y only and $z = f(a, y)$.

The vanishing of the two first partials of z is a necessary condition for the existence of extrema. Indeed, the extrema of many functions can be located exclusively by the use of these two relations, that is,

$$\boxed{\left.\frac{\partial z}{\partial x}\right|_{(a,\, b)} = 0 = \left.\frac{\partial z}{\partial y}\right|_{(a,\, b)}}\,.$$

However, this necessary condition is not a *sufficient condition* for locating extrema of many other functions. Sufficient conditions are included in the following theorem, which describes the **Second Derivative Test**. This is a basic test for finding relative maxima and minima for functions of two variables.

THEOREM 1 The Second Derivative Test

Suppose:

(1) f *is a function of two variables*, x *and* y;

(2) f, *its first partials, and its second partials are continuous at all points of a "neighborhood" about* (a, b);

(3) $f_x(a, b) = f_y(a, b) = 0$.

Then, if $D = f_{xx}(a, b) \cdot f_{yy}(a, b) - f_{xy}^2(a, b)$:

(i) f *has a* ***relative minimum*** *at* (a, b) *if* $D > 0$ *and* $f_{xx}(a, b) > 0$;

(ii) f *has a* ***relative maximum*** *at* (a, b) *if* $D > 0$ *and* $f_{xx}(a, b) < 0$;

(iii) f *has* ***no extrema*** *(a saddle point exists) if* $D < 0$;

(iv) *the* ***test fails*** *if* $D = 0$.

The point (a, b) is called the critical point for extreme values of f.

The following example illustrates one aspect of the procedure outlined in Theorem 1 for finding the extrema of a function of two variables.

EXAMPLE 1 Let a function representing profit (gain) in hundreds of dollars be defined by $G(x, y) = 100 - 2y^2 - 6y + 3xy - 2x^2 + 15x$, where x denotes the amount of advertising in thousands of dollars and y is the quantity of sales in thousands of units. Find values of x and y for which G is a maximum, and find $G_{\max}$.

Solution

$$\begin{cases} G_x = 3y - 4x + 15 = 0: & R_1 \qquad (1) \\ G_y = -4y + 3x - 6 = 0: & R_2. \qquad (2) \end{cases}$$

Solving Eqs. (1) and (2) simultaneously, we perform the row operations, $4R_1$ plus $3R_2$, and obtain the system:

$$\begin{aligned} 12y - 16x + 60 &= 0 \\ -12y + 9x - 18 &= 0 \\ \hline -7x + 42 &= 0 \\ -7x &= -42 \\ x &= 6. \end{aligned}$$

Solving:

Substituting $x = 6$ in Eq. (1) we get

$$\begin{aligned} 3y - 24 + 15 &= 0 \\ 3y &= 9 \\ y &= 3. \end{aligned}$$

Therefore, (6, 3) is the critical point.

Continuing with the procedure outlined in Theorem 1 we have

$$G_{xx} = -4, \quad G_{yy} = -4, \quad G_{xy} = 3 \qquad \text{at } (6, 3).$$

Then

$$G_{xx} \cdot G_{yy} - G_{xy}^2 = (-4)\cdot(-4) - (3)^2 = 16 - 9 > 0. \qquad (3)$$

Since $G_{xx} < 0$ and the result in Eq. (3) is positive, we conclude from (ii) of Theorem 1 that G has a relative maximum at the critical point (6, 3). Furthermore, G_{max} is defined by

$$\begin{aligned} G(6, 3) &= 100 - 2(9) - 6(3) + 3(18) - 2(36) + 15(6) \\ &= 136 \quad \text{or} \quad \$13{,}600. \end{aligned}$$

Thus, for $x = 6$ or an advertising layout of \$6000, and for $y = 3$ or a sales output of 3000 units, there is a maximum profit of \$13,600.

EXAMPLE 2 Find the dimensions of an open, rectangular container having a volume of 32 ft^3 if the least amount of lumber is to be used in its manufacture.

Solution With the dimensions of the container as shown in the figure, we have volume $V = xyz = 32$ and surface area $S = xy + 2xz + 2yz$. Solving V for z we get $z = 32/xy$. Substituting this value of z into the formula for S yields

$$S = xy + 2x \cdot \frac{32}{xy} + 2y \cdot \frac{32}{xy}$$

or

$$S = xy + \frac{64}{y} + \frac{64}{x}.$$

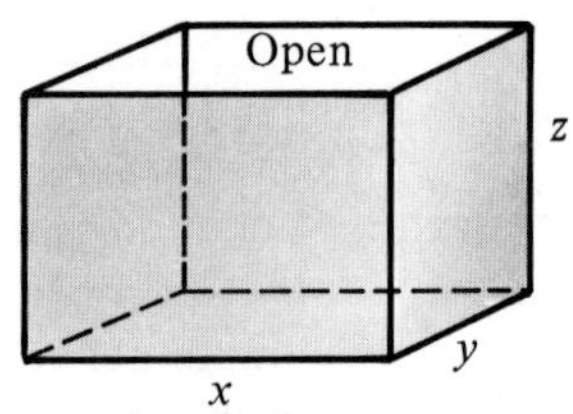

Then the partial of S with respect to x is

$$S_x = y + 64(-x^{-2}) = y - \frac{64}{x^2}$$

and similarly,

$$S_y = x + 64(-y^{-2}) = x - \frac{64}{y^2}.$$

We set these two first partials equal to zero and solve simultaneously:

$$\left.\begin{aligned} y - \frac{64}{x^2} &= 0 \\ x - \frac{64}{y^2} &= 0 \end{aligned}\right\}$$

$$\left.\begin{aligned} x^2y &= 64 \qquad (4) \\ xy^2 &= 64 \qquad (5) \end{aligned}\right\}.$$

Therefore,

$$x^2y = xy^2$$

or

$$xy(x - y) = 0$$

from which $x = y = 0$ or $x = y$. Since $x = y = 0$ is not applicable, then $x = y$ is the only valid solution. Using this solution in Eq. (4) gives us

$$\begin{aligned} y^3 &= 64 \\ y &= 64^{1/3} = 4 = x. \end{aligned}$$

It follows that $z = 32/xy = 32/(4)(4) = 2$.

We now use the Second Derivative Test to check for optimality:

$$S_{xx} = -64(-2x^{-3}) = \frac{128}{x^3}\bigg|_{x=4} = 2$$

$$S_{yy} = -64(-2y^{-3}) = \frac{128}{y^3}\bigg|_{y=4} = 2$$

$$S_{xy} = 1.$$

Then

$$S_{xx} \cdot S_{yy} - S_{xy}^2 = 2 \cdot 2 - 1^2$$
$$= 4 - 1 > 0. \tag{6}$$

Since $S_{xx} > 0$ and the result in Eq. (6) is positive, it follows that S is a minimum value for the dimensions x ft by y ft by z ft $\to$ 4 ft by 4 ft by 2 ft. The least amount of lumber is given by

$$S = xy + 2xz + 2yz\bigg|_{\substack{x=y=4\text{ ft}\\ z=2\text{ ft}}}$$
$$= 16 + 16 + 16 = 48\text{ ft}^2.$$

The theory of extrema for functions of two variables may be extended to functions of three or more variables. However, we choose not to pursue the extended theory in this text.

PROBLEM SET 5.5

Test the function f for extrema, if there are any:

1. $f(x, y) = 2x^2 + y^2 - 2xy - 4x + 3.$

2. $f(x, y) = 3xy - 4x^2 - y^2 + x - 3y.$

3. $f(x, y) = \dfrac{x^3}{3} + x^2y - xy^2 + \dfrac{y^3}{3}.$

4. $f(x, y) = x^3 + y^3 - 3xy + 10.$

5. $f(x, y) = xy - \dfrac{2}{x} + \dfrac{4}{y} + 5.$

6. $f(x, y) = xe^{-x} + ye^{-y} + 1.$

7. $f(x, y) = x \ln x + y \ln y - 2x - y.$

8. Work Example 1 of this section if $G(x, y) = 14x - y + 2xy - (3x^2/2) - (3y^2/2) + 200.$

9. Given the cost function $C(u, p) = 3u^2 - 4up - 28u + 3p^2 - 8p + 200$ where u denotes the expense of upkeeping plant facilities and p represents the amount of payroll disbursements. Find the critical point (u, p) and determine $C_{\min}$ in thousands of dollars.

10. Find the dimensions of a closed, rectangular container if the least amount of lumber is to be used in its manufacture. The volume is fixed at 27 m^3. Also find the minimum surface area.

11. Work problem 10 if the volume is fixed at 125 m^3.

12. Under conditions of monopoly a company manufactures two products, P_1 and P_2, whose demand equations are $x = 4 - p + q$ for P_1 and $y = 2 - p - q$

for P_2. Each of the x units of P_1 and y units of P_2 is in hundreds. The price per unit of x is p dollars and the price per unit of y is q dollars. P_1 costs \$2 per unit and P_2 costs \$1 per unit. Find the demands, x and y, and the prices, p and q, so that profit G (in hundreds of dollars) is maximized. *Hint*: If $100x$ units of P_1 and $100y$ units of P_2 are manufactured and sold, then total dollar revenue R is $100px + 100qy$. Furthermore, total cost C of production is $200x + 100y$. So,

$$\begin{aligned} G &= R - C \\ &= (100px + 100qy) - (200x + 100y) \\ &= (100p - 200)x + (100q - 100)y \\ &= (100p - 200)(4 - p + q) + (100q - 100)(2 - p - q). \end{aligned}$$

13. A certain profit function P in dollars is defined by $P = 18(100z) - 4(100x) - 2(100y) - 800$ where $100z$ represents amount of output priced at \$18 per unit, $100x$ denotes amount of an input costing \$4 per unit, and $100y$ denotes amount of a second input costing \$2 per unit. If the productivity function is given by $z = 8 - (2/x) - (1/y)$, find maximum profit. [*Hint*: $P = 1800(8 - 2/x - 1/y) - 400x - 200y - 800$.]

14. Work exercise 13 above if $z = 10 - (4/x) - (2/y)$ and $P = 20(100z) - 5(100x) - 2(100y) - 1000$.

15. A rectangular bin of volume of 48 ft^3 is to be constructed in a basement corner by using the floor and the two walls, thus minimizing cost. Find the minimum cost if the lid of the bin costs 15¢ per ft^2 and the sides cost 20¢ per ft^2.

16. Work exercise 15 if the sides cost 24¢ per ft^2 and the volume is 40 ft^3.

5.6 LAGRANGE MULTIPLIERS (SINGLE CONSTRAINT)

The problem of finding extrema for the function

$$f(x, y) = x^2 + y^2 \tag{1}$$

is different from that of finding extrema for

$$f(x, y) = x^2 + y^2 \quad \text{subject to the condition} \quad 3x + 2y = 13. \tag{2}$$

In Eq. (1) we are optimizing a function with no added condition, while in Eq. (2) there is a side condition (or constraint) imposed on the function. In the second case we say that the problem is one of **constrained maxima and minima**. In (1) the problem is one of unconstrained maxima and minima. The presence of a constraint is usually significant in that the domain is restricted. The function in Eq. (1) with no side condition has a minimum whose value is clearly seen to be zero. The minimum value in Eq. (2) is not so easily seen. However, it can be readily found by an interesting and powerful technique called the **method of Lagrange multipliers**. This method involves the introduction of an artificial variable l, called the Lagrange multiplier, into a newly formed auxiliary function F. We illustrate in Eq. (2) above.

Let the constraint be

$$g(x, y) = 3x + 2y - 13$$

and let

$$F(x, y, l) = f(x, y) + lg(x, y).$$

Substituting from Eq. (2) we have

$$F(x, y, l) = x^2 + y^2 + l(3x + 2y - 13).$$

The necessary conditions for extrema are that

$$F_x = 0, \quad F_y = 0 \quad \text{and} \quad F_l = 0.$$

We perform these operations and get

$$\left.\begin{aligned} F_x &= 2x + 3l = 0 \\ F_y &= 2y + 2l = 0 \\ F_l &= 3x + 2y - 13 = 0 \end{aligned}\right\}. \qquad \begin{aligned} &(3) \\ &(4) \\ &(5) \end{aligned}$$

Note that the last equation is precisely the constraint equation. Thus, any critical point (x, y) of F will necessarily satisfy the side condition $g(x, y) = 0$. That is, the two functions F and f have the same values at the critical points of F.

The introduction of the Lagrange multiplier l enables us to have a system of three equations in three unknowns, allowing effective computation. However, it is not implied that the computations are always fast and easy. Returning to the system of necessary conditions, we have from Eq. (3) that

$$l = -\frac{2x}{3}$$

and from Eq. (4),

$$l = -\frac{2y}{2} = -y.$$

Therefore,

$$-\frac{2x}{3} = -y$$

or

$$x = \frac{3y}{2}.$$

Substituting this value of x into Eq. (5) we get

$$\frac{9y}{2} + 2y = 13$$

or

$$\frac{13y}{2} = 13$$

and

$$y = \frac{2}{13} \cdot 13 = 2$$

Since $x = \frac{3}{2} y$ then

$$x = \frac{3}{2} \cdot 2 = 3$$

critical point (3, 2)

We now find sufficient conditions for extrema as defined in the Second Derivative Test for functions of two variables. Accordingly,

$$F_{xx} = 2, \quad F_{yy} = 2, \quad F_{xy} = 0.$$

Then
$$F_{xx} \cdot F_{yy} - F_{xy}^2 = 2 \cdot 2 - 0^2 > 0.$$

Since $F_{xx} > 0$, there exists a minimum for F, and thus for f, at the critical point (3, 2). The value of $f_{\min}$ is $f(3, 2) = 3^2 + 2^2 = 13$. We found that $l = -y$, and since $y = 2$ then $l = -2$. Now $F(x, y, l)$ has the value

$$F(3, 2, -2) = 9 + 4 - 2(9 + 4 - 13) = 13$$

which equals the value of $f(3, 2)$.

In General:

The method of Lagrange multipliers may be expressed as follows for a function of two variables:

(1) Given a function $f(x, y)$ to be optimized subject to the constraint $g(x, y) = 0$.

(2) Form the Lagrangian expression

$$\boxed{F(x, y, l) = f(x, y) + lg(x, y)}\,.$$

(3) Determine the critical points (a, b) which satisfy $F_x(a, b) = 0$, $F_y(a, b) = 0$, and $F_l(a, b) = 0$.

(4) Evaluate the difference function $D = F_{xx}(a, b) \cdot F_{yy}(a, b) - F_{xy}^2(a, b)$. The points (a, b) for which $D > 0$ are extreme points.

(5) If $F_{xx}(a, b) > 0$, then f has a relative minimum at (a, b). If $F_{xx}(a, b) < 0$, then f has a relative maximum at (a, b).

The power of the Lagrange multiplier method asserts itself in cases where the side condition is complicated. In particular, if it is inconvenient or impossible to solve the constraint for one variable in terms of another, then the value of the method becomes apparent. For example, if we were to optimize

$$f(x, y) = x^2 + y^2 \quad \text{subject to the constraint} \quad x^3 + y^3 = xy \tag{6}$$

it would be difficult to solve Eq. (6) for y as a function of x if we wished to optimize by former methods. In this case, the method of Lagrange multipliers is useful. For, even though the constraint cannot be expressed as $y = f(x)$, still the system of first partials under this method is usually solvable. Consider the following example:

EXAMPLE 1 Given total cost $C = 3w^2 + 2k^2 - 2wk - 2w - 5k + 5$ (in millions of dollars) where w is the total number of man-hours of work (labor) and k is the total capital outlay in millions of dollars. Minimize cost subject to the constraint $4w + 3k = 17$.

Solution It is inconvenient if not difficult to solve the side condition for k or w and substitute into the many-termed cost function. Since the multiplier method makes it unnecessary to do this, we find the method advantageous. And so we let

$$g(w, k) = 4w + 3k - 17$$

and let $F(w, k, l) = C(w, k) + lg(w, k)$

or $$F = 3w^2 + 2k^2 - 2wk - 2w - 5k + 5 + l(4w + 3k - 17).$$

Next we find the necessary conditions $F_w = 0$, $F_k = 0$, and $F_l = 0$ and solve simultaneously the system of first partials. Thus,

$$\begin{cases} F_w = 6w - 2k - 2 + 4l = 0 & (7) \\ F_k = 4k - 2w - 5 + 3l = 0 & (8) \\ F_l = 4w + 3k - 17 = 0. & (9) \end{cases}$$

Equation (7) minus Eq. (8) yields

$$8w - 6k + 3 + l = 0$$

from which

$$l = 6k - 8w - 3.$$

Substituting this value of l into Eq. (8) we have

$$4k - 2w - 5 + 18k - 24w - 9 = 0.$$

Simplifying results in

$$22k - 26w - 14 = 0$$

or $$11k - 13w - 7 = 0. \qquad (10)$$

We now solve Eqs. (9) and (10) simultaneously:

$$\begin{cases} 11k - 13w - 7 = 0 & (10) \\ 3k + 4w - 17 = 0. & (9) \end{cases}$$

The row operations $3 \cdot$ Eq. (10) $- 11 \cdot$ Eq. (9) give us

$$\begin{cases} 33k - 39w - 21 = 0 & (11) \\ -33k - 44w + 187 = 0. & (12) \end{cases}$$

Now the sum of Eqs. (11) and (12) yields

$$-83w + 166 = 0$$

or $$w = 2.$$

Placing this value of w into Eq. (10), we obtain

$$11k - 26 - 7 = 0$$

or $$k = 3.$$

Therefore, $(w, k) = (2, 3)$ is a critical point. Applying the tests for sufficiency we get

$$F_{ww} = 6, \quad F_{kk} = 4, \quad F_{wk} = -2.$$

Then $$F_{ww} \cdot F_{kk} - F_{wk}^2 = 6 \cdot 4 - (-2)^2 > 0. \qquad (13)$$

Since $F_{ww} > 0$ and the difference in Eq. (13) is positive, then the cost function has a minimum at (2, 3). This value $C_{\min}$ is given by

$$C(2, 3) = 12 + 18 - 12 - 4 - 15 + 5 = 4.$$

That is, when $w = 2$ or 20,000 man-hours, and $k = 3$ or \$3,000,000, then minimum cost $C = 4$ or \$4,000,000.

The method of Lagrange multipliers is applicable to functions having three or more variables and having more than one constraint. A multiplier is used for each constraint when two or more constraints are present. When optimizing a function of three variables, the Lagrange multiplier technique fails to tell us whether the critical points give a maximum, a minimum, or neither. In such a case, other criteria such as geometric analysis must be used in decision-making for optimality.

As a final example of this section we work Example 2 of the previous section by the multiplier method.

EXAMPLE 2 Minimize $S = xy + 2xz + 2yz$ subject to the constraint $V = xyz = 32$.

Solution Let $g = xyz - 32$ and introduce the Lagrangian form

$$F = xy + 2xz + 2yz + l(xyz - 32).$$

Then

$$\left.\begin{aligned} F_x &= y + 2z + lyz = 0 \quad (14)\\ F_y &= x + 2z + lxz = 0 \quad (15)\\ F_z &= 2x + 2y + lxy = 0 \quad (16)\\ F_l &= xyz - 32 = 0. \quad (17)\end{aligned}\right\}$$

Equation (14) minus Eq. (15) yields

$$y - x + (y - x)lz = 0$$

or

$$(y - x)(1 + lz) = 0.$$

Therefore,

$$y = x \tag{18}$$

and

$$lz = -1. \tag{19}$$

From Eq. (19) we get

$$l = -\frac{1}{z}. \tag{20}$$

Using $l = -1/z$ in Eq. (14) we have

$$y + 2z - y = 0$$

or

$$z = 0.$$

This value of z is not relevant since, by the nature of the problem, z is a dimension such that $z > 0$. Using $y = x$ from Eq. (18) in Eq. (16) we get

$$2x + 2x + lx^2 = 0.$$

Then

$$4x + lx^2 = 0$$

$$x(4 + lx) = 0$$

$$x = 0, \quad \text{not applicable,}$$

and $$l = -\frac{4}{x}.$$

This value of l, substituted in Eq. (15), yields

$$x + 2z - 4z = 0$$

Solving for z: $$x - 2z = 0$$

$$z = \frac{x}{2}. \tag{21}$$

Using $y = x$ from Eq. (18) and $z = x/2$ from Eq. (21) in Eq. (17), we obtain

$$x \cdot x \cdot \frac{x}{2} = 32.$$

Then $$x^3 = 64$$

or $$x = 4.$$

It follows that $$y = 4$$

and $$z = \frac{4}{2} = 2.$$

Thus, the critical point is $(x, y, z) = (4, 4, 2)$, at which point S is a minimum as was shown in Section 5.5.

PROBLEM SET 5.6

1. Total cost of output is given by $C = k^2 + 2w^2 - 2kw + 1$ (millions of dollars) where k and w are the amounts of inputs, capital and labor, respectively. If cost C is subject to the constraint $k - w = 1$, find the value of each input that results in minimum cost.
2. Minimize $H = x^2 + 3y^2 + 8$ subject to the side condition $2x + 4y = 14$.
3. Maximize sales $S = 94c + 106w - 5c^2 - 5w^2$, where expenses are planned so that $2c + 3w = 22$. Cost of materials is represented by c and the amount of man-hours of work by w.
4. Find the maximum of $h(x, y) = -4x^2 - 4y^2 - 32$ subject to the constraint $8x + 8y + 16 = 0$.
5. Determine the minimum value of the expense function $E(x, y) = 4.5x^2 + 4y^2 - 6xy$ subject to the constraint $3x + 2y = 30$. Assume that x denotes the cost of raw materials and y is the cost of shipping and handling.
6. Minimize the distance d from the origin (0, 0) to the line $x + 3y = 5$. [*Hint*: $d = \sqrt{x^2 + y^2}$; minimizing d is the equivalent of minimizing d^2.]
7. A profit function P is defined by

$$P(x, y) = \frac{5x}{4 + x} + \frac{4y}{5 + y} - x - y + 16$$

(in millions of dollars), where x is quantity of items sold and y is cost of production. If P is subject to the side condition $x + y = 15$, find P_{max}.

8. Work exercise 7 when

$$P(x, y) = \frac{2x}{6 + x} + \frac{3y}{4 + y} - 2x - 2y + 30$$

and the side condition is $2x + 2y = 28$.

9. Find the point on the plane $x + 2y - z = 3$ that is nearest the origin (0, 0, 0) and find the minimum distance. [*Hint*: Minimize $d^2 = x^2 + y^2 + z^2$ subject to $x + 2y - z = 3$. The critical point (x, y, z) obtained will furnish the coordinates for which $d_{\min} = \sqrt{x^2 + y^2 + z^2}$.]

10. Work problem 9 if the plane has the equation $2x - 3y + 4z = 22$.

11. A rectangular box is to be manufactured with no top and a volume of 27 ft^3. Find dimensions that will result in the most economical box if the base costs 20¢ per ft^2 and the sides cost 10¢ per ft^2. That is, form a cost equation from the surface-area equation and then minimize cost subject to the constraint $V = xyz = 27$. Assume that cost will have a minimum value at the critical point.

12. Work problem 11 if the box is to have a lid and a volume of 128 ft^3.

13. A company's total output (production) function is given by $P(x, y, z) = x^2 + y^2 + z^2 - 2x - 3y - 4z + 10^6$, where x, y, and z are elements of input. If P is constrained by $2x + 3y + 4z = 29$, find values for the input factors which will maximize output and find that maximum value. Assume that $P_{\max}$ occurs at the critical point obtained.

5.7 LEAST SQUARES (LINEAR REGRESSION)

Another important application of optimizing a function of two variables is the method of least squares. This application deals with the problem of fitting a linear function

$$y = mt + b$$

to a set of experimental readings

$$(t_1, y_1),\quad (t_2, y_2),\quad (t_3, y_3), \ldots, (t_n, y_n).$$

The line that best fits a set of points serves as a trend line which is easy to read and interpret. One may extrapolate beyond the line and thus predict future sales or profits or other needed information. Forecasts would be more likely to be valid for short terms only. We illustrate by example.

EXAMPLE 1 Consider a company which had profits (in millions of dollars) of 4 in 1970, 4 in 1971, 7 in 1972, and 6 in 1973. These data are shown in Figure 5.12 where $t = 1$ is associated with 1970, $t = 2$ with 1971, $t = 3$ with 1972, and $t = 4$ with 1973. A free-hand line (L in the figure) is drawn as a good "guess-timate" of the best-fitting line. The lengths of the segments AB, CD, EF, and GH represent deviations of points B, D, E, and H from the line. Since line L has the equation $y = mt + b$ and point B has co-

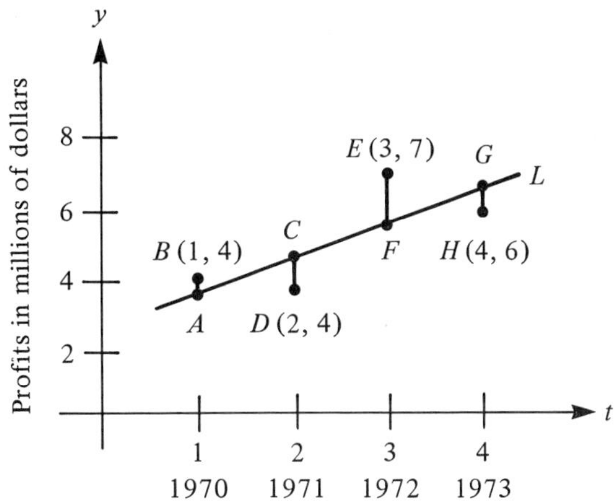

Figure 5.12

ordinates (1, 4) we may write the **deviation** AB as

$$d_{AB} = [4 - m(1) - b] = 4 - m - b.$$

Similarly, the deviations CD, EF, and GH are written as

$$d_{CD} = 4 - 2m - b$$
$$d_{EF} = 7 - 3m - b$$

and

$$d_{GH} = 6 - 4m - b.$$

In other words, for each value of t there is a pair of values of y, namely, y_{exp} and $mt_{\text{exp}} + b$ whose difference

$$y_{\text{exp}} - (mt_{\text{exp}} + b)$$

or

$$y_{\text{exp}} - mt_{\text{exp}} - b$$

we call a deviation. The sub-notation "exp" denotes a value obtained experimentally. The total of all these deviations is a measure of the "goodness of fit" of line L to the given experimental data. A perfect fit (generally not possible) results when the sum of these deviations is zero. Since some deviations could be positive and some negative, thus adding up to zero even when the line is a "bad fit," we resort to the following technique: We form a sum of the squares of the deviations. Then our objective is to find the line L for which the sum of the squares of the deviations is a minimum value. This sum will be nonnegative and can be zero only if line L is a perfect fit. The following tabular form exhibits the basic ingredients of the problem.

t_{exp}	y_{exp}	$y_{\text{exp}} - mt_{\text{exp}} - b$: Deviation	$(\text{Deviation})^2$
1	4	$4 - m - b$	$(4 - m - b)^2$
2	4	$4 - 2m - b$	$(4 - 2m - b)^2$
3	7	$7 - 3m - b$	$(7 - 3m - b)^2$
4	6	$6 - 4m - b$	$(6 - 4m - b)^2$

The method of least squares tells us to form the function of two variables

$$f(m, b) = d_{AB}^2 + d_{CD}^2 + d_{EF}^2 + d_{GH}^2$$

and to minimize f (see Figure 5.12). Accordingly, we have

$$f(m, b) = (4 - m - b)^2 + (4 - 2m - b)^2 + (7 - 3m - b)^2 + (6 - 4m - b)^2.$$

First, we find and solve the necessary conditions

$$f_m = 0 \quad \text{and} \quad f_b = 0.$$

And so we have

$$f_m = -2(4-m-b) - 4(4-2m-b) - 6(7-3m-b) - 8(6-4m-b) = 0$$

or

$$-114 + 60m + 20b = 0 \quad (1)$$

$$f_b = -2(4-m-b) - 2(4-2m-b) - 2(7-3m-b) - 2(6-4m-b) = 0$$

or

$$-42 + 20m + 8b = 0. \quad (2)$$

Then 3 times Eq. (2) minus Eq. (1) results in the system

$$\left.\begin{aligned} -126 + 60m + 24b &= 0 \\ 114 - 60m - 20b &= 0 \end{aligned}\right\}.$$

Adding and solving, we obtain

$$-12 + 4b = 0$$

or

$$b = 3.$$

Substituting this value of b into Eq. (2) we get

$$\begin{aligned} -42 + 20m + 24 &= 0 \\ 20m &= 18 \\ m &= \frac{18}{20} = \frac{9}{10}. \end{aligned}$$

Therefore the desired line L has the equation

$$y = \frac{9}{10}t + 3. \quad (3)$$

According to the method of least squares, Eq. (3) represents the line that best fits the given experimental data. We can check this result by invoking tests for sufficiency as follows:

$$f_{mm} = 60, \quad f_{bb} = 8, \quad f_{mb} = 20.$$

Then

$$\begin{aligned} f_{mm} \cdot f_{bb} - f_{mb}^2 &= 60 \cdot 8 - (20)^2 \\ &= 480 - 400 > 0. \end{aligned}$$

This result plus the fact that $f_{mm} > 0$ indicates a minimum for f at the critical point $(m, b) = (9/10, 3)$. The sufficiency test is not necessary, however, since the theory underlying the method of least squares offers proof that in problems of fitting a straight line the results always imply a minimum.

To predict the amount of profits for 1975, we substitute $t = 6$ in Eq. (3) and obtain

$$P(6) = \frac{9}{10}(6) + 3$$
$$= 5.4 + 3 \quad \text{or} \quad 8.4 \text{ million dollars.}$$

Keep in mind that we have minimized a function which represents the sum of the squares of the deviations. We have not minimized profits. Let us consider one more example in order to gain more familiarity with the method of least squares.

EXAMPLE 2 During 1949, the year of the start of a new "bull market" according to some analysts, the Dow-Jones Averages stood at 1.6 (hundreds). In 1953 the price reached a low of 2.6 (hundreds) and in 1956 it attained a high of 5.2 (hundreds). Find the trend line L that best fits these data and use the equation of L to predict prices for the averages in 1961 and 1966.

Solution Let $t = 1$ correspond to 1949, $t = 5$ to 1953 and $t = 8$ to 1956 (see Figure 5.13). We then have the three points P (1, 1.6), Q (5, 2.6) and R (8, 5.2). The deviations relative to P, Q, and R are, respectively,

$$d_1 = 1.6 - 1m - b$$
$$d_2 = 2.6 - 5m - b$$
and
$$d_3 = 5.2 - 8m - b.$$

We form the sum of the squares of the deviations, call it $f(m, b)$, and proceed to minimize f. Thus,

$$f(m, b) = (1.6 - m - b)^2 + (2.6 - 5m - b)^2 + (5.2 - 8m - b)^2.$$

Then
$$f_m = -2(1.6 - m - b) - 10(2.6 - 5m - b) - 16(5.2 - 8m - b)$$
$$= -112.4 + 180m + 28b = 0$$
or
$$-28.1 + 45m + 7b = 0. \tag{4}$$
$$f_b = -2(1.6 - m - b) - 2(2.6 - 5m - b) - 2(5.2 - 8m - b)$$
$$= -18.8 + 28m + 6b = 0$$
or
$$-9.4 + 14m + 3b = 0. \tag{5}$$

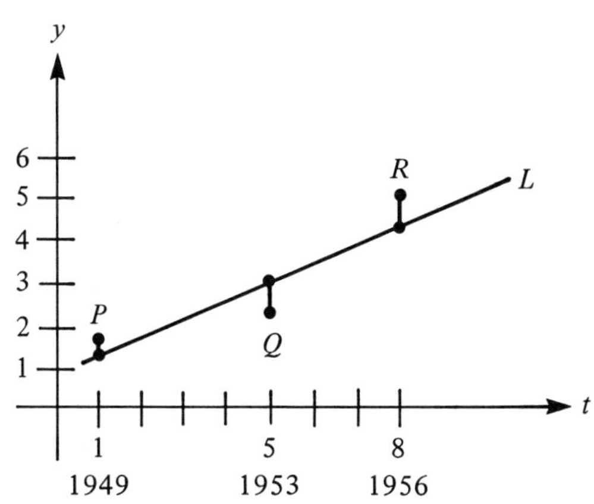

Figure 5.13

Now 3 times Eq. (4) minus 7 times Eq. (5) leads to the system

$$\left.\begin{aligned} -84.3 + 135m + 21b &= 0 \\ 65.8 - 98m - 21b &= 0 \end{aligned}\right\}.$$

Solving, we get

$$-18.5 + 37m = 0$$

$$m = \frac{18.5}{37} = \frac{1}{2}.$$

Substituting in Eq. (5) yields

$$-9.4 + 7 + 3b = 0$$

$$-2.4 + 3b = 0$$

$$b = \frac{2.4}{3} = 0.8.$$

Therefore, the line of best fit has the equation

$$y = \frac{1}{2}t + 0.8. \tag{6}$$

For 1961, $t = 13$, so that Eq. (6) becomes

$$y = \frac{1}{2}(13) + 0.8 = 7.3 \text{ hundreds}$$

or 730. A check of the records reveals that the averages did record a price of 730 in late 1961. For 1966, $t = 18$ and Eq. (6) becomes

$$y = \frac{1}{2}(18) + 0.8 = 9.8 \text{ hundreds}$$

or 980. The Dow-Jones Averages actually attained a price of 980 in early February, 1966, and rose to an all-time high close of about 995 a very short time later. Of course, there were times when extrapolations did not conform to the explicit behavior of the Dow-Jones Averages.

In the method of least squares the line which best fits the experimental data (t_i, y_i), $i = 1, 2, 3, \ldots$, is often referred to as the **line of regression** of y on t. In line with this designation the least squares method is called **simple regression analysis**. The method is being applied extensively and is used as a basis for certain problems in the comparatively new field of cybernetics.

PROBLEM SET 5.7

1. The following table shows the increase in price which a firm had to contend with in the period 1970–1972:

Year t	1970	1971	1972
Price y	\$2	\$3	\$5

(a) Find the line L of best fit to these data using the method of least squares.
(b) Use the equation of L to forecast the price in 1973.

2. The OTS Corp. had total sales of 3 million dollars in 1970, 5 million in 1971, and 4.5 million in 1972.

   (a) Find the line L that best fits these data by use of the least squares method.
   (b) Use the equation of L to predict a sales value for 1973.

3. Use the method of least squares to obtain the line $y = mx + b$ which best fits the given points (x, y): (2, 20), (2, 40), (3, 40), (4, 30).

4. The average price per ounce of gold on the world market was about \$40 in 1971, about \$60 in 1972, and about \$90 in 1973. What should be the average price per ounce of gold in 1974 according to the method of least squares?

5. In problem 4, if the price is \$60 in 1972, \$90 in 1973, and \$125 in 1974, use the least squares method to predict the price in 1975.

6. Find the line $y = mx + b$ of best fit for the data: (1, 20), (2, 15), (3, 20), and (4, 30).

7. If the cost of T-bone steaks was \$1.10 per lb. in 1970, \$1.50 in 1971, and \$1.80 in 1972, find the cost per lb. of T-bone steaks in 1973 by the method of least squares.

8. Find the line of regression for the data: (1, 1), (2, 3), (3, 2), (4, 5).

9. If the population of a country was 10 million in 1950, 14 million in 1960, and 16 million in 1970, find the country's population in 1980 by regression analysis.

10. Out of a possible 100 points, Rick scores 78 on Test I, 88 on Test II, and 92 on Test III. According to regression analysis, what score should Rick make on Test IV?

11. In problem 10 assume that for the following semester Rick makes 82 on Test 1, 90 on Test 2, and 88 on Test 3. Using the method of least squares, what score should Rick make on Test 4?

12. If profits P are 1 million (dollars) when advertising expense A is 100,000 (dollars). P is 1.4 million when A is 200,000, P is 1.8 million when A is 300,000, and P is 2 million when A is 400,000, find P when A is 500,000. Use the method of least squares.

13. In problem 12 find P when A is 500,000 (dollars) given that P is 1 million (dollars) when A is 100,000, P is 1.2 million when A is 200,000, P is 1.1 million when A is 300,000, and P is 1 million when A is 400,000.

5.8 REVIEW PROBLEMS

1. Sketch graphs of the following planes in (x, y, z) space:

   (a) $x + y = 1$ (b) $x + z = 2$
   (c) $y + z = 3$ (d) $x + y + z = 1$
   (e) $2x + 3y + 4z = 12$

2. Sketch graphs of the following functions in E^3:

   (a) $x^2 + y^2 = 4$ (b) $y = 4x^2$ (c) $z = x^2$
   (d) $y^2 = 4x$ (e) $z = 1 - x^2$ (f) $y = 4 - x^2$

3. Given a production function $P(k, l) = 2k^2 + 3l^2 - 2kl$ where k and l represent amounts of capital and labor inputs, respectively.

   (a) Find $P_k(2, 1)$ and $P_l(2, 1)$ (b) Find P_{kk}, P_{ll} and P_{lk}.

4. It is found that total revenue per acre of barley is given by $R(w, s) = 24w + 28s + 6ws$ and total cost per acre is $Q(w, s) = 6w^2 + 3s^2$, where w and s are expenditures per acre for labor and soil improvements, respectively.

   (a) Find the profit function $P(w, s)$
   (b) Find P_w and P_s at (2, 4)
   (c) Determine P_{ww}, P_{ss}, and P_{ws}.

5. Work problem 4 when $R(w, s) = w^2 + s^2 + ws - 5w + 2s - 10$ and $Q(w, s) = 2ws + 3w + 4s - 20$.

6. Find F_x and F_y for each of the following:

   (a) $F(x, y) = 2x^2 + \dfrac{1}{y} - \dfrac{1}{x}$ (b) $F(x, y) = x^2y^2 + \sqrt{xy}$

   (c) $F(x, y) = 2x^3y + 3xy^3$ (d) $F(x, y) = xy^2 + x^2y - \dfrac{1}{xy}$.

7. Determine G_{uu}, G_{vv}, and G_{uv} for:

   (a) $G(u, v) = ue^v + ve^u$ (b) $G(u, v) = e^{u-v} + (v/u)$
   (c) $G(u, v) = (2u + 3v)^3$ (d) $G(u, v) = \ln(1 - uv)$.

8. Find dy/dx if:

   (a) $xy^2 + x^2y^3 = 1$ (b) $2x^2y - 3xy^2 = xy + 1$

   (c) $\sqrt{x} + \sqrt{y} = xy$ (d) $\dfrac{y+1}{y} = \dfrac{x+1}{x}$

9. Find an equation of the line tangent to the graph of the given equation at the indicated points:

   (a) $x^2 + y^2 = 3xy - 1$; (1, 1) (b) $\ln(x + y) - xy = 2$; (2, −1)
   (c) $x^{2/3} + y^{2/3} = 5$; (1, 8) (d) $\sqrt{2x + 4y + 7} = 5$; (3, 3)

10. If the total sales of the TBR Co. are defined by $S(x, y) = x^2y^3$ and $x = t^2 + 1$, $y = t^3 - 2$, find ds/dt by the chain rule.

11. Work problem 10 when $S(x, y) = e^{x^2+y^2}$ and $x = 1/t$, $y = 1/t^2$.

12. Work problem 10 when $S(x, y) = \ln(2x - 3y)$ and $x = e^{2t}$, $y = e^{-3t}$.

13. Maximize the profit function defined by $P = 100 - 2u^2 + 4uv - 3v^2 + 4u + 6v$ where u and v are inputs of production.

14. Work problem 13 if $P = 24x + 10y - x^3 - y^2 - 27$.

15. Examine the function $z = 2xy - 5x^2 - 2y^2 + 4x + 4y - 4$ for maxima or minima.

16. A total cost function is given by $Q(x, y) = x^3 + y^3 - 3xy + 11$, where x is the number of man-hours of labor (in tens) and y is the number of pounds of type A steel (in hundreds) per run of production. What amount of each input should be used to minimize the cost per run?

17. For each of the following functions find the total derivative dz/dx (recall that $\left.\dfrac{dz}{dx} = \dfrac{\partial z}{\partial x} \cdot \dfrac{dx}{dx} + \dfrac{\partial z}{\partial y} \cdot \dfrac{dy}{dx}\right)$:

   (a) $z = x^2 + y^2 + 2$ and $y = 1 - x^2$
   (b) $z = \sqrt{x + y}$ and $y = x^2 + 2$
   (c) $z = e^{x-2y}$ and $y = \ln x$
   (d) $z = \ln(2x + 4y)$ and $y = \ln 2x$

18. A closed rectangular box is to have fixed volume of 8 ft^3. Find dimensions for minimum surface area.

19. Work problem 18 if the volume is 27 ft^3.

20. A rectangular box to contain 270 cm^3 is to be manufactured from two types of material. The cost of the material in the top and bottom is 10¢ per cm^2 and the cost of the material in the four sides is 8¢ per cm^2. Find the dimensions of the box so that the cost of materials is minimized.

21. A firm allocates \$30,000 for the manufacture and promotion of a new product. If x thousand dollars are spent on manufacturing and y thousand dollars on promotion, then the firm estimates $S(x, y) = 5x^2y$ units of the product will be sold. Find values of x and y so that sales are maximized.

22. Given the profit function $P(x, y) = 9 - \frac{2}{x} - \frac{3}{y} - \frac{x}{2} - \frac{y}{3}$ where x denotes the amount of radio advertising (in thousands of dollars), y is the amount of television advertising (in thousands of dollars), and P is in tens of thousands of dollars. Find maximum profit.

23. Use the method of Lagrange multipliers to find the critical point of $f(x, y) = 2x^2 + 3xy + 2y^2$ subject to the constraint $x - y = 8$.

24. Maximize the product of three positive numbers subject to the constraint that their sum is 60.

25. Let $C(x, y) = 4x^2 + 7y^2 + 2xy - 700$ define the total cost of producing x units of model A radios and y units of model B radios. A total of 66 units of radios are to be produced. How should production allocations be made in order to minimize total cost?

26. An epidemic of flu virus affected 2.5 million people in January, 3.5 million in February, and 3.1 million in March. Use the method of least squares to fit a line to the given data and predict the number of people affected in April, other factors remaining unchanged.

27. The Gross National Product (GNP) of a certain country was $\$(10)^8$ in 1960, $\$1.6 \times 10^8$ in 1965, and $\$1.8 \times 10^8$ in 1970. Predict the GNP for 1975 by the least squares method.

28. Find the least squares straight-line fit to the data: $(\frac{1}{2}, \frac{1}{2})$, $(\frac{7}{4}, \frac{9}{4})$, $(3, 3)$.

29. A firm has annual sales in a four-year period as shown in the table:

x: year	y: sales in thousands
1971	5
1972	10
1973	8
1974	12

Let 1971 be 1, 1972 be 2, etc. Find the least squares regression line and use it to predict the firm's sales for 1975 and 1976.

5.9 ENRICHMENT TOPICS AND PROBLEMS

1. Use L'Hospital's Rule to evaluate the following limits:

(a) $\lim_{x \to \infty} \frac{x}{e^x}$ **(b)** $\lim_{x \to 9} \frac{x^2 - 81}{x - 9}$

(c) $\lim_{x \to \infty} \frac{x^2}{e^x}$ (d) $\lim_{x \to 2} \frac{x^3 - 5x + 2}{x^4 + 6x^2 - 40}$

(e) $\lim_{x \to 0} \frac{e^x - 1 - x}{x^2}$.

2. The demand for product A is $p = 72 - 6x$ and for product B is $q = 80 - 10y$. The joint-cost function is $C(x, y) = x^2 + 2xy + 2y^2$. Maximize the profit. [*Hint*: profit P = revenue R − cost $C = px + qy - C$.]
3. Work problem 2 if $p = 20 - 2x$, $q = 15 - 2y$, and $C = x^2 + xy + y^2$.
4. If the demand functions are $p = 6 - 2x$ and $q = 7 - (y^2/2)$ for products A and B, and cost is $C(x, y) = 5 + 2x + y$, find maximum profit (use the method of problem 2).
5. Find the point on the plane $z = x$ that is closest to the point (1, 0, 0). [*Hint*: minimize $F(x, y, z) = (x - 1)^2 + (y - 0)^2 + (z - 0)^2$.]
6. Use Lagrange multipliers to find the critical point of $f(x, y, z) = x^2 + y^2 + z^2$ subject to the constraint $x - 2y + 3z - 14 = 0$.
7. For a certain firm, output is $1000z$ units priced at \$4 per unit. One input totals $1000x$ units at \$1 per unit and a second input is $1000y$ units at \$4 per unit. The profit function is $P = 4000z - 1000x - 4000y$ where the production function is $z = 5 - 1/x - 1/y$. Maximize P and evaluate $P_{\max}$. [*Hint*: By the multiplier method form the Lagrangian expression $F(x, y, z, l) = 4000z - 1000x - 4000y + l\left(\frac{1}{x} + \frac{1}{y} + z - 5\right)$. Assume that max P occurs at the critical point.]
8. Work problem 7 when $P = 900z - 200x - 100y$ and $z = 6 - (2/x) - (4/y)$.
9. The production function for the QMD Corp. is $z = \frac{9}{2} - \frac{4}{xy}$. One input totals $100x$ units priced at \$8 per unit and a second input totals $100y$ units priced at \$4 per unit. The amount of output is $100z$ units priced at \$32 each. Find $P_{\max}$ by methods of problem 7.
10. A certain type of house cost \$24,000 in 1950, \$32,000 in 1960, and \$46,000 in 1970. Find an equation of the line of best fit using the method of least squares, and use the equation to forecast the cost of the house in 1980.
11. Don makes a score of 81 on Test 1, 92 on Test 2, and 93 on Test 3. According to the least squares method, what should he score on Test 4?
12. Find the least squares straight-line fit to the data: (1, 1), (2, 2), (3, 1), (4, 2).
13. In a certain economy the consumer index was 0.35 in 1955, 0.55 in 1960, 1.05 in 1965, and 1.45 in 1970. Find the line of regression and use it to predict the consumer index for 1975.
14. Analogous to the differential of a function of one variable (if $y = f(x)$, then $dy = f'(x)\,dx$), there is the **differential of a function of two variables**: If $z = f(x, y)$, then $dz = f_x\,dx + f_y\,dy$. Just as $dy = f'(x)\,dx$ is useful as an approximation to Δy, so

$$dz = f_x\,dx + f_y\,dy$$

can be useful as an approximation to Δz. "Good" approximations result when the changes (or increments) of the independent variables are "sufficiently" small. Consider the rectangular container of length 2 meters having a square end of side 1 meter. If these measurements are accurate to within 0.01 m, what error can we expect in calculating the volume of the container?

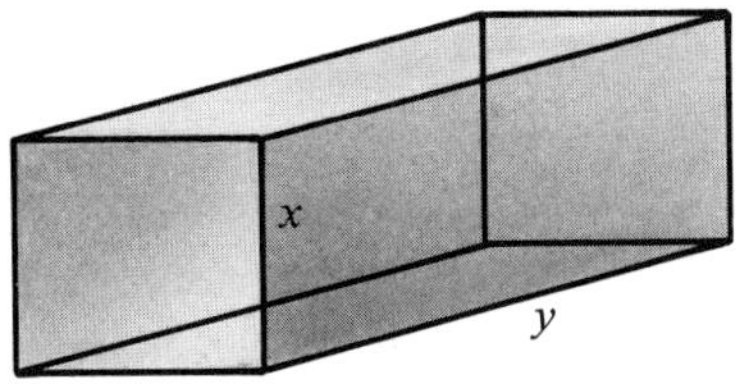

Figure 5.14

Solution From the figure, volume V is given by

$$V(x, y) = x^2 y.$$

To find the *actual* amount of possible error in the volume, we compute

$$\Delta V = V(x \pm dx, y \pm dy) - V(x, y).$$

The $\pm$ is used since we don't know whether the errors in measurement are positive or negative. When positive signs are used we get

$$\begin{aligned}\Delta V &= (1 + 0.01)^2(2 + 0.01) - (1)^2(2) \\ &\cong 0.0504 \text{ m}^3.\end{aligned}$$

When negative signs are chosen we obtain

$$\begin{aligned}\Delta V &= (1 - 0.01)^2(2 - 0.01) - (1)^2(2) \\ &\cong 0.0496 \text{ m}^3.\end{aligned}$$

We choose the signs in a way that results in the maximum absolute value for ΔV. Accordingly,

$$|\Delta V| \leq 0.0504.$$

The *approximate* amount of possible error in the volume is found by use of the differential. Thus, the differential of $V = x^2 y$ is

$$dV = 2xy\, dx + x^2\, dy.$$

Evaluated at $x = 1$, $dx = +0.01$, $y = 2$, $dy = +0.01$, this becomes

$$\begin{aligned}dV &= 2(1)(2)(0.01) + (1)^2(0.01) \\ &= 0.04 + 0.01 = 0.05 \text{ m}^3.\end{aligned}$$

Therefore, the approximate amount of error in the volume differs from the actual amount by only $(0.0504 - 0.05)$ or 0.0004 m^3.

15. Use the ideas and methods of problem 14 to find ΔV and dV if a side of the square end is 2 ft, the length is 1 ft, and measurements are accurate to within 0.02 ft.

16. Use differentials to approximate the value of $f(2.02, 1.01)$ if $f(x, y) = x^2 + 3xy + y^2$. [*Hint*: Let $x = 2$, $dx = 0.02$, $y = 1$, $dy = 0.01$. Then $f(2.02, 1.01) \cong f(x, y) + df(x, y)$.

17. Profits for the LTP Co. in 1975 are forecast to be $P(x, y) = x^{3/2}y^{3/2}$ subject to $x = 9$, $dx = 0.02$, $y = 4$, $dy = 0.01$. Use differentials to approximate the amount of profits in 1975.

18. Work problem 17 if $x = 4$, $dx = 0.01 = dy$, $y = 1$.

19. The PCR Corp. has a contract to make closed rectangular crates with dimensions 3 ft by 4 ft by 5 ft. The materials to be used in the construction cost 10¢ per ft^2. If each dimension has a possible error of 0.05 ft, use differentials to approximate the error in the cost per crate. [*Hint*: Cost $C = 20xy + 20xz + 20yz$, $dC = C_x\,dx + C_y\,dy + C_z\,dz$.]

20. Work problem 19 when the possible error in measuring each dimension is 0.02 ft.

21. Given that the approximate relative error in cost Q is defined by dQ/Q and the approximate percentage error is $(dQ/Q)(100)$. A certain cost function is defined by $Q = 4x^{3/2}y^{1/2}$ where x is the amount of man-hours of labor and y is the cost of raw materials. A 2% increase in x and a 4% decrease in y will produce what relative change and what percentage change in Q? [*Hint*: $\ln|Q| = \ln 4 + \frac{3}{2}\ln|x| + \frac{1}{2}\ln|y|$. Then $\frac{dQ}{Q} = \frac{3}{2}\cdot\frac{dx}{x} + \frac{1}{2}\frac{dy}{y}$ evaluated at the given conditions.]

22. Work problem 21 if the increase in x is 3% and the decrease in y is 5%.

23. In the manufacture of a ball bearing (spherical in shape) the diameter is controlled to within an error of not more than 0.08%. What relative error and percentage error in the volume will this produce?

24. Work problem 23 if the diameter can be controlled to within an error of not more than 0.10%.

25. If $\lim_{x\to a} f(x) = \infty$ and $\lim_{x\to a} g(x) = \infty$, then

$$\lim_{x\to a}\frac{f(x)}{g(x)} = \lim_{x\to a}\frac{f'(x)}{g'(x)}$$

by L'Hospital's Rule and the process may be repeated until a result, finite or infinite, is obtained (see section 2.9). For example,

$$\begin{aligned}\lim_{x\to\infty}\frac{x^2}{e^x}\left(\frac{\infty}{\infty}\right) &= \lim_{x\to\infty}\frac{D_x x^2}{D_x e^x}\\ &= \lim_{x\to\infty}\frac{2x}{e^x}\left(\frac{\infty}{\infty}\right)\\ &= \lim_{x\to\infty}\frac{D_x 2x}{D_x e^x}\\ &= \lim_{x\to\infty}\frac{2}{e^x} = 0.\end{aligned}$$

Apply L'Hospital's Rule to evaluate:

(a) $\lim_{x\to+\infty}\frac{\ln x}{x}$ **(b)** $\lim_{x\to 1}\frac{\ln(x-1)}{\ln(x^2-1)}$ **(c)** $\lim_{x\to\infty}\frac{e^{3x}}{x^3}$.

6 INTEGRAL CALCULUS

Antidifferentiation as a Practical Tool

IMPORTANT EQUATIONS

$$\int f(x)\,dx = F(x) + C$$

$$A = f(x_{i-1})\,\Delta x, \qquad i = 1, 2, 3, \ldots, n$$

$$\lim_{n\to\infty} \sum_{i=1}^{n} f(t_i)\,\Delta x_i = L = \int_a^b f(x)\,dx$$

$$\int_a^b f(x)\,dx = \int_a^b F'(x)\,dx = F(x)\Big]_a^b = F(b) - F(a)$$

$$\sum_{i=0}^{n} a_i x^i = a_0 + a_1 x + a_2 x^2 + \cdots + a_n x^n$$

$$(p + q)^n = \sum_{r=0}^{n} \binom{n}{r} p^r q^{n-r}$$

So far we have introduced and discussed some basic concepts of the differential calculus. In this chapter we introduce and discuss many fundamental ideas of the other main branch of calculus, namely, the **integral calculus**. We found that differential calculus deals with slopes of tangent lines. We will now find that integral calculus deals with areas of regions. The concept of slope is a "close-in" idea, focusing on a small part of a curve in the neighborhood of a point. In contrast, area is a "far-out" notion, concerned with boundaries and regions. Yet we will learn that these two main branches of calculus are "close relatives." Furthermore, the algebraic analyses of each have geometric interpretations.

At the start of our study of differentiation we may have entertained doubts as to how derivatives could be used in a practical sense. As we progressed our doubts were dispelled when we learned to apply derivatives in marginal analysis and in the optimization of sales, cost, profits, and other functions through max-min theory. Similarly, in this chapter we may at first wonder what the connection is between certain important real-world problems and the concept of area. It will be interesting to see how the solutions to these problems will depend on the ideas underlying the calculation of an area "under a curve."

6.1 THE INDEFINITE INTEGRAL

The evaluation of an integral involves a process called integration. We will use the process to evaluate (1) indefinite integrals, and (2) definite integrals. Let us peruse the ideas underlying (1) first, reserving the notions in (2) for later sections.

To integrate a function means to find an antiderivative of the function. Thus, indefinite integration and antidifferentiation are synonymous terms and will be used interchangeably. The connotation of the word "antidifferentiate" suggests a process that is the reverse of "differentiate." It suggests the reconstruction of a function whose rate of change is given.

In many business situations it often happens that the rate of change of a function (with respect to some independent variable) is known or can be easily found. The problem may then be to find the function whose rate of change, i.e., its derivative, is known. For example, a marginal cost function or a marginal revenue function may be known to a firm. Given this information, it might be necessary to reconstruct the total cost function and total revenue function for purposes of intelligent planning and decision-making. In line with this reasoning, suppose a marginal revenue function is given by

$$\frac{dR}{dx} = 2x$$

where x denotes the amount of production units and R is the total revenue. In order to find the revenue function R, we use our knowledge of derivatives plus our diagnostic ability and write

$$R = x^2.$$

However, we soon realize that this is only one of many possible answers such as

$$R = x^2 + 1,$$
$$R = x^2 - 2,$$
$$R = x^2 + 3,$$

and so on, since the derivative of each of these answers is $2x$. Accordingly, we write the answer as

$$R = x^2 + C \tag{1}$$

where C is any constant in the real number system. Thus,

$$\text{If} \quad \frac{dR}{dx} = 2x \quad \text{then} \quad R = x^2 + C \quad \text{because} \quad \frac{d}{dx}(x^2 + C) = 2x.$$

We say that x^2 is an **antiderivative** of $2x$. The answer in Eq. (1) denotes a whole family of functions since C may assume any real value. The derivatives of any two members of this family are identical, since any two functions have the same derivatives when the functions differ only by a constant. Graphs of R are shown in Figure 6.1 for the cases when the additive constant C equals -1, 0, and 1. The graphs have the same shape but different y-intercepts.

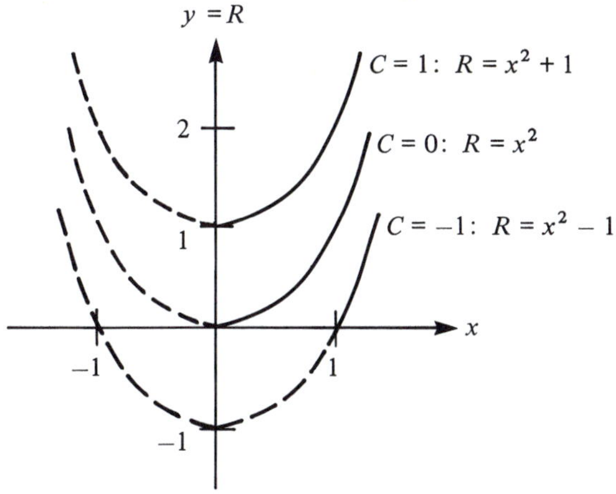

Figure 6.1

We can find a specific member R of the family $(x^2 + C)$ if we know the value of R for a certain value of x. Suppose that $R = 10$ when $x = 2$. Then

boundary condition

$$R = x^2 + C$$

becomes

$$10 = (2)^2 + C$$

or

$$C = 6.$$

Thus, for the initial (or boundary) condition $R = 10$ when $x = 2$, we have the unique solution

$$R = x^2 + 6.$$

The dotted portions of the graphs in Figure 6.1 indicate where revenue R is not relevant. For a particular value of the additive constant C we write other examples as follows:

An antiderivative of x^2 is $(x^3/3) + C$. (2)

An antiderivative of $2x^3$ is $(x^4/2) + C$. (3)

An antiderivative of $6x^{1/2}$ is $4x^{3/2} + C$. (4)

An antiderivative of $(2x + 1)^5$ is $\dfrac{(2x+1)^6}{12} + C$. (5)

To verify the correctness of each of these answers we need only differentiate the answer and compare the result with the given function. Thus, to verify the answer in Eq. (4) we perform the operations

$$\begin{aligned} D_x(4x^{3/2} + C) &= 4 \cdot (3/2)x^{1/2} + 0 \\ &= 6x^{1/2}, \quad \text{the given function.} \end{aligned}$$

In Eqs. (2) and (3) it is not very difficult to guess the answers. In Eq. (4) the task would be more challenging, while in Eq. (5) it would probably be impossible. This latter fact points to the need for formal methods that yield antiderivatives conveniently. In each case it is important to add the arbitrary constant C so that the most general form of the answer is exhibited. This additive constant is often referred to as the "constant of integration." This leads to a definition of the indefinite integral.

DEFINITION 1 *The **indefinite integral** of a function f with respect to x is given by*

$$\boxed{\int f(x)\,dx = F(x) + C}$$

where

(a) *$F(x)$ is an elementary function with the property $F'(x) = f(x)$;*

(b) *the symbol $\int$ is called the integral sign;*

(c) *$f(x)$ is called the integrand;*

(d) *dx symbolizes integration (antidifferentiation) with respect to x;*

(e) *C is the constant of integration.*

To check the result of an integration, we differentiate the result and show that the answer corresponds to the integrand. That is,

$$\frac{d}{dx}\left[\int f(x)\,dx\right] = f(x).$$

In Eqs. (1)–(5) we have by the integral notation:

$$\int 2x\,dx = x^2 + C \tag{1'}$$

$$\int x^2\,dx = (x^3/3) + C \tag{2'}$$

$$\int 2x^3\,dx = (x^4/2) + C \tag{3'}$$

$$\int 6\sqrt{x}\,dx = 4x^{3/2} + C \tag{4'}$$

$$\int (2x + 1)^5\,dx = \frac{(2x + 1)^6}{12} + C. \tag{5'}$$

Take special note of the fact that the antiderivative is a function whose derivative is the integrand. We do not imply that all integrands are functions of x. For example, we may write

$$\int q^t \, dq = \frac{q^{t+1}}{t+1} + C, \qquad t \neq -1,$$

or

$$\int [r(t) + s(t)] \, dt = \int r(t) \, dt + \int s(t) \, dt + C$$

or

$$\int \frac{dw}{w} = \ln|w| + C, \quad \text{and so on.}$$

For complicated integrands we need more than the ability to guess the answer. Certain formulas are at our disposal for finding antiderivatives of a large class of functions. We list some of them.

I. The integral of a power n of $u(x)$ is u to the $(n + 1)$ power all divided by $n + 1$, provided $n \neq -1$:

$$\int u^n \, du = \frac{u^{n+1}}{n+1} + C, \qquad n \neq -1.$$

EXAMPLE 1 The rate of supply is given by $u(x) = x^3$. Find the supply function $S(x)$.

Solution

$$S(x) = \int x^3 \, dx$$
$$= \frac{x^4}{4} + C.$$

II. The integral of a constant k times a function is k times the integral of the function:

$$\int ku(x) \, dx = k \int u(x) \, dx + C$$

EXAMPLE 2

$$\int \frac{3}{2} x^2 \, dx = \frac{3}{2} \int x^2 \, dx$$
$$= \frac{3}{2} \cdot \frac{x^3}{3} + C$$
$$= \frac{x^3}{2} + C.$$

III. The integral of the sum or difference of two (or more) functions is the sum or difference of the integrals of the functions:

$$\int [u(x) \pm v(x)] \, dx = \int u(x) \, dx \pm \int v(x) \, dx + C$$

where the two constants of integration are incorporated as one.

EXAMPLE 3 If marginal cost is known to be $MQ = 2x^2 - 4x + 3$, find $Q(x)$ if $Q = 11/3$ when $x = 1$.

Solution

$$Q = \int MQ\,dx$$
$$= \int (2x^2 - 4x + 3)\,dx$$
$$= \frac{2x^3}{3} - 2x^2 + 3x + C \qquad \text{[by I, II, and III]}.$$

For the given boundary conditions,

$$\frac{11}{3} = \frac{2}{3} - 2 + 3 + C$$
$$\frac{11}{3} = \frac{5}{3} + C$$

or
$$C = 2.$$

Therefore,
$$Q(x) = \frac{2x^3}{3} - 2x^2 + 3x + 2,$$

the desired cost function.

IV. The integral of a power n of $u(x)$ is the natural logarithm of u if $n = -1$:

$$\int u^n\,du = \ln|u| + C \quad \text{if } n = -1$$

or
$$\int \frac{du}{u} = \ln|u| + C$$

EXAMPLE 4 $\int \frac{2\,dx}{3x} = \frac{2}{3}\int x^{-1}\,dx = \frac{2}{3}\int \frac{dx}{x} = \frac{2}{3}\ln|x| + C.$

Warning: Do not write $\int x^{-1}\,dx = x^{-1+1}/(-1+1) + C$ and do not compound your error by calling this answer 1 or 0 or some other misinterpretation!

At this stage the integral $\int(2x + 1)^5\,dx$ in Eq. (5) may still be haunting the reader. Let us clarify the situation.

EXAMPLE 5 Evaluate $\int(2x + 1)^5\,dx$.

Solution We are immediately aware that none of formulas I–IV explains the integral of a binomial raised to a power. What to do? As in many other mathematical dilemmas, we revert to a known procedure. Accordingly, we make a **single variable substitution** for $(2x + 1)$ and apply formula I:

Let
$$u = 2x + 1.$$

Then the differential of u is

$$du = 2\,dx \qquad [\text{recall } du = f'(x)\,dx.]$$

and
$$\frac{du}{2} = dx.$$

Making the appropriate substitution and evaluating, we get

$$\begin{aligned}\int (2x+1)^5\,dx &= \int u^5 \cdot \frac{du}{2}\\ &= \frac{1}{2}\int u^5\,du\\ &= \frac{1}{2}\cdot\frac{u^6}{6}+C\\ &= \frac{1}{12}u^6+C\\ &= \frac{1}{12}(2x+1)^6+C.\end{aligned}$$

Observe that the dx in the given integral is useful for effecting a change of variables. To verify that our answer is indeed the correct answer, we differentiate it:

$$\begin{aligned}D_x \frac{1}{12}(2x+1)^6 + C &= \frac{1}{12}\cdot(6)(2x+1)^5(2)+0\\ &= (2x+1)^5\end{aligned}$$

which is the integrand of the original integral, as we wished to show.

PROBLEM SET 6.1

1. Evaluate each indefinite integral:

(a) $\int (3x-4)\,dx$ **(b)** $\int \left(\frac{x}{3}+\frac{5}{3}\right)dx$

(c) $\int (x^2+2)\,dx$ **(d)** $\int (1-2x^2)\,dx$

(e) $\int (4x^2-x+1)\,dx$ **(f)** $\int (1+2x-6x^2)\,dx$

2. If marginal revenue is given by $MR = 100x - 8x^2$ and total revenue $R =$ \$600 when $x = 3$, find $R(x)$.

3. The marginal cost function for a commodity is given by $MQ = 10 - 12x + 15x^2$ where x is the amount of output. Find the total cost function $Q(x)$ if $Q =$ \$136 when $x = 2$.

4. If the rate of change of u with respect to t is $\frac{t}{2} - \frac{t^2}{3}$ and $u = 0$ when $t = 0$, find u as a function of t.

5. A certain marginal profit function is known to be defined by $f(q) = 4q^3 + 3q^2$. Find profit $P(q)$ if $P =$ \$44 when $q = 2$ units.

Evaluate the following indefinite integrals:

6. $\int (2x-1)\,dx$ **7.** $\int \left(2-\frac{1}{u}\right)du$

8. $\int (x^3+x^2)\,dx$ **9.** $\int (2-3t^2+4t^3)\,dt$

10. $\int \left(h^4 + \frac{2}{h}\right) dh$

11. $\int \frac{x^2 - x + 3}{x}\, dx$

12. $\int \frac{2x^3 - 3x^2}{2x}\, dx$

13. $\int (x - 2)^2\, dx$

14. $\int (2t + 1)^2\, dt$

15. $\int (1 - 2v)^3\, dv$

16. $\int (u^{1/2} - 2)^2\, du$

17. $\int \left(x^4 - x^3 + \frac{3}{\sqrt{x}}\right) dx.$

18. Assume that marginal selling expense is defined by $MS = \sqrt{2q + 1}$ where q denotes the quantity of units sold. Find the selling expense function $S(q)$ if $S = 20$ when $q = 4$.

19. In problem 18 find $S(q)$ if $MS = \sqrt{3q - 2}$ and $S = 10$ when $q = 2$.

20. A certain marginal demand function is given by $MD = 10/(x + 1)$ where x denotes amount of a commodity demanded. Find the demand function $D(x)$ if $D = 1$ when $x = 0$. [*Hint*: Let $u = x + 1$, then evaluate $\int(10/u)\, du$.]

21. Work problem 20 when $MD = 10/(5x + 1)$.

22. If marginal demand is defined by $MD = 4/x^2, x \neq 0$, where x denotes quantity demanded, find the demand function $D(x)$ if $D = 10$ when $x = 2$.

23. A marginal supply function is given by $MS = q^2/2 + 1$ where q is the quantity of a commodity. Find $S(q)$ if $S = 52$ when $q = 6$.

6.2 SIMPLE DIFFERENTIAL EQUATIONS

In this section we discuss simple differential equations and their solutions. Applications to real-world business and economic situations will be introduced wherever feasible. But first we need a definition.

DEFINITION 1 *Equations of the form $dy/dx = h(x)$ and $dy/dx = g(x, y)$, which specify the derivative as a function of x or as a function of x and y, are called* ***differential equations****. For example, $dy/dx = 2x$ and $dy/dx = x/\sqrt{y}$ are differential equations. These are examples of* ***first-order*** *differential equations.*

Second-order and third-order (and higher) differential equations exist also, but we will not pursue them in this section. The equation

$$\frac{d^2y}{dx^2} + 2x\frac{dy}{dx} + 3xy = 0$$

is an example of a differential equation of second order. The highest-order derivative in the equation is the second; thus the name, second-order differential equation.

THEOREM 1 *The differential equation $dy/dx = f(x)$ has a solution $y = F(x)$ provided*

(a) *$F(x)$ is a differentiable function;*

(b) $\frac{d}{dx}F(x) = f(x)$.

That is, to solve for y we write

$$dy = f(x)\,dx.$$

Integrating,

$$\int dy = \int f(x)\,dx$$

$$y = F(x) + C.$$

If two or more integrations are involved in a solution, all constants of integration are "collected" into one and written by convention at the end of the right-hand expression.

Interpretations:

(1) The integral of $1 \cdot dy$ is y because the derivative of y with respect to y is 1.

(2) The integral of $f(x)$ is $F(x) + C$ because the derivative of $[F(x) + C]$ with respect to x is $f(x)$.

Or:

(1) The integral of dy is y because the differential of y is dy.

(2) The integral of $f(x)$ is $F(x) + C$ because the differential of $[F(x) + C]$ is $f(x)\,dx$.

EXAMPLE 1 Solve the differential equation $dy/dx = 4x^3/3$.

Solution Multiplying both sides of the given equation by dx ($\neq 0$) we get differential y equal to:

$$dy = \frac{4}{3}x^3\,dx.$$

Integrating,

$$\int dy = \frac{4}{3}\int x^3\,dx$$

or

$$y = \frac{4}{3} \cdot \frac{x^4}{4} + C$$

$$y = \frac{x^4}{3} + C.$$

When x and y occur in a differential equation in such a way that we can separate the variables effectively, a solution can be obtained. The next example illustrates.

EXAMPLE 2 Find y if $dy/dx = 2xy$ and $y = 5$ when $x = 0$.

Solution We effect a "separation of variables" by multiplying both sides of the given equation by dx,

$$dy = 2xy\,dx,$$

then dividing both sides by y:

$$\frac{dy}{y} = 2x\,dx, \qquad y \neq 0.$$

Thus we have the y term with the dy and the x term with the dx. Integrating both sides of the equation, we have

$$\int \frac{dy}{y} = 2\int x\,dx$$

or

$$\ln|y| = x^2 + C.$$

To solve for y we apply the definition

$$\log_b N = x \leftrightarrow N = b^x.$$

Accordingly,

$$\begin{aligned} y &= e^{x^2+C} \\ &= e^{x^2} \cdot e^C && [\text{since } a^{m+n} = a^m \cdot a^n] \\ &= Ae^{x^2} && [A = e^C]. \end{aligned}$$

Substituting the given initial condition $y = 5$ when $x = 0$, we obtain

$$5 = Ae^0 = A.$$

Therefore, $\quad y = 5e^{x^2}$, the desired solution.

EXAMPLE 3 A company's marginal cost function is defined by $dQ/dq = 1000/\sqrt{q-1}$ where $q\ (>1)$ denotes the quantity of items manufactured. Find the total cost function $Q(q)$ if $Q = \$13{,}000$ when $q = 26$.

Solution We multiply both sides of the given equation by dq and express $1/\sqrt{q-1}$ as $(q-1)^{-1/2}$. This yields the integrable form

$$dQ = 1000(q-1)^{-1/2}\,dq.$$

Integrating, we have

$$\int dQ = 1000\int (q-1)^{-1/2}\,dq.$$

Let $u = q - 1$. Then $du = dq$ and

$$\begin{aligned} Q &= 1000\int u^{-1/2}\,du \\ &= 1000\,\frac{u^{1/2}}{1/2} + C \\ &= 2000\sqrt{u} + C. \end{aligned}$$

Resubstituting $(q - 1)$ for u, we get

$$Q = 2000\sqrt{q-1} + C. \tag{1}$$

When the boundary conditions $Q = \$13{,}000$ and $q = 26$ are substituted in Eq. (1), we have

$$\begin{aligned} 13{,}000 &= 2000\sqrt{26-1} + C \\ &= 10{,}000 + C \end{aligned}$$

or

$$C = 3000.$$

This value of C is placed in Eq. (1), yielding

$$Q(q) = 2000\sqrt{q-1} + 3000,$$

the desired total cost function.

EXAMPLE 4 A mathematical model for predicting the national debt was developed by a Mr. Domar. The underlying hypothesis stated that marginal national income is a constant,

$$\frac{dI}{dt} = A \tag{2}$$

where A is a constant representing an amount of dollars per year. A second assumption stipulated that an annual rate of increase in the national debt N varies directly as the national income. Using k as the constant of proportionality we write

$$\frac{dN}{dt} = kI. \tag{3}$$

From Eq. (2) we have

$$dI = A\,dt.$$

Integration of both sides results in

$$\int dI = A\int dt$$

or

$$I = At + C_1.$$

Applying the initial condition $I = I_0$ when $t = 0$, we have

$$I_0 = A(0) + C_1$$

or

$$C_1 = I_0$$

and

$$I = At + I_0.$$

This value of I is substituted in Eq. (3) so that

$$\frac{dN}{dt} = k(At + I_0)$$

or

$$dN = k(At + I_0)\,dt.$$

Integrating, we get

$$\int dN = k\int (At + I_0)\,dt$$

or

$$N = k\left(\frac{At^2}{2} + I_0t\right) + C_2. \tag{4}$$

Let us assume that $N = N_0$ when $t = 0$. Then

$$N_0 = k(0) + C_2$$

or

$$C_2 = N_0.$$

Substituting this value of C_2 in Eq. (4) gives

$$N = k\left(\frac{At^2}{2} + I_0 t\right) + N_0.$$

This model is a good example of a practical application of differential equations in economic analysis.

PROBLEM SET 6.2

Solve the differential equations 1–6:

1. $\dfrac{dy}{dx} = x^2 - 2$

2. $\dfrac{dy}{du} = 2u + \dfrac{3}{u^2}$

3. $\dfrac{dy}{dt} = \dfrac{4t}{y}$

4. $\dfrac{du}{dt} = (t + 1)^3$

5. $\dfrac{dy}{dx} = \sqrt{xy}$

6. $\dfrac{dy}{dz} = \dfrac{2z - 1}{3y^2 + 1}$

7. If the gross national product G grows at a rate proportional to the size of the national capital stock s, then $dG/ds = ks$, k is constant. Find $G(s)$ if $G = 10k$ when $s = 4$.

8. A company finds that the rate of decrease of sales with respect to price varies directly with sales volume and inversely with sales price. That is, $ds/dp = -2s/p$, where s is sales volume and p is the unit sales price. Find $\ln|s|$ if $s = 1$ when $p = 2$.

9. Suppose that marginal profit P is a constant M. That is, $dP/dq = M$ where q is quantity of production. Let $P = 1$ when $q = 1$. Furthermore, suppose that marginal sales vary directly with profits, i.e., $dS/dq = P$. Let $S = 2$ when $q = 0$. Find S as a function of q.

10. A marginal profit function is defined by

$$\frac{dP}{dc} = -\frac{500}{c^2}$$

where c denotes the total cost of items produced.

(a) Find the total profit function $P(c)$ if $P = \$120$ when $c = \$25$.
(b) Find c in (a) if $P = \$150$.

11. In problem 10 let $dP/dc = -800/(4c + 10)^2$. Find $P(c)$ if $P = \$400$ when $c = \$10$.

Solve the following differential equations subject to the initial conditions as indicated.

12. $\dfrac{dy}{dx} = \dfrac{1 + x}{1 + y}$, $\quad y = 0$ when $x = 1$.

13. $\dfrac{dy}{dt} = \dfrac{1 + 3t}{1 + 3y}$, $\quad y = 1$ when $t = 0$.

14. $\dfrac{du}{dv} = v(u + 1)$, $\quad u = 0$ when $v = 0$.

15. $\dfrac{dy}{dx} = 2x(1 - y)^2, \qquad y = 0$ when $x = 1$.

16. $\dfrac{dy}{dx} = \left(\dfrac{2 + y}{3 - x}\right)^2, \qquad y = 0$ when $x = 2$.

17. $\dfrac{dy}{dt} = \dfrac{2t - 1}{1 - 4y}, \qquad y = 1$ when $t = 2$.

6.3 MORE APPLICATIONS

Antidifferentiation can be used to advantage in finding functions whose second derivatives are known. The procedure involves two steps: (1) reconstructing the first derivative from the second derivative, and (2) reconstructing the desired function from its first derivative.

EXAMPLE 1 The rate of change of a marginal cost function is defined by $d^2Q/dx^2 = 2x + 2$, where x is the number of units manufactured. Find the total cost function $Q(x)$ if $dQ/dx = 1$ and $Q = 7/3$ when $x = 1$.

Solution

$$\frac{dQ}{dx} = \int (2x + 2)\, dx$$
$$= x^2 + 2x + C_1.$$

When $dQ/dx = 1$ and $x = 1$, we have

$$1 = 1 + 2 + C_1 \quad \text{or} \quad C_1 = -2.$$

Then
$$\frac{dQ}{dx} = x^2 + 2x - 2$$
or
$$dQ = (x^2 + 2x - 2)\, dx.$$

Integrating, we get

$$\int dQ = \int (x^2 + 2x - 2)\, dx$$
or
$$Q = (x^3/3) + x^2 - 2x + C_2.$$

When $Q = 7/3$ and $x = 1$, we have

$$\frac{7}{3} = \frac{1}{3} + 1 - 2 + C_2 \quad \text{or} \quad C_2 = 3.$$

So $\quad Q(x) = (x^3/3) + x^2 - 2x + 3,\quad$ the desired equation.

EXAMPLE 2 A certain acceleration is given by $a(t) = 3t^2 + 2t$, where t is the number of units of time. Find the velocity $v(t)$ and the law of motion $s(t)$ if $v = 2$ and $s = 1$ when $t = 1$.

Solution

$$a = \frac{dv}{dt} = 3t^2 + 2t$$
$$dv = (3t^2 + 2t)\,dt$$
$$\int dv = \int (3t^2 + 2t)\,dt$$
$$v = t^3 + t^2 + C.$$

Substituting the initial condition $v = 2$ at $t = 1$, we have

$$2 = 1 + 1 + C$$

or

$$C = 0$$

Then

$$v(t) = t^3 + t^2 = \frac{ds}{dt}$$

and

$$\int ds = \int (t^3 + t^2)\,dt$$
$$s = \frac{t^4}{4} + \frac{t^3}{3} + C_1.$$

When $s = 1$ and $t = 1$, we obtain

$$1 = \frac{1}{4} + \frac{1}{3} + C_1$$

or

$$C_1 = 1 - \frac{7}{12} = \frac{5}{12}.$$

Therefore

$$s(t) = \frac{t^4}{4} + \frac{t^3}{3} + \frac{5}{12},$$

the desired law of motion.

EXAMPLE 3 The rate of change of a marginal profit function is given by $P''(x) = (1 + x)^3$. Find the profit function $P(x)$ if $P' = 5$ and $P = 18/5$ when $x = 1$.

Solution

$$P'(x) = \int (1 + x)^3\,dx$$
$$= \int u^3\,du \qquad \text{where } u = 1 + x.$$

Then

$$P'(x) = \frac{u^4}{4} + C_1$$
$$= \frac{(1 + x)^4}{4} + C_1 \qquad [\text{replacing } u \text{ by } 1 + x].$$

If $P'(x) = 5$ when $x = 1$, we have

$$5 = \frac{(1 + 1)^4}{4} + C_1$$
$$= 4 + C_1 \quad \text{or} \quad C_1 = 1.$$

So

$$P'(x) = \frac{(1 + x)^4}{4} + 1.$$

To find $P(x)$ we integrate $P'(x)$,

$$\begin{aligned} P(x) &= \int P'(x)\,dx \\ &= \int \left[\frac{(1+x)^4}{4} + 1\right] dx \\ &= \int \frac{u^4}{4}\,du + \int 1 \cdot dx \qquad [u = 1 + x] \\ &= \frac{u^5}{4(5)} + x + C_2 \end{aligned}$$

or $$P(x) = \frac{(1+x)^5}{20} + x + C_2 \qquad [\text{replacing } u \text{ by } 1 + x].$$

When $P = 18/5$ and $x = 1$ we obtain

$$\begin{aligned} \frac{18}{5} &= \frac{(1+1)^5}{20} + 1 + C_2 \\ &= \frac{8}{5} + 1 + C_2 \quad \text{or} \quad C_2 = 1. \end{aligned}$$

Therefore, $$P(x) = \frac{(1+x)^5}{20} + x + 1,$$

the desired profit function.

EXAMPLE 4 The rate of change of a certain marginal demand function is defined by $d^2D/dx^2 = 10/x^2$, $x > 0$. Find dD/dx if $dD/dx = -5$ when $x = 1/2$. Then continue the antidifferentiation and find $D(x)$ if $D = 5$ when $x = 1$.

Solution

$$\begin{aligned} \frac{dD}{dx} &= \int \frac{10}{x^2}\,dx \\ &= 10 \int x^{-2}\,dx \\ &= 10\,\frac{x^{-1}}{-1} + C_1 \\ &= -\frac{10}{x} + C_1. \end{aligned}$$

At the given boundary condition $dD/dx = -5$ when $x = 1/2$ we get

$$\begin{aligned} -5 &= -\frac{10}{1/2} + C_1 \\ &= -20 + C_1 \quad \text{or} \quad C_1 = 15. \end{aligned}$$

Then $$\frac{dD}{dx} = -\frac{10}{x} + 15.$$

Continuing, $$\begin{aligned} D(x) &= \int \left(-\frac{10}{x} + 15\right) dx \\ &= -10 \ln|x| + 15x + C_2. \end{aligned}$$

When $D = 5$ and $x = 1$, we have

$$\begin{aligned} 5 &= -10 \ln 1 + 15(1) + C_2 \\ &= 15 + C_2 \quad \text{or} \quad C_2 = -10. \end{aligned}$$

Therefore, $$D(x) = -10 \ln|x| + 15x - 10,$$

the desired demand function.

In each example above the second derivative was given. We integrated twice in order to find the function with the known second derivative. After the first integration the initial (or boundary) conditions were used to find a particular member of the family of answers. This procedure was repeated in going from the first derivative of f to f itself.

PROBLEM SET 6.3

Find y if its second derivative is given as follows in problems 1–6. In each case $dy/dx = y = 1$ when $x = 1$.

1. $\dfrac{d^2y}{dx^2} = 2$
2. $\dfrac{d^2y}{dx^2} = 3 - 4x$
3. $\dfrac{d^2y}{dx^2} = 2x^2 - x$
4. $\dfrac{d^2y}{dx^2} = x^3 + x^2$
5. $\dfrac{d^2y}{dx^2} = \dfrac{6}{x^2}$
6. $\dfrac{d^2y}{dx^2} = \dfrac{x + 1}{x^3}$
7. The rate of change of a marginal profit function is known to be $d^2P/dx^2 = 2x - 3x^2$. Find the profit function $P(x)$ if $dP/dx = P = 1$ when $x = 1$.
8. If the rate of change of the slope of a price function is $d^2S/dx^2 = -1/(x + 1)^2$ and $dS/dx = 2$, $S = 1$ when $x = 0$, find $S(x)$.
9. Work problem 8 if $dS/dx = 1$ and $S = 2$ when $x = 0$.
10. A marginal stock price is changing at the rate $P''(t) = 9t^{-3/2}$ where t is time in years. Let $P'(t) = 41$ when $t = 4$.
    - **(a)** Find $P(t)$ if $P = 18$ when $t = 4$.
    - **(b)** What is the price of the stock at the end of 9 years?
11. Find the answer to (b) of problem 10 if $P'(t) = 31$ and $P = 8$ when $t = 4$.

Determine R if its second derivatives are as follows:

12. $\dfrac{d^2R}{dx^2} = \sqrt{x + 2}$
13. $\dfrac{d^2R}{dt^2} = 3t^2 + 4t + 5$
14. $R''(u) = \dfrac{1}{(1 - 2u)^2}$
15. $R''(y) = (3y - 2)^4$
16. $T''(r) = 4r^3 - 6r^2$
17. $H''(q) = 27(1 - 3q)^8$

6.4 OTHER SUBSTITUTIONS

There are many different types of rates of change. Since integrands of integrals represent rates of change, we must investigate these types if we wish to perform the integration. We will introduce only those types whose antiderivatives are not very complicated.

EXAMPLE 1 Evaluate $\int x(x^2 + 1)^3\, dx$.

Solution We could cube the $(x^2 + 1)$, multiply the result by x, and integrate the four terms. But this is a lengthy procedure. Besides, this approach would be awful if the power were 33 instead of 3! So we employ a convenient, single-variable substitution whose acquaintance we made in previous sections. Accordingly, let $u = x^2 + 1$. Then $du = 2x\,dx$ and $du/2 = x\,dx$. The given integral now assumes the form:

$$\int u^3 \cdot \frac{du}{2} = \frac{1}{2}\int u^3\, du.$$

Integrating, we get
$$= \frac{1}{2}\frac{u^4}{4} + C.$$

Simplifying, we have:
$$= \frac{1}{8}(x^2 + 1)^4 + C.$$

EXAMPLE 2 Find $\displaystyle\int \frac{2t\,dt}{\sqrt{1 - 3t^2}}$.

Solution Write the integral as $2\int(1 - 3t^2)^{-1/2} t\, dt$. Letting $u = 1 - 3t^2$, then $du = -6t\,dt$ and $-du/6 = t\,dt$. Substituting and integrating, we have

$$2\int u^{-1/2}\left(-\frac{du}{6}\right) = -\frac{2}{6}\int u^{-1/2}\, du$$
$$= -\frac{1}{3}u^{1/2}(2) + C$$
$$= -\frac{2}{3}\sqrt{1 - 3t^2} + C \qquad [u = 1 - 3t^2].$$

EXAMPLE 3 A marginal cost function is represented by $MQ = x\sqrt{x - 1}$. Find cost $Q(x)$.

Solution I
$$Q = \int x\sqrt{x - 1}\, dx.$$

Let $u = \sqrt{x - 1}$. Then

$$u^2 = x - 1, \qquad 2u\,du = dx, \qquad x = u^2 + 1.$$

Substituting in the integral we get

$$Q = \int (u^2 + 1) \cdot u \cdot 2u\, du$$
$$= 2\int (u^4 + u^2)\, du.$$

Integrating and simplifying yields:

$$Q = 2\left(\frac{u^5}{5} + \frac{u^3}{3}\right) + C.$$

Since $u = (x - 1)^{1/2}$, then $u^5 = (x - 1)^{5/2}$ and $u^3 = (x - 1)^{3/2}$. So

$$Q(x) = 2\left[\frac{(x - 1)^{5/2}}{5} + \frac{(x - 1)^{3/2}}{3}\right] + C.$$

Solution II Let $u = x - 1$. Then $du = dx$ and $x = u + 1$. Making the appropriate substitutions we get

$$\begin{aligned} Q &= \int (u + 1) \cdot u^{1/2}\, du \\ &= \int (u^{3/2} + u^{1/2})\, du \\ &= u^{5/2} \cdot (2/5) + u^{3/2} \cdot (2/3) + C \\ &= (2/5)(x - 1)^{5/2} + (2/3)(x - 1)^{3/2} + C \end{aligned}$$

(replacing u by its equivalent $(x - 1)$), which agrees with the answer obtained by method I. Observe that the answer is a family of curves when no initial conditions are given.

EXAMPLE 4 Suppose that marginal supply is defined by $MS = t/(t + 1)^3$. Find the general supply function $S(t)$.

Solution

$$S = \int \frac{t}{(t + 1)^3}\, dt.$$

Let $u = t + 1$. Then $du = dt$ and $t = u - 1$. Using these substitutions in the integral, we have

$$\begin{aligned} S &= \int \frac{u - 1}{u^3}\, du \\ &= \int (u^{-2} - u^{-3})\, du \\ &= \frac{u^{-1}}{-1} - \frac{u^{-2}}{-2} + C \\ &= -\frac{1}{u} + \frac{1}{2u^2} + C \\ &= -\frac{1}{t + 1} + \frac{1}{2(t + 1)^2} + C \qquad \text{[replacing } u \text{ by } t + 1\text{]}. \end{aligned}$$

EXAMPLE 5 A sales function is changing at the rate $MS = 1/(1 + \sqrt{x})$. Find $S(t)$, the general sales function.

Solution

$$S = \int \frac{dx}{1 + \sqrt{x}}.$$

Let $u = \sqrt{x}$. Then $u^2 = x$, $2u\, du = dx$. The integral now becomes a function of u:

$$S = \int \frac{2u\,du}{1 + u}$$
$$= 2\int \left(1 - \frac{1}{1 + u}\right) du$$

$\left(\text{since the quotient } \frac{u}{u + 1} = 1 - \frac{1}{1 + u} \text{ by long division}\right).$

Note: When a quotient of two polynomials exhibits the degree of the numerator as greater than or equal to the degree of the denominator, we then divide so as to try to obtain an integrable form. In this case, the degree of the numerator equals that of the denominator, and the degree is 1.

Integrating, we get

$$S = 2(u - \ln|1 + u|) + C$$

or $S(x) = 2(\sqrt{x} - \ln|1 + \sqrt{x}|) + C$ [replacing u by its equivalent $\sqrt{x}$].

EXAMPLE 6 The marginal productivity of a certain item is given by $MP = x^2/(x - 2)$. Find productivity $P(x)$ without boundary conditions.

Solution

$$P = \int \frac{x^2}{x - 2}\,dx.$$

Since the integrand is a quotient of two polynomials and the degree of the numerator is greater than that of the denominator, we divide and get

$$\frac{x^2}{x - 2} = x + 2 + \frac{4}{x - 2}.$$

Then
$$P = \int \left(x + 2 + \frac{4}{x - 2}\right) dx$$

or
$$P(x) = \frac{x^2}{2} + 2x + 4\ln|x - 2| + C.$$

PROBLEM SET 6.4

Given the following marginal functions, find the indicated function by anti-differentiation:

1. $MP = x(x^2 - 1)^3;\ P(x)$

2. $MQ = \dfrac{t}{\sqrt{1 - 2t^2}};\ Q(t)$

3. $MR = x\sqrt{x + 1};\ R(x)$

4. $MS = \dfrac{t}{(t + 1)^2};\ S(t)$

5. $MT = \dfrac{1}{1 - \sqrt{x}};\ T(x)$

6. $MU = \dfrac{x^2}{x - 1};\ U(x)$

7. $MV = t(t + 1)^5;\ V(t)$

8. $MW = \dfrac{x^2 + 2}{x + 1};\ W(x)$

9. $MX = \dfrac{u}{u^2 + 1};\ X(u)$

10. $MZ = (x^2 + 2x)(x + 1);\ Z(x)$

11. $MF = (x^2 - 6x)^2(x - 3); F(x)$
12. $MG = (x^2 - 4x)^3(x - 2); G(x)$
13. $MH = (t^3 - 3t)^3(t^2 - 1); H(t)$
14. $MP = x^2(3x^3 + 4)^5; P(x)$
15. $MP = \dfrac{1 + \sqrt{x}}{\sqrt{x}}; P(x)$
16. $MR = \dfrac{t - 1}{t^2 - 2t - 1}; R(t)$

6.5 AREA

In this section we will consider the specific problem of calculating the areas of regions bounded above by the graph of a simple curve, below by the horizontal axis, and laterally by two vertical lines. Figure 6.2 illustrates such a region R, whose area will be referred to as "area under a curve." The calculation of an area under a curve has so many interesting and practical applications that it is worthwhile to discuss the underlying ideas at some length and depth.

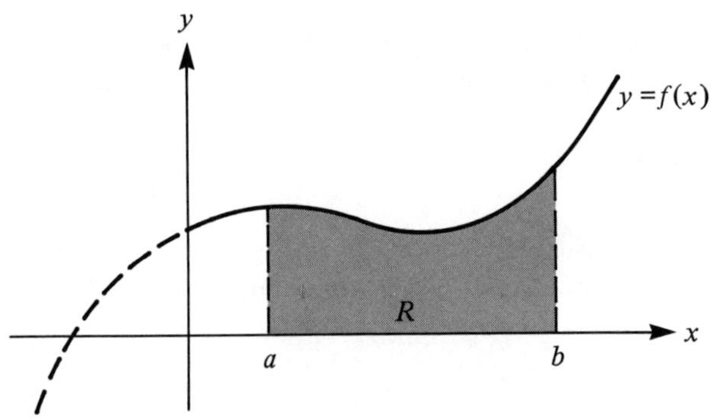

Figure 6.2

6.5.1 Approximating Area by Rectangles

In Figure 6.2, if region R is bounded above by some linear function $y = mx + K$, then R could be a triangle, a rectangle, or a trapezoid (see Figure 6.3). The area in these three cases can be easily calculated by methods learned in elementary geometry. However, if f is not a linear

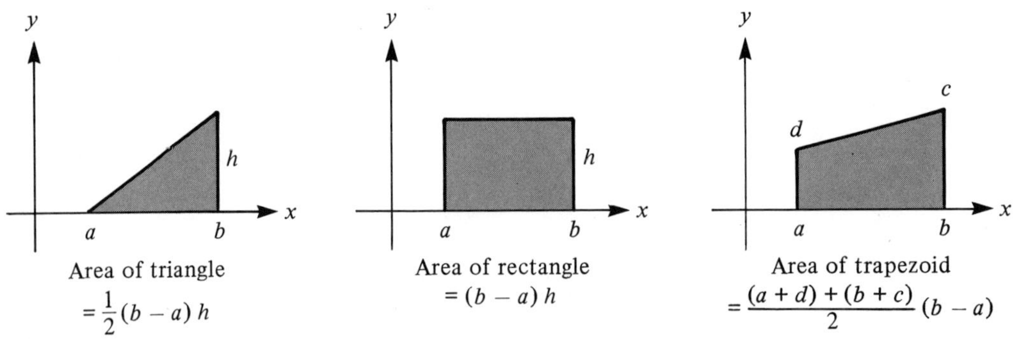

Figure 6.3

function, how do we find the area of region R? We proceed initially with approximation methods.

The Method of Inscribed Rectangles

Suppose f is continuous and nonnegative on the interval $a \le x \le b$ and R is a bounded region as shown in Figure 6.2. We divide the interval $[a, b]$ into subintervals by inserting $(n - 1)$ points between a and b. These points are labeled $x_1, x_2, x_3, \ldots, x_{n-1}$. Points a and b are represented by x_0 and x_n, respectively (see Figure 6.4). The subintervals $[x_i, x_{i-1}]$, $i = 1, 2, 3, \ldots, n$, need not be equal, but for convenience, we will work with equal subintervals. That is, we will use

$$x_1 - x_0 = x_2 - x_1 = \cdots = x_n - x_{n-1} = \Delta x = \frac{b - a}{n}$$

where $\quad a = x_0 < x_1 < x_2 < \cdots < x_{n-1} < x_n = b.$

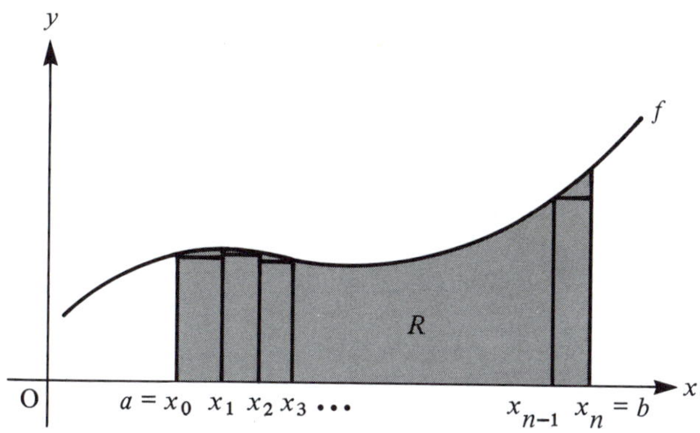

Figure 6.4

On each subinterval we construct a rectangle whose height is the *minimum* value of f on that subinterval and whose width is Δx. We say that rectangles constructed in such a manner are **inscribed rectangles**. The sum of their areas is *less* than the true area of R. A typical inscribed rectangle (enlarged) is shown in Figure 6.5. The area A of each inscribed

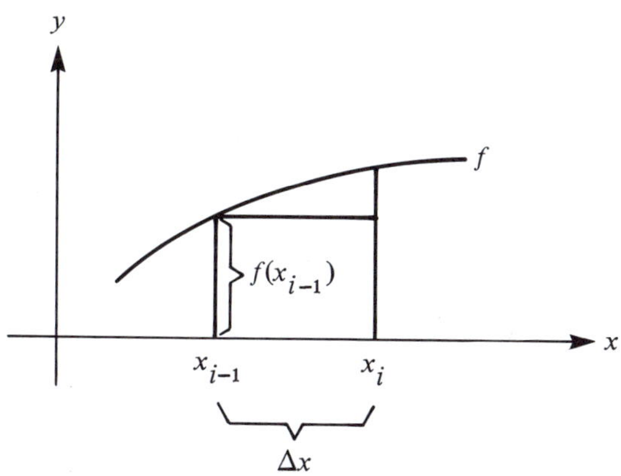

Figure 6.5

rectangle is its height times its width or

$$\boxed{A = f(x_{i-1}) \cdot \Delta x, \qquad i = 1, 2, 3, \ldots, n}.$$

The sum of all such areas approximates (underestimates) the true area of R.

If we make the widths of the subintervals smaller and smaller, it should be intuitively clear that the approximation of area will get better and better. This is so because more rectangles would cover a larger part of the given region R. Thus, a limiting process seems to be involved here; and letting the widths of the subintervals approach zero seems to be a logical move to improve the approximation. Equivalently, we can make the approximation as close as we want by using a sufficiently large number, n, of rectangles. In other words, the area "under a curve" should be the limit of the sum of the areas of approximating rectangles as their number n increases without bound, that is, as $n \to \infty$. Later, we will see that this is so. This limiting process is defined to be a **definite integral** and is written thus:

$$\lim_{n\to\infty} [f(x_0)\,\Delta x + f(x_1)\,\Delta x + \cdots + f(x_{n-1})\,\Delta x] = \int_a^b f(x)\,dx.$$

Many practical applications follow from this unique characterization of definite integrals as limits of sums.

Let us illustrate the method of inscribed rectangles by an example.

EXAMPLE 1 Consider the region R bounded by the graph of the marginal sales function $f(x) = x^2/10$, $x = 5$, $x = 10$, and $y = 0$ (see Figure 6.6). We wish to approximate the area of R using inscribed rectangles and $n = 5$ subintervals. Then

$$\Delta x = \frac{b-a}{n} = \frac{10-5}{5} = 1.$$

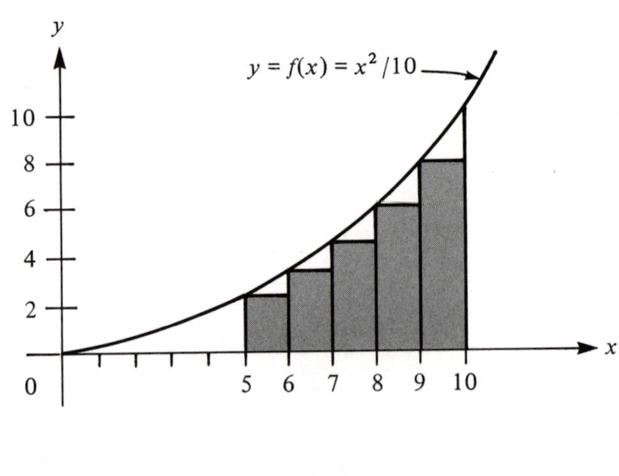

Figure 6.6

The sum of the areas of the five inscribed rectangles is given by

$$\begin{aligned} A_{\text{I}} &= f(5)\cdot 1 + f(6)\cdot 1 + \cdots + f(9)\cdot 1 \\ &= \frac{5^2}{10} + \frac{6^2}{10} + \cdots + \frac{9^2}{10} \\ &= \frac{1}{10}[25 + 36 + 49 + 64 + 81] \\ &= \frac{1}{10}(255) = 25.5 \text{ units}^2. \end{aligned}$$

The true area of R can be shown to be $29\frac{1}{6}$ units^2. Thus, our answer underestimates the true area by about $29.167 - 25.5 = 3.667$ units^2.

By a similar procedure we can estimate the area by use of circumscribed rectangles.

The Method of Circumscribed Rectangles

EXAMPLE 2 We apply this method to region R of Example 1. Figure 6.7 exhibits region R with circumscribed rectangles constructed on the 5 subintervals.

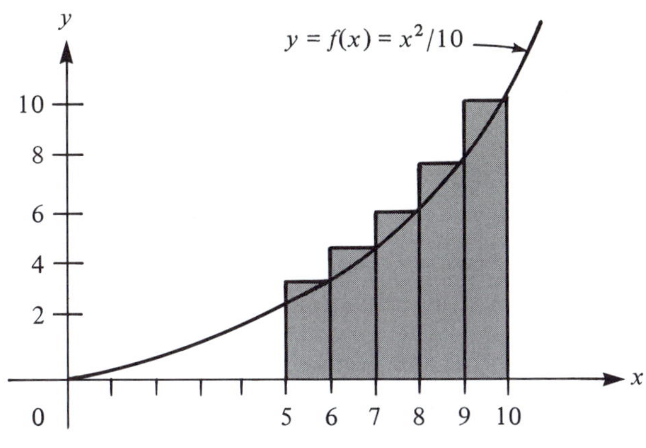

Figure 6.7

On each subinterval we construct a rectangle whose height is the *maximum* value of f on that subinterval and whose width is $\Delta x = (b - a)/n$ or $(10 - 5)/5 = 1$ in this case. Such rectangles are called **circumscribed rectangles**. The area of each rectangle is given by

$$A = f(x_i)\,\Delta x, \qquad i = 1, 2, 3, \ldots, n.$$

A typical circumscribed rectangle (enlarged) is shown in Figure 6.8. The sum of all such areas *overestimates* the true area of R and, in this example,

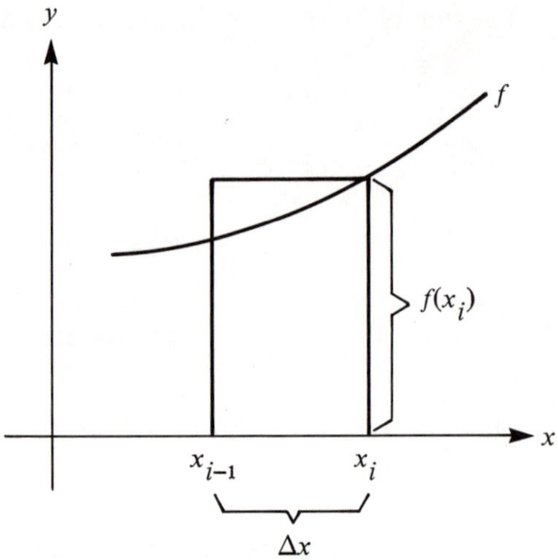

Figure 6.8

is given by

$$A_C = f(6)\cdot 1 + f(7)\cdot 1 + \cdots + f(10)\cdot 1 \qquad \left[f(x) = \frac{x^2}{10}\right]$$

$$= \frac{6^2}{10} + \frac{7^2}{10} + \cdots + \frac{10^2}{10}$$

$$= \frac{1}{10}(36 + 49 + 64 + 81 + 100)$$

$$= \frac{1}{10}(330) = 33 \text{ units}^2.$$

This answer is about $33 - 29.167 = 3.833$ units^2 more than the true area of R.

There is another method for approximating areas which makes use of the midpoints of subintervals.

The Midpoint Method

EXAMPLE 3 Let us apply this method to region R of Examples 1 and 2. In so doing, we will have a result to compare with the answers in the two previous methods. Figure 6.9 gives us a picture of region R with the heights of rectangles at the midpoints of the subintervals. From inspection it appears that the approximation of area by this method should be better than the two previous methods. This turns out to be correct.

On each subinterval we construct a rectangle whose height is $f\left(\frac{x_i + x_{i-1}}{2}\right)$ and whose width is $\Delta x = (b - a)/n$ or $(10 - 5)/5 = 1$ in this case. A typical rectangle (enlarged) is shown in Figure 6.10. The area of each rectangle is

$$A = f\left(\frac{x_i + x_{i-1}}{2}\right)\cdot \Delta x$$

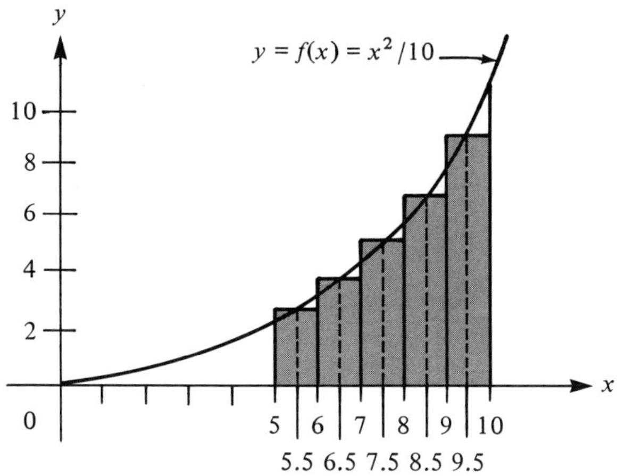

Figure 6.9

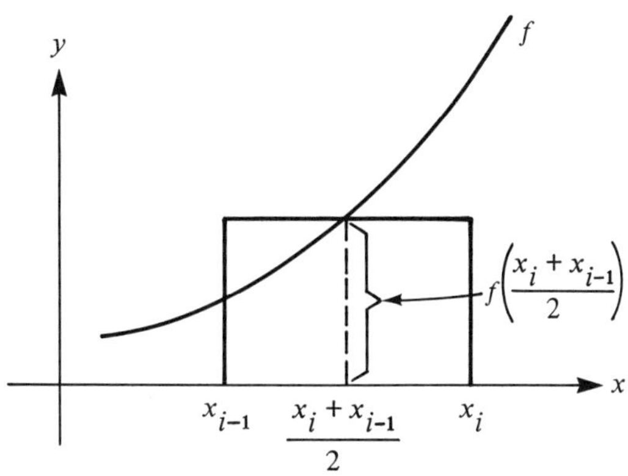

Figure 6.10

and the sum of all such areas in this example is given by

$$\begin{aligned} A_M &= f(5.5) \cdot 1 + f(6.5) \cdot 1 + \cdots + f(9.5) \cdot 1 \\ &= \frac{(5.5)^2}{10} + \frac{(6.5)^2}{10} + \cdots + \frac{(9.5)^2}{10} \\ &= \frac{1}{10}(30.25 + 42.25 + 56.25 + 72.25 + 90.25) \\ &= \frac{1}{10}(291.25) = 29.125 \text{ units}^2 \end{aligned}$$

This answer underestimates the true area by about $29.167 - 29.125 = 0.042$ units^2, a much better estimate of the true area than that of either of the two previous methods.

In this section we dealt with the approximation of area by the use of a finite number of rectangles. This approach will help motivate the subsequent definition of area "under a curve" as the limit of a sum.

PROBLEM SET 6.5.1

1. A firm's marginal cost function is represented by $f(x) = x^2$.
   (a) Sketch a graph of f.
   (b) Consider the interval $[1, 2]$ and divide the interval into four subintervals, each of length $(2 - 1)/4 = 1/4$.
   (c) Draw the inscribed rectangles and calculate the sum of their areas.
   (d) Find by what amount the answer in (c) underestimates the true area, $7/3$ units^2.
2. Work problem 1 using circumscribed rectangles.
3. A company estimates that the annual rate of savings accruing from the use of new equipment will be given by $MS = 1/t$, where t denotes the number of years after purchase of the equipment. Savings S is in thousands of dollars.
   (a) Sketch a graph of MS.
   (b) Consider the interval $[1, 4]$ and divide the interval into five equal subintervals.
   (c) Draw the circumscribed rectangles and calculate the sum of their areas.
   (d) By what amount does the result in (c) overestimate the true area $2 \ln 2 \cong 1.38$?
4. Work problem 3 using inscribed rectangles.
5. Work problem 1 by the midpoint method.
6. Work problem 3 using the midpoint method.
7. A marginal demand function is defined by $MD = 1/x^2$, $x > 0$.
   (a) Sketch a graph of MD.
   (b) Consider the interval $[1, 2]$ and divide the interval into four equal subintervals.
   (c) Draw the inscribed rectangles and calculate the sum of their areas.
   (d) By what amount does the answer in (c) underestimate the true area, $1/2$ units^2?
8. Work problem 7 using the midpoint method.
9. The rate of change of demand for a certain commodity is given by the marginal demand function $MD = 4 - x^2$.
   (a) Sketch a graph of MD.
   (b) In the interval $[0, 2]$ construct four inscribed rectangles of equal bases and calculate the sum of their areas.
   (c) How does the answer in (b) compare with the true area ($16/3$ units^2)?
10. Work problem 9 using circumscribed rectangles.
11. Work problem 9 by the midpoint method.
12. A marginal supply function is defined by $S(x) = \sqrt{x}$. Use six subintervals in $[0, 6]$ to find area by the method of:
    (a) inscribed rectangles
    (b) circumscribed rectangles
    (c) midpoint method.
13. The rate of change of demand for a commodity is given by $MD = 10/(x + 1)$. Use four subintervals in $[0, 8]$ to find the area by the method of:
    (a) inscribed rectangles
    (b) midpoint method.

6.5.2 Finding Exact Areas by Calculus. The Definite Integral

In the following discussion we attempt to develop an intuitive understanding of how the concept of limit plays a vital role in the calculation of true area (as opposed to approximate area).

You have at some time probably worked with the area of a circle, whose formula is $A = \pi r^2$, r denoting the circle's radius. This area was found by the ancient Greeks in the following ingenious manner (Figure 6.11 illustrates geometrically).

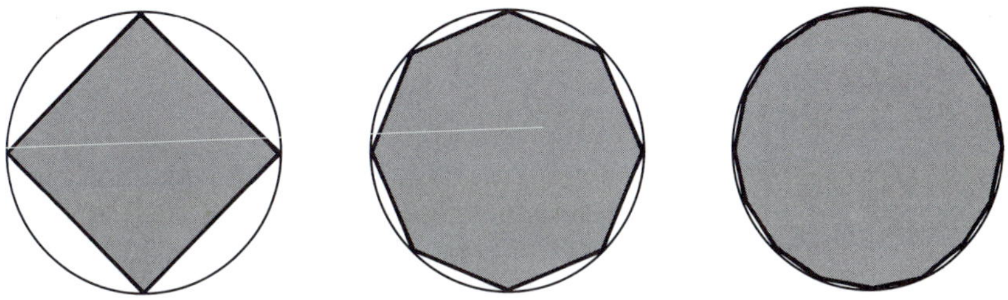

Figure 6.11 Inscribed polygon of $n = 4$ sides; Inscribed polygon of $n = 8$ sides; Inscribed polygon of $n = 16$ sides

We consider a sequence of circles with inscribed polygons successively having an increasing number of sides. The area of each inscribed polygon underestimates the true area of the circle. However, as the number of sides becomes larger and larger, the approximation to the area of the circle becomes better and better. If we imagine the number of sides n approaching infinity, the limit of the area of the inscribed polygon will be the area of the circle. We will not go through the mechanics of calculating this limiting value. The process is tedious and involves certain concepts not yet used in this text. Besides, we have more convenient methods at our disposal which we will soon introduce.

The area of the circle discussed above can also be shown to be the limiting value of the area of circumscribed polygons. In this case the area overestimates the area of the circle by smaller and smaller values as the number of sides gets larger and larger. As the number of sides n of the circumscribed polygon is increased, the polygon fits tighter and tighter about the circle. If $n \to \infty$ then the area of the circumscribed polygon $\to$ the area of the circle.

Thus, the concept of a limit is basic to the calculation of area. We extend the limit concept to the "area under a curve" by developing the ideas which underlie the definite integral. Area is one of many important applications of the notion of **definite integral**.

(1) Let a function $f \geq 0$ be continuous on the closed interval $[a, b]$.
(2) Select the points

$$a = x_0 < x_1 < x_2 < \cdots < x_n = b,$$

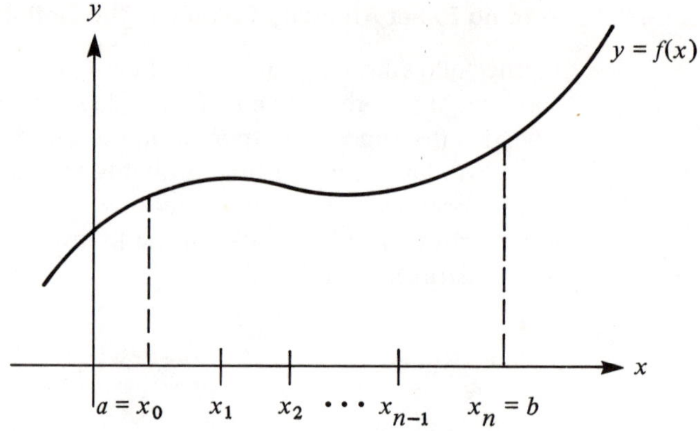

Figure 6.12

as illustrated in Figure 6.12. These points constitute a subdivision of $[a, b]$ into n subintervals, not necessarily all equal, each subinterval having length

$$\Delta x_i = x_i - x_{i-1}, \qquad i = 1, 2, 3, \ldots, n.$$

The largest of the Δx's is called the *mesh* of the subdivision and is denoted by

$$\text{mesh} = \max \Delta x_i = \|\Delta\|, \qquad i = 1, 2, 3, \ldots, n.$$

(3) Let t_i (read "tee sub i") be an arbitrary number in a typical subinterval such that

$$x_{i-1} \le t_i \le x_i, \qquad i = 1, 2, 3, \ldots, n$$

(refer to Figure 6.13). Next form the sum

$$S_n = f(t_1)\,\Delta x_1 + f(t_2)\,\Delta x_2 + \cdots + f(t_n)\,\Delta x_n,$$

which can be written in more compact notation as

$$S_n = \sum_{i=1}^{n} f(t_i)\,\Delta x_i. \tag{1}$$

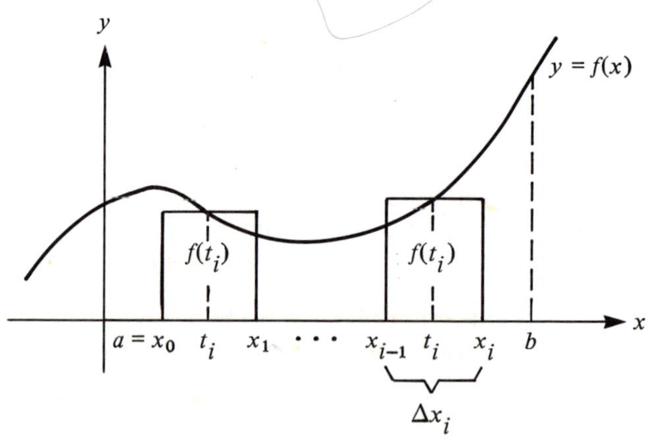

Figure 6.13

The symbol Σ is the Greek letter "sigma," commonly used to conveniently abbreviate the sum of many terms. In the Greek alphabet the letter Σ has the same connotation as the letter S (for the first letter in "sum") in the English alphabet. A brief explanation of the mechanics of "sigma notation" is in order. When

$$\begin{aligned} i &= 1, \text{ we have } & f(t_1)\,\Delta x_1, & \\ i &= 2, \text{ we have } & f(t_2)\,\Delta x_2, & \\ i &= 3, \text{ we have } & f(t_3)\,\Delta x_3, & \quad \text{and so on.} \end{aligned}$$

We will have more to say about sigma notation in a later section.

The sum S_n in Eq. (1) is called a **Riemann sum** for the given function $f(x)$ and the closed interval $[a, b]$. The name is due to the eminent mathematician G. B. F. Riemann, 1826–1866. This Riemann sum is clearly an approximation to the area under the curve of $f(x)$ over $[a, b]$, $f \geq 0$. Each subinterval Δx_i may be considered as the width of a rectangle of height $f(t_i)$, so that the typical product

$$f(t_i)\,\Delta x_i$$

represents the area of the rectangle (see Figure 6.13 again). In developing the steps leading to the concept of a definite integral we are (a) employing approximation methods which tie in with previous discussion; (b) continuing geometric interpretations which exercise our intuitive senses; and (c) developing a base from which the limit concept follows naturally as the basic feature underlying the concept of definite integral.

As n increases, the sum S_n of all these areas becomes an increasingly better approximation to the area under the curve of f in the interval $a \leq x \leq b$. If we denote the exact area by the number L, then the limit of S_n as n becomes infinite is defined to be the number L provided the mesh $\|\Delta\|$ approaches zero. In symbols we write this as

$$\lim_{n \to \infty} S_n = L, \qquad \text{provided } \|\Delta\| \to 0.$$

That is, the Riemann sum $\sum_{i=1}^{n} f(t_i)\,\Delta x_i$ approaches a limit which is *defined to be the area under the curve* of $f \geq 0$ from $x = a$ to $x = b$.

Note: The subdivision of the interval $[a, b]$ becomes finer and finer as n increases in size. In this process of "refinement," $n \to \infty$ as the mesh $\|\Delta\| \to 0$, but the converse is not necessarily true. This fact is proven in calculus texts which are concerned with a rigorous treatment of the integral calculus. Since our treatment is less theoretical and thus more practical, we will usually state significant ideas without detailed proofs.

Let us generalize the basic notions discussed above.

DEFINITION 1 *If there exists a number L such that the Riemann sum approaches L when the mesh approaches zero for all possible patterns of subdivision of the interval*

*$[a, b]$, then the limiting value L is the **definite integral** of the continuous function $f \geq 0$ from $x = a$ to $x = b$. Symbolically, we write*

$$\boxed{\lim_{n\to\infty} \sum_{i=1}^{n} f(t_i)\,\Delta x_i = L = \int_a^b f(x)\,dx}\ .$$

The symbol $\int_a^b f(x)\,dx$ is read "the integral from a to b of f of x dee x." The numbers a and b are, respectively, the **lower and upper limits of integration**. The word "limit" in connection with limits of integration is not the same as the limit used in $\lim_{x\to a} f(x) = f(a)$. Limits of integration should be considered as values for the independent variable which (1) bound a region whose area is desired, or (2) simply act as the domain of definition of an integrand. Moreover, the dx tells us that the integration is being performed with respect to x, so that the a and b are values of x. As explained previously in the discussion of the indefinite integral, the symbol $\int$ is the integral sign and $f(x)$ is the integrand. The symbols x and dx are dummy variables and can be replaced by other variables. Thus,

$$\int_a^b f(x)\,dx = \int_a^b f(t)\,dt = \int_a^b f(u)\,du,$$

and so on.

Proponents of the abstract believe that the definite integral is concerned only with a function (f in our case) and two constants (a and b in our situation). However, we will emphasize the *practical aspects* of the definite integral in those situations directly and indirectly related to business and economics. In this section we are particularly interested in the application to area under a curve. This leads us to a statement of a theorem which provides us with a surprisingly simple procedure for finding areas. It relieves us of the burden of approximating areas by long and tedious arithmetical computations and it is a powerful tool that in essence relates differentiation and integration. This very important theorem was evolved almost simultaneously by Newton of England and Leibniz of Germany in the mid-1600's. It is a fascinating truth that the antiderivative, the reverse of finding tangents, lends itself as a handy tool for conquering the problems of summation. The applications of this famous theorem are found extensively in problems other than areas.

THEOREM 1 The Fundamental Theorem of Integral Calculus

If

(1) *$f(x)$ is a continuous function on $[a, b]$ and*
(2) *$F(x)$ is a function whose derivative is $f(x)$,*
then

(3)
$$\boxed{\begin{aligned}\int_a^b f(x)\,dx &= \int_a^b F'(x)\,dx \\ &= F(x)\Big]_a^b \\ &= F(b) - F(a).\end{aligned}}$$

It is not necessary to include the constant of integration when evaluating a definite integral. The reason for this is that the constant vanishes in the process. For instance,

$$\begin{aligned}\int_a^b f(x)\,dx &= F(x) + C \Big]_a^b \\ &= F(b) + C - [F(a) + C] \\ &= F(b) + C - F(a) - C \\ &= F(b) - F(a).\end{aligned}$$

Note 1: The integral $\int_a^b f(x)\,dx$ is the definite integral of f from $x = a$ to $x = b$. This integral is a symbol which represents the limit

$$\lim_{n\to\infty} \sum_{i=1}^{n} f(t_i)\,\Delta x_i.$$

The Fundamental Theorem provides us with an amazing shortcut for computing the definite integral. It tells us to find the antiderivative F of the integrand f, evaluate F at the upper limit b of integration, evaluate F at the lower limit a of integration, then calculate the difference $F(b) - F(a)$.

Note 2: If the integrand f is a nonnegative function ($f \geq 0$) on $[a, b]$, then the answer $F(b) - F(a)$ is a number representing the area under the curve of f. However, if f is such that it is negative for all $x \in [a, b]$, then the result $F(b) - F(a)$ would be a negative number. Since area A is a positive concept, we would need to write

$$\begin{aligned}A = -\int_a^b f(x)\,dx &= -F(x)\Big]_a^b \\ &= -[F(b) - F(a)],\end{aligned}$$

in order to calculate area. Thus, if $y = f(x)$ has its graph below the x-axis we find the area "under" f by evaluating the definite integral $\int_a^b f(x)\,dx$ and negating the algebraic sign (negative) obtained.

EXAMPLE 4 Find the area of region R bounded by the graph of $f(x) = x^2 - 2x$ and $y = 0$.

Solution From the figure we see that the graph of f lies below the x-axis for the interval $[0, 2]$. Therefore, the area of region R (see figure) will turn out to

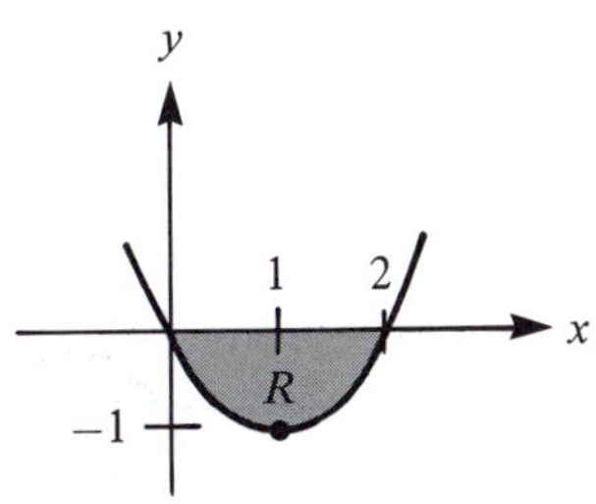

be a negative number by definite integration. Since area A is a positive number, we supply the negative sign before the integral and evaluate

$$\begin{aligned} A &= -\int_0^2 (x^2 - 2x)\,dx \\ &= -\left[\frac{x^3}{3} - x^2\right]_0^2 \\ &= -\left[\frac{2^3}{3} - 2^2 - \frac{1}{3}(0^3) + 0^2\right] \\ &= -\left(\frac{8}{3} - 4\right) = -\left(-\frac{4}{3}\right) = \frac{4}{3} \text{ units}^2. \end{aligned}$$

If the function f of this example represented a marginal profit function, then the negative algebraic sign of the evaluated integral would indicate a negative profit, that is, a loss. Consider a function f such that its graph crosses the horizontal axis at a point c where $a \leq c \leq b$. Furthermore, one of the regions in question is above and the other is below the horizontal axis. Then each part must be evaluated separately and the results added to obtain the total area "under" f.

EXAMPLE 5 Calculate the total area "under" the graph of $f(x) = x(x + 1)(x - 2)$.

Solution From the graph of f we see that the x-intercepts occur at $(0, 0)$, $(-1, 0)$, and $(2, 0)$. The region under f from -1 to 0 is above the x-axis and the region from 0 to 2 is below the x-axis. Then the total area A is the algebraic sum of the two definite integrals

$$\begin{aligned} A &= \int_{-1}^0 f(x)\,dx + \left[-\int_0^2 f(x)\,dx\right] \\ &= \int_{-1}^0 x(x + 1)(x - 2)\,dx - \int_0^2 x(x + 1)(x - 2)\,dx \\ &= \int_{-1}^0 (x^3 - x^2 - 2x)\,dx - \int_0^2 (x^3 - x^2 - 2x)\,dx \\ &= \left[\frac{x^4}{4} - \frac{x^3}{3} - x^2\right]_{-1}^0 - \left[\frac{x^4}{4} - \frac{x^3}{3} - x^2\right]_0^2 \\ &= -\frac{1}{4}(-1)^4 + \frac{1}{3}(-1)^3 + (-1)^2 - \left[\frac{2^4}{4} - \frac{1}{3}(2)^3 - (2)^2\right] \\ &= -\frac{1}{4} - \frac{1}{3} + 1 - 4 + \frac{8}{3} + 4 = \frac{37}{12} \text{ units}^2. \end{aligned}$$

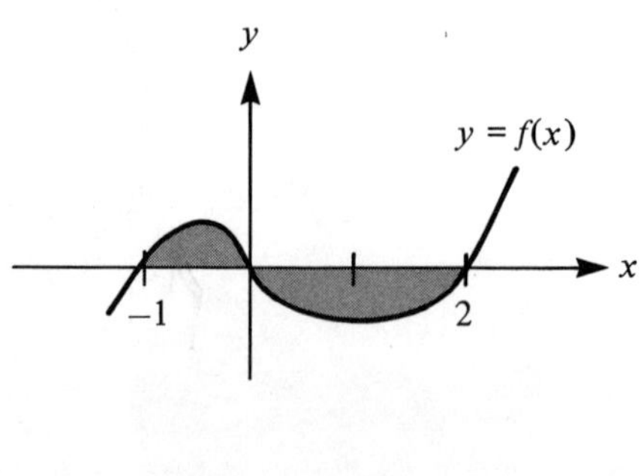

The properties of definite integrals are similar to those of indefinite integrals except for the evaluative characteristic defined by the Fundamental Theorem. We list some of these important properties.

Property 1 $\int_a^b kf(x)\,dx = k\int_a^b f(x)\,dx$, *k a constant.*

Property 2 $\int_a^b f(x)\,dx = -\int_b^a f(x)\,dx.$

Property 3 $\int_a^a f(x)\,dx = 0.$

Property 4 $\int_a^b [f(x) \pm g(x)]\,dx = \int_a^b f(x)\,dx \pm \int_a^b g(x)\,dx.$

Property 5 $\int_a^c f(x)\,dx + \int_c^b f(x)\,dx = \int_a^b f(x)\,dx$ *if c is between a and b (see Figure 6.14).*

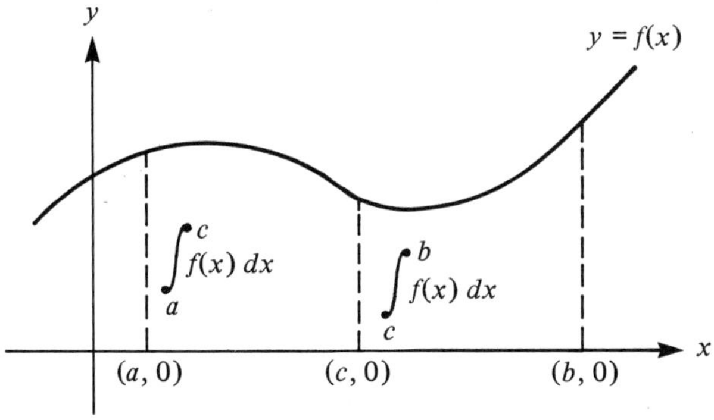

Figure 6.14

If $c = a$ in Property 5, then $\int_a^a f(x)\,dx = 0$ by Property 3. If $c \notin [a, b]$, as in Figure 6.15, then Property 5 is written

$$\int_a^c f(x)\,dx - \int_b^c f(x)\,dx = \int_a^b f(x)\,dx.$$

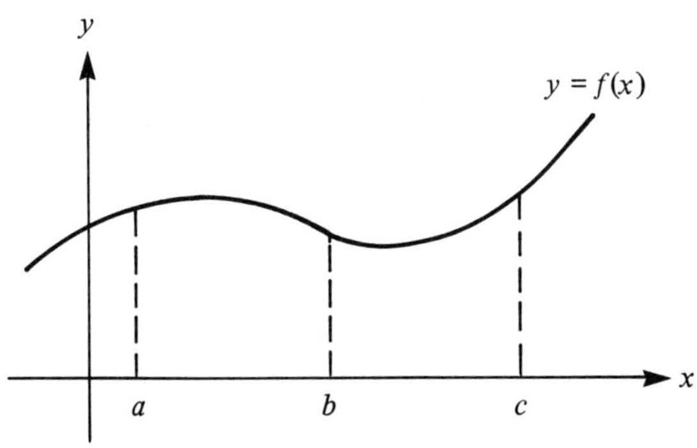

Figure 6.15

Property 6 $\int_a^b u^n\,du = \left.\frac{u^{n+1}}{n+1}\right]_a^b = \frac{1}{n+1}\left[b^{n+1} - a^{n+1}\right]$ *where* $n \neq -1$.

Property 7 $\int_a^b u^{-1}\,du = \ln|u|\Big]_a^b = \ln|b| - \ln|a|$.

PROBLEM SET 6.5.2

Find the area under the curves defined in problems 1–6:

1. $f(x) = 2x + 3$ from $x = 0$ to $x = 1$.
2. $f(x) = 3x - 3$ from $x = 1$ to $x = 2$.
3. $g(x) = x^2 + 2$ from $x = 0$ to $x = 3$.
4. $g(x) = x^2 - x$ from $x = 1$ to $x = 3$.
5. $h(x) = 4 - x^2$; a graph of h is helpful.
6. $h(x) = 9 - x^2$; a graph of h is helpful.
7. Marginal profit is given by $MP = 4x - x^2$.
   (a) Sketch a graph of MP. (b) Find the area under MP.
   (c) The numerical value in (b) represents profit. Write this value in terms of dollars.
8. Work problem 7 when marginal profit $MP = 12x - 3x^2$.
9. Assume that a marginal selling expense is given by $MS = 5p^{3/2}$. Find the selling expense $S(p)$ if p increases from 0 to 4.
10. Work problem 9 if p increases from 0 to 9.
11. Work problem 9 if p increases from 0 to 16.

Calculate the area "under" the curves defined in problems 12–16:

12. $f(x) = x^2 - 3x$
13. $g(x) = x^2 - 2x$
14. $h(x) = \frac{3}{5}x - 1$
15. $f(t) = \frac{2}{3}t - \frac{1}{2}$.
16. The management of Colodyne learns that its marginal cost of production is given by $MQ = 3.2x - 9.2$ where x denotes the number of units produced. Find the total cost function $Q(x)$ if fixed cost is \$10. [*Hint*: The initial conditions are $Q = 10$ when $x = 0$.]
17. The management in problem 16 decides on a price of \$6.80 per unit of sales. This implies that the marginal sales revenue is given by $MS = 6.8$. Find the total sales revenue function $S(x)$ if $S = 0$ when $x = 0$.
18. Refer to problems 16 and 17 and find maximum profits P_{max}, recalling that P_{max} occurs when marginal cost equals marginal revenue.
19. Continue with problem 18 and find the increase in profits when sales units increase from $x = 3$ to $x = 4$. [*Hint*: $P = \int_3^4 P'(x)\,dx = \int_3^4 (MS - MQ)\,dx$, and so on.]
20. In each of the following problems marginal cost MQ and marginal sales MS are given functions. Find the total cost and total sales functions if fixed cost is given as \$8 in each case:
    (a) $MQ = 4x - 8$, $MS = 8$ (b) $MQ = 4.8x - 10.6$, $MS = 13.4$
    (c) $MQ = 2.25x - 6$, $MS = 12$.
21. Find the profit function and maximum profit for each part of problem 20. Refer to problem 18.

6.6 SIGMA NOTATION APPLIED

Let us examine some of the ideas which are basic to the notion of definite integral. One of these basic concepts is that of summation, whose abbreviation involves sigma notation. For example, the sum of the first ten positive integers can be written as

$$\sum_{i=1}^{10} i$$

and read as "the summation of i from $i = 1$ through $i = 10$." This shorthand form indicates that we replace i with 1, then with 2, then with 3, and so on, and form their sum. Accordingly, we have the sum of the first ten positive integers:

$$\sum_{i=1}^{10} i = 1 + 2 + 3 + \cdots + 10.$$

We state without proof that the sum of the first n consecutive positive integers is given by

$$\sum_{i=1}^{n} i = 1 + 2 + 3 + \cdots + n$$
$$= \frac{n(n+1)}{2}.$$

With the aid of this formula we can quickly and conveniently find, for example, the sum of the first 400 natural numbers, as follows.

EXAMPLE 1

$$\sum_{k=1}^{400} k = \frac{400(400+1)}{2}$$
$$= 200(401) = 80{,}200.$$

Other examples of sigma notation are:

$$\sum_{i=1}^{5} c = c + c + c + c + c, \quad \text{where } c \text{ is a constant.}$$

$$\sum_{i=1}^{n} k = kn, \quad \text{where } k \text{ is a constant.}$$

$$\sum_{i=1}^{6} a_i = a_1 + a_2 + a_3 + \cdots + a_6.$$

$$\sum_{j=1}^{5} b^j = b + b^2 + b^3 + b^4 + b^5.$$

$$\sum_{k=1}^{n} \frac{1}{k} = 1 + \frac{1}{2} + \frac{1}{3} + \cdots + \frac{1}{n}.$$

$$\sum_{i=2}^{4} (2n - 3) = [2(2) - 3] + [2(3) - 3] + [2(4) - 3]$$
$$= 1 + 3 + 5.$$

Observe that the answer contains three terms in this last example. This suggests that the evaluation of $(4 - 2 + 1)$ will yield the correct number of terms. In general we may write

$$\sum_{i=m}^{n} F(i) = F(m) + F(m+1) + F(m+2) + \cdots + F(n)$$

where

(a) m and n are integers,

(b) $m \leq n$,

(c) $(n - m + 1)$ is the number of terms in the summation,

(d) $F(i)$ is a formula involving i,

(e) i is the index of summation,

(f) m and n are the lower and upper limits of summation, respectively.

The following theorem contains six fundamental properties of the sigma notation. Proofs are omitted.

THEOREM 1 **(1)** $\sum_{i=1}^{n} ka_i = k \sum_{i=1}^{n} a_i, \quad k \text{ a constant}$

(2) $\sum_{i=1}^{n} (a_i + b_i) = \sum_{i=1}^{n} a_i + \sum_{i=1}^{n} b_i$

(3) $\sum_{i=1}^{n} (a_i - b_i) = \sum_{i=1}^{n} a_i - \sum_{i=1}^{n} b_i$

(4) $\sum_{i=1}^{n} (a_i - a_{i-1}) = a_n - a_0.$

(5) $\sum_{i=m}^{n} a_i = \sum_{i=m+1}^{n+1} a_{i-1}$

(6) $\sum_{i=k}^{n} a_i = \sum_{i=m}^{n} a_i - \sum_{i=m}^{k-1} a_i.$

Some examples of item (5) are:

$$\sum_{i=-1}^{1} (i+1) = \sum_{i=0}^{2} i$$

$$\sum_{i=2}^{4} (i-2)^2 = \sum_{i=0}^{2} i^2$$

A few examples of item (6) are:

$$\sum_{i=5}^{10} i^2 = \sum_{i=0}^{10} i^2 - \sum_{i=0}^{4} i^2$$

$$\sum_{k=15}^{25} (k-2) = \sum_{k=0}^{25} (k-2) - \sum_{k=0}^{14} (k-2).$$

As we have already learned, sigma notation and the limit concept, taken together, play vital roles in the development of the definite integral. We illustrate by concrete example.

EXAMPLE 2 A marginal sales revenue function is given by $MS = 2x$, where x denotes the amount of production units. Find total sales revenue S using summation of rectangles in conjunction with the limit concept when x increases from 1 to 3. Check the answer by evaluating the appropriate definite integral.

Solution

$$\begin{aligned} MS &= f(x) = 2x, \qquad 1 \le x \le 3, \\ \Delta x &= \frac{3-1}{n} = \frac{2}{n}, \\ x_1 &= 1 + \Delta x, \\ x_2 &= 1 + 2\,\Delta x, \\ x_3 &= 1 + 3\,\Delta x, \\ &\cdots \\ x_{n-1} &= 1 + (n-1)\,\Delta x \end{aligned}$$

(see Figure 6.16). Also,

$$\begin{aligned} f(1) &= 2 \\ f(x_1) &= f(1 + \Delta x) = 2(1 + \Delta x) \\ f(x_2) &= f(1 + 2\,\Delta x) = 2(1 + 2\,\Delta x) \\ f(x_3) &= f(1 + 3\,\Delta x) = 2(1 + 3\,\Delta x) \\ &\cdots \\ f(x_{n-1}) &= f[1 + (n-1)\,\Delta x] = 2[1 + (n-1)\,\Delta x]. \end{aligned}$$

So the sum of the areas of the inscribed rectangles is

$$\begin{aligned} S_n &= f(1)\,\Delta x + f(x_1)\,\Delta x + f(x_2)\,\Delta x + \cdots + f(x_{n-1})\,\Delta x \\ &= [2 + 2 + 2\,\Delta x + 2 + 4\,\Delta x + 2 + 6\,\Delta x + \cdots + 2 + 2(n-1)]\,\Delta x \\ &= \Delta x[(2 + 2 + 2 + \cdots + 2) + \Delta x(2 + 4 + 6 + \cdots + 2(n-1))] \\ &= \Delta x[2n + 2\,\Delta x(1 + 2 + 3 + \cdots + (n-1))] \end{aligned}$$

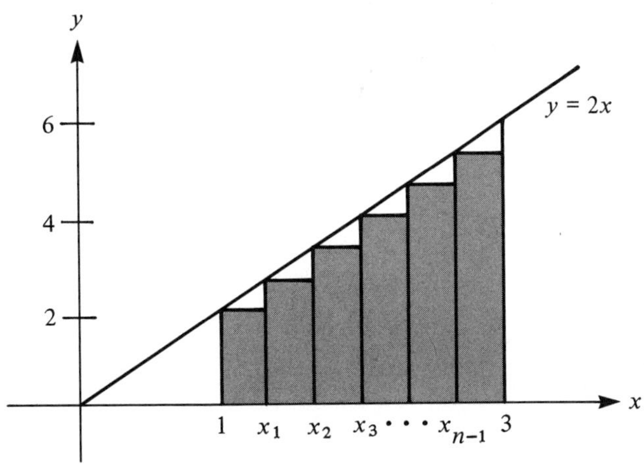

Figure 6.16

since $(2 + 2 + 2 + \cdots + 2)$ means there are n of the 2's, or $2n$ [i.e., $\sum_{i=1}^{n} 2 = 2n$]. So

$$S_n = \frac{2}{n}\left[2n + 2\cdot\frac{2}{n}(1 + 2 + 3 + \cdots + n - 1)\right] \qquad \left[\text{since } \Delta x = \frac{2}{n}\right];$$
$$= \frac{2}{n}\left[2n + \frac{4}{n}\cdot\frac{(n-1)\cdot n}{2}\right]$$

because $1 + 2 + 3 + \cdots + n - 1$ is the sum of the first $(n - 1)$ positive integers, indicating that the formula $n(n + 1)/2$ becomes $(n - 1)n/2$ when n is replaced by $n - 1$ [i.e., $\sum_{i=1}^{n-1} i = (n - 1)n/2$]. Continuing,

$$S_n = \frac{2}{n}[2n + 2n - 2] \qquad [\text{simplifying}]$$
$$= \frac{2}{n}[4n - 2]$$
$$= 8 - \frac{4}{n}.$$

Then
$$\lim_{n\to\infty} S_n = \lim_{n\to\infty}\left(8 - \frac{4}{n}\right)$$
$$= 8 \qquad \left[\lim_{n\to\infty}\frac{4}{n} = 0\right].$$

Checking by definite integration, we obtain total sales revenue

$$S = \int_1^3 MS\,dx$$
$$= \int_1^3 2x\,dx$$
$$= x^2\Big]_1^3$$
$$= 3^2 - 1^2 \doteq \$8$$

which agrees with the previous result. This is one method of applying the limit concept to finding area under a curve. The area in our particular situation represents the total amount of sales revenue S as the number of production units increases from $x = 1$ to $x = 3$.

Another important example of sigma notation is

$$\sum_{i=1}^{4} i^2 = 1^2 + 2^2 + 3^2 + 4^2$$

which denotes the sum of the squares of the first four natural numbers. The notation may be generalized to indicate the sum of the squares of the first n consecutive positive integers, thus:

$$\sum_{i=1}^{n} i^2 = 1^2 + 2^2 + 3^2 + \cdots + n^2.$$

This sum can be formulated as

$$\sum_{i=1}^{n} i^2 = \frac{n(n+1)(2n+1)}{6}.$$

We now exhibit the usefulness of this formula by calculating cost using the limit process.

EXAMPLE 3 Given the marginal cost function $MQ = f(x) = x^2/2$, find cost Q when x increases from $x = 0$ to $x = 1$.

Solution Our approach will be basically similar to that of Example 1 of Section 6.5, the chief difference being the use of circumscribed rectangles (see Figure 6.17). Then $0 \le x \le 1$, $\Delta x = (1 - 0)/n = 1/n$ and

$$\begin{aligned} S_n &= f\left(\frac{1}{n}\right)\cdot\frac{1}{n} + f\left(\frac{2}{n}\right)\cdot\frac{1}{n} + f\left(\frac{3}{n}\right)\cdot\frac{1}{n} + \cdots + f\left(\frac{n}{n}\right)\cdot\frac{1}{n} \\ &= \frac{1}{n}\left[\frac{1}{2}\cdot\frac{1}{n^2} + \frac{1}{2}\cdot\frac{4}{n^2} + \frac{1}{2}\cdot\frac{9}{n^2} + \cdots + \frac{1}{2}\cdot\frac{n^2}{n^2}\right] \\ &= \frac{1}{2}\cdot\frac{1}{n}\left[\frac{1}{n^2}(1^2 + 2^2 + 3^2 + \cdots + n^2)\right] \\ &= \frac{1}{2n^3}\cdot\frac{n}{6}(n+1)(2n+1) \\ &= \frac{1}{12n^2}(2n^2 + 3n + 1) = \frac{1}{6} + \frac{1}{4n} + \frac{1}{12n^2}. \end{aligned}$$

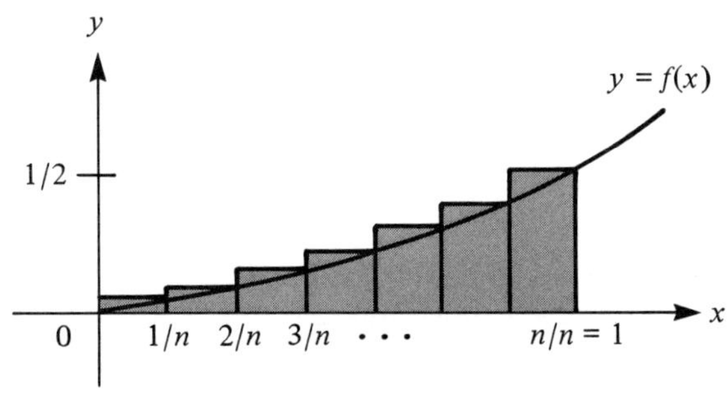

Figure 6.17

As we let the number of rectangles increase to infinity, we have

$$\begin{aligned} \lim_{n\to\infty} S_n &= \lim_{n\to\infty}\left(\frac{1}{6} + \frac{1}{4n} + \frac{1}{12n^2}\right) \\ &= \frac{1}{6}. \end{aligned}$$

This number should equal the definite integral of f from $x = 0$ to $x = 1$.

Let us see if it does:

$$Q = \int_0^1 MQ\,dx = \int_0^1 \frac{x^2}{2}\,dx$$
$$= \frac{x^3}{6}\bigg]_0^1 = \frac{1}{6},$$

which agrees with the previous answer.

It should be clear at this stage that if the limit exists then the definite integral exists also. Moreover, their common answer is a unique number. In advanced calculus courses it is proved that if a function f is continuous on the closed interval $[a, b]$ then the definite integral $\int_a^b f(x)\,dx$ always exists and has a unique value. There are some functions which are not continuous on $[a, b]$ but whose definite integral still exists. However, we will not pursue these functions at this point of our discussion.

Let us now evaluate an indicated sum by use of the properties of sigma notation.

EXAMPLE 4

$$\sum_{i=1}^{n} (i+1)(i-3) = \sum_{i=1}^{n} (i^2 - 2i - 3)$$
$$= \sum_{i=1}^{n} i^2 - 2\sum_{i=1}^{n} i - \sum_{i=1}^{n} 3$$
$$= \frac{n}{6}(n+1)(2n+1) - 2\cdot\frac{n}{2}(n+1) - 3n$$
$$= \frac{n(n+1)(2n+1) - 6n(n+1) - 18n}{6}$$
$$= \frac{n}{6}[(n+1)(2n+1) - 6(n+1) - 18]$$
$$= \frac{n}{6}[2n^2 - 3n - 23].$$

If we wish to find the sum of the first 30 terms, we use $n = 30$ in the formula obtained above and calculate as follows:

$$\sum_{i=1}^{30} (i+1)(i-3) = \frac{30}{6}[2(30)^2 - 3(30) - 23]$$
$$= 5(1800 - 90 - 23)$$
$$= 5(1687) = 8435.$$

In the next example we make use of summation and a geometric progression to compute the value of an annuity.

EXAMPLE 5 Mrs. Kay deposits \$1000 in a savings account at the end of each year for 5 years. The account earns 8% compounded annually. What is the value V of the account at the end of 5 years?

Solution Keeping in mind that the last deposit will earn no interest, we need to sum up all terms of the form

$$\$1000(1 + 0.08)^{t-1}$$

for $t = 1$ to $t = 5$. That is,

$$\begin{aligned} V &= 1000 \sum_{t=1}^{5} (1.08)^{t-1} \\ &= 1000[1 + (1.08) + (1.08)^2 + \cdots + (1.08)^4] \\ &= 1000 + 1000[1.08 + (1.08)^2 + \cdots + (1.08)^4] \qquad (1) \end{aligned}$$

The expression on the right in square brackets is a **geometric progression** whose first term $a_1 = 1.08$ and common ratio $r = 1.08$. Substituting these values in the formula for the sum S_n of a geometric progression, we have

$$\begin{aligned} S_n &= a_1 \frac{(r^n - 1)}{r - 1} \qquad [n = 4] \\ &= 1.08\left[\frac{(1.08)^4 - 1}{1.08 - 1}\right] \\ &= 1.08\left[\frac{1.36048896 - 1}{0.08}\right] \\ &= 27\frac{(0.36048896)}{2} \\ &= 4.86660096 \end{aligned}$$

The value of $(1.08)^4$ can be found by applying either logarithms or the binomial theorem or a desk calculator or compound interest tables. From Eq. (1):

$$\begin{aligned} V &= 1000 + 1000S_n \\ &= 1000 + 1000(4.86660096) \\ &= 1000 + 4866.60 = \$5866.60. \end{aligned}$$

This total annuity value agrees very closely with the value found from annuity tables.

To express the sum of an **infinite** number (or **series**) of terms we write

$$S_\infty = \sum_{i=1}^{\infty} s_i.$$

In expanded form this is an infinite sequence of partial sums:

$$\begin{aligned} S_1 &= s_1 \\ S_2 &= s_1 + s_2 \\ S_3 &= s_1 + s_2 + s_3 \\ &\cdots \\ S_n &= s_1 + s_2 + s_3 + \cdots + s_n \\ &\cdots \\ S_\infty &= s_1 + s_2 + s_3 + \cdots + s_n + \cdots. \end{aligned}$$

The sum S_∞ of an **infinite geometric progression**

$$a + ar + ar^2 + \cdots + ar^n + \cdots,$$

where $|r| < 1$, is defined by

$$\begin{aligned} S_\infty &= \lim_{n\to\infty} S_n \\ &= \lim_{n\to\infty} \frac{a_1(1 - r^n)}{1 - r} = \frac{a_1}{1 - r}. \end{aligned} \tag{2}$$

As a matter of explanation of the result in Eq. (2),

$$\lim_{n\to\infty} (1 - r^n) = 1 \qquad \text{when } |r| < 1.$$

For instance, for $r = 1/2$,

$$\begin{aligned} \lim_{n\to\infty} \left[1 - \left(\frac{1}{2}\right)^n\right] &= \lim_{n\to\infty} \left(1 - \frac{1}{2^n}\right) \\ &= 1 - 0 = 1. \end{aligned}$$

Remark: It is indeed amazing that mathematicians have been able to invent a method for calculating the sum of a never-ending series of terms!

EXAMPLE 6 Find the sum of the infinite series

$$1 + \frac{1}{2} + \frac{1}{4} + \frac{1}{8} + \cdots + \frac{1}{2^{n-1}} + \cdots.$$

Solution Since $r = 1/2$, Eq. (2) is applicable and we have

$$\begin{aligned} S_\infty &= \frac{a_1}{1 - r} \qquad [\text{where } a_1 = 1], \\ &= \frac{1}{1 - \frac{1}{2}} = 2. \end{aligned}$$

EXAMPLE 7 Calculate the sum $\sum_{i=1}^{\infty} (-1/3)^i$.

Solution

$$\sum_{i=1}^{\infty} \left(-\frac{1}{3}\right)^i = -\frac{1}{3} + \frac{1}{9} - \frac{1}{27} + \cdots$$

where $|r| = |-1/3| = 1/3 < 1$ and $a_1 = -1/3$. Therefore,

$$\begin{aligned} S_\infty &= \frac{a_1}{1 - r} \\ &= \frac{-\frac{1}{3}}{1 - \left(-\frac{1}{3}\right)} \\ &= -\frac{1}{3}\left(\frac{3}{4}\right) = -\frac{1}{4}. \end{aligned}$$

Formula (2) also finds application in expressing **repeating decimals** as common fractions.

EXAMPLE 8 Express the repeating decimal $0.8333\overline{3}$ as a common fraction.

Solution Since $\overline{3}$ means that the 3's are repeated, we have

$$\begin{aligned}
0.833\overline{3} &= 0.8 + [0.03 + 0.003 + 0.0003 + \cdots] \\
&= 0.8 + \sum_{k=1}^{\infty} (0.03)(0.1)^{k-1} \\
&= 0.8 + \frac{a_1}{1-r} \qquad [a_1 = 0.03, r = 0.1] \\
&= 0.8 + \frac{0.03}{1-0.1} \\
&= 0.8 + \frac{0.03}{0.90} \\
&= \frac{4}{5} + \frac{1}{30} \\
&= \frac{25}{30} = \frac{5}{6}.
\end{aligned}$$

Remember that the bar over the last 3 in the decimal $0.8333\overline{3}$ indicates that the 3 is repeated indefinitely; hence the name "repeating decimal."

EXAMPLE 9 In sigma notation the following infinite series may be written thus:

$$2x - 3x^2 + 4x^3 - 5x^4 + \cdots = \sum_{k=2}^{\infty} (-1)^k k x^{k-1}.$$

Observe: *1.* The coefficient of each term is one more than the power of x. So, if we let k represent the coefficient, then the power of x must necessarily be denoted by $k - 1$.

2. The symbol $(-1)^k$ serves the purpose of producing the correct algebraic sign in the right place.

3. If $k = 2$, then $(-1)^2 \cdot 2x^{2-1} = 2x$, the first term; if $k = 3$, then $(-1)^3 \cdot 3x^{3-1} = -3x^2$, the second term, and so on.

EXAMPLE 10 If p and q are any numbers, the **binomial theorem** states that

$$(p + q)^n = \binom{n}{0} p^n + \binom{n}{1} p^{n-1}q + \binom{n}{2} p^{n-2}q^2 + \cdots + \binom{n}{n} q^n$$

where

$$\binom{n}{r} = \frac{n!}{r!(n-r)!}$$

and n is a natural number, $r \geq 0$. For instance,

$$\binom{5}{3} = \frac{5!}{3!(5-3)!} = \frac{5 \cdot 4 \cdot 3!}{3! \cdot 2} = 10.$$

In the case of the binomial expansion, the sigma notation is highly useful in lending brevity and compactness. Thus,

$$\boxed{(p + q)^n = \sum_{r=0}^{n} \binom{n}{r} p^{n-r}q^r} \quad (3)$$

EXAMPLE 11

$$(2x - y)^5 = \sum_{r=0}^{5} \binom{5}{r} (2x)^{5-r}(-y)^r$$

where $p = 2x$, $q = (-y)$, and $n = 5$.

EXAMPLE 12 Apply the binomial expansion to evaluate $(0.97)^4$ to three decimal places.

Solution

$$\begin{aligned}(0.97)^4 &= (1 - 0.03)^4\\ &= 1^4 + 4(1^3)(-0.03) + 6(1^2)(-0.03)^2 + 4(1)(-0.03)^3 + (-0.03)^4\\ &= 1 - 0.12 + 0.0054 - 0.000108 + 0.00000081\\ &= 0.88529281 \quad \text{or } 0.885 \text{ to three decimal places.}\end{aligned}$$

The coefficients 1, 4, 6, 4, 1 are the result of computing $\binom{4}{0}$, $\binom{4}{1}$, $\binom{4}{2}$, $\binom{4}{3}$, and $\binom{4}{4}$. Since these computations are rather time-consuming, we recommend an easier and faster method of finding the coefficients in a binomial expansion. This method is called Pascal's Triangle, and its development is as follows:

$(a+b)^0 = 1 \longrightarrow$ 1

$(a+b)^1 = 1a + 1b \longrightarrow$ 1 1

$(a+b)^2 = 1a^2 + 2ab + 1b^2 \longrightarrow$ 1 2 1

$(a+b)^3 = 1a^3 + 3a^2b + 3ab^2 + 1b^3 \longrightarrow$ 1 3 3 1

. . . 1 4 6 4 1

1 5 10 10 5 1

1 6 15 20 15 6 1

. . .

The coefficients form a triangle, hence the name Pascal's Triangle (first found by the great French mathematician and philosopher Blaise Pascal, 1623–1662). Observe how the coefficients are formed. To obtain the 4, 6, 4, for example, we simply add the 1 and 3, 3 and 3, and 3 and 1 from above and to each side of the 4, 6, 4; and similarly for the others. Thus, the coefficients of $(a + b)^5$ are 1, 5, 10, 10, 5, 1.

PROBLEM SET 6.6

Express each formula of problems 1–4 in sigma notation:

1. $1 + 3 + 5 + \cdots + (2n - 1) = n^2$

2. $2 + 4 + 6 + \cdots + 2n = n(n + 1)$

3. $4 + 8 + 12 + \cdots + 4n = 2n(n + 1)$

4. $1 + 8 + 27 + \cdots + n^3 = \left[\dfrac{n(n + 1)}{2}\right]^2.$

Use the methods of this section to find the total cost from the marginal cost function and the corresponding interval indicated:

5. $MQ = 3x$ for $x = 0$ to $x = 2$.

6. $MQ = 2x + 1$ for $x = 1$ to $x = 2$.

Write each sum in problems 7–10 in expanded form:

7. $\sum_{i=1}^{n} (2i + 5)$

8. $\sum_{j=1}^{n} (j^2 + 1)$

9. $\sum_{k=2}^{n} (k^3 - 3k)$

10. $\sum_{i=1}^{n} H(t_i)\,\Delta t_i$

Using methods of this section, calculate the total sales revenue from the given marginal sales revenue and its corresponding interval:

11. $MS = x^2$ from $x = 0$ to $x = 3$.

12. $MS = 2x^2$ from $x = 1$ to $x = 2$.

Express problems 13 and 14 in sigma notation:

13. $y_1\,\Delta x_1 + y_2\,\Delta x_2 + y_3\,\Delta x_3 + \cdots + y_n\,\Delta x_n$

14. $g(t_1)\,\Delta u_1 + g(t_2)\,\Delta u_2 + g(t_3)\,\Delta u_3 + \cdots + g(t_n)\,\Delta u_n$.

Write problems 15 and 16 in expanded form:

15. $\sum_{n=1}^{5} \frac{1}{n!}$

16. $\sum_{i=1}^{n} \pi[f(t_i)]^2\,\Delta x_i$.

A marginal output function is defined in problems 17 and 18. Find the total output for the interval of input indicated:

17. $MR = 1 + 2x$ for $[1, 3]$.

18. $MT = 1 - x^2$ for $[0, 4]$.

Find an expression for the general term in problems 19–20 and write in sigma notation:

19. $x + x^3 + x^5 + \cdots$

20. $\frac{x}{2} + \frac{x^2}{3} + \frac{x^3}{4} + \cdots$

21. Work Example 5 when \$1000 is invested at the end of each year at 6% compounded annually for 5 years.

22. Write in expanded form: $\sum_{j=1}^{n} b_j x_j = c$.

Find the sum of the infinite geometric progression in problems 23–24:

23. $4 + 3 + \frac{9}{4} + \cdots$

24. $\sum_{k=1}^{\infty} \left(\frac{2}{3}\right)^k$.

Write in sigma notation after finding the nth term for problems 25–26:

25. $1 - 2 + 4 - 8 + \cdots$

26. $x^3 - x^6 + x^9 - x^{12} + \cdots$

In problems 27–28 expand each indicated sum and compute the total value:

27. $\sum_{r=0}^{2} \binom{2}{r} \left(\frac{1}{3}\right)^{2-r} \left(\frac{2}{3}\right)^r$

28. $\sum_{r=1}^{3} \binom{3}{r} (0.2)^{3-r}(0.8)^r$.

Express the following binomial expansions for problems 29–31 in sigma notation:

29. $(x + y)^6$

30. $(x - 2y)^7$

31. $(3a + 2b)^4$

Expand each of problems 32–34:

32. $\sum_{k=1}^{4} \frac{[1 + (-1)^k]}{2}$

33. $\sum_{k=2}^{\infty} \frac{(-1)^{k+1}}{(x + 1)^k}$

34. $\sum_{i=3}^{\infty} \frac{i}{(i - 2)^i}$.

35. Express each repeating decimal as a common fraction:
(a) $0.444\overline{4}$ (b) $1.666\overline{6}$ (c) $9.333\overline{3}$

36. Express each repeating decimal as a common fraction:
(a) $2.1414\overline{14}$ (b) $5.03636\overline{36}$ (c) $7.180180\overline{180}$.

37. Use the binomial expansion to evaluate the following numbers to three decimal places:
(a) $(0.99)^4$ (b) $(1.02)^5$ [*Hint*: $1.02 = 1 + 0.02$]

6.7 REVIEW PROBLEMS

Evaluate the following functions:

1. $\int (4x - 5)\, dx$
2. $\int \left(x^2 - \frac{1}{2}\right) dx$
3. $\int (2t^2 - 3t)\, dt$
4. $\int (x^2 - 2x^3)\, dx$
5. $\int (x^3 - 3x + 1)\, dx$
6. $\int (x^4 - x^2)\, dx$
7. $\int (t^4 - 2t^2 + 2)\, dt$
8. $\int (t^5 + t^4 - 1)\, dt$
9. If marginal cost is defined by $MQ = \sqrt{2q + 1}$, find the total cost function $Q(q)$ if $Q = 3$ when $q = 4$.
10. A certain marginal demand function is given by $MD = 1/(2x - 1)$. Find the total demand function $D(x)$ if $D = 1$ when $x = 1$.
11. Evaluate:
(a) $\int \frac{2\sqrt{u} - 3u}{u^2}\, du$ (b) $\int (1 - 3t)^2\, dt$
(c) $\int (2\theta + 1)^3\, d\theta$ (d) $\int \frac{2\, dv}{\sqrt{1 - 3v}}$
12. Given a marginal selling expense function $MS = (\frac{1}{5}q + 1)^4$. Find the total selling expense function $S(q)$ if $S = 1$ when $q = 0$.
13. The marginal supply for q units of a commodity is given by $MS = 3(2q + 1)^2$. Find total supply $S(q)$ if $S = 1$ when $q = 1$.

Solve the differential equations in problems 14–19:

14. $\frac{dy}{dx} = xy$
15. $\frac{dy}{dx} = \frac{x + 1}{y + 1}$
16. $\frac{dy}{dt} = ty^2$
17. $\frac{dy}{dt} = \frac{2y}{\sqrt{t - 1}}$
18. $\frac{du}{dx} = \sqrt{\frac{x}{u}}$
19. $\frac{du}{dx} = \frac{1 - u}{x + 1}$
20. Assume that the gross national product grows at a rate proportional to the size of the national capital assets a as follows: $dG/da = \sqrt{a}$. Find $G(a)$ if $G = 54$ when $a = 9$.
21. The marginal profit for x items of production is defined by $dP/dx = 50x^4$. Find profit $P(x)$ if $P = \$400$ when $x = 2$ items.

22. Solve the following differential equations subject to the indicated boundary conditions:

(a) $\dfrac{dy}{dx} = x^2y^2, \qquad y = 1 \text{ when } x = 3$

(b) $\dfrac{dy}{dx} = x(x + 1)^3, \qquad y = 2 \text{ when } x = 0$

(c) $\dfrac{du}{dt} = \dfrac{t(t^2 + 1)^3}{u(u^2 - 1)^3}, \qquad u = 1 \text{ when } t = 1$

(d) $\dfrac{dz}{dt} = 6\sqrt{t} + 4t, \qquad z = 7 \text{ when } t = 1.$

23. The rate of change of a marginal profit function is defined by $d^2P/dx^2 = 4(1 + 2x)^2$ where x denotes quantity of items sold. Find the total profit function $P(x)$ if $dP/dx = 1$ when $x = 1$ and $P = 0$ when $x = 0$.

24. Work problem 23 if $d^2P/dx^2 = 6(1 + x)^3$.

25. Find $T(x)$ if its second derivatives are as follows:

(a) $T''(x) = (x - 1)^2$ (b) $T''(x) = \dfrac{1}{\sqrt{2x}}$

(c) $T''(x) = \dfrac{1}{\sqrt{3x + 2}}$ (d) $T''(x) = \dfrac{1}{(2x - 1)^2}$

26. A marginal supply function is changing at the rate $MS = x/(x - 1)$. Find the general supply function $S(x)$.

27. Work problem 26 if $MS = x(1 + x^2)^4$.

28. The marginal productivity of q items of production is given by $MP = \frac{1}{10}(1 + 2q)^2$. Find the productivity $P(q)$ without boundary conditions.

29. Work problem 28 if $MP = \frac{1}{20}(1 + 3q)^2$.

For each given marginal cost function in problems 30–33:

(a) Sketch a graph
(b) Divide the interval [1, 2] into four equal subintervals, draw appropriate inscribed rectangles, and calculate the sum of their areas.
(c) By what amount does the approximate area underestimate the true area under the curve on [1, 2]?

30. $MQ = 2x + 1$ 31. $MQ = \dfrac{x^2}{3}$

32. $MQ = x^2 + 1$ 33. $MQ = x$

34. Work problems 30–33 using circumscribed rectangles.

35. Work problems 30–33 using the midpoint method.

36. State the Fundamental Theorem of Integral Calculus and discuss its importance.

Calculate the areas under the given curves:

37. $f(x) = 3x + 5$ on [0, 2] 38. $g(x) = 2x^2 - 8$ on [0, 3]

39. $h(t) = 16 - t^2$ 40. $F(x) = 25 - x^2$

41. $F(x) = x^2 - 2x$ 42. $G(t) = t^2 - 3t$

43. $H(q) = \sqrt{q}$ on [1, 4] 44. $F(q) = \dfrac{1}{q}$ on [2, 3]

45. If marginal profit is defined by $MP = 4q - 2q^2$ find profit $P(q)$ if $P = 3$ when $q = 0$.

46. A certain marginal cost of production function is found to be $MQ = 0.3x^2 - 0.2x$ for each production run, $10 \leq x \leq 20$, x denoting the size per run. Find total cost Q.

47. Work problem 46 if $MQ = 0.6x^2 - 2x - 100$.

48. If the rate of change of sales revenue is $MS = 0.09p^2 + 2p + 24$ (in thousands of dollars) for $2 \leq p \leq 4$, where \$$p$ is the price per unit, find total sales revenue $S(p)$.

49. Work problem 48 when $MS = 0.04p^3 - 10p + 60$.

50. A marginal sales revenue function is defined by $MS = 3x$. Find total sales revenue, using sigma notation together with the limit concept, over the interval [0, 3]. Use circumscribed rectangles. Check your answer by definite integration.

51. Work problem 50 using inscribed rectangles.

52. The amount of \$500 is deposited into an account at the end of each year for 6 years. Find the total value V of the account if money is worth 6% compounded annually.

53. Work problem 52 if \$1000 is deposited at the end of each year.

54. Find the sum of the infinite series:

(a) $1 + \frac{1}{3} + \frac{1}{9} + \cdots + \frac{1}{3^{n-1}} + \cdots$

(b) $0.333\overline{3}$ **(c)** $1.2727\overline{27}$

(d) $16 + 4 + 1 + \cdots + \frac{16}{4^{n-1}} + \cdots$

55. Calculate the sum of each of the following finite series:

(a) $1 + (1.05) + (1.05)^2 + \cdots + (1.05)^6$

(b) $1 + (1.01) + (1.01)^2 + \cdots + (1.01)^{10}$

(c) $\frac{1}{2} + \frac{1}{4} + \frac{1}{8} + \cdots + \frac{1}{64}$.

56. Write the following binomial expansions in sigma notation:

(a) $(2a + b)^5$ **(b)** $\left(\frac{x}{2} - 1\right)^6$ **(c)** $(3t - 2)^4$.

57. Expand and compute the total value:

(a) $\sum_{r=0}^{2} \binom{2}{r}\left(\frac{1}{4}\right)^{2-r}\left(\frac{3}{4}\right)^r$ **(b)** $\sum_{r=2}^{3} \binom{3}{r}(0.1)^{3-r}(0.9)^r$.

58. Assuming that the binomial theorem is valid for rational powers, calculate to four decimal places:

(a) $(1.1)^5$ **(b)** $(0.98)^6$ **(c)** $(1.04)^{1/2}$ **(d)** $(0.95)^{1/3}$

[*Hint*: $1.1 = 1 + 0.1$, $0.98 = 1 - 0.02$, and so on.]

59. Find the area under each curve for the domain indicated or implied:

(a) $f(x) = 1 + (x + 1)^3$ for $[-2, 0]$

(b) $f(x) = -2 + (x - 1)^3$ for $[0, 2]$

(c) $y = (x + 1)(x - 1)(x - 3)$.

60. Find the area under the graph of:

(a) $f(x) = 1 + \sqrt{x}$ from $x = 1$ to $x = 9$
(b) $f(x) = x^{1/3} + 1$ from $x = 0$ to $x = 8$
(c) $f(x) = 6x - 3x^2$ from $x = 1$ to $x = 2$

61. Write the following expressions using sigma notation:

(a) $H(q_1)\,\Delta q_1 + H(q_2)\,\Delta q_2 + \cdots + H(q_n)\,\Delta q_n$.
(b) $P(z) = a_0 + a_1 z + a_2 z^2 + \cdots + a_n z^n$.

62. Express in sigma notation:

(a) $\dfrac{1}{1\cdot 2} + \dfrac{1}{2\cdot 3} + \dfrac{1}{3\cdot 4} + \cdots + \dfrac{1}{n(n+1)}$

(b) $\left(\dfrac{1}{n}\right)^2 + \left(\dfrac{2}{n}\right)^2 + \left(\dfrac{3}{n}\right)^2 + \cdots + \left(\dfrac{n}{n}\right)^2$

63. Expand:

(a) $\sum_{i=1}^{n} G(u_i)$ (b) $\sum_{k=1}^{3} k(k+1)$

64. Expand and describe the resulting expression:

(a) $\sum_{k=1}^{n} ar^{k-1}$ (b) $\sum_{k=0}^{n} \binom{n}{k} x^{n-k}y^k$

65. Integrate:

(a) $\int (3x + 4)^2\,dx$ (b) $\int (\sqrt{x} + 1)^2\,dx$

66. Solve for y if:

(a) $\dfrac{dy}{dx} = \dfrac{x^3}{y^4}$ (b) $\dfrac{dy}{dx} = \dfrac{3x + 1}{\sqrt{2y + 1}}$

(c) $\dfrac{d^2y}{dx^2} = \sqrt{x + 1}$ (d) $\dfrac{d^2y}{dx^2} = \dfrac{1}{(3x + 1)^2}$

6.8 ENRICHMENT TOPICS AND PROBLEMS

1. Consider a function f continuous on the closed interval $a \le x \le b$, where $f(x)$ represents the number of units of force (variable) acting in the direction of motion on an object moving to the right along the x-axis from a point a to a point b. If W is the number of units of **work** done by the force acting on the object in $[a, b]$, then W is defined by

$$W = \lim_{n\to\infty} \sum_{i=1}^{n} f(t_i)\,\Delta x_i = \int_a^b f(x)\,dx.$$

EXAMPLE A company needs a component whose performance specification fulfills the required work to be done under the following condition: It must be able to do the work necessary to stretch a spring 2 inches. A force of $6x$ pounds is required to hold the spring stretched x inches beyond its natural length.

Solution $f(x) = 6x$ for $0 \le x \le 2.$

Then
$$W = \int_0^2 6x\,dx$$
$$= 3x^2 \Big]_0^2$$
$$= 3(2)^2 - 3(0)^2$$
$$= 12 \text{ inch-pounds.}$$

With an understanding of this example, work the following problems:

(a) A spring has a natural length of 10 inches. If a force of 5 pounds stretches the spring 1/2 inches, find the work done in stretching the spring from 10 inches to 12 inches.

(b) A 6-pound force can stretch a spring 2/3 feet. How much work is needed to stretch the spring 2 feet? [*Hint*: $f(x) = kx$. Then $6 = k \cdot 2/3$ or $k = 9$. Therefore, $f(x) = 9x$ for $0 \le x \le 2$.]

(c) A spring has a natural length of 16 inches. A 10,000-pound force compresses the spring to 12 inches. Find the work done in compressing it from 16 inches to 10 inches. [*Hint*: $f(x) = kx$ (called *Hooke's Law*) holds for compressing as well as extending a spring. Thus, $10{,}000 = k \cdot 4$, $k = 2500$, and $f(x) = 2500x$, $W = \int_0^6 2500x\,dx$.]

2. A sack of wheat weighing 800 pounds is to be unloaded. As the sack is lifted vertically, wheat spills out so that, after having been lifted x feet, the sack weighs $(800 - 4x)$ pounds. What is the amount of work done by a force applied to the base of the sack to lift it 10 feet? [*Hint*: Refer to problem 1 and let $f(x) = 800 - 4x$. Then work $W = \int_0^{10} f(x)\,dx$.]

3. A cylindrical tank of diameter 6 feet and height 10 feet is half full of a certain chemical. Calculate the work required to pump all the chemical up over the rim of the tank. [*Hint*: A "typical element" of the chemical, for example the ith one, will have thickness Δx_i and will be at a distance of x_i, say, from the bottom of the tank. To get this typical element moved to the top rim of the tank, it must traverse a distance of $(10 - x_i)$ feet (see figure). Since work is force times distance, then the work done in pumping the element up will be defined by

$$W_i = \underbrace{(10 - x_i)}_{\text{distance}}\underbrace{[(62.4)(\pi \cdot 3^2)\,\Delta x_i}_{\text{force}} \text{ foot-pounds,}$$

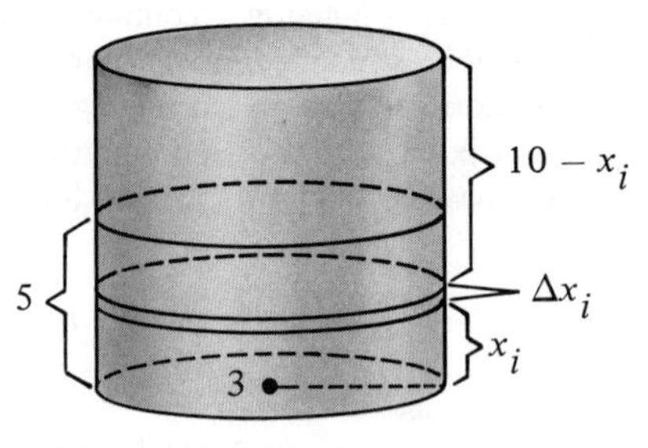

where $\pi \cdot 3^2$ is the area of the base, 62.4 pounds per foot^3 is the density of the liquid, and the product $(62.4)(\pi \cdot 3^2)\,\Delta x_i$ pounds is the weight or force of the liquid. We sum these typical products, take the limit, and obtain the integral which represents the required work:

$$W = \int_0^5 (10 - x)(62.4)(9\pi)\,dx = ?$$

4. Work problem 3 if the tank is 12 feet high and has a radius of 6 feet.
5. The total cost of a new piece of equipment is \$30,000. It is estimated that cost savings from the use of the equipment will allow a rate of depreciation defined by $MD = 2000x - 1000$. How long (t is in years) will it take for the equipment to pay for itself, assuming a scrap value of zero? [*Hint*: $D = \int_0^t MD\,dx = \$30{,}000$. Find t.]
6. Given the following marginal cost and marginal sales functions. Find the total cost and total sales functions if fixed cost is \$10 in each case:

   (a) $MQ = 2x + 1,\quad MS = 12$
   (b) $MQ = 3x + 2,\quad MS = 10$
   (c) $MQ = x^2 + (1/2).\quad MS = 3x - 10.$
7. If you could possibly save 1 cent the first day of November, 2 cents the second day of November, 4 cents the third day, 8 cents the fourth day, and so on, each day after the first saving twice as much as you saved the preceding day, what accumulated amount would you have saved at the end of the month?
8. Mr. Jay's salary was \$6000 the first year. Each year after the first his salary was increased 10% over the amount of the preceding year.

   (a) Find the amount of his salary the fifth year.
   (b) Find his total salary for the first five years.

   [*Hint*: For this geometric progression the common ratio $r = 1.1$.]
9. Work problem 8 if the starting salary was \$8000.
10. A house costing \$20,000 initially is depreciated at year's end by 10% of its value at the beginning of the year. What is the value of the house at the end of 5 years? [*Hint*: The common ratio of the geometric progression is $r = 0.9$.]
11. Work problem 10 if the initial cost is \$30,000.
12. Write each repeating decimal as a common fraction:

    (a) $0.3666\overline{6}$ **(b)** $1.255\overline{5}$ **(c)** $2.7474\overline{74}$
13. Show that:

    (a) $\int_0^1 t(1 - \sqrt{t})^2\,dt = \dfrac{1}{30}$

    (b) $\int_4^8 \dfrac{3u\,du}{\sqrt{u^2 - 15}} = 18$

    (c) $\int_0^a \sqrt{a^2 - x^2}\,dx = a^2\pi/4$, a is constant [*Hint*: $x^2 + y^2 = a^2$ is the equation of a circle.]
14. Evaluate:

    (a) $\int_0^1 x(x - 1)^7\,dx$ **(b)** $\int_0^{\sqrt{2}} \sqrt{4x^2 + 1}\,x\,dx$
15. Determine

    (a) $\int \dfrac{1 + \sqrt{x}}{\sqrt{x}}\,dx$ **(b)** $\int \dfrac{1 + t^{1/3}}{t^{2/3}}\,dt$

16. Solve for y:

(a) $\frac{dy}{dx} = \frac{1 - 2x^2}{1 - 2y^2}$ (b) $\frac{dy}{dx} = \frac{1}{\sqrt{xy - x}}$

17. Evaluate $\int_0^2 \frac{x^2}{x + 1}\,dx$. Divide first.

18. Find $\int x^3(2x^4 + 1)^{-2}\,dx$.

19. Find the area of the region bounded by $y = x^3$ and the x-axis on the interval $[-1, 1]$.

20. Determine $\sum_{i=2}^{16} (i - 2)^2$.

21. Evaluate $\sum_{i=20}^{30} (i^2 + i)$

22. Use the limiting process to find the area under $y = 2x^2 + 3$ on $[0, 2]$.

23. In problem 22, find the area on $[1, 4]$.

24. Expand:

(a) $\sum_{j=1}^{5} \frac{1}{j(j + 1)}$ (b) $\sum_{i=2}^{10} \sqrt{i - 2}$

25. Describe the sum:

(a) $\sum_{k=1}^{n} k = \frac{n(n + 1)}{2}$ (b) $\sum_{k=1}^{n} k^2 = \frac{n}{6}(n + 1)(2n + 1)$

26. Solve the differential equations:

(a) $\frac{dy}{dx} = \sqrt{xy - x + y - 1}$ (b) $\frac{dy}{dt} = t + ty - y - 1$

27. Integrate:

(a) $\int (2t^4 + 1)^4 t^3\,dt$ (b) $\int (3x^2 - x^3)x(2 - x)\,dx$

(c) $\int (x^4 + 8x^2 + 9)(x^3 + 4x)\,dx$

(d) $\int (4t^3 - 3t^2 - 2)(4t^2 - 2t)\,dt$

28. Given the marginal sales function $MS = 2x^2$, find the sales function $S(x)$ when x increases from 2 units to 4 units.

29. Expand by the binomial expansion:

(a) $(x + 1)^8$ (b) $(1 - 2x)^5$ (c) $(ab + c)^3$

30. Find the area under the curve:

(a) $F(x) = (x - 3)^2$ from $x = 0$ to $x = 3$
(b) $G(x) = 16 - x^2$ from $x = 0$ to $x = 4$
(c) $H(x) = x^2 - x^3$ from $x = 0$ to $x = 2$

31. Use the binomial theorem to compute the following to three decimal places:

(a) $(1.02)^{10}$ (b) $(0.97)^{1/3}$ (c) $(24)^{3/2}$
(d) $(35)^{-1/2}$ (e) $(1002)^{2/3}$ (f) $(103)^{-1/2}$

7 OTHER APPLICATIONS AND TECHNIQUES OF INTEGRATION

More Tools for the Decision-Maker

IMPORTANT EQUATIONS

$$\int e^u\,du = e^u + C$$

$$CS = \int_0^{\bar{q}} D(q)\,dq - (\bar{q})(\bar{p})$$

$$PS = (\bar{q})(\bar{p}) - \int_0^{\bar{q}} S(q)\,dq$$

$$\bar{y} = \frac{1}{b-a}\int_a^b f(x)\,dx$$

$$\int u\,dv = uv - \int v\,du$$

$$\int_a^{\infty} f(x)\,dx = \lim_{h\to\infty}\int_a^h f(x)\,dx$$

$$\int_{-\infty}^{b} f(x)\,dx = \lim_{h\to-\infty}\int_h^b f(x)\,dx$$

$$p(E) = \int_a^b f(x)\,dx$$

In the last chapter the basic ideas underlying integration were presented along with certain techniques of integration and some applications. In the present chapter we will extend many of these basic notions to include further applications and techniques of integration. The intuitive approach together with the computational aspects of the calculus will be favored and emphasized. Wherever feasible the practical side of the calculus will find expression in business situations. We will present examples and problems that illustrate applications of the integral calculus to decision-making in business operations. We hope that you will develop a conceptual understanding and gain insight into the relation between a particular mathematical model and the real-world (empirical) business situation. The presentation will include such topics as the area between two curves, consumers' and producers' surplus, average value of a function, integration by parts, improper integrals, integration by partial fractions, the integral as a probability distribution, and others.

7.1 AREA BETWEEN TWO CURVES

Decision-making in business may be involved with areas between curves as well as "under" curves. Consider two functions f and g, where $f \geq g$ and both functions are continuous on the closed interval $a \leq x \leq b$. A typical "element of area" has height $[f(t_i) - g(t_i)]$ and width Δx_i (see Figure 7.1), where $\Delta x_i = (b - a)/n$, a constant. The product

$$[f(t_i) - g(t_i)]\,\Delta x_i$$

represents the area of a typical rectangular element. Our objective is to find the area of the entire region R bounded above by f and below by g and laterally by $x = a$ and $x = b$. Accordingly, let there be n elements comprising R and take the sum of all these elements of area from $i = 1$ to $i = n$ as follows:

$$\sum_{i=1}^{n} [f(t_i) - g(t_i)]\,\Delta x_i.$$

This sum is called a Riemann sum and approximates the true area A of region R. Expanded, the sigma notation means:

$$A \cong [f(t_1) - g(t_1)]\,\Delta x_1 + [f(t_2) - g(t_2)]\,\Delta x_2 + \cdots + [f(t_n) - g(t_n)]\,\Delta x_n.$$

As $n \to \infty$ we say the limit of the Riemann sum is a number which repre-

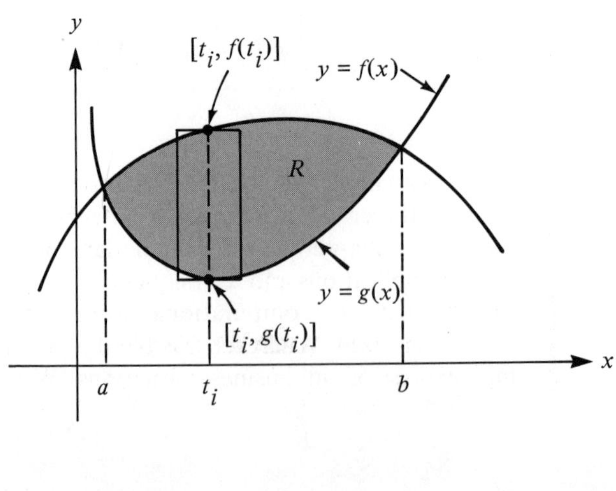

Figure 7.1

sents the area A of R. This number is the definite integral of $(f - g)$ from $x = a$ to $x = b$. We formalize thus:

$$A = \lim_{n \to \infty} \sum_{i=1}^{n} [f(t_i) - g(t_i)] \, \Delta x_i = \int_a^b [f(x) - g(x)] \, dx. \tag{1}$$

The values $x = a$ and $x = b$ will usually be obtained by solving the functions $f(x)$ and $g(x)$ simultaneously. That is, points of intersection will be found such that one of the points will furnish the coordinate of the lower limit of integration and the other point will furnish the coordinate of the upper limit of integration.

EXAMPLE 1 Find the area of the region bounded by the curves $f(x) = 6x - x^2$ and $g(x) = 10 - x$.

Solution The limits of integration are determined by the points of intersection of f and g. We find these points by solving f and g simultaneously:

$$\begin{aligned} 10 - x &= 6x - x^2 \\ x^2 - 7x + 10 &= 0 \\ (x - 2)(x - 5) &= 0, \quad x = 2 \quad \text{or} \quad x = 5 \\ \text{when } x &= 2, \quad \text{then } g(2) = 10 - 2 = 8 \\ \text{when } x &= 5, \quad \text{then } g(5) = 10 - 5 = 5. \end{aligned}$$

Therefore the points of intersection are (2, 8) and (5, 5) (Figure 7.2 illustrates). In the interval [2, 5] the graph of f is above the graph of g, so that for any $x \in [2, 5]$ the difference $f - g$ is a positive number. This is as it should be, since area is a positive concept.

According to Eq. (1), the area A of R (shaded) is defined by

$$\begin{aligned} A &= \lim_{n \to \infty} \sum_{i=1}^{n} [f(t_i) - g(t_i)] \, \Delta x_i \\ &= \int_a^b [f(x) - g(x)] \, dx \\ &= \int_2^5 [6x - x^2 - (10 - x)] \, dx \\ &= \int_2^5 [-x^2 + 7x - 10] \, dx \\ &= \left[\frac{-x^3}{3} + \frac{7x^2}{2} - 10x \right]_2^5 \\ &= \left[-\frac{125}{3} + \frac{7}{2}(25) - 10(5) + \frac{8}{3} - \frac{28}{2} + 10(2) \right] \\ &= \frac{9}{2} \text{ units}^2. \end{aligned}$$

It can be shown that the definition of area in Eq. (1) is valid when region R is located in other quadrants. However, most of our examples will be exhibited in quadrant I, since it is there that real-world situations are defined mathematically.

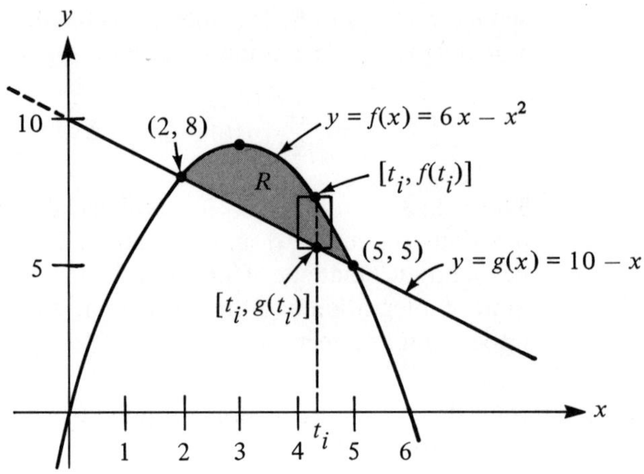

Figure 7.2

EXAMPLE 2 Since profit $P(x)$ = revenue $R(x)$ minus cost $Q(x)$, it follows that

$$P'(x) = R'(x) - Q'(x).$$

Integrating both sides of the equation we obtain

$$\int P'(x)\,dx = \int [R'(x) - Q'(x)]\,dx. \tag{2}$$

Let $R'(x)$, the marginal revenue, be given by $(x + 1)$ and $Q'(x)$, the marginal cost, by $x^{-1/2}$ where x items of a commodity are produced and sold. To find total profit P (in hundreds of dollars) when x increases from $x = 0$ to $x = 4$, we substitute in Eq. (2) above and proceed thus:

$$\begin{aligned} P = \int_a^b P'(x)\,dx &= \int_a^b [R'(x) - Q'(x)]\,dx \\ &= \int_0^4 [(x+1) - x^{-1/2}]\,dx \\ &= \left(\frac{x^2}{2} + x - x^{1/2}\cdot 2\right)\Bigg]_0^4 \\ &= 8 + 4 - 2\sqrt{4} = 8 \quad \text{or } \$800. \end{aligned}$$

In this example, for convenience we neglected fixed costs.

We have already learned that the derivative and differential of $e^{u(x)}$ are given, respectively, by

$$D_x e^{u(x)} = e^{u(x)} \cdot D_x u(x)$$

and

$$de^{u(x)} = e^{u(x)}\,du(x). \tag{3}$$

It follows from Eq. (3) that the indefinite integral of e^u is

$$\boxed{\int e^u\,du = e^u + C}. \tag{4}$$

In Eq. (4) we abbreviated $u(x)$ as u for brevity's sake. If e^u is defined on the interval $[a, b]$, we may write the definite integral as

$$\int_a^b e^{u(x)}\,du = e^{u(x)}\Big]_a^b = e^{u(b)} - e^{u(a)}. \tag{5}$$

EXAMPLE 3 Find the area of the region bounded by the graphs of $y = e^x$, $y = 1$, and $x = 2$.

Solution

$$\begin{aligned}\text{Area } A &= \lim_{n\to\infty} \sum_{i=1}^{n} (e^{t_i} - 1)\,\Delta x_i \qquad \text{[see Figure 7.3].}\\ &= \int_0^2 (e^x - 1)\,dx\\ &= \int_0^2 e^x\,dx - \int_0^2 dx\\ &= [e^x - x]_0^2\\ &= e^2 - 2 - e^0 + 0\\ &= (e^2 - 3)\text{ units}^2.\end{aligned}$$

In the next example we show how an area between two curves may be calculated by using horizontal rectangular elements of area as well as vertical elements.

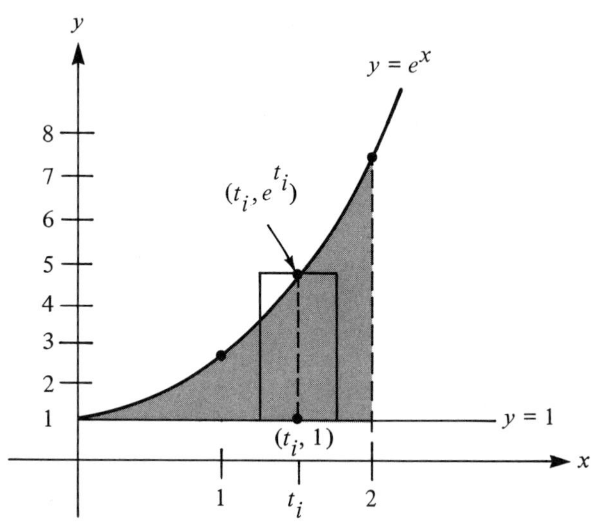

Figure 7.3

EXAMPLE 4 Given a marginal revenue function $MR = x^2$ and a marginal cost function $MQ = x^3$, find profit P. Consider the answer as representing thousands of dollars.

Solution As pointed out in Example 2, we may write

$$\int_a^b MP\,dx = \int_a^b [MR - MQ]\,dx.$$

To get the limits of integration we solve MR and MQ simultaneously:

$$\begin{aligned} x^2 &= x^3 \\ x^2 - x^3 &= 0 \\ x^2(1 - x) &= 0 \\ x &= 0, \quad x = 1. \end{aligned}$$

At $x = 0$, $MR = MQ = 0$. At $x = 1$, $MR = MQ = 1$. It is clear that the two curves intersect at the points (0, 0) and (1, 1) (see Figure 7.4).

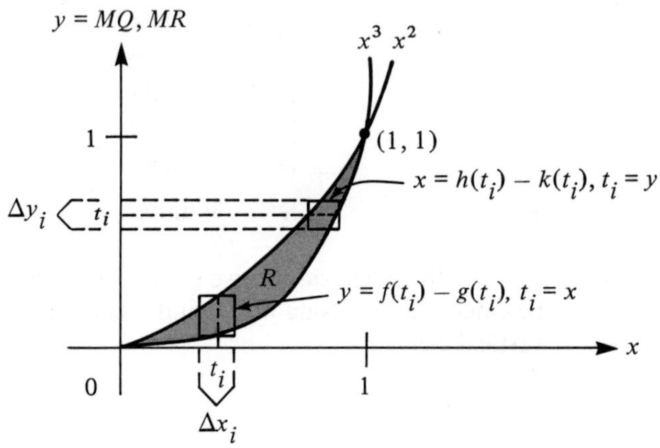

Figure 7.4

Using vertical elements of area we have area

$$A = \lim_{n \to \infty} \sum_{i=1}^{n} [f(t_i) - g(t_i)] \, \Delta x_i = \int_a^b [f(x) - g(x)] \, dx$$

where a and b are 0 and 1, respectively, $f(x) = x^2$, and $g(x) = x^3$. Substituting these values and performing the integration, we obtain profit

$$\begin{aligned} P &= \int_0^1 (x^2 - x^3) \, dx \\ &= \left[\frac{x^3}{3} - \frac{x^4}{4} \right]_0^1 \\ &= \frac{1}{3} - \frac{1}{4} = \frac{1}{12} \quad \text{thousands of dollars.} \end{aligned}$$

When using horizontal elements of area, we think of the y-axis as being partitioned into subintervals between $y = 0$ and $y = 1$. In the ith subinterval we choose a point t_i and designate the width of the subinterval as $\Delta y_i = y_i - y_{i-1}$. The length of the ith element is $[h(t_i) - k(t_i)]$ and its area is

$$[h(t_i) - k(t_i)] \, \Delta y_i.$$

We form a Riemann sum of these typical elements, take the limit as

$n \to \infty$, and define the area of R to be this limit:

$$A = \lim_{n \to \infty} \sum_{i=1}^{n} [h(t_i) - k(t_i)] \Delta y_i = \int_c^d [h(y) - k(y)]\,dy.$$

In this case c and d are the y values 0 and 1, respectively. To obtain $h(y)$, we solve $y = x^3$ for x. Then

$$x = y^{1/3} = h(y).$$

Similarly,

$$x = y^{1/2} = k(y).$$

It follows that profit P is equivalent to the numerical value of the area given by

$$\begin{aligned} P &= \int_0^1 [y^{1/3} - y^{1/2}]\,dy \\ &= y^{4/3} \cdot \frac{3}{4} - y^{3/2} \cdot \frac{2}{3}\Big]_0^1 \\ &= \frac{3}{4}(1) - \frac{2}{3}(1) \\ &= \frac{9}{12} - \frac{8}{12} = \frac{1}{12} \quad \text{thousands of dollars.} \end{aligned}$$

This answer agrees with that obtained when using vertical elements of area. In some instances the area between two curves can be calculated more conveniently by the horizontal-element approach.

In some situations it is possible (or necessary) to evaluate two definite integrals in order to obtain the total area of a region. The next example illustrates.

EXAMPLE 5 Find the area of the triangular region whose vertices are (0, 0), (1, 2), and (3, 0).

Solution Figure 7.5 exhibits the triangular region. The area A that we desire is the sum of the areas of R_1 and R_2. That is,

$$A = (\text{area of } R_1) + (\text{area of } R_2).$$

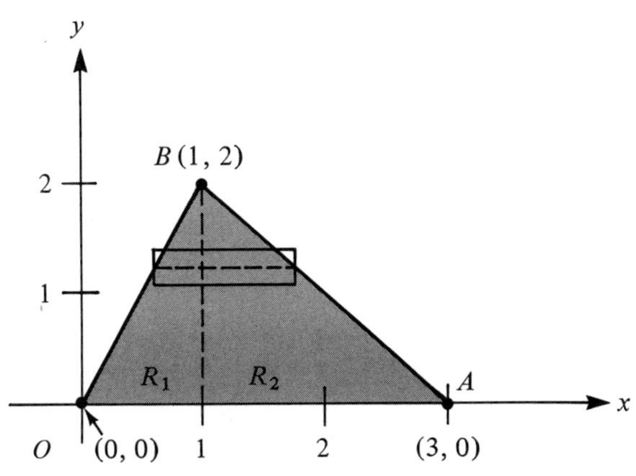

Figure 7.5

First, we find equations of lines OB and AB. The slope of OB is

$$\begin{aligned} m &= \frac{\Delta y}{\Delta x} \\ &= \frac{2-0}{1-0} = 2. \end{aligned}$$

An equation of line OB is of the form

$$y - y_1 = m(x - x_1)$$

or

$$\begin{aligned} y - 0 &= 2(x - 0) \\ y &= 2x \end{aligned}$$

where we used (0, 0) for (x_1, y_1). The slope of line AB is

$$m = \frac{2-0}{1-3} = -1.$$

An equation of AB is

$$y - 0 = -1(x - 3)$$

or

$$y = -x + 3,$$

where we used (3, 0) for (x_1, y_1). Of course, we could use (1, 2) for (x_1, y_1) and get the same equation for line AB. The area of R_1 is

$$\begin{aligned} A_1 &= \int_0^1 2x\,dx \\ &= x^2 \Big]_0^1 = 1 \text{ unit}^2. \end{aligned}$$

To find the area of R_2 we proceed thus:

$$\begin{aligned} A_2 &= \int_1^3 (-x + 3)\,dx \\ &= \left[-\frac{x^2}{2} + 3x\right]_1^3 \\ &= -\frac{9}{2} + 9 + \frac{1}{2} - 3 \\ &= -4 + 6 = 2 \text{ units}^2. \end{aligned}$$

Then the required area A is

$$\begin{aligned} A &= A_1 + A_2 \\ &= 1 + 2 = 3 \text{ units}^2. \end{aligned}$$

We can use a horizontal element to check this answer. To this end, we solve $y = 2x$ for x and obtain

$$x = y/2 = h(y).$$

Similarly, we solve $y = -x + 3$ for x and get

$$x = 3 - y = k(y).$$

The required area is

$$\begin{aligned} A &= \int_0^2 [k(y) - h(y)]\,dy \\ &= \int_0^2 \left[(3 - y) - \frac{y}{2}\right] dy \\ &= \int_0^2 \left(3 - \frac{3}{2}y\right) dy \\ &= 3y - \frac{3y^2}{4}\Big]_0^2 \\ &= 6 - 3 = 3 \text{ units}^2, \quad \text{as before.} \end{aligned}$$

A simpler check can be made without the calculus. The area of a triangle is $\frac{1}{2}$(base)(height) or

$$\frac{1}{2}(1)(2) + \frac{1}{2}(2)(2) = 1 + 2 = 3 \text{ units}^2.$$

PROBLEM SET 7.1

Find the area of the region bounded by the graphs of the given functions and relations. Check your answer by use of horizontal elements wherever feasible. Sketch graphs.

1. $y = x$ and $y = x^2$

2. $y = 2x + 4$ and $y = 2x^2$

3. $y = 2x$ and $y = x^2 - 3$

4. $y = x/2$ and $y = (7x/2) - x^2$

5. $y = x$ and $y = \sqrt{x}$

6. $y = x^2$ and $y = 4x - x^2$

7. $y = 4x$ and $y = x^3$

8. $y = 2x^2$ and $y = 4x^3$

9. $y = e^x$, $y = 2$, and $x = 0$

10. $y = e^x$, $y = e^{-x}$, and $y = 2$

11. $y = e^x$, $y = e^{-x}$, and $y = 1/2$

12. $y = e^{-x}$, $y = 1$, and $x = 2$

Given the following marginal revenue and marginal cost functions, find the profit by methods of this section. Interpret the answer as thousands of dollars.

13. $MR = 4x - x^2$ and $MQ = 4 - x$.

14. $MR = 16 - x^2$ and $MQ = 19 - 4x$.

15. $MR = x$ and $MQ = x^2 - 4x + 6$.

16. $MR = 2 - x^2$ and $MQ = x^3$ on $[0, 1]$.

17. $MR = 12 - x^2$ and $MQ = x^3$ on $[0, 2]$.

18. $MR = e^{x/2}$ and $MQ = e^{-x}$ on $[0, 2]$.

19. $MR = 3/2$ and $MQ = e^x/2$

20. $MR = x$ and $MQ = x^4$.

21. $MR = 6x$ and $MQ = 3x^2$.

Find the area of the region bounded by the following curves:

22. $3y = 2x$, $y = 5 - x$, and $y = 0$.

23. $y = 3x$, $y = -3x + 6$, and $y = 0$.

24. $y = \sqrt{x}$, $y = 4 - x/2$, and $y = 0$.

25. $y = 5 - x^2$ and $y = x^2 - 3$.

26. Find the area of the triangular region having vertices (0, 0), (1, 1), and (0, 3).

7.2 CONSUMERS' AND PRODUCERS' SURPLUS

Assume that the demand D and the supply S for a certain commodity X are defined by the system (called a demand-supply system)

$$\left.\begin{aligned} D:&\quad p = 900 - q^2 \\ S:&\quad p = 200 + 15q \end{aligned}\right\}$$

where price p is a function of q, the quantity of X demanded (or, equivalently, the quantity supplied). In an economic system of pure competition (or monopoly), suppose that a quantity $\overline{q}$ represents the market demand which corresponds to a price $\overline{p}$. The point $(\overline{q}, \overline{p})$ is often referred to as a "break-even" point, that is, a point of market equilibrium.

We find the point $(\overline{q}, \overline{p})$ for the above demand-supply system by solving equations D and S simultaneously. Accordingly,

$$\begin{aligned} 200 + 15q &= 900 - q^2 \\ q^2 + 15q - 700 &= 0 \\ (q - 20)(q + 35) &= 0 \\ q &= 20 \end{aligned}$$

is the only solution since $q = -35$ is not empirically valid (that is, the amount of a commodity cannot be negative). Substituting in D we have

$$p = 900 - (20)^2 = 500.$$

The resulting point of intersection (20, 500) represents $(\overline{q}, \overline{p})$, the point of equilibrium (see Figure 7.6).

It is reasonable to assume that some shoppers would be willing to buy commodity X even if the price were higher than \$500, the equilibrium price. For price $p =$ \$644, the demand can be found by solving

$$\begin{aligned} 644 &= 900 - q^2 \\ q^2 &= 900 - 644 = 256 \\ q &= \sqrt{256} = 16 \text{ units.} \end{aligned}$$

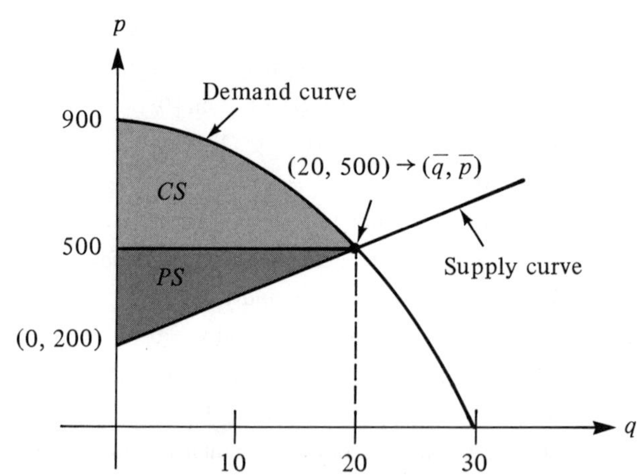

Figure 7.6

Even at a price of $p = \$899$ there would still be a demand of

$$q^2 = 900 - 899 = 1$$

or

$$q = \sqrt{1} = 1 \text{ unit of commodity } X.$$

Assuming certain economic conditions, we define the consumers' gain as the **consumers' surplus**, CS, and calculate it by evaluating the difference of the two areas:

$$(\text{Area under } D) \quad \text{minus} \quad (\text{Area under } \overline{p}).$$

That is, we find the area between the two curves $p = 900 - q^2$ and $p = 500$. The area under D is given by the definite integral:

$$\int_0^{20} (900 - q^2)\, dq$$

and the area under $\overline{p}$ is simply the area of a 20 by 500 rectangle. Combining these two features as a difference of two areas, we find consumers' surplus to be

$$\begin{aligned} CS &= \int_0^{20} (900 - q^2)\, dq - (20)(500) \\ &= 900q - q^3/3 \Big]_0^{20} - 10{,}000 \\ &= 900(20) - \frac{8000}{3} - 10{,}000 \cong \$5333.33. \end{aligned}$$

In a similar vein, it is reasonable to think that there would be some producers who would be willing to supply units of commodity X at a price below the market equilibrium price. The producers' gain is defined as the **producers' surplus** and is represented by the area between the supply curves and the line $\overline{p} = 500$. The area under $\overline{p} = 500$ was shown to be the product of a 20×500 rectangle or (20)(500). The area under S is given by the definite integral

$$\int_0^{20} (200 + 15q)\, dq.$$

The difference of these two areas is a value which defines producers' surplus:

$$\begin{aligned} PS &= (20)(500) - \int_0^{20} (200 + 15q)\, dq \\ &= 10{,}000 - \left[200q + \frac{15q^2}{2}\right]_0^{20} \\ &= 10{,}000 - 200(20) - \frac{15}{2}(400) \\ &= \$3000. \end{aligned}$$

In general, consumers' surplus is a difference of two areas given by

$$\boxed{CS = \int_0^{\overline{q}} D(q)\, dq - (\overline{q})(\overline{p})}$$

and producers' surplus is defined as

$$PS = (\bar{q})(\bar{p}) - \int_0^{\bar{q}} S(q)\, dq$$

Figure 7.7 illustrates these concepts. Observe that the demand curve is a nonincreasing function, whereas the supply curve is a nondecreasing function.

We now present one more example in order to increase our understanding of these important economic concepts.

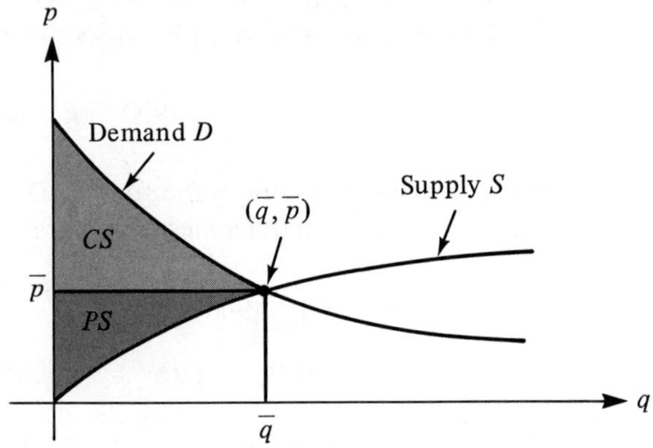

Figure 7.7

EXAMPLE 1 Find consumers' surplus CS and producers' surplus PS for the given demand-supply system:

$$\left.\begin{aligned} D:&\quad p = 7.5 - q \\ S:&\quad 10p = q^2 + 2p + 11 \end{aligned}\right\}$$

Solution Solving D and S simultaneously, we get

$$q^2 + 2q + 11 = 10p = 10(7.5 - q) = 75 - 10q.$$

Then

$$q^2 + 12q - 64 = 0$$
$$(q - 4)(q + 16) = 0$$

where

$$q = 4$$

is the only relevant solution. When $q = 4$, we obtain

$$p = 7.5 - 4 = 3.5.$$

Thus, the equilibrium point $(\bar{q}, \bar{p})$ is (4, 3.5). Figure 7.8 exhibits this example geometrically.

To find consumers' surplus we evaluate

$$\begin{aligned} CS &= \int_0^4 D(q)\, dq - (\bar{q})(\bar{p}) \\ &= \int_0^4 (7.5 - q)\, dq - (4)(3.5) \\ &= 7.5q - q^2/2 \Big]_0^4 - 14 \\ &= 30 - 8 - 14 = \$8. \end{aligned}$$

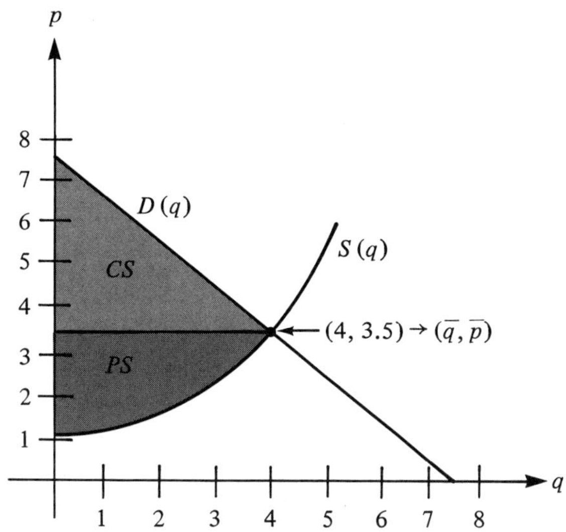

Figure 7.8

We calculate the producers' surplus by evaluating the difference of two areas as follows:

$$\begin{aligned} PS &= (\bar{q})(\bar{p}) - \int_0^4 S(q)\,dq \\ &= (4)(3.5) - \int_0^4 \frac{(q^2 + 2q + 11)}{10}\,dq \\ &= 14 - \frac{1}{10}[q^3/3 + q^2 + 11q]_0^4 \\ &= 14 - \frac{1}{10}\left(\frac{64}{3} + 16 + 44\right) \\ &= 14 - \frac{64}{30} - 6 \cong \$5.87. \end{aligned}$$

PROBLEM SET 7.2

Find the consumers' and producers' surplus for each given demand-supply system. Sketch graphs of the odd-numbered problems.

1. $\left.\begin{aligned} D&: \ p = 11 - q \\ S&: \ p = q + 5 \end{aligned}\right\}$

2. $\left.\begin{aligned} D&: \ p = -2q + 8 \\ S&: \ p = q + 2 \end{aligned}\right\}$

3. $\left.\begin{aligned} D&: \ p = -q + 3 \\ S&: \ p = q^2/4 \end{aligned}\right\}$

4. $\left.\begin{aligned} D&: \ p = \frac{-3q}{2} + 8 \\ S&: \ p = q^2 + 1 \end{aligned}\right\}$

5. $\left.\begin{aligned} D&: \ p = 6 - q^2 \\ S&: \ p = \frac{q}{2} + 1 \end{aligned}\right\}$

6. $\left.\begin{aligned} D&: \ 2p = 22 - 2q^2 \\ S&: \ 2p = q + 1 \end{aligned}\right\}$

7. $\left.\begin{aligned} D&: \ p = 9 - q^2 \\ S&: \ p = q^2 + 1 \end{aligned}\right\}$

8. $\left.\begin{aligned} D&: \ p = 12 - q^2 \\ S&: \ p = \frac{q^2}{3} \end{aligned}\right\}$

9. $\left.\begin{array}{l} D\colon\ p = -q^2 - 2q + 5 \\ S\colon\ p = q^2 + 1 \end{array}\right\}$

10. $\left.\begin{array}{l} D\colon\ p = -q^2 - 4q + 14 \\ S\colon\ p = 5 + 4q \end{array}\right\}$

11. $\left.\begin{array}{l} D\colon\ p = 14 - q^2 \\ S\colon\ p = 2 + 2q^2 \end{array}\right\}$

12. $\left.\begin{array}{l} D\colon\ p = -q^3 + 34 \\ S\colon\ p = q^3 - 20 \end{array}\right\}$

13. $\left.\begin{array}{l} D\colon\ p = \dfrac{2}{q+1} \\ S\colon\ p = q \end{array}\right\}$

14. $\left.\begin{array}{l} D\colon\ p = \dfrac{4}{q+1} \\ S\colon\ p = q + 1 \end{array}\right\}$

15. $\left.\begin{array}{l} D\colon\ p = 5.5 - 1.5q \\ S\colon\ p = 1.5 + 0.5q \end{array}\right\}$

16. $\left.\begin{array}{l} D\colon\ p = \dfrac{25}{2} - \dfrac{3q}{8} \\ S\colon\ p = \dfrac{16}{3} + \dfrac{2q}{9} \end{array}\right\}$

7.3 AVERAGE VALUE OF A FUNCTION

If a set of data (finite in number) is given by $y = f(x)$, where the function f is defined and continuous on the closed interval $[a, b]$, it may be of interest in some situations to find the average value of these data.

Most of us are familiar with the method of calculating the average value of a set of data representing employees' wages, for example, y_1, y_2, $y_3, \ldots, y_n$ where there are n of the y's. Then the arithmetic mean or average value of the wages can be computed by

$$\bar{y} = \frac{y_1 + y_2 + y_3 + \cdots + y_n}{n} = \frac{1}{n}\sum_{i=1}^{n} y_i.$$

If $\lim_{n\to\infty}(1/n)\sum_{i=1}^{n} y_i$ exists, then we have the average value of $f(x)$ on $[a, b]$. The next definition follows naturally.

DEFINITION 1 *The **average value** of a continuous function $f(x)$ with respect to x on $[a, b]$ is given by*

$$\boxed{\bar{y} = \frac{1}{b-a}\int_a^b f(x)\,dx}\,. \qquad (1)$$

Our geometric intuition is satisfied that Eq. (1) makes sense, for it is reasonable to expect that there exists a value of $x = t \in [a, b]$ such that the area of the rectangle of height $\bar{y}$ and base length $(b - a)$ is equal to the area under f on $[a, b]$ (see Figure 7.9). That is,

$$\bar{y}\cdot(b - a) = \int_a^b f(x)\,dx$$

or

$$\bar{y} = \frac{1}{b-a}\int_a^b f(x)\,dx.$$

Actually, there is a theorem which establishes the validity of Eq. (1). We state the theorem without proof.

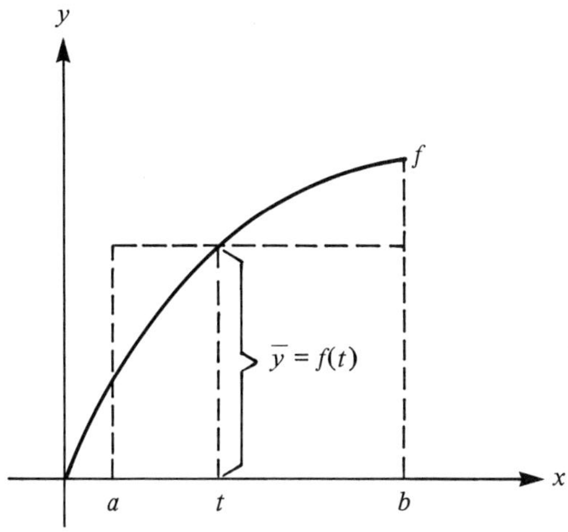

Figure 7.9

THEOREM 1 A Mean Value Theorem for Integrals

If $y = f(x)$ is continuous in the interval $[a, b]$, then there exists a point $x = t$ in $[a, b]$ such that

$$\int_a^b f(x)\,dx = f(t)(b - a).$$

If, for instance, $y = Q(x)$ defines the cost of production per month of x units of input, then Eq. (1) would represent the average cost of production for that month. The average cost is illustrated geometrically in Figure 7.10.

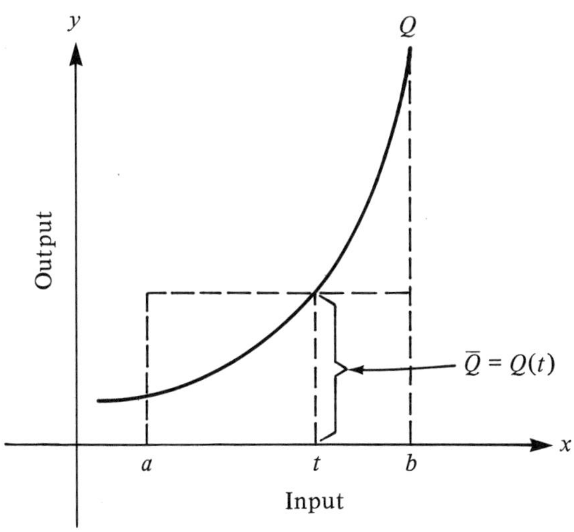

Figure 7.10

EXAMPLE 1 Find the average total cost if marginal cost is given by $MR = (x^2/2) + 1$ thousands of dollars, where x is in hundreds of units of production and x increases from 1 to 4.

Solution Average total cost is found by evaluating

$$\begin{aligned}\bar{Q} &= \frac{1}{4-1}\int_1^4 \left(\frac{x^2}{2}+1\right)dx \\ &= \frac{1}{3}\left[\frac{x^3}{6}+x\right]_1^4 \\ &= \frac{1}{3}\left[\frac{64}{6}+4-\frac{1}{6}-1\right] \\ &= \frac{9}{2} \quad \text{or} \quad \$4500 \qquad \left[\text{since } \frac{9}{2}\times \$1000 = \$4500\right].\end{aligned}$$

There may be occasions when we wish to know the value of $x = t$ at which $\bar{y}$ occurs. The Mean Value Theorem for Integrals does not stipulate what this value of t is. However, it can be found in certain situations. One such case is Example 1. Let $\bar{y} = f(t)$ in Eq. (1) so that

$$f(t) = \frac{1}{b-a}\int_a^b f(x)\,dx.$$

If we multiply both sides by $(b-a)$ we obtain

$$f(t)\cdot(b-a) = \int_a^b f(x)\,dx.$$

From Example 1 we have $(b-a) = 4 - 1 = 3$, $f(x) = x^2/2 + 1$, and $f(t) = t^2/2 + 1$. Substituting, we have

$$\left(\frac{t^2}{2}+1\right)(3) = \int_1^4\left(\frac{x^2}{2}+1\right)dx = \frac{27}{2}.$$

Then

$$\begin{aligned}\frac{t^2}{2}+1 &= \frac{1}{3}\left(\frac{27}{2}\right) = \frac{9}{2} \\ \frac{t^2}{2} &= \frac{9}{2} - 1 = \frac{7}{2} \\ t^2 &= 7 \\ t &= \pm\sqrt{7} \quad \text{or} \quad \sqrt{7} \qquad [\text{since } -\sqrt{7} \notin [1, 4]].\end{aligned}$$

Thus, the average total cost occurs when $x = t = \sqrt{7} \cong 2.6457$, or about 265 units are produced ($2.6457 \times 100 \cong 265$ units).

In this section we learned how the definite integral acts as an ideal tool for generalizing from the concept of average (or mean) to the concept of the average value of a continuous function f with respect to a variable x.

PROBLEM SET 7.3

Find the average value of the given function on the indicated interval. In problems 1–8, solve for the value of x at which the average value occurs.

1. Cost $Q(x) = (2x/3) + 10$ on $[0, 5]$.

2. Demand $D(x) = 20 - (x/5)$ on $[0, 30]$.

3. Demand $D(x) = 50 - (x/2)$ on $[0, 20]$.
4. Cost $Q(x) = (3x/5) + 5$ on $[5, 10]$.
5. Revenue $R(x) = x^2/2$ on $[0, 10]$.
6. Revenue $R(x) = (x^2/5) - 1$ on $[0, 20]$.
7. Profit $P(x) = 10x - x^2$ on $[0, 3]$.
8. Profit $P(x) = 40x - 3x^2$ on $[0, 10]$.
9. Sales $S(x) = (x + 1)^2/10$ (in thousands) on $[0, 3]$.
10. Sales $S(x) = x^3/5$ on $[10, 20]$.
11. Price $P(x) = (x - 1)^3/25$ (in hundreds) on $[2, 6]$.
12. Price $P(x) = (x^2 + 1)(x + \frac{1}{2})$ on $[1, 4]$.
13. Demand $D(x) = 100 - x^3/10$ on $[5, 10]$.
14. Supply $S(x) = 2/x^2$ (in thousands) on $[1, 3]$
15. Cost $Q(x) = x(x^2 + 1)^3$ (in thousands) on $[0, 1]$.
16. Cost $Q(x) = 900\sqrt{x}$ on $[0, 9]$.
17. Profit $P(x) = 750\sqrt{x}$ on $[0, 25]$.
18. Demand $D(x) = 200/(10 + 2x)$ on $[0, 10]$.
19. Demand $D(x) = 300/(10 + 3x)$ on $[10, 30]$.
20. Depreciation $D(x) = 100e^{-x/2}$ on $[0, 5]$.

7.4 INTEGRATION BY PARTS

In this section we continue the development of computational skills by the introduction and application of a method called **integration by parts**. Actually, this is one of the two more important methods of integration, the other being the method of substitution. Most problems of integration can be solved by an application of either of these two main techniques.

Integration by parts finds its basis in the formula for the differential of the product:

$$d(uv) = u\,dv + v\,du$$

where u and v are continuously differentiable functions of the independent variable x. We say that a function $f(x)$ is continuously differentiable on (a, b) if $f'(x)$ is defined and continuous on (a, b). Solving for $u\,dv$ and integrating, we obtain

$$\boxed{\int u\,dv = uv - \int v\,du + C}. \tag{1}$$

Equation (1) is called the formula for integration by parts. The integral to be evaluated is expressed in terms of another integral, $\int v\,du$. Since this latter integral must be integrable if we are to make progress, we should choose u and dv in such a way that this aim is accomplished. If, for instance, we are attempting to apply Eq. (1) to the integral

$$\int g(x) \cdot h(x)\,dx,$$

we might let $u = g(x)$ and $dv = h(x)\,dx$. Then we would find

$$du = g'(x)\,dx \quad \text{and} \quad v = \int h(x)\,dx.$$

If the latter equation were not integrable we would back up and try again by letting $u = h(x)$ and $dv = g(x)\,dx$. In the process we obviously must be able to recognize whether a given integral is integrable. In some cases our knowledge and diagnostic ability can be put to a severe test.

EXAMPLE 1 Evaluate $\int xe^x\,dx$.

Solution Let $u = x$ and $dv = e^x\,dx$. Then $du = dx$ and

$$dv = \int e^x\,dx$$

or

$$v = e^x.$$

We will not include the constant of integration after each integration but will rather write the "$+C$" at the end of the final integrated equation. This is standard procedure and can be proven to be mathematically valid. Substituting in the given integral we have

$$\begin{aligned}\int xe^x\,dx = \int u\,dv &= uv - \int v\,du \\ &= xe^x - \int e^x\,dx \\ &= xe^x - e^x + C.\end{aligned}$$

Caution: If we had let $u = e^x$, $du = e^x\,dx$ and $dv = x\,dx$, $v = x^2/2$, then we would be faced with a worsening situation in the form of:

$$\begin{aligned}\int xe^x\,dx &= uv - \int v\,du \\ &= \frac{x^2e^x}{2} - \int \frac{x^2e^x}{2}\,dx.\end{aligned}$$

It is apparent that $\int(x^2e^x/2)\,dx$ is more difficult to cope with than is the original integral. Thus, we have retrogressed and not progressed. The discerning student will check himself (or herself) before wasting much time on the wrong path and return to the right course.

EXAMPLE 2 Find $\displaystyle\int \frac{\ln|x|\,dx}{x}$.

Solution

$$\text{Let } u = \ln|x|. \qquad \text{Then } du = \frac{1}{x}\,dx$$

$$\text{Let } dv = \frac{dx}{x}. \qquad \text{Then } v = \int \frac{dx}{x} = \ln|x|.$$

Observe that letting $dv = \ln|x|\,dx$ would have resulted in finding v by integration by parts, thus compounding the problem. Therefore,

$$\int \frac{\ln|x|\,dx}{x} = \ln|x| \cdot \ln|x| - \int \frac{\ln|x|\,dx}{x}.$$

Solving for the required integral we get

$$2\int \frac{\ln|x|\,dx}{x} = (\ln|x|)^2$$

or

$$\int \frac{\ln|x|\,dx}{x} = \frac{(\ln|x|)^2}{2} + C.$$

The given integral can be calculated by a previous method of substitution. You should apply the substitution method and satisfy yourself that your answer agrees with the one above.

EXAMPLE 3 If marginal profit is defined by $MP = x\sqrt{9 - x}$ (in thousands of dollars), where x denotes quantity of production, find profit P when x increases from $x = 0$ to $x = 5$.

Solution

$$P = \int_0^5 MP\,dx = \int_0^5 x(9 - x)^{1/2}\,dx.$$

Let

$$\begin{aligned} u &= x, \quad du = dx, \\ dv &= (9 - x)^{1/2}\,dx \\ v &= \int (9 - x)^{1/2}\,dx \\ &= -(9 - x)^{3/2} \cdot \frac{2}{3} \\ &= -\frac{2}{3}(9 - x)^{3/2}. \end{aligned}$$

Then

$$\begin{aligned} P &= -\frac{2x}{3}(9 - x)^{3/2}\Big]_0^5 - (-2/3)\int_0^5 (9 - x)^{3/2}\,dx \\ &= -\frac{10}{3}(4^{3/2}) + \frac{2}{3}\left[-(9 - x)^{5/2} \cdot \frac{2}{5}\right]_0^5 \\ &= -\frac{80}{3} - \frac{4}{15}[4^{5/2} - 9^{5/2}] \\ &= -\frac{80}{3} - \frac{4}{15}[32 - 243] \\ &= \frac{444}{15} \text{ thousands of dollars} \\ &= \$29{,}600. \end{aligned}$$

EXAMPLE 4 Given the marginal cost function $MQ = \ln(t + 1)$, where time t is the number of quarters. Find cost Q when t increases from 0 to 2.

Solution

$$Q = \int_0^2 \ln(t + 1)\, dt.$$

Let $u = \ln(t + 1),$

then $du = \dfrac{dt}{t + 1},$

$$dv = dt,$$

$$v = t.$$

Then
$$\begin{aligned} Q &= t \ln(t + 1) \Big]_0^2 - \int_0^2 \frac{t\, dt}{t + 1} \\ &= 2 \ln 3 - \int_0^2 \left[1 - \frac{1}{t + 1}\right] dt \qquad \text{[by long division]} \\ &= 2 \ln 3 - [t - \ln|t + 1|]_0^2 \\ &= 2 \ln 3 - [2 - \ln 3 + \ln 1] \\ &= 3 \ln 3 - 2 \\ &= 3(1.0986) - 2 \cong \$1.30. \end{aligned}$$

PROBLEM SET 7.4

Integrate by parts:

1. $\int xe^{-x}\, dx$ **2.** $\int xe^{-x/2}\, dx$ **3.** $\int te^{2t}\, dt$

4. $\int \ln u\, du$ **5.** $\int y \ln y\, dy$ **6.** $\int \ln(2x + 1)\, dx$

7. $\int x(x - 1)^3\, dx$ **8.** $\int t\sqrt{t + 5}\, dt$ **9.** $\int ze^{3z}\, dz$

Compute the area of the region R bounded by the given curves:

10. $y = \ln x$, $y = 0$, and $x = e^2$.

11. $y = xe^{x/3}$, $x = 0$, and $x = 3$.

12. $y = x(2 - x)^2$, $x = 0$, and $x = 2$.

13. $y = \dfrac{x^2}{x - 1}$, $x = 2$, and $x = 3$.

14. $y = (3^x)$, $x = 0$, and $x = 1$. [*Hint*: use $3 = e^{\ln 3}$.]

15. If total cost Q is changing at the rate $MQ = xe^{0.02x}$, find Q when x increases from 0 to 50.

16. Work problem 15 if $MQ = xe^{0.01x}$ and the domain of integration is $0 \leq x \leq 100$.

17. A certain gold stock grows at the rate $f(t) = 10 + \ln t$ where t is time in months. Estimate the increase in growth as t increases from 1 to 2.

18. Work problem 17 if $f(t) = 5 + \ln 5t$ in the time period $1 \leq t \leq 5$.

19. Use integration by parts to evaluate:

(a) $\int \ln(x^3)\, dx$ **(b)** $\int 2xe^{2x-1}\, dx$

(c) $\int_0^{0.1} 10te^{-10t}\, dt$ **(d)** $\int_2^6 \ln(2 + t)\, dt$

7.5 PARTIAL FRACTIONS: LINEAR FACTORS

If $f(x)$ and $g(x)$ are polynomial functions, then their quotient $f(x)/g(x)$ is called a rational function. When the degree of the numerator $f(x)$ is less than the degree of the denominator $g(x)$, then $f(x)/g(x)$ is a *proper* rational function. When integrating a rational function, we first perform a "decomposition" of the fraction into a sum of fractions whose denominators are **distinct linear factors**, none of them repeated. Each fraction in the sum will be of the form $A/(ax + b)$ where the A, a, and b are real numbers. We assume the denominator has factors of the form $ax + b$.

The following example illustrates the decomposition or resolution of a fractional integrand into its equivalent **partial fractions**.

EXAMPLE 1 Resolve into partial fractions: $\dfrac{2x - 1}{x^2 - x - 6}$.

Solution I The denominator factors into a product of two linear factors, none of them repeated:

$$x^2 - x - 6 = (x - 3)(x + 2).$$

By a theorem in algebra, there exist constants A and B such that

$$\frac{2x - 1}{x^2 - x - 6} = \frac{2x - 1}{(x - 3)(x + 2)} = \frac{A}{x - 3} + \frac{B}{x + 2} \tag{1a}$$

$$= \frac{A(x + 2) + B(x - 3)}{(x - 3)(x + 2)}. \tag{1b}$$

Clearing fractions (that is, multiplying both sides of Eq. (1) by the common denominator), we obtain

$$2x - 1 = A(x + 2) + B(x - 3). \tag{2}$$

This form defines the equality of two polynomials. In fact, an algebraic theorem goes farther and declares that the two polynomials in Eq. (2) are *identically* equal. In other words, Eq. (2) is valid for all real values of x. (We shall not pursue these useful theorems from algebra, but instead we shall exploit their practical implications.)

One method of expediting the calculation of constants A and B in Eq. (2) is to select convenient values of x, substitute in Eq. (2), and perform the algebraic manipulations:

Let $x = -2$; then $2(-2) - 1 = A(0) + B(-5)$

or $-5 = -5B \rightarrow B = 1.$

Let $x = 3$; then $2(3) - 1 = A(5) + B(0)$

$5 = 5A \rightarrow A = 1.$

Substituting these values of A and B in Eq. (1a), we have

$$\frac{2x - 1}{x^2 - x - 6} = \frac{1}{x - 3} + \frac{1}{x + 2},$$

the resolved form.

If we wished to integrate the original fraction, we would resolve it into partial fractions in this way, obtaining a form clearly integrable. We would then have

$$\begin{aligned}\int \frac{2x-1}{x^2-x-6}\,dx &= \int \left[\frac{1}{x-3} + \frac{1}{x+2}\right] dx \\ &= \ln|x-3| + \ln|x+2| + C \\ &= \ln|(x-3)(x+2)| + C.\end{aligned}$$

Solution II Since Eq. (2) is an identity in x, then like powers of x on the two sides of Eq. (2) are equal. That is, corresponding coefficients are the same. This fact prompts us to equate the coefficients of like powers of x, thus:

(a) Coefficients of x: $2 = A + B$

(b) Coefficients of x^0: $-1 = 2A - 3B$ (or equating constants)

The result is a system of two equations in the two unknowns A and B. Solving the system, we get

$$\left.\begin{aligned} 4 &= 2A + 2B \\ -1 &= 2A - 2B \end{aligned}\right\}.$$

Subtraction yields

$$5 = 5B \quad \text{or} \quad B = 1.$$

Substituting, we get

$$\begin{aligned} 4 &= 2A + 2(1) \\ 2 &= 2A \quad \text{or} \quad A = 1. \end{aligned}$$

These answers check those in Solution I.

The denominator of an integrand may contain more than two distinct linear factors. The technique for resolving into partial fractions is the same as in the previous example except for additional algebraic manipulations. The next example illustrates.

EXAMPLE 2 Find the total cost Q (in hundred thousands of dollars) if the marginal cost is defined by

$$MQ = \frac{x^2+1}{x^3-x} \qquad \text{on } [2, 4].$$

Solution

$$Q = \int_2^4 \frac{x^2+1}{x^3-x}\,dx.$$

We have $\dfrac{x^2+1}{x^3-x} = \dfrac{x^2+1}{x(x-1)(x+1)} = \dfrac{A}{x} + \dfrac{B}{x-1} + \dfrac{C}{x+1}$

and $x^2 + 1 = A(x^2-1) + Bx(x+1) + Cx(x-1)$ (3)

is an identity in x. The form in Eq. (3) is the result of clearing fractions in the step immediately above. We wish to find values of A, B, and C so as to have a sum of fractions which is integrable. Toward this end we choose convenient values of x in Eq. (3) and solve:

Let $x = 1$; then $1 + 1 = 0 + B(2) = 2$, $B = 1$
Let $x = -1$; then $1 + 1 = 0 - B(0) - C(-2) = 2C = 2$, $C = 1$.

To find the value of A we may equate coefficients of like powers of x as follows:

$$\text{Equating constants:} \quad 1 = -A \quad \text{or} \quad A = -1.$$

Substituting $A = -1$, $B = 1$, $C = 1$ in the original resolved form, we get

$$\frac{x^2 + 1}{x^3 - x} = \frac{-1}{x} + \frac{1}{x - 1} + \frac{1}{x + 1} \tag{4}$$

where the right side of Eq. (4) is an integrable form. Integrating, we have total cost

$$\begin{aligned}
Q &= \int_2^4 \frac{x^2 + 1}{x^3 - x}\,dx \\
&= \int_2^4 \left[-\frac{1}{x} + \frac{1}{x - 1} + \frac{1}{x + 1}\right] dx \\
&= [-\ln|x| + \ln|x - 1| + \ln|x + 1|]_2^4 \\
&= -\ln|x| + \ln|(x - 1)(x + 1)| \Big]_2^4 \\
&= \ln\left|\frac{(x - 1)(x + 1)}{x}\right| \Bigg]_2^4 \\
&= \ln\frac{3(5)}{4} - \ln\frac{3}{2} \\
&= \ln\frac{15}{4} - \ln\frac{6}{4} \\
&= \ln\frac{15}{4}\cdot\frac{4}{6} = \ln\frac{5}{2} \\
&\cong 0.91629 \quad \text{hundred thousands of dollars}
\end{aligned}$$

or \$91,629.

If we elected to find A, B, and C solely by the method of equating coefficients of like powers of x, we would obtain from Eq. (3):

(a) Coefficients of x^2: $1 = A + B + C$

(b) Coefficients of x: $0 = B - C$ or $B = C$

(c) Constants: $1 = -A$ or $A = -1$.

Substituting $A = -1$ in (a) we get

$$1 = -1 + B + C$$

or

$$2 = B + C. \tag{5}$$

From (b) we have $B = C$, so Eq. (5) becomes

$$2 = C + C = 2C$$

or

$$C = 1 = B.$$

These results clearly agree with those obtained above.

When a rational function of the form $f(x)/g(x)$ is such that the degree of the numerator $f(x)$ is *greater* than the degree of the denominator $g(x)$, we say that the fraction is *improper*. In order to integrate improper fractions, we first divide the denominator into the numerator using long division until a remainder is obtained that is of lower degree than the denominator. Let us illustrate.

EXAMPLE 3 Integrate $\int \frac{x^4 + x^3 - 1}{x^2 + 1}\,dx$.

Solution We first perform the following long division:

$$\begin{array}{lrl}
\text{Divisor} \rightarrow \ x^2 + 1 & x^4 + x^3 - 1 & \underline{|x^2 + x - 1} \leftarrow \text{Quotient} \\
& \underline{x^4 + x^2 } & \\
& x^3 - x^2 - 1 & \\
& \underline{x^3 + x } & \\
& - x^2 - x - 1 & \\
& \underline{- x^2 - 1} & \\
& - x & \leftarrow \text{Remainder.}
\end{array}$$

So
$$\frac{x^4 + x^3 - 1}{x^2 + 1} = x^2 + x - 1 - \frac{x}{x^2 + 1}.$$

Then
$$\int \frac{x^4 + x^3 - 1}{x^2 + 1}\,dx = \int \left[x^2 + x - 1 - \frac{x}{x^2 + 1}\right] dx$$
$$= \frac{x^3}{3} + \frac{x^2}{2} - x - \frac{1}{2}\ln(x^2 + 1) + C.$$

We bring this section to a close by considering the case where the denominator has only linear factors, *some* of which *are repeated*.

EXAMPLE 4 Find the area under the curve

$$f(x) = \frac{3x - 2}{(x + 1)(x + 2)^2} \qquad \text{on } [0, 1].$$

Solution
$$\text{Area} = \int_0^1 \frac{3x - 2}{(x + 1)(x + 2)^2}\,dx$$

where
$$\frac{3x - 2}{(x + 1)(x + 2)^2} = \frac{A}{x + 1} + \frac{B}{x + 2} + \frac{C}{(x + 2)^2}. \tag{6}$$

Again we have invoked the aid of an algebraic theorem which states:

For each factor $(x - a)$ there corresponds a term $\frac{A}{x - a}$.

For each factor $(x - a)^2$ there correspond two terms $\frac{B}{x - a} + \frac{C}{(x - a)^2}$.

For each factor $(x - a)^3$ there correspond three terms $\dfrac{D}{x - a} + \dfrac{E}{(x - a)^2} + \dfrac{F}{(x - a)^3}$, and so on.

Clearing fractions in Eq. (6) results in the form

$$3x - 2 = A(x + 2)^2 + B(x + 1)(x + 2) + C(x + 1). \tag{7}$$

We equate coefficients of like powers of x and get:

(a) Coefficient of x^2: $0 = A + B$ or $B = -A$

(b) Coefficient of x: $3 = 4A + 3B + C$

(c) Constants: $-2 = 4A + 2B$.

Using $B = -A$ from (a) in (b) and in (c), we have

(d) $3 = A + C$

(e) $-2 = 2A$ or $A = -1$ and $B = -(-1) = 1$.

Putting $A = -1$ in (d) we get $C = 4$. Therefore,

$$\frac{3x - 2}{(x - 1)(x + 2)^2} = \frac{-1}{x + 1} + \frac{1}{x + 2} + \frac{4}{(x + 2)^2}.$$

An evaluation of the definite integral follows:

$$\begin{aligned}
\int_0^1 \frac{3x - 2}{(x + 1)(x + 2)^2}\,dx &= \int_0^1 \left[-\frac{1}{x + 1} + \frac{1}{x + 2} + 4(x + 2)^{-2}\right] dx \\
&= -\ln|x + 1| + \ln|x + 2| + \frac{4(x + 2)^{-1}}{-1}\Bigg]_0^1 \\
&= -\ln 2 + \ln 3 - 4\left(\frac{1}{3}\right) - \ln 2 + 4\left(\frac{1}{2}\right) \\
&= \ln 3 - 2\ln 2 - \frac{4}{3} + 2 \\
&= \ln 3 - \ln 4 + \frac{2}{3} \\
&= \left(\ln\frac{3}{4} + \frac{2}{3}\right) \text{units}^2.
\end{aligned}$$

PROBLEM SET 7.5

Integrate by methods of this section:

1. $\displaystyle\int \frac{dx}{x^2 + x - 2}$

2. $\displaystyle\int \frac{x + 3}{x^2 + x}\,dx$

3. $\displaystyle\int \frac{x^3 + 5}{x^2 - 25}\,dx$

4. $\displaystyle\int \frac{x\,dx}{x^2 + 2x - 8}$

5. $\displaystyle\int \frac{(x + 2)\,dx}{x^3 + x^2 - 2x}$

6. $\displaystyle\int \frac{(x^2 + 9)\,dx}{x(x - 3)^2}$

7. $\displaystyle\int \frac{(5x + 3)\,dx}{x^3 - 2x^2 - 3x}$

8. $\displaystyle\int_0^1 \frac{x\,dx}{x^2 - 3x - 4}$

9. $\displaystyle\int_2^3 \frac{(t + 1)\,dt}{t^2(t - 1)}$

10. Compute the area of the region bounded by the graphs of $y = \dfrac{x - 1}{x^2 - 5x + 6}$, $y = 0$, $x = 4$, and $x = 6$.

11. Find the average value of the demand function

$$D(x) = \frac{x^2 - 3}{(x - 1)(x - 2)}$$

on the interval $[3, 6]$.

12. Determine the total cost Q if marginal cost is defined by $MQ = \dfrac{x^2 + 1}{x^3 - x^2}$ on the interval $[3, 4]$.

13. Evaluate $\displaystyle\int \frac{x^3 + x^2 + 3}{x^2 + 3}\,dx$.

14. Assume that a cost function changes at a rate given by $MQ = 1/x(x - 5)^2$, $x > 2$. Find the value of Q when x increases from $x = 2$ to $x = 3$.

15. Evaluate the indefinite integral $\displaystyle\int \frac{x + 1}{(x - 1)^2}\,dx$.

16. Suppose a marginal price function is given by $MP = x^2/(x^2 - 9)$. Find price P in millions of dollars when x increases from 4 to 5.

7.6 IMPROPER INTEGRALS

The importance to us of improper integrals stems from their applicability in such fast-growing fields as probability theory and statistical decision theory.

The notation of definite integrals is useful in extending our study of definite integrals to improper integrals. In contrast to this new name we will refer to definite integrals as proper integrals. The latter were introduced and discussed previously with the assumption that

$$\int_a^b f(x)\,dx \tag{1}$$

involved an integrand $f(x)$ which was continuous for all values of x in the interval from $x = a$ (finite) to $x = b$ (finite), including the end points. In this section we will study the two cases for which (1) is an improper integral. In Case I, a or b (the "or" is inclusive) is infinite, so that the interval of integration is an infinite one. In Case II the function $f(x)$ is discontinuous at an end point of the interval or within the interval. In order to clarify, we introduce the geometric notion of **boundedness**.

We will agree that the continuity of f on $[a, b]$ implies that f is bounded on $[a, b]$. Furthermore, if f is bounded on $[a, b]$ then the integral $\int_a^b f(x)\,dx$ may exist. (We say "may" because more precise statements of necessary and sufficient conditions for the existence of the Riemann integral are beyond the scope of this text.) However, if f is unbounded on $[a, b]$ then $\int_a^b f(x)\,dx$ does not exist. Figure 7.11 illustrates.

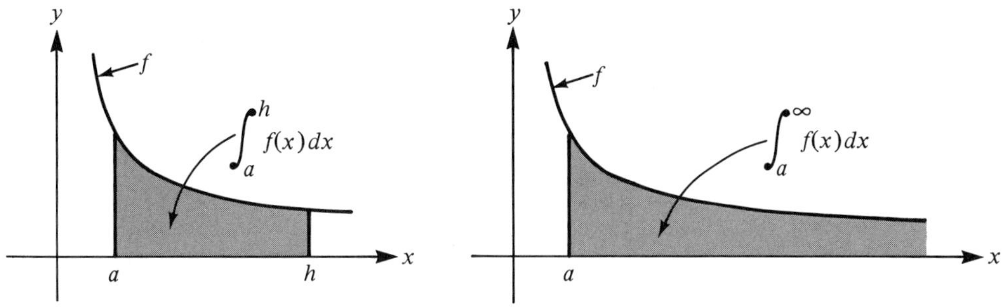

Figure 7.11

EXAMPLE 1 The function $f(x) = 1/(x - 2)$ is unbounded on any interval containing 2; but f is bounded on any interval $a \le x \le b$ where $a < b < 2$ or $2 < a < b$.

We now exhibit a function that is defined and bounded on the interval $0 \le x \le 1$ but $\int_0^1 f(x)\,dx$ does not exist.

EXAMPLE 2

$$\text{If } f(x) = \begin{cases} 0 & \text{when } 0 \le x \le 1, \quad x \text{ is rational} \\ 1 & \text{when } 0 \le x \le 1, \quad x \text{ is irrational} \end{cases}$$

then $\int_0^1 f(x)\,dx$ is not defined. However, we will not work with such exceptional cases.

Case I The notation $\int_a^\infty f(x)\,dx$ indicates that the values of $f(x)$ start at $x = a$ and continue to "infinity." This integral is not the ordinary or proper integral we have been studying. We seek to make it consistent with the meanings of the definite integral. To do so we offer the following definition.

DEFINITION 1 *If a function f is continuous on the closed interval $a \le x \le h$ for all values of h no matter how large and if*

$$\lim_{h \to \infty} \int_a^h f(x)\,dx$$

exists, then

$$\boxed{\int_a^\infty f(x)\,dx = \lim_{h \to \infty} \int_a^h f(x)\,dx}.$$

The integral $\int_a^\infty f(x)\,dx$ is known as an ***improper integral****.*

The symbol "$\lim_{x\to\infty}$" is read "the limit as x tends to infinity" or "the limit as x increases without bound." Similar to finite cases, we have basic properties of cases involving infinite limits:

Given that

$$\lim_{x\to\infty} f(x) = L,$$

$$\lim_{x\to\infty} g(x) = M,$$

and

$$\lim_{x\to\infty} h(x) = N,$$

then

Property 1 $\lim_{x\to\infty} kf(x) = k \lim_{x\to\infty} f(x) = kL$ *for any constant k*

Property 2 $\lim_{x\to\infty} [f(x) \pm g(x) \pm h(x)] = L \pm M \pm N$

Property 3 $\lim_{x\to\infty} [f(x) \cdot g(x) \cdot h(x)] = LMN$

Property 4 $\lim_{x\to\infty} \frac{h(x)}{g(x)} = \frac{N}{M}$

These properties are also valid if $x \to -\infty$.

The improper integral $\int_a^\infty f(x)\,dx$ is said to be **convergent** if

$$\lim_{h\to\infty} \int_a^h f(x)\,dx$$

exists. If $\lim_{h\to\infty} \int_a^h f(x)\,dx$ does not exist, then the improper integral is said to be **divergent**. Some explanation of the terms convergent and divergent is in order.

Initially, we say that an infinite series of terms is convergent if it has a sum. Otherwise, it is divergent. That is, if

$$S_n = u_1 + u_2 + u_3 + \cdots + u_n$$

represents the sum of the first n terms of an infinite series

$$u_1 + u_2 + u_3 + \cdots + u_n + \cdots$$

and if

$$\lim_{n\to\infty} S_n = S,$$

that is, the limit of S_n exists as n approaches infinity, then this limit S is called the sum of the series. In other words, the series is said to converge to S, or to have the value S. Otherwise, the series is said to be divergent, which means it has no sum.

For example, the series

$$1 + 3 + 5 + 7 + \cdots$$

is divergent because $\lim_{n\to\infty} S_n$ is positively infinite. As another example, the series

$$1 - 1 + 1 - 1 + \cdots$$

is also divergent because $\lim_{n\to\infty} S_n$ alternately increases and decreases without reaching a finite limit. This series is said to oscillate. An example of a convergent series is the geometric series

$$1 + \frac{1}{2} + \frac{1}{4} + \frac{1}{8} + \cdots$$

The sum of this series can be proven to be

$$S = \frac{a}{1 - r} \tag{2}$$

where $a = 1$, the first term of the series, and $r = 1/2$, the common ratio. Substituting in Eq. (2) we get

$$S = \frac{1}{1 - \frac{1}{2}} = 2.$$

In general, if $|r| < 1$, the geometric series

$$a + ar + ar^2 + \cdots + ar^{n-1} + \cdots$$

approaches a finite limit as n becomes infinite. This limit is the sum

$$S = \lim_{n\to\infty} \frac{a(1 - r^n)}{1 - r} = \frac{a}{1 - r},$$

the result being due to the fact that $|r| < 1$ so that

$$\lim_{n\to\infty} r^n = 0.$$

Thus,

$$S = \frac{a(1 - 0)}{1 - r} = \frac{a}{1 - r}.$$

Since we are interested in practical uses we are therefore concerned more with convergent series. Hence, we want to be able to distinguish between convergence and divergence. Let us illustrate the ideas of Definition 1.

EXAMPLE 3 Examine I for convergence or divergence if $I = \int_0^\infty e^{-x}\,dx$.

Solution

$$\begin{aligned} I &= \lim_{h\to\infty} \int_0^h e^{-x}\,dx \\ &= \lim_{h\to\infty} (-e^{-x})_0^h \\ &= \lim_{h\to\infty} (-e^{-h} + e^0) = 1 \end{aligned}$$

since

$$\lim_{h\to\infty} (-e^{-h}) = \lim_{h\to\infty} \left(-\frac{1}{e^h}\right) = 0.$$

Thus, the limit exists and so I converges and its value is 1.

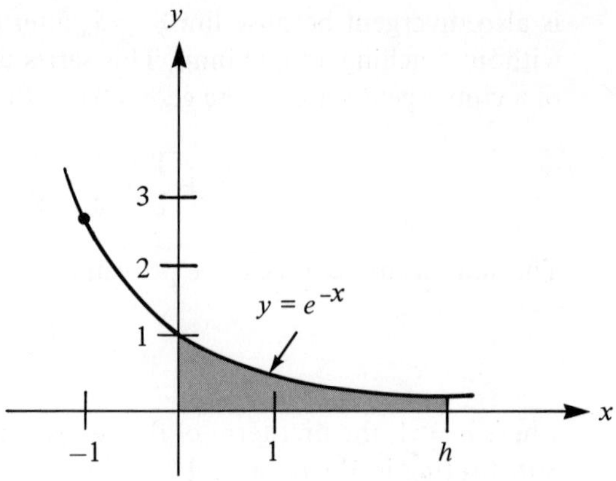

Figure 7.12

Let us illustrate $I = \int_0^\infty e^{-x}\,dx$ geometrically. To this end we consider the proper integral $\int_0^h e^{-x}\,dx$ as representing the area of the region bounded by $y = e^{-x}$, $y = 0$, $x = 0$, and $x = h$ (see Figure 7.12).

EXAMPLE 4 Evaluate $I = \displaystyle\int_0^\infty \frac{dx}{x+1}$ and exhibit geometrically.

Solution

$$\begin{aligned} I &= \lim_{h\to\infty} \int_0^h \frac{dx}{x+1} \\ &= \lim_{h\to\infty} \ln|x+1| \Big]_0^h \\ &= \lim_{h\to\infty} [\ln|h+1| - \ln|0+1|] \\ &= \infty \end{aligned}$$

since $\ln|h+1|$ increases without bound as $h \to \infty$. Figure 7.13 illustrates. The limit does not exist and so I diverges.

In both of these examples we pictured a region extending to infinity. And yet in the previous example the improper integral converged to a finite value. What is the explanation? How can an infinite region have finite area? The reason is: The existence of convergence basically stems from the fact that the curve in question approaches the horizontal axis

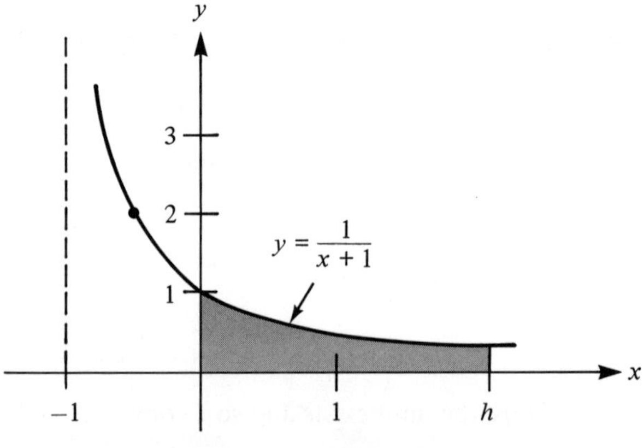

Figure 7.13

"sufficiently fast" as the independent variable approaches infinity, thus keeping the area from increasing indefinitely.

Similar to Definition 1 we have

DEFINITION 2 *If $f(x)$ is continuous on the interval $h \leq x \leq b$ for all values of h less than or equal to b, and if*

$$\lim_{h \to -\infty} \int_h^b f(x)\, dx$$

exists, then

$$\boxed{\int_{-\infty}^b f(x)\, dx = \lim_{h \to -\infty} \int_h^b f(x)\, dx}.$$

The symbol $\int_{-\infty}^b f(x)\, dx$ is also called an ***improper integral*** *and is said to* ***converge*** *if*

$$\lim_{h \to -\infty} \int_h^b f(x)\, dx$$

exists. Otherwise, it ***diverges.***

EXAMPLE 5 Evaluate $I = \int_{-\infty}^1 \frac{dx}{(x-2)^2}$.

Solution

$$\begin{aligned} I &= \lim_{h \to -\infty} \int_h^1 (x-2)^{-2}\, dx \\ &= \lim_{h \to -\infty} \frac{(x-2)^{-1}}{-1}\Bigg]_h^1 \\ &= -\lim_{h \to -\infty} \frac{1}{x-2}\Bigg]_h^1 \\ &= -\lim_{h \to -\infty} \left[\frac{1}{1-2} - \frac{1}{h-2}\right] \\ &= -[-1-0] = 1. \end{aligned}$$

Thus I converges to the value 1. The graph of the integrand $y = 1/(x-2)^2$ is shown in Figure 7.14 together with the area (shaded) represented by the integral $\int_h^1 \frac{dx}{(x-2)^2}$.

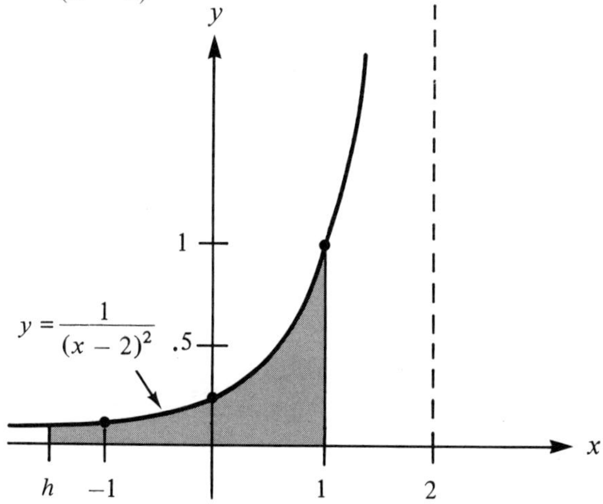

Figure 7.14

As a final example for Case I, there is the situation where both limits of integration are infinite. Improper integrals of such form also may be defined and evaluated by means of definite integration.

EXAMPLE 6 Find the value of $I = \int_{-\infty}^{\infty} xe^{-x^2}\,dx$ if I converges.

Solution Assuming that h, 0, and k are points in a finite interval on which the integrand is continuous, we integrate I as follows:

$$
\begin{aligned}
I &= \int_{-\infty}^{0} xe^{-x^2}\,dx + \int_{0}^{\infty} xe^{-x^2}\,dx \\
&= \lim_{h\to-\infty} \int_{h}^{0} xe^{-x^2}\,dx + \lim_{k\to\infty} \int_{0}^{k} xe^{-x^2}\,dx \\
&= -\frac{1}{2} \lim_{h\to-\infty} e^{-x^2}\Big]_{h}^{0} + \left(-\frac{1}{2}\right) \lim_{k\to\infty} e^{-x^2}\Big]_{0}^{k} \\
&= -\frac{1}{2} \lim_{h\to-\infty} \left[e^{0} - \frac{1}{e^{h^2}}\right] - \frac{1}{2} \lim_{k\to\infty} \left[\frac{1}{e^{k^2}} - e^{0}\right] \\
&= -\frac{1}{2}[1] - \frac{1}{2}[-1] = 0 \quad \to \quad I \text{ converges.}
\end{aligned}
$$

If we interpret the preceding equations geometrically, as in Figure 7.15, then this last example involves a region that extends to the right and left indefinitely, and yet its area is shown to be finite. Convergence then hinges on whether or not the graph of $y = xe^{-x^2}$ approaches the x-axis with sufficient rapidity as $h \to -\infty$ and $k \to \infty$ to prevent the area from "blowing up" beyond all bounds. The choice of 0 as one of the intermediate limits of integration was arbitrary. The same answer would result if any real number b had been used, in which case we would have had, after some manipulation,

$$
\begin{aligned}
I &= -\frac{1}{2} \lim_{h\to-\infty} \left[e^{b} - \frac{1}{e^{h^2}}\right] - \frac{1}{2} \lim_{k\to\infty} \left[\frac{1}{e^{k^2}} - e^{b}\right] \\
&= -\frac{1}{2}[e^{b}] - \frac{1}{2}[-e^{b}] \\
&= -\frac{e^{b}}{2} + \frac{e^{b}}{2} = 0, \quad \text{as before.}
\end{aligned}
$$

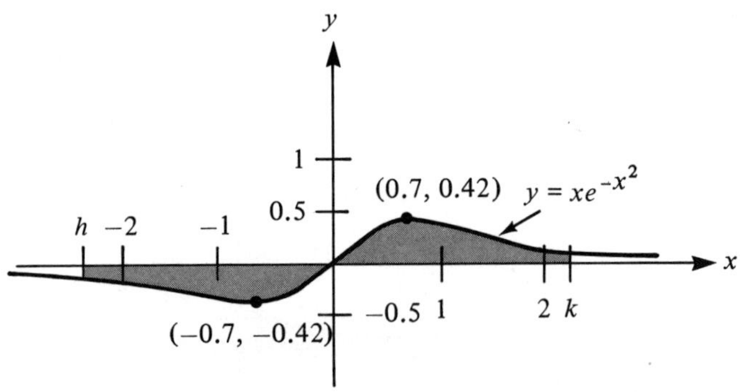

Figure 7.15

Case II In certain cases the notions underlying the definite integral can be extended to functions with infinite discontinuities on the interval of integration. We first define and illustrate the discontinuity of an integrand $f(x)$ at one of the end points of the range of integration.

DEFINITION 3 *If f is continuous on the interval $a \leq x < b$, then*

$$\int_a^b f(x)\,dx = \lim_{h \to b^-} \int_a^h f(x)\,dx, \tag{3}$$

provided the limit exists in Eq. (3). The symbol $\lim_{h \to b^-}$ indicates that b is being approached through values less than b. If the limit exists, the improper integral converges; otherwise, it diverges.

EXAMPLE 7 Evaluate $\int_0^1 \frac{dx}{\sqrt{1-x}}$ if a value exists.

Solution Since the integrand is discontinuous at the upper limit of integration, $x = 1$, we apply the results of Definition 3 and obtain

$$\begin{aligned}
\int_0^1 \frac{dx}{\sqrt{1-x}} &= \lim_{h \to 1^-} \int_0^h (1-x)^{-1/2}\,dx \\
&= -\lim_{h \to 1^-} (1-x)^{1/2} \cdot 2 \Big]_0^h \\
&= -2 \lim_{h \to 1^-} [\sqrt{1-h} - \sqrt{1-0}] \\
&= -2[0-1] = 2 \quad \to \quad \text{convergence.}
\end{aligned}$$

Similarly, if f is discontinuous at the lower limit of integration, we have the following definition.

DEFINITION 4 *If f is continuous on the interval $a < x \leq b$, then*

$$\int_a^b f(x)\,dx = \lim_{h \to a^+} \int_h^b f(x)\,dx, \tag{4}$$

provided the limit exists. Here the limiting value results from approaching the value a through values greater than a.

Figure 7.16 illustrates Definition 4.

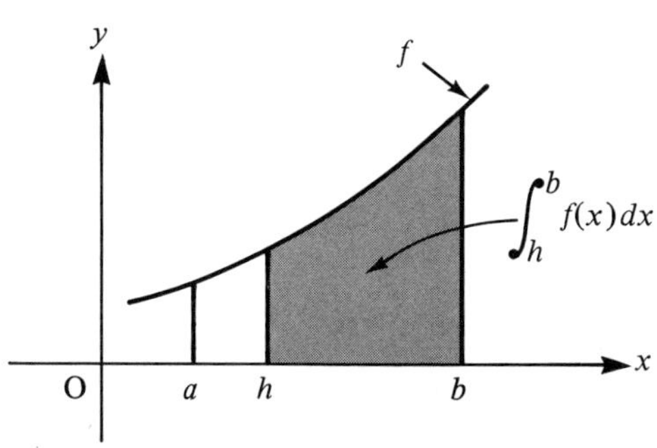

Figure 7.16

EXAMPLE 8 Find the area A under the graph of $y = 1/\sqrt{x}$ for the domain $0 < x \leq 4$. Illustrate geometrically.

Solution The geometric interpretation is shown in Figure 7.17.

$$\begin{aligned} A = \int_0^4 \frac{dx}{\sqrt{x}} &= \lim_{h\to 0^+} \int_h^4 x^{-1/2}\,dx \\ &= \lim_{h\to 0^+} x^{1/2}\cdot 2\,\Big]_h^4 \\ &= 2\lim_{h\to 0^+} [\sqrt{4}-\sqrt{h}] \\ &= 2[2] = 4. \end{aligned}$$

Thus, the area exists and has the value 4 units^2.

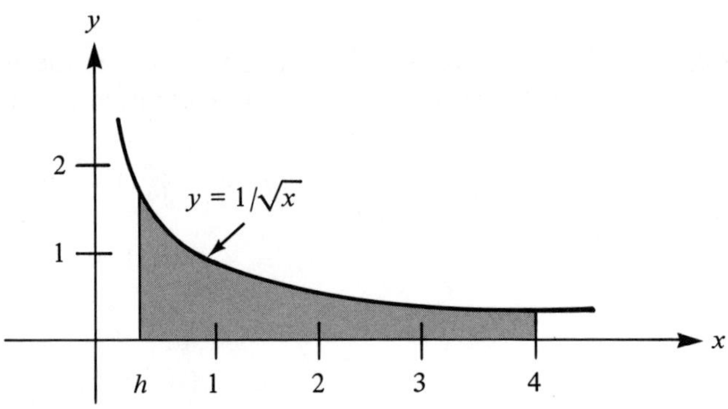

Figure 7.17

Finally, if f is discontinuous at an interior point of the range of integration, say at some $x = c$ where $a < c < b$, but is continuous for all other values of x in $[a, b]$, then we define the integral

$$\int_a^b f(x)\,dx = \int_a^c f(x)\,dx + \int_c^b f(x)\,dx. \tag{5}$$

The right side of Eq. (5) is evaluated according to the methods of Eqs. (3) and (4).

EXAMPLE 9 Find the area A under the curve $f(x) = 1/(x-8)^{2/3}$ for $0 \leq x \leq 16$.

Solution Upon careful scrutiny we see that f is not defined at the interior point $x = 8$ of the interval $[0, 16]$. Therefore, we proceed according to Eq. (5):

$$\begin{aligned} A &= \int_0^{16} \frac{dx}{(x-8)^{2/3}} \\ &= \lim_{h\to 8^-} \int_0^h (x-8)^{-2/3}\,dx + \lim_{k\to 8^+} \int_k^{16} (x-8)^{-2/3}\,dx \\ &= \lim_{h\to 8^-} (x-8)^{1/3}\cdot 3\,\Big]_0^h + \lim_{k\to 8^+} (x-8)^{1/3}\cdot 3\,\Big]_k^{16} \\ &= 3\lim_{h\to 8^-} [(h-8)^{1/3} - (-8)^{1/3}] + 3\lim_{k\to 8^+} [8^{1/3} - (k-8)^{1/3}] \\ &= 3[0+2] + 3[2-0] = 12 \text{ units}^2. \end{aligned}$$

In problems of this last type, both integrals on the right side of Eq. (5) must converge independently in order for the original integral $\int_a^b f(x)\,dx$ to converge. If either one of the integrals on the right side fails to exist, i.e., diverges, then the original integral diverges.

PROBLEM SET 7.6

In problems 1–20 find the value of the given improper integral or show that it diverges.

1. $\int_1^\infty \frac{dx}{x^3}$
2. $\int_1^\infty \frac{dx}{x^4}$
3. $\int_1^\infty \frac{dx}{\sqrt{x}}$
4. $\int_0^\infty \frac{dt}{e^t}$
5. $\int_0^\infty \frac{dz}{2e^{2z}}$
6. $\int_1^\infty \frac{dx}{\sqrt{1+x}}$
7. $\int_0^\infty xe^{-x^2}\,dx$
8. $\int_1^\infty \frac{x\,dx}{x^2+1}$
9. $\int_1^\infty \ln x\,dx$
10. $\int_0^\infty xe^{-x}\,dx$
11. $\int_{-\infty}^\infty xe^{-x^2}\,dx$
12. $\int_{-\infty}^\infty \frac{dx}{x+1}$
13. $\int_0^1 \frac{dx}{\sqrt{x}}$
14. $\int_0^1 \frac{dx}{x^{1/3}}$
15. $\int_0^3 \frac{x\,dx}{(9-x^2)^2}$
16. $\int_0^9 \frac{dx}{(x-1)^{1/3}}$
17. $\int_1^3 \frac{dx}{(x-2)^2}$
18. $\int_0^8 \frac{dx}{(8-x)^3}$
19. $\int_0^4 \frac{dx}{\sqrt{4-x}}$
20. $\int_0^2 \frac{x\,dx}{(x^2-1)^2}$.
21. Find the area of the region bounded by the graph of $y = 1/(x-8)^{2/3}$, the x-axis, and the lines $x = 0$, $x = 8$. Sketch the region.
22. Given the demand curve $D(x) = 1/\sqrt{x}$ and the supply curve $S(x) = \sqrt{x}$. Find the area between D and S for the interval $(0, 1]$.
23. Find the area, if it exists, under the curve $xy = 9$ and to the right of $x = 1$.
24. A certain rate of depreciation is given by $MD(t) = 10e^{-2t}$ where t is time in years. Find D for $[0, \infty)$.
25. Work problem 24 when $MD(t) = 20e^{-4t}$.
26. Suppose the inflation index changes at a rate defined by $MI(t) = \dfrac{1}{10\sqrt{t-4}}$ where t is the number of years. Find the index I over the interval $(4, 8]$ where 4 and 8 denote the years 1974 and 1978, respectively.

7.7 PROBABILITY DENSITY FUNCTIONS

There will be no attempt here to present a thorough outline of the theory of probability. Rather, we will develop those few concepts and procedures needed for our immediate use in applying some of the mathematics developed in previous sections. Furthermore, we will discuss infinite sample spaces as well as sample spaces with a finite number of elements. At this point the reader is probably asking: What are sample spaces and elements of sample spaces? Let us attempt to clarify the situation.

DEFINITION 1 *A **sample space** S for an experiment is the set of all possible outcomes of the experiment. Each outcome is the result of an action or trial, and the totality of these trials constitutes an **experiment**.*

Consider the experiment which involves a toss of two "fair" coins. This experiment consists of one trial. All the possible outcomes of this experiment can be indicated by the ordered pairs (h, h), (h, t), (t, h), and (t, t), where h and t denote heads and tails, respectively. The set of these possible outcomes defines a sample space S for the experiment of tossing two coins once. The formal designation of this sample space is

$$S = \{(h, h), (h, t), (t, h), (t, t)\}. \tag{1}$$

If all the outcomes of an experiment are "equally likely" to occur, we say the experiment is *fair*. It is hoped that the term "equally likely" is intuitively clear.

A **subset** A of a sample space S is called an **event** E. The event E is said to occur if any one of its elements occurs. Specifically, for the experiment of tossing two coins once, the event

$$E = \{(h, h), (h, t), (t, h)\} \tag{2}$$

describes the event that at least one head turns up. From our definition of an event, we see that the event E in Eq. (2) is a subset of the sample space S in Eq. (1). Since any one of the outcomes, say (h, t), is an element of E, we say that the event E has occurred. By the definition of a subset, we write $\{(h, h)\}$, $\{(h, t)\}$, $\{(t, h)\}$, and $\varnothing$ as the subsets of E. The symbol $\varnothing$ denotes the empty (or null) set and is defined to be a subset of any set. The sample spaces in Eqs. (1) and (2) contain 4 elements and 3 elements, respectively, and accordingly are defined as finite sample spaces.

For the experiment of tossing two coins once, we say that the probability of two heads occurring is 1/4, meaning that one of the four possible equally likely outcomes is (h, h). It follows that the probability of *not* getting (h, h) is

$$1 - \frac{1}{4} = \frac{3}{4}.$$

Symbolically, if event $E = \{(h, h)\}$ then the probability

$$P(E) = \frac{1}{4}$$

and

$$P(E') = \frac{3}{4}.$$

In other words,

$$P(E') = 1 - P(E)$$

where E' denotes the complement of E. Observe that

$$P(E) + P(E') = 1.$$

Let us investigate some elementary properties of the probability concept.

Property 1 *The probability P of an event E is a nonnegative number. That is,* $P(E) \geq 0$ *for a subset E of a given sample space S.*

Property 2 *The probability of either of two mutually exclusive events occurring equals the sum of the probabilities that the individual events occur. That is,* $P(E_1 \cup E_2) = P(E_1) + P(E_2)$ *where* E_1 *and* E_2 *are any two mutually exclusive events. Events* E_1 *and* E_2 *are said to be* ***mutually exclusive*** *if their intersection is the empty set. Symbolically, we write*

$$E_1 \cap E_2 = \varnothing.$$

Property 3 *The sum of the probabilities of all the possible outcomes in a sample space S equals 1. In other words,*

$$P(S) = 1 \quad or \quad \sum_{i=1}^{n} P(E_i) = 1$$

for any S where E is any event in S. An alternative expression is $P(S) = \sum_{i=1}^{n} P(E_i) = 1$ *if and only if the* $E_1, E_2, \ldots, E_n$ *are a* ***partition*** *of the sample space S.*

Let us clarify the meaning of partition. If a probability function is defined on a sample space S and that space is divided into n subspaces, then the set of all such subspaces of the sample space S is called a **partition** of S. Figure 7.18 illustrates one such partition of S.

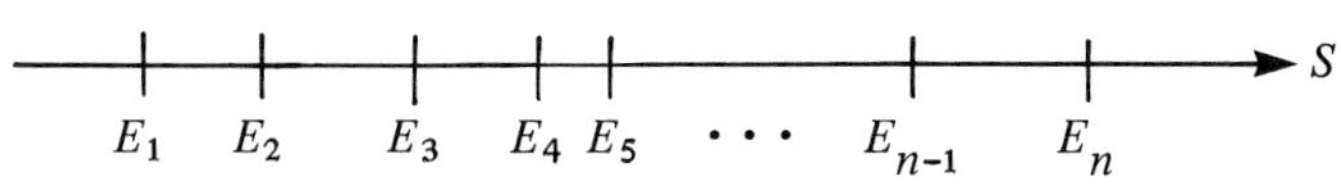

Figure 7.18

Thus, a partition is a set of points

$$E_1, E_2, E_3, \ldots, E_n$$

where $E_1 < E_2 < E_3 < \cdots < E_n$

and the subintervals (subspaces) need not be of equal length.

When a real number of the form $P(E_i)$ is assigned to each event in S such that properties 1, 2, and 3 are satisfied, we say that the set of pairs of the form $\{(E_i), P(E_i)\}$ defines a **probability function**. This is reminiscent of the definition of a function of one variable, which states that the set of ordered pairs (x, y) defines a function $y = f(x)$.

This brief introduction to probability was confined to sample spaces which have only a finite number of elements. Let us continue the discussion with sample spaces containing infinitely many elements. In contrast to probabilities of events associated with points of a finite sample space, we will be considering those probabilities associated with events defined as intervals or subintervals or a union of subintervals in a "continuous" sample space. As a case in point, consider the experiment of selecting a light bulb at random and measuring the intensity of its light. (We say that an item is selected *at random* from a list of items if the selection is such that each item of the list is equally likely to be chosen.) The device measuring the intensity of light of the bulb may record values ranging, let's say, from 1.0000 to 9.9999. Knowing that there are infinitely many numbers in the interval $\{1.0000 \leq x \leq 9.9999\}$ we find it feasible and practical to work with the interval as a continuous sample space. The latter is often referred to as a **continuum** as opposed to a discrete sample space which has only a finite number of elements.

We will be discussing sample spaces (continuums) containing intervals from the set of real numbers. This implies that events can be open or closed subintervals, some of which may be of infinite extent. Since there will be a continuum of possible outcomes, we may need to assign a probability to each of an unbounded number of subintervals. In the example of the light bulb above, the variable x is a measure of the light intensity. We will refer to x as a random variable and will assume, for the sake of example, that x takes on the values $1, 2, 3, \ldots, 9$ for a certain experiment. In Figure 7.19 the random variable x is graphed on the horizontal x-axis and $f(x)$, the frequency of occurrence of each value of x, is plotted in the vertical direction.

Since the measurements could be made to any degree of accuracy, in theory at least, it is mathematically feasible to consider all real numbers in the specified interval. Thus, we will be considering measurements

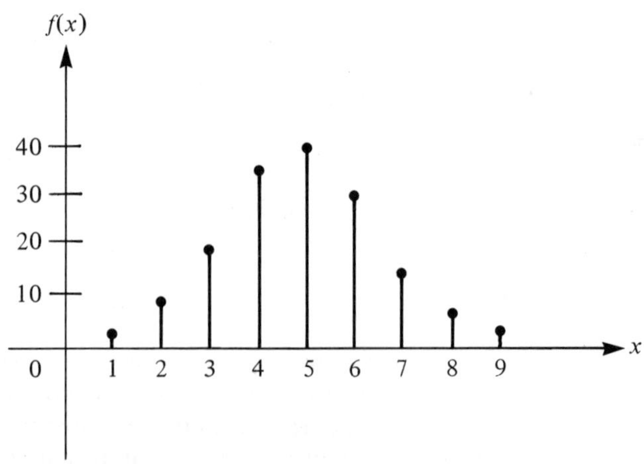

Figure 7.19

which are continuous and dense in a continuum. (Between any two rational numbers a and b of a sample space, there is another rational number $(a + b)/2$. This implies that there are infinitely many rational numbers. Because of this we define the set of rational numbers as being *dense.*) The discrete graph of Figure 7.19 has its counterpart in Figure 7.20, where the same data are interpreted in terms of a continuous sample space. The graph is a smooth curve drawn through the tops of the frequency ordinates shown in the previous graph. However, the important idea suggested by the graph in Figure 7.20 is that *area can be viewed as a probability.*

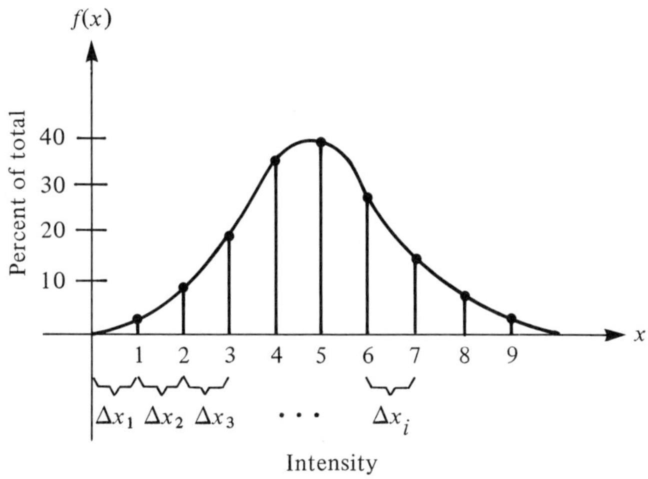

Figure 7.20

If, for the sake of simplicity, the device accurately measures light intensity to within tenths only, we may wish to know the probability that the intensity of the light of a bulb (selected at random) will be between 1.1 and 1.3. To this end, let us assume that the total number of bulbs tested is 500 and that the graph of Figure 7.20 represents a distribution of light intensities of bulbs from this "population" of 500. Intervals of intensity are measured along the horizontal while the percentage of the total number of bulbs is shown along the vertical. A typical interval of intensity is indicated by Δx_i (see Figure 7.20) and the associated percentage by p_i so that

$$\sum_{i=1}^{10} p_i \, \Delta x_i = 1.$$

This result is due to the following reasons:

1. The probability that a bulb (selected at random) will have an intensity in a given interval is assumed to be the percentage value (shown in Figure 7.20) based on experimental computations.
2. Probability functions for sample spaces of infinitely many elements have the same properties as they do for finite sample spaces. In other

words, for any event E_i belonging to a sample space containing the real numbers $\mathbf{R}$, where $p(E_i)$ is the associated probability value, then

(a) $p(E_i) \geq 0$,

(b) $p(E_i \cup E_j) = p(E_i) \cup p(E_j), \qquad i \neq j$,

(c) $\sum_{i=1}^{n} p(E_i) = 1, E_i \in \mathbf{R}$.

From our knowledge of limits and their role in the development of definite integrals, we assume that as the size of the sample gets bigger and the interval of measurement gets smaller, the accuracy of the distribution gets better. Accordingly, we may write

$$\lim_{n \to \infty} \sum_{i=1}^{n} p_i \, \Delta x_i = 1.$$

This formulation essentially resembles that of the definite integral. We are thus led to consider p_i as some functional value $f(t_i)$ where $x_{i-1} \leq t_i \leq x_i$ and $\Delta x_i = x_i - x_{i-1}$. Applying these notions to our preceding example, we conclude that the probability p that a bulb will have a light intensity between 1.1 and 1.3 will be

$$p = \int_{1.1}^{1.3} f(x) \, dx.$$

It is now apparent that the calculation of a probability will be the same as the calculation of an area.

Note: The function $f(x)$ must be such that it appropriately defines the distribution (of intensities, in our example). In this context f is called a **probability density function**, or simply pdf.

DEFINITION 2 *If f is a function defined for all real numbers x in a sample space S of an experiment, then f is a* ***probability density function (pdf)*** *provided*

(1) $f(x) \geq 0, \quad x \in \mathbf{R}$ *(the real numbers),*

(2) $p(E) = \int_a^b f(x) \, dx$, *$E$ is any event in S such that* $E = \{x | a \leq x \leq b\}$,

(3) $\int_0^\infty f(x) \, dx = 1$.

From this definition it should be clear that the integral of the pdf determines the probability function $p(E)$.

EXAMPLE 1 Suppose that an appropriate pdf for our light bulb experiment is given by $f(x) = \frac{1}{2}e^{-x/2}$ and we wish to find the probability previously indicated by

$$p(E) = \int_{1.1}^{1.3} f(x) \, dx.$$

where E is the event that a bulb will have a light intensity between 1.1 and 1.3.

First we must check that the properties for a pdf hold true in this case. It follows immediately that Properties 1 and 2 of Definition 2 are satisfied, so that only Property 3 remains to be checked. To this end we evaluate the improper integral

$$\int_0^\infty \frac{1}{2} e^{-x/2}\,dx = \lim_{h\to\infty} \frac{1}{2}(-2)e^{-x/2}\Big]_0^h$$
$$= -\lim_{h\to\infty}\left(\frac{1}{e^{h/2}}\right) - 1$$
$$= -(0-1) = 1, \quad \text{as required.}$$

Thus all three conditions for a pdf are fulfilled, so we proceed with calculations as follows:

$$p(E) = \int_{1.1}^{1.3} \frac{1}{2} e^{-x/2}\,dx = \frac{1}{2}(-2)e^{-x/2}\Big]_{1.1}^{1.3}$$
$$= -e^{-1.3/2} + e^{-1.1/2}$$
$$= -e^{-0.65} + e^{-0.55}$$
$$= -0.5220 + 0.5769 \quad \text{[from exponential tables]}$$
$$= 0.0549, \quad \text{the desired probability.}$$

Because we are finding probabilities by integrating over intervals, it follows that the probability that the intensity is exactly 1.1 is the value zero, since

$$\int_{1.1}^{1.1} f(x)\,dx = 0.$$

EXAMPLE 2 Assume that the "life" of a certain type of battery is defined by $f(x) = 1/(2x+1)^2$, where x denotes the number of years. Find the probability of the event E that such a battery will last anywhere from 4 to 5 years.

Solution $f(x) \geq 0$ for all x in the domain (the sample space) of f, so Property 1 of Definition 2 holds. Property 2 also holds. We check to see if Property 3 is satisfied:

$$\int_0^\infty \frac{dx}{(2x+1)^2} = \lim_{h\to\infty}\int_0^h (2x+1)^{-2}\,dx$$
$$= \lim_{h\to\infty} \frac{1}{2}\frac{(2x+1)^{-1}}{-1}\Big]_0^h$$
$$= -\frac{1}{2}\lim_{h\to\infty} \frac{1}{2x+1}\Big]_0^h$$
$$= -\frac{1}{2}\lim_{h\to\infty}\left[\frac{1}{2h+1} - \frac{1}{1}\right]$$
$$= -\frac{1}{2}[0-1] = \frac{1}{2}.$$

Since $\frac{1}{2} \neq 1$, as it should if f is to be a suitable pdf, we effect a conformation of f to a pdf by letting a function $g(x) = 2f(x)$, that is, $g(x) = 2/(2x+1)^2$.

To find $p(E)$ we integrate $g(t)$ as t increases from $t = 4$ to $t = 5$:

$$\begin{aligned} p(E) &= \int_4^5 g(t)\,dt \\ &= \int_4^5 \frac{2\,dt}{(2t+1)^2} \\ &= 2\int_4^5 (2t+1)^{-2}\,dt \\ &= \left.\frac{2(2t+1)^{-1}}{-1}\right]_4^5 \\ &= \left.-\frac{2}{2t+1}\right]_4^5 \\ &= -\frac{2}{11} + \frac{2}{9} \\ &= \frac{2}{99} \cong 0.0202, \quad \text{the desired probability.} \end{aligned}$$

If $g(t)$ is integrated on the interval from 3 to 4 years, we obtain, after preliminary steps,

$$\begin{aligned} p(E) &= \left.-\frac{2}{2t+1}\right]_3^4 \\ &= -\frac{2}{9} + \frac{2}{7} \\ &= \frac{4}{63} \cong 0.0635, \end{aligned}$$

a larger probability value than that obtained when integrating on the interval from 4 to 5 years. The probabilities are different even though the intervals are of the same length. Thus, when the intervals of integration are of the same length, the values of the corresponding probabilities are not necessarily the same.

From the previous example we see that a function f can be transformed into a pdf by letting

$$g(x) = \frac{a}{b} f(x),$$

provided

$$\int_0^\infty f(x)\,dx = \frac{b}{a}.$$

This process helps us to make Property 3 valid, a necessary step before $p(E)$ can be found.

PROBLEM SET 7.7

1. Assume that a filament in an electronic tube has a useful lifetime of t hours and that the associated pdf is given by $f(t) = \frac{1}{20}e^{-t/20}$.

   (a) Verify that $f(t)$ is a pdf.

(b) Find the probability $p(E_1)$ if E_1 is the event that a filament (selected at random) will last more than 1000 hours. [*Hint*: $P(E_1) = \int_{1000}^{\infty} f(t)\,dt$.]
(c) If E_2 is the event that the filament will last at least 500 hours, find $p(E_2)$. [*Hint*: $P(E_2) = \int_{500}^{\infty} f(t)\,dt$.]
(d) Find the probability that the filament will last at least 100 but no more than 500 hours.

2. Show that f is a probability density function if:

$$f(x) = \begin{cases} \dfrac{6x - 3x^2}{4}, & x \in [0, 2] \\ 0, & x \notin [0, 2]. \end{cases}$$

3. In a certain nuts and bolts warehouse the frequency function $f(x) = \frac{4}{7}(6x^2 - x^3)$ on $[0, 1]$ defines the proportion of orders filled per day.

(a) Show that f is a pdf.
(b) Find the probability that no more than 50% of the orders will be processed in one day.
(c) Determine the probability that more than 50% of the orders will be processed in a day.
(d) What is the probability that between 1/3 and 2/3 of the orders will be processed in a day?

Find the constants K in each of problems 4–8 such that the given function is a pdf:

4. $$f(x) = \begin{cases} \dfrac{K}{\sqrt{x}} & \text{for } 0 < x < 1 \\ 0 & \text{elsewhere} \end{cases}$$

5. $$f(x) = \begin{cases} \dfrac{K}{(1 + x)^2} & \text{for } x \in [0, 1] \\ 0 & \text{for } x \notin [0, 1] \end{cases}$$

6. $$f(x) = \begin{cases} K\left(\dfrac{13}{4} - \dfrac{3x}{2}\right) & \text{for } 0 \le x \le 4 \\ 0 & \text{otherwise} \end{cases}$$

7. $$f(x) = \begin{cases} \dfrac{K}{3} xe^{-x^2/2} & \text{for } x \ge 0 \\ 0 & \text{elsewhere} \end{cases}$$

8. $$f(x) = \begin{cases} K(4x - x^4) & \text{for } [0, 1] \\ 0 & \text{elsewhere} \end{cases}$$

9. Work problem 3 if $f(x) = \frac{4}{3}(3x^2 - x^3)$ on $[0, 1]$.

10. A certain type of light bulb in burning for t months is found to have a probability density function given by

$$f(t) = \begin{cases} 0.05e^{-0.05t} & \text{for } t \ge 0 \\ 0 & \text{for } t < 0. \end{cases}$$

(a) Find the probability that a bulb chosen at random will burn for less than 10 months.
(b) Compute the probability that a bulb will burn for between 8 and 10 months.
(c) Calculate the probability that a bulb will burn for at least 18 months.

11. The number of units x produced by the ABC Co. in one hour is found to have the pdf $f(x) = 2(1 + x)^{-3}$. Find the probability that the number of items produced will be:

    (a) No more than 1 **(b)** At least 2.

12. The probability density function for the useful life of a battery is defined by $f(t) = \frac{1}{100}e^{-t/100}$ where t is the number of hours. What is the probability that the battery will have a useful life of:

    (a) 100 to 110 hours **(b)** 500 to 510 hours.

13. The probability density function for a person dying at age t is given by $f(t) = 1/2(1 + t)^{3/2}$. Find the probability of a person surviving to more than 63 years of age. [*Hint*: Find the probability p of dying at age 63. Then $1 - p$ is the probability desired.]

14. The fraction of orders processed within a 9-hour shift in a certain mail-order firm has the probability density function $f(x) = 12(x^2 - x^3), 0 \leq x \leq 1$.

    (a) Find the probability that over an extended time period at least 80% of all orders are handled within 9 hours.

    (b) Find the probability that less than 50% of all orders are processed within 9 hours. [*Hint*: Use the decimal equivalent of percentages in the limits of integration.]

15. The annual income per person in a certain country has the probability density function $f(x) = 4x(2 + x^2)^{-2}$, $x \geq 0$, and x denotes thousands of dollars. Find the probability that a person has an income of:

    (a) less than \$2000.

    (b) anywhere from 1 to 4 thousand dollars.

    (c) more than \$4000. [*Hint*: $p(>\$4000) = 1 - p(\leq\$4000)$.]

7.8 INTEGRATION BY USE OF TABLES

Many real-world situations present problems whose integrals are difficult or impossible to evaluate by formulas and techniques introduced thus far. Time and labor can be substantially reduced by use of tables of integrals. A very brief table of integrals is given below for supplementary use in working certain types of integrals. A comprehensive table would contain many more formulas than shown here. This is simply a brief introduction to integration by use of tables. The letters a, b, and C are constants in the real number system.

Forms Containing ($a + bu$)

1. $$\int \frac{du}{u(a + bu)^2} = \frac{1}{a(a + bu)} - \frac{1}{a^2} \ln \left|\frac{a + bu}{u}\right| + C$$

2. $$\int \frac{du}{u^2(a + bu)} = -\frac{1}{au} + \frac{b}{a^2} \ln \left|\frac{a + bu}{u}\right| + C$$

3. $$\int \frac{du}{u^2(a + bu)^2} = -\frac{a + 2bu}{a^2u(a + bu)} + \frac{2b}{a^3} \ln \left|\frac{a + bu}{u}\right| + C$$

EXAMPLE 1 $$\int \frac{dx}{x^2(1+2x)^2} = -\frac{1+4x}{x(1+2x)} + 4\ln\left|\frac{1+2x}{x}\right| + C$$

using Formula 3 with $x = u$, $a = 1$, and $b = 2$.

Exercise I Find: (a) $\displaystyle\int \frac{dx}{x^2(4+3x)}$ (b) $\displaystyle\int \frac{dt}{t(3+4t)^2}$

Forms Containing $\sqrt{a+bu}$

4. $$\int \frac{\sqrt{a+bu}}{u}\,du = 2\sqrt{a+bu} + \sqrt{a}\ln\left|\frac{\sqrt{a+bu}-\sqrt{a}}{\sqrt{a+bu}+\sqrt{a}}\right| + C$$

5. $$\int \frac{u^2\,du}{\sqrt{a+bu}} = \frac{2(8a^2 - 4abu + 3b^2u^2)\sqrt{a+bu}}{15b^3} + C$$

EXAMPLE 2 $$\int \frac{x^2\,dx}{\sqrt{3+2x}} = \frac{2(72 - 24x + 12x^2)\sqrt{3+2x}}{120} + C$$

using Formula 5 with $x = u$, $a = 3$, and $b = 2$.

Exercise II Find: (a) $\displaystyle\int \frac{\sqrt{4+5x}}{x}\,dx$ (b) $\displaystyle\int \frac{z^2\,dz}{\sqrt{2+z}}$

Forms Containing $(u^2 + a^2)$ or $(u^2 - a^2)$

6. $$\int \frac{du}{u(u^2 \pm a^2)} = \pm\frac{1}{2a^2}\ln\left|\frac{u^2}{u^2 \pm a^2}\right| + C$$

EXAMPLE 3 $$\int \frac{dt}{t(t^2+7)} = \frac{1}{14}\ln\left|\frac{t^2}{t^2+7}\right| + C$$

using Formula 6 with $t = u$, $a^2 = 7$.

Exercise III Find: (a) $\displaystyle\int \frac{dw}{w(w^2-5)}$ (b) $\displaystyle\frac{dt}{t(4t^2+9)}$

Forms Containing $\sqrt{a^2 - u^2}$, $\sqrt{a^2 + u^2}$, or $\sqrt{u^2 - a^2}$

7. $$\int \frac{du}{u^2\sqrt{a^2-u^2}} = -\frac{\sqrt{a^2-u^2}}{a^2u} + C$$

8. $$\int \sqrt{u^2 \pm a^2}\,du = \frac{u}{2}\sqrt{u^2 \pm a^2} \pm \frac{a^2}{2}\ln\left|u + \sqrt{u^2 \pm a^2}\right| + C$$

9. $$\int u^3\sqrt{u^2+a^2}\,du = \left(\frac{u^2}{5} - \frac{2a^2}{15}\right)(u^2+a^2)^{3/2} + C$$

10. $\int u^3\sqrt{u^2 - a^2}\, du = \frac{(u^2 - a^2)^{5/2}}{5} + \frac{a^2}{3}(u^2 - a^2)^{3/2} + C$

11. $\int \frac{du}{u^3\sqrt{u^2 + a^2}} = -\frac{\sqrt{u^2 + a^2}}{2a^2u^2} + \frac{1}{2a^3}\ln\left|\frac{a + \sqrt{u^2 + a^2}}{u}\right| + C$

EXAMPLE 4

$$\int \frac{dx}{x^3\sqrt{x^2 + 2}} = -\frac{\sqrt{x^2 + 2}}{4x^2} + \frac{1}{4\sqrt{2}}\ln\left|\frac{\sqrt{2} + \sqrt{x^2 + 2}}{x}\right| + C$$

using Formula 11 with $x = u$, $a^2 = 2$.

Exercise IV Find:

(a) $\int \frac{d\theta}{\theta^2\sqrt{6 - \theta^2}}$ (b) $\int \sqrt{x^2 + 9}\, dx$ (c) $\int z^3\sqrt{z^2 + 4}\, dz$

(d) $\int t^3\sqrt{t^2 - 10}\, dt$ (e) $\int \frac{dv}{v^3\sqrt{v^2 + 16}}$

7.9 REVIEW PROBLEMS

Find the area of the region bounded by the given relations in problems 1–6:

1. $y = 3x/2$, $y = 6 - \frac{3x}{2}$, $x = 0$.
2. $y = (x - 2)^2$, $y = x$.
3. $y = x^2$, $y = 2x - x^2$.
4. $y = x^2 + 2$, $y = 4 - x^2$.
5. $y = x^3$, $y = 4x - 3x^2$, in Quadrant I.
6. $y = x^4$, $y = 8x$.

Given marginal cost MQ and marginal revenue MR, find the profit P if $P = R - Q$ (neglect fixed costs).

7. $MR = \sqrt{x}$, $MQ = x/2$, x increases from 0 to 4.
8. $MR = e^{x-1}$, $MQ = e^{2x+1}$, x increases from 1 to 3.
9. $MR = \frac{x + 6}{2}$, $MQ = \frac{x^2 + 4}{4}$, x increases from 0 to 2.
10. $MR = 3x$, $MQ = x^2(x + 2)$, x increases from 0 to 1.

For each given demand-supply system find the consumers' and producers' surplus:

11. $\left.\begin{array}{l} D:\ p = 20 - x^2 \\ S:\ p = x \end{array}\right\}$

12. $\left.\begin{array}{l} D:\ p = 40 - q^2 \\ S:\ p = 2q + q^2 \end{array}\right\}$

13. $\left.\begin{array}{l} D:\ p = 100 - q^2 \\ S:\ p = 2 + q^2 \end{array}\right\}$

14. $\left.\begin{array}{l} D:\ p = \frac{15}{q + 1} \\ S:\ p = 2q + 1 \end{array}\right\}$

Find the average value of the given function on the indicated interval:

15. Marginal cost $MQ(x) = 51 - 2x + x^2$ on $[1, 3]$.
16. Profit $P(x) = \sqrt{4x - 3}$ on $[1, 3]$.

17. Sales $S(x) = \sqrt[3]{x^2}$ (in hundreds) on $[0, 8]$.

18. Temperature $T(t) = 2 + (t - 1)^3$ on $[0, 3]$.

19. Marginal productivity $MP(q) = 0.64q^{1/3}$ (in hundreds) on $[0, 8]$.

Evaluate using integration by parts in 20–21:

20. $\int 2x(1 + 2x)^3\, dx$

21. $\int \frac{t\, dt}{(1 - 2t)^3}$

22. The marginal consumer price index for a certain economy is defined by $MI(t) = \frac{t}{10} e^{t/10}$. Using 1971 for $t = 1$, 1972 for $t = 2$, etc., find the increase in the price index when t increases

(a) from 1 to 4 (b) from 4 to 5.

23. If marginal demand is defined by $MD = x/(x^2 + 1)$, find the increase in demand D (in thousands) on $[0, 1]$.

Find:

24. $\int \frac{\ln \theta\, d\theta}{\theta}$

25. $\int \frac{\ln \sqrt{t}\, dt}{\sqrt{t}}$

26. $\int_0^4 x\sqrt{9 + x^2}\, dx$

27. $\int \sqrt{1 - e^t}e^t\, dt.$

Integrate by partial fractions in problems 28–31:

28. $\int \frac{(u + 1)\, du}{u^2(u - 1)}$

29. $\int \frac{16\, dt}{t(t - 4)^2}$

30. $\int \frac{x\, dx}{x^2 - 2x + 1}$

31. $\int \frac{2x^2 + 8x - 12}{x^2 - 4}\, dx$

32. A certain marginal demand function is given by $MD(x) = (x^2 + 1)/(x - 1)^2$ (in thousands) where x denotes the quantity demanded. Find the increase in demand when x increases from 2 to 3.

33. Find the area under the curve $f(x) = 1/e^{2x}$ on $[0, \frac{1}{2}]$.

34. Find the average value of the function $f(x) = \ln\sqrt{x}/\sqrt{x}$ on the interval $[0, 4]$.

35. Evaluate:

(a) $\int (9t^2 + 1)^{1/2}t\, dt$ (b) $\int x(16x^2 - 9)^3\, dx$

36. A certain marginal function is given by $MD = 20/3x^{4/3}$. Find the increase in D (hundreds of units) when x increases from 1 to 8.

37. Find the area, if it exists, under the curve $y = 1/10e^{10x}$ on the infinite interval $[0, \infty)$.

38. Establish convergence or divergence of the following integrals:

(a) $\int_0^1 \frac{3dx}{2x^2}$ (b) $\int_0^3 \frac{x\, dx}{\sqrt{9 - x^2}}$

(c) $\int_{-1}^1 \frac{dx}{x^{5/2}}$ (d) $\int_{-\infty}^{\infty} \frac{dx}{2x^3}.$

39. If the filament in an electronic tube has a useful lifetime of t hours and the associated probability density function (pdf) is given by

$$f(t) = \frac{1}{100} e^{-t/100}$$

(a) Find the probability $p(E_1)$ where E_1 is the event "the filament will last more than 300 hours."

(b) Find the probability $p(E_2)$ if E_2 is the event "the filament will last between 400 and 500 hours."

40. Convert $f(x) = \sqrt{x} - \dfrac{1}{\sqrt{x}}$ to a probability density function over $[1, 4]$; then find the probability that a value of x chosen at random will lie in the interval $[1, 2]$.

7.10 ENRICHMENT TOPICS AND PROBLEMS

The **volume of a solid** generated by revolving about the x-axis the region bounded by $y = f(x)$, $y = 0$, $x = a$, and $x = b$ is given by

$$V = \pi \int_a^b [f(x)]^2\,dx.$$

Figure 7.21 illustrates this region. For example, if $f(x) = x^2$ and the region under f between $x = 1$ and $x = 2$ is revolved about the x-axis, then the volume V of the solid generated is found by evaluating

$$\begin{aligned} V &= \int_1^2 \pi[f(x)]^2\,dx \\ &= \pi \int_1^2 (x^2)^2\,dx \\ &= \pi \int_1^2 x^4\,dx \\ &= \pi \cdot \frac{x^5}{5}\Big]_1^2 \\ &= \frac{\pi}{5}[2^5 - 1^5] = \frac{31}{5}\pi \text{ units}^3. \end{aligned}$$

Use the notions of this discussion to find the volume V of the solid generated when the regions bounded by the given curves and lines are rotated about the x-axis:

1. $f(x) = x^3$, $x = 0$, $x = 1$.
2. $3y = y - 2x$, $x = 0$, $x = 2$.
3. $f(x) = x - x^2$, $y = 0$.
4. $f(x) = \sqrt{x}$, $y = 0$, $x = 4$.

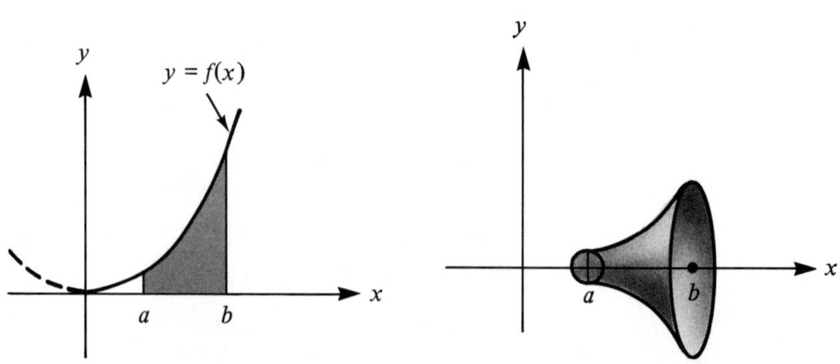

Figure 7.21

5. $f(x) = e^x$, $y = 0$, $x = 1/2$.

6. $f(x) = \sqrt{x - 1}$, $y = 0$, $x = 2$.

7. $f(x) = x^2 + 2$, $x = 0$, $x = 2$.

8. $f(x) = x^2$, $g(x) = x^3$ [*Hint*: $V = \pi\int_0^1[f^2(x) - g^2(x)]\,dx$.]

9. $f(x) = 4$, $g(x) = x^2$ (See the hint in prob. 8 and make use of symmetry.)

10. A toy manufacturer wishes to make a toy having the volume of a solid generated when the region bounded by the graphs of $y = \sqrt{x}$ and $y = x$ is revolved about the x-axis. Find the volume of the solid. [*Hint*: Refer to the notions in problem 8.]

11. The concept of a **moment** occurs in the study of statistical decision theory, particularly in relation to the notions of variance and standard deviation. The first moment M_y of f about the y-axis is defined by the definite integral

$$M_y = \int_a^b xf(x)\,dx$$

where f is a function, continuous and nonnegative on $[a, b]$. The word "first" in "first moment" refers to the power of x. If the integrand has x in powers of two and three, then the moments are called second and third moments, respectively. The above integral is sometimes defined as the first moment of region R about the y-axis where R is bounded by $y = f(x)$, $x = a$, $x = b$, and the x-axis. Other symbols used for the first three moments are μ_1, μ_2, and μ_3. For instance, if region R is bounded by $y = x^2$, $y = 0$, $x = 1$, $x = 2$, then the first three moments of R about the y-axis are

$$\mu_1 = \int_1^2 x \cdot x^2\,dx = \int_1^2 x^3\,dx$$
$$\mu_2 = \int_1^2 x^2 \cdot x^2\,dx = \int_1^2 x^4\,dx$$
$$\mu_3 = \int_1^2 x^3 \cdot x^2\,dx = \int_1^2 x^5\,dx.$$

Find the first three moments of region R about the y-axis if R is defined by:

(a) $f(x) = 2 - x$, $x = 0$, $y = 0$

(b) $f(x) = x^2 + 1$, $x = 0$, $x = 1$

(c) $f(x) = x + 4/x$, $y = 0$, $x = 2$, $x = 3$.

12. The **length of an arc** of a curve. For purposes of planning the cost budget for mining a certain vein of gold, a mining firm needs to know the length of the vein. Its geological survey indicates the shape (contour) of the vein to be approximated by the curve $y = x^{3/2}/3$ on the interval $0 \le x \le 4$ miles. The length of an arc of a curve, when $y = f(x)$, can be found by evaluating the integral

$$\int_a^b \sqrt{1 + \left(\frac{dy}{dx}\right)^2}\,dx \tag{1}$$

or

$$\int_a^b \sqrt{1 + \left(\frac{dx}{dy}\right)^2}\,dy. \tag{2}$$

To apply the integral in Eq. (1) we find

$$\frac{dy}{dx} = \frac{1}{3} \cdot \frac{3}{2} x^{1/2} = \frac{1}{2}\sqrt{x}.$$

and

$$\left(\frac{dy}{dx}\right)^2 = \frac{x}{4}.$$

Then

$$\int_0^4 \sqrt{1 + \frac{x}{4}}\,dx = 4 \int_0^4 \left(1 + \frac{x}{4}\right)^{1/2} \cdot \frac{1}{4}\,dx$$
$$= 4\left(1 + \frac{x}{4}\right)^{3/2} \cdot \frac{2}{3}\Bigg]_0^4$$
$$= \frac{8}{3}[2^{3/2} - 1]$$
$$= \frac{8}{3}[2.828 - 1]$$
$$\cong 4.874 \text{ miles,}$$

the estimated length of the vein of gold.

Use the integrals in Eqs. (1) or (2) to find the length of the given curve on the indicated interval:

(a) $3y = (x^2 + 2)^{3/2}$ from $x = 0$ to $x = 1$
(b) $4y^3 = 9x^2$ from $(0, 0)$ to $(2\sqrt{3}, 3)$
(c) $3y^2 = (2x + 8)^3$ from $x = 0$ to $x = 4$ in quadrant I
(d) $4y^2 = x^3$ from $x = 0$ to $x = 1$.

13. Assume that the present value of all dividend payments made in the future by the ACP Co. is defined by the improper integral

$$\int_0^\infty D_0 e^{(RE-K)t}\,dt$$

where t = time in years,
R = the fraction of income retained,
E = the fraction of the book value earned on ACP common stock,
D_0 = amount of dividend paid per share at time $t = 0$
and $RE - K < 0$.

Show that the value of the integral is $D_0/(K - RE)$.

14. Let the present value of the total outlay for promoting and advertising product P_{13} be defined by $P = \int_0^\infty Q(t)e^{-rt}\,dt$ where P is in millions of dollars, $Q(t) = t/10$ denotes the total cost of producing, marketing, and promoting P_{13}, and r is the required rate of return on all capital invested in activities related to P_{13}. Find the value of P when $r = 10\%$. [*Hint*: $\lim_{h\to\infty}(h/e^h) = \lim_{h\to\infty}(D_h h/D_h e^h) = \cdots$.]

15. The arithmetic mean of a frequency distribution is defined by

$$\bar{x} = \frac{\int_a^b xf(x)\,dx}{\int_a^b f(x)\,dx}$$

where the function f denotes the frequency distribution on the interval $[a, b]$. If the frequency distribution of hourly wages between \$2 and \$4 is given by $f(x) = 4x - x^2$, find the arithmetic mean of the wages.

16. The function $f(x) = e^{-x}$ defines the frequency distribution of consumer index values between 0.5 and 2. Find the arithmetic mean of the index values. (Refer to problem 15.)

17. If $f(t) = 6/(t + 27)^{5/3}$ defines the probability that a light bulb will fail in t months, then:

(a) Verify that f is a pdf.
(b) Find the probability that a similar bulb picked at random will last for more than 37 months.

18. Find the area between the given curves:

(a) $y = x^2 + x + 1$ and $y = 2x^2 + 2x - 5$
(b) $y = e^x$, $y = e^{-x}$, and $y = e$
(c) $y = 3x$, $y = -\frac{3}{2}x + \frac{9}{2}$, and $y = 0$.

19. Find maximum profit if:

(a) Revenue $R(x) = 5 + 2x$ and cost $C(x) = 5 + x^2$
(b) Revenue $R(x) = 1 + 32x$ and cost $C(x) = 1 + x^4$
(c) Revenue $R(x) = 10x + 1$ and cost $C(x) = (x^2/80) + 1$.

20. Integrate by parts:

(a) $\int xe^{-6x}\,dx$ (b) $\int te^{t-2}\,dt$.

21. Find the consumers' and producers' surplus if demand D is $y = 7 - x$ and supply S is $y = x^2 + 1$.

22. Determine the average value of $f(x) = 6x - x^2$ on the interval $[0, 6]$.

23. Integrate: $\int \frac{(x^2 + 3x)\,dx}{x^2 - 6x + 9}$.

24. Find the value, if it exists, of $\int_0^\infty x^2e^{-4x^3}\,dx$.

25. Determine the area of the region between the given curves:

(a) $y = x^2 - 2x$ and $y = 4x - 2x^2$
(b) $y = -1 + e^x$, $y = -1 + e^{-x}$, and $y = 1$.
(c) $y = \ln 2x$, $y = 0$, and $x = 2$
(d) $y = x^3 - 2$ and $y = x^3 - x^2 + 2$.

26. Integrate:

(a) $\int \frac{9dt}{t(t^2 - 9)}$ (b) $\int \frac{4z^2dz}{4z^2 - 8z + 3}$.

27. Find the average value of $g(t) = t^4 - 54t^2$ on the interval $[-1, 1]$.

28. Given demand D: $y = 20 - x^2$ and supply S: $y = (x + 2)^2$; find consumers' and producers' surplus.

29. Use integration by parts twice to determine:

(a) $\int x^2e^x\,dx$ (b) $\int t^2e^{2t}\,dt$

(c) $\int (\ln z)^2\,dz$ (d) $\int x^2(x + 1)^4\,dx$

30. Find the area of the triangular region with vertices (0, 0), (2, 3), and (4, 0).

31. Evaluate:

(a) $\int_0^1 \frac{1 - \sqrt{x}}{1 + \sqrt{x}}\,dx$; let $u = \sqrt{x}$, $u^2 = x$, $2u\,du = dx$, etc.

(b) $\int_1^8 \frac{dx}{1 + \sqrt[3]{x}}$; let $u = x^{1/3}$, $u^3 = x$, $3u^2\,du = dx$, etc.

(c) $\displaystyle\int_0^1 \frac{x^2\,dx}{(1+x)^3}$; let $u = 1 + x$, etc.

(d) $\displaystyle\int_0^1 \frac{dt}{e^t + 1}$. [*Hint*: Multiply numerator and denominator by e^{-t}.]

(e) $\displaystyle\int_0^3 \frac{x^3 + x^2 - x - 1}{x + 1}\,dx$

8 DIFFERENTIAL EQUATIONS

Additional Calculus Tools

IMPORTANT EQUATIONS

$$\frac{dy}{dx} + P(x)y = Q(x)$$

$$\text{integration factor} = e^{\int P\,dx}$$

$$y = vx + Cu(x)$$

$$f(tx, ty) = t^k f(x, y)$$

$$M(x, y)\,dx + N(x, y)\,dy = 0$$

8.1 INTRODUCTION

The subject of differential equations is a broad one, having extensive applications. This chapter will be a very short introduction to this interesting and important branch of mathematics.

A differential equation is an equation relating certain independent variables with certain functions and with derivatives of these functions with respect to the independent variable(s). Thus

$$F\left(x, y, \frac{dy}{dx}, \frac{d^2y}{dx^2}, \ldots, \frac{d^ny}{dx^n}\right) = 0 \tag{1}$$

which contains the independent variable x, the function y, and derivatives of y with respect to x, is called an **ordinary differential equation**. If an equation involves partial derivatives it is called a **partial differential equation**. We will confine our discussion to ordinary differential equations.

A relation $y = g(x)$ is a solution of Eq. (1) above if

$$F[x, g(x), g'(x), g''(x), \ldots, g^{(n)}(x)] \equiv 0. \tag{2}$$

For example, the differential equation

$$\frac{3\,d^2y}{dx^2} - \frac{dy}{dx} - 2y = 0$$

has the solution $y = e^x$ since, by substitution, we get

$$3e^x - e^x - 2e^x = 0,$$

which is valid for all real values of x.

The **order** of a differential equation is determined by the order of the derivative (or differential) of highest order in that equation. For example,

$$dy = 2x\,dx \tag{3}$$

$$x\left(\frac{dy}{dx}\right)^2 = 3 \tag{4}$$

$$\frac{d^2y}{dx^2} - 4y = 0 \tag{5}$$

Thus, Eqs. (3) and (4) are first-order differential equations because of the presence of the first derivative only; Eq. (5) is of the second order because the second derivative is the highest-order derivative in that equation.

The **degree** of an ordinary differential equation is the algebraic degree of the highest-order derivative appearing in the equation. Accordingly, Eqs. (3) and (5) are of degree one because of the $(dy)^1$ and $(d^2y/dx^2)^1$. Similarly,

$$\left(\frac{d^2y}{dx^2}\right)^3 + \left(\frac{dy}{dx}\right)^4 - x^2y = e^x$$

is of degree three because d^2y/dx^2 is the highest-order derivative and its degree is three.

If y and its derivatives occur to the first degree, then the equation is defined to be linear in y. Thus,

$$x\frac{d^2y}{dx^2} + \frac{dy}{dx} + ye^x = 0$$

is linear in y whereas

$$\frac{d^2y}{dx^2} + \left(\frac{dy}{dx}\right)^2 - x = 0$$

is not linear.

8.2 SOLUTIONS

It is possible to test whether a given relation is a solution of a given differential equation by the process of differentiating and substituting in the given equation. This section is involved with that process. However, a large part of our later discussion will consist of devising methods for finding solutions.

Some differential equations have no solutions, some have one or more.

EXAMPLE 1 $(dy/dx)^2 + 2y^2 = -3$ has no solution since the left side of the equation is the sum of two terms raised to even powers and the right side is negative. Hence, no function $y(x)$ will satisfy the given differential equation.

EXAMPLE 2 $(y')^2 + y^2 = 0$ has $y = 0$ as its only solution.

EXAMPLE 3 $dy = 2\,dx$ has the general solution $y = 2x + C$. Since C represents any real number, the given equation has infinitely many solutions.

It should be understood that a general solution cannot always be indicated by a single formula. For instance, $y' - y^2 = 0$ has $y = -1/x$ and $y = 0$ as two particular solutions.

EXAMPLE 4 Verify that $y = \dfrac{2}{x} - 1$ is a solution of

$$x\frac{d^2y}{dx^2} + 2\frac{dy}{dx} = 0.$$

Solution Differentiating, we get

$$\frac{dy}{dx} = \frac{-2}{x^2}$$

and

$$\frac{d^2y}{dx^2} = \frac{4}{x^3}.$$

Substituting these values in the given differential equation, we get

$$x\left(\frac{4}{x^3}\right) + 2\left(\frac{-2}{x^2}\right) = \frac{4}{x^2} - \frac{4}{x^2} = 0.$$

Thus, $y = (2/x) - 1$ is a solution.

EXAMPLE 5 Show that $x^2 + y^2 = x^2y^2 + C$ is a solution of

$$(x - xy^2)\,dx + (y - x^2y)\,dy = 0.$$

Solution Differentiating implicitly, we obtain

$$2x + 2y\frac{dy}{dx} = 2x^2y\frac{dy}{dx} + 2xy^2$$

or

$$(x - xy^2) = (x^2y - y)\frac{dy}{dx},$$

which simplifies to the given differential equation.

PROBLEM SET 8.1, 8.2

1. Show that $xy = C(x - y)$ is a general solution of $x^2y' - y^2 = 0$.
2. Verify that $y = Ce^{y/x}$ is a general solution of $(xy - y^2)\,dy/dx - y^2 = 0$.
3. Show that $y = 2/(x^2 - 1)$ is a particular solution of $y' + xy^2 = 0$ on $-1 < x < 1$.
4. Find whether $y = e^x + e^{2x}$ is a solution of $y'' - 4y' + 4y - e^x = 0$.
5. Verify that $y = e^{-x} + xe^{-x} + 1$ is a solution of $y'' + 2y' + y - 1 = 0$.
6. Show that $y + \ln y + x^2 = C$ is a general solution of $(y + 1)\,dy + 2xy\,dx = 0$.
7. Find values of C_1 and C_2 such that $f(x) = x^2 + C_1x + C_2 - 1$ satisfies $f(1) = 2$ and $f'(1) = 1$.
8. Determine values of C_1 and C_2 such that $g(x) = C_1(C_2x - 1)^4$ satisfies $g'(0) = -4$ and $g''(0) = 24$.
9. Verify that $x^4 = Cy^3$ is a general solution of $4y\,dx = 3x\,dy$.
10. Show that $y(x) = C_1e^{3x} + C_2e^{-3x}$ is a solution of $y'' - 9y = 0$.
11. Find the particular solution for $y' = \sqrt{x}/2$ which satisfies the condition $y = 2$ when $x = 9$.
12. Verify that $2x^2 + 3xy^2 - 4y^3 = C$ is an implicit solution of $4x + 3y^2 + 6y(x - 2y)\,dy/dx = 0$.
13. Show whether $y = C_1e^x + C_2xe^x$ is a solution of $\dfrac{d^2y}{dx^2} - 2\dfrac{dy}{dx} + y = 0$.
14. Find the particular solution satisfying $dy/dx = x\sqrt{1 - y^2}/y\sqrt{1 - x^2}$ if $y = 3/4$ when $x = 7/16$.
15. Show that $y = 3e^x$, $y = \frac{1}{4}e^{-x}$, and $y = 5$ are all solutions of $y''' - y' = 0$.
16. Show that $s = \frac{1}{2}(2t^2 + 4t + 3)$ satisfies $s'' - 4s' + 4s = 4t^2$.

17. Verify that $4e^{-x} - e^{-4y} = C$ is a solution of $e^{x+y}\,dx = e^{2x-3y}\,dy$.

18. Show whether $1/y = Ce^x + x + 1$ is a solution of $y' + y - xy^2 = 0$. [*Hint*: Use $(y'/y^2) + (1/y) - x = 0$ for convenience.]

8.3 SEPARATION OF VARIABLES

In business and economics many practical problems involving rates of change, for example, marginal cost and marginal profit, can be described by differential equations. Certain of these equations have been introduced earlier. Some repetition is unavoidable as we progress. However, this should not hinder us but rather should help us get a better foothold.

The law of normal population growth is an example of the application of differential equations to natural phenomena. Consider the differential equation

$$\frac{dy}{dt} = ky,$$

which implies that a population y reproduces at a rate proportional to its size. This equation does not include those factors which restrain or augment population growth. The solution of this equation involves the following steps.

$$\begin{aligned} \frac{dy}{y} &= k\,dt && \text{[separating the variables]} \\ \ln|y| &= kt + C \\ |y| &= e^{kt+C} = e^{kt} \cdot e^{C} \\ y &= y_0 e^{kt} && [y_0 = e^C] \end{aligned}$$

where y_0 is a constant denoting the size of the initial population at time $t = 0$, and e^{kt}, $k > 0$, indicates exponential growth. This solution exhibits the population growth pattern and also furnishes us the means with which to predict the amount of future growth. For instance, if the population of a village is 1000 and doubles in 10 years, in order to find the population after 20 years we proceed as follows:

$$\begin{aligned} 2000 &= 1000e^{k(10)} \\ 2 &= e^{10k} \\ 10k &= \ln 2 \\ k &= \frac{\ln 2}{10} \cong 0.06931. \end{aligned}$$

Then $$y = 1000e^{0.06931t}$$

At $t = 20$, $$\begin{aligned} y &= 1000e^{1.3862} \\ &\cong 1000(3.7026) \\ &\cong 3703 \text{ people.} \end{aligned}$$

EXAMPLE 1 Separate the variables and solve:

$$y\sqrt{3x^2 + 4}\,dy + x\sqrt{5 - y^2}\,dx = 0.$$

Solution Separating the variables, we obtain

$$\frac{y}{\sqrt{5 - y^2}}\,dy + \frac{x}{\sqrt{3x^2 + 4}}\,dx = 0.$$

Then $\quad (5 - y^2)^{-1/2}y\,dy + (3x^2 + 4)^{-1/2}x\,dx = 0.$

Integrating, $\quad -\frac{1}{2}(5 - y^2)^{1/2} \cdot 2 + \frac{1}{6}(3x^2 + 4)^{1/2} \cdot 2 = C.$

Simplifying, $\quad -\sqrt{5 - y^2} + \frac{1}{3}\sqrt{3x^2 + 4} = C,$

is a general solution. To find a particular solution if $y = 1$ when $x = 2$, then

$$-\sqrt{5 - 1} + \frac{1}{3}\sqrt{12 + 4} = C = -2 + \frac{4}{3} = -\frac{2}{3}$$

Thus, $\quad -\sqrt{5 - y^2} + \frac{1}{3}\sqrt{3x^2 + 4} = -\frac{2}{3}$

or $\quad 3\sqrt{5 - y^2} - \sqrt{3x^2 + 4} = 2$

is a particular solution.

PROBLEM SET 8.3

Find the general solutions of problems 1–8:

1. $(2 - x)\,dy + (1 + y)\,dx = 0$

2. $dy + e^{x+y}\,dx = 0$

3. $xy\,dx + (1 - x)\,dy = 0$

4. $xy\,dx + \sqrt{1 + x^2}\,dy = 0$

5. $(3 + e^{2x})\,dy - ye^{2x}\,dx = 0$

6. $\dfrac{dy}{dt} - 3\sqrt{t}(1 - y)^{1/3} = 0$

7. $(1 + \ln y)\,dy + (1 + \ln x)\,dx = 0$

8. $y(x - 1)\,dy + x^2\,dx = 0$

Find particular solutions satisfying the given conditions in problems 9–13:

9. $\sqrt{y + 1}\,dx + x\,dy = 0; \quad y = 3$ when $x = 1$

10. $y(x - 1)\,dx = x(y + 1)\,dy; \quad y = 1$ when $x = 2$

11. $2(1 + e^x)\,dy + ye^{2x}\,dx = 0; \quad y = 1$ when $x = 0$

12. $x\sqrt{5 - y^2}\,dx - y\sqrt{10 - x^2}\,dy = 0; \quad y = 2$ when $x = 3$

13. $(2 - e^t)\,dy - e^{t-y}\,dt = 0; \quad y = 0$ when $t = 0$

14. A certain population grows at a rate proportional to its size P. Due to a persistently unfavorable environment, the population also grows at a rate inversely proportional to time t. Write the law of population growth.

15. The growth rate of a population is directly proportional to its size P and inversely proportional to the cube root of P (because of limitations on space). Find the law of population growth.

16. A deposit is made in a credit union paying 7% interest compounded continuously. How long will it take to double the amount deposited? Use $A = Pe^{it}$.

17. In problem 16, how long will it take to double the amount deposited if the interest rate is 8%?

18. A certain amount deposited in a savings firm doubles every 10 years. What is the rate of interest if it is compounded continuously?

19. The GNP (gross national product) of a country grows at an annual rate proportional to 4% of the GNP. In 1970 the GNP was 10^7 dollars. Find the GNP in 1980.

20. Work problem 19 if the GNP *decreases* at an annual rate proportional to 4% of GNP.

21. Assume the population P in the world was 2200 million in 1930 and 3300 million in 1960. Find a formula for the value of P at any time t and use it to determine:
    (a) the population of the world in the year 2030;
    (b) the year in which each inhabitant of the world will have only one square yard of land for himself (use total "useable" land area as $1.6(10^{14})$ square yards).

8.4 SEPARATION OF VARIABLES: OTHER APPLICATIONS

There could be factors involved with population growth other than the size of the population. The lack of food, for instance, might cause the **rate of growth** to decrease as the size of that population increases. Such a restraint could be expressed by letting the growth rate be proportional to both the size and the remaining amount of growth $S - y$:

$$\frac{dy}{dt} = ky(S - y)$$

where S is the limiting size caused by the shortage of food. From the equation it is clear that the growth rate approaches zero as y approaches S. The equation is a reasonable mathematical model for interpreting the given conditions. To solve it, we proceed as follows:

$$\frac{dy}{y(S - y)} = k\,dt \qquad \text{[separating the variables]}$$

$$\int\left(\frac{A}{y} + \frac{B}{S - y}\right)dy = k\int dt \qquad \left[\frac{1}{y(S-y)} = \frac{A}{y} + \frac{B}{S-y} \text{ for some constants } A \text{ and } B\right]$$

$$\int\left(\frac{1/S}{y} + \frac{1/S}{S - y}\right)dy = kt + C_1 \qquad \left[A = \frac{1}{S} = B, \text{ explained below}\right]$$

$$\frac{1}{S}\ln|y| - \frac{1}{S}\ln|S - y| = kt + C_1$$

$$\begin{aligned}
\ln\left|\frac{y}{S-y}\right| &= Skt + C && [C = SC_1] \\
\frac{y}{S-y} &= e^{Skt+C} = e^{Skt} \cdot e^{C} \\
&= De^{Skt} && [D = e^{C}] \\
y &= SDe^{Skt} - yDe^{Skt} \\
&= SDe^{Bt} - yDe^{Bt} && [B = Sk] \\
(1 + De^{Bt})y &= SDe^{Bt} \\
y &= \frac{SDe^{Bt}}{1 + De^{Bt}} \\
y &= \frac{SD}{D + e^{-Bt}} && (1)
\end{aligned}$$

where we multiplied numerator and denominator by e^{-Bt} so as to have a simpler form of the general solution. Let us review the method of partial fractions in order to clarify that $A = 1/S = B$ above. Thus,

$$\frac{1}{y(S-y)} = \frac{A}{y} + \frac{B}{S-y} = \frac{A(S-y) + By}{y(S-y)}$$

for some constants A and B. Since the denominators are equal, then so are the numerators. Accordingly,

$$1 = A(S - y) + By.$$

This is valid for all real values of y, so we choose $y = 0$ and get

$$1 = AS \quad \text{or} \quad A = \frac{1}{S}.$$

We next choose $y = S$ and obtain

$$1 = BS \quad \text{or} \quad B = \frac{1}{S}.$$

Thus,

$$A = \frac{1}{S} = B,$$

as we wished to show.

We return to the general solution (1) and draw solution curves, shown in Figure 8.1. Notice that these curves are confined to the interval $0 < y < S$, $t \geq 0$, since $\lim_{t\to\infty} y = S$ and $y > 0$.

Some common problems involve differential equations of the form

$$\frac{du}{dt} = ku + a \qquad (k \text{ and } a \text{ constants}).$$

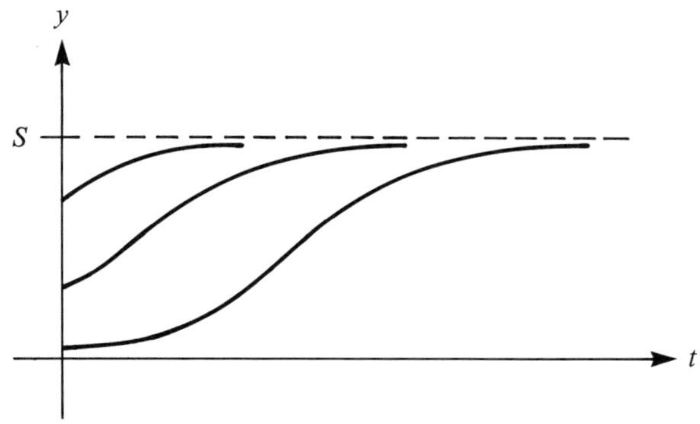

Figure 8.1

An alternative form is

$$\frac{du}{dt} = k(u + b) \qquad \left(b = \frac{a}{k}\right).$$

We introduce and solve a typical problem.

The rate of change of the temperature of a body is proportional to the difference in temperature between the body and the surrounding medium. (This is Newton's Law of Cooling.) Suppose the temperature in a warehouse is 70°F and an employee carries in a cup of coffee whose temperature is 180°F. After 6 minutes the coffee's temperature is found to be 120°F. If H denotes the temperature of the coffee at any time $t \geq 0$ and M is the temperature (constant) of the surrounding medium, then the initial differential equation can be formulated as

$$\frac{dH}{dt} = k(H - M)$$

or

$$\frac{dH}{dt} = k(H - 70) \qquad \text{(for this example).}$$

This is subject to the conditions $H = 180$ at $t = 0$ and $H = 120$ at $t = 6$. Separating the variables, we get

$$\frac{dH}{H - 70} = k\,dt.$$

Integrating,

$$\ln|H - 70| = kt + C$$

Then

$$H - 70 = e^{kt+C} = e^{kt} \cdot e^{C}$$

$$H = 70 + Be^{kt} \qquad [B = e^{C}].$$

For $H = 180$, $t = 0$,

$$180 = 70 + B$$

$$B = 110.$$

Therefore,

$$H = 70 + 110e^{kt} \tag{2}$$

Putting $H = 120$, $t = 6$ in Eq. (2),

$$120 = 70 + 110e^{6k}$$

Solving for k, $$e^{6k} = \frac{5}{11}$$
$$6k = \ln 5 - \ln 11$$
$$\cong -0.7885$$
$$k = -0.1314$$

Therefore, $$H = 70 + 110e^{-0.1314t}$$

is the desired formula giving the temperature of the coffee for any time $t \geq 0$. The graph in Figure 8.2 shows us pictorially the variation of temperature with time. Observe that $H \to 70°\text{F}$ as $t \to \infty$.

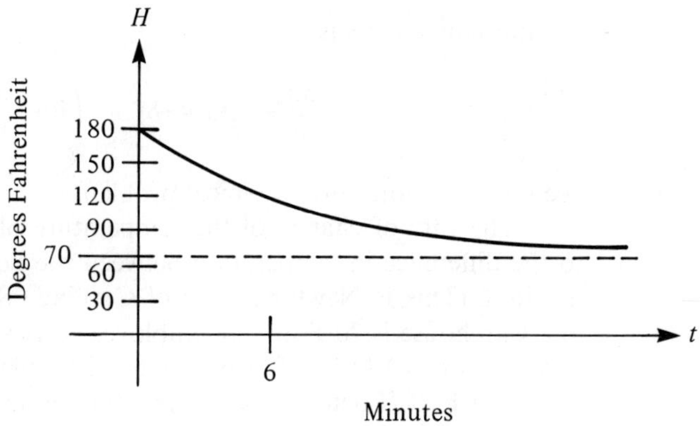

Figure 8.2

PROBLEM SET 8.4

In problems 1–7, find the particular solution of the given differential equation that satisfies the given initial conditions.

1. $dy = 2(y - 100)\, dx$; $y = 101$ when $x = 1$.
2. $dy = 3(2y - 55)\, dx$; $y = 28$ when $x = 2$.
3. $du = \frac{1}{2}\left(\frac{3}{4} - u\right) dt$; $u = \frac{5}{4}$ when $t = 2$
4. $dw = \frac{t^2}{w}(w^2 - 1)\, dt$; $w = \frac{1}{2}$ when $t = 3$
5. $dz = \frac{\sqrt{t}}{z}(2z^2 - 1)\, dt$; $z = 1$ when $t = 4$
6. $dx = \frac{\sqrt{5 - x^2}}{tx}\, dt$; $x = 1$ when $t = 1$.
7. $dx = \frac{4}{x}(y + xy)\, dy$; $y = 2$ when $x = 0$.

Solve problems 8–13:

8. The growth rate of a population is proportional to its size P and also inversely proportional to time t (due to a persistently deteriorating environment). Find the law of population growth.

9. The growth rate of a population is proportional to its size P and also to the remaining amount of growth $L - P$. Write the pertinent differential equation and determine its general solution.

10. Let 75°F be the temperature of a room and 200°F be the temperature of a cup of coffee brought into the room. After 8 minutes, the coffee has a temperature of 130°F. Find a formula for the coffee's temperature at any time t while in the room.

11. Work problem 10 by first replacing 75°F, 200°F, 8 minutes, and 130°F with 74°F, 194°F, 7 minutes, and 144°F, respectively.

12. For a certain firm the rate at which profit grows is given by $dP/dx = 100(1 - e^{1-0.01x})$ where x is the amount of products sold. If $P = \$10{,}000$ when $x = 100$, find $P(x)$ when $x = 200$. Sketch a graph of dP/dx. What is the upper limit for dP/dx?

13. A certain population grows at a rate dy/dt proportional to its size y and inversely proportional to the quantity $y - 1$. Using $\ln C$ as the constant of integration, show that the general solution can be written as $ye^{-y} = \dfrac{1}{C}e^{-kt}$.

14. A firm is concerned with the spreading of an epidemic in a town where most of the firm's employees live. It is known that the rate at which the epidemic spreads is jointly proportional to the number of residents who have already caught the disease and also to the number of residents S who are susceptible but have not yet gotten the disease. The firm wishes to find a formula for the number of residents $h(t)$ who have caught the disease by the time t (weeks). To find this formula we let $S - h$ be the number of susceptibles who have not yet contracted the disease at time t, then form the differential equation $dh/dt = kh(S - h)$. Show that the formula desired by the firm, that is, the general solution, is $h(t) = AS/(A + e^{-Bt})$ where A and B are certain constants.

15. In periods of worldwide inflation and depression, many small business firms, say T of them, are threatened with bankruptcy. If we let $x(t)$ be the number which have become bankrupt by time t, then $T - x$ represents the number of threatened firms which have not yet become bankrupt at time T. Assuming that $dx/dt = kx(T - x)$ is a practical mathematical model to use in this instance, show that $x(t) = DT/(D + e^{-Bt})$ where C is the constant of integration and $B = Tk$, $D = e^{TC}$ are also constants.

16. Assume that the growth of a certain population is defined by

$$N(x) = \frac{BN_0}{N_0 + (B - N_0)e^{-0.0001Bx}}$$

where B is the upper limit of the population, N_0 is the initial number of the population, and t is the time in months. If $B = 10^4$ and $N_0 = 10^3$, find $N(10)$.

8.5 LINEAR EQUATIONS

A differential equation of the form

$$A(x)\frac{dy}{dx} + B(x)y + C(x) = 0, \qquad A(x) \neq 0,$$

is called a **first-order linear differential equation**. This reduces to

$$\boxed{\frac{dy}{dx} + P(x)y = Q(x)} \tag{1}$$

upon division by $A(x)$. Observe in Eq. (1) that the coefficient of y is a function of x and that the right side is some other function of x. Some illustrations of linear differential equations are:

$$\frac{dy}{dx} + e^x y = 2x$$

$$\frac{dy}{dx} - \frac{2}{x} y = 4$$

$$\frac{dy}{dx} + (1 - x)y = e^{-2x}$$

The following are nonlinear differential equations:

$$\frac{dy}{dx} + xe^y = e^x \qquad \text{(because of the } e^y\text{)}$$

$$\frac{dy}{dx} - y^2 = 4x \qquad \text{(because of the } y^2\text{)}$$

If $Q(x) = 0$, then Eq. (1) can be solved by a separation of variables. Thus,

$$\frac{dy}{dx} + Py = 0 \tag{2}$$

$$\frac{dy}{y} + P\,dx = 0$$

$$\ln|y| + \int P\,dx = C$$

$$|y| = e^{-\int P\,dx + C} = e^{-\int P\,dx} \cdot e^C$$

$$y = ke^{-\int P\,dx}, \quad k = e^C \tag{3}$$

[There will be later references to the important equations (2) and (3).] For $k \neq 0$, we can find a particular solution $y = u(x)$. The solution in Eq. (3) offers us a clue for developing a formula that gives the solution. To this end, we note that

$$\frac{d}{dx} e^{\int P\,dx} y = e^{\int P\,dx} \frac{dy}{dx} + yPe^{\int P\,dx}$$

$$= e^{\int P\,dx}\left(\frac{dy}{dx} + Py\right).$$

This form suggests that we multiply Eq. (1) by $e^{\int P\,dx}$. We then obtain

$$e^{\int P\,dx}\left(\frac{dy}{dx} + Py\right) = e^{\int P\,dx} Q.$$

Integrating, we get

$$e^{\int P\,dx}y = \int e^{\int P\,dx}Q\,dx + C$$

or

$$y = e^{-\int P\,dx}\int e^{\int P\,dx}Q\,dx + Ce^{-\int P\,dx}.$$

The student is advised not to memorize this cumbersome formula but to keep in mind the inherent implication, which is:

> *To solve the linear equation (8), multiply by the* ***integrating factor*** *$e^{\int P\,dx}$ and integrate.*

EXAMPLE 1 Solve: $\dfrac{dy}{dx} - y = x$

Solution Comparing this to Eq. (1), we see that $P = -1$, so

$$e^{\int P\,dx} = e^{\int(-dx)} = e^{-x}$$

Multiplying the given equation by e^{-x}, we obtain

$$e^{-x}\left(\frac{dy}{dx} - y\right) = xe^{-x}$$

or

$$\frac{d}{dx}(e^{-x}y) = xe^{-x}$$

Integrating, $\qquad e^{-x}y = \int xe^{-x}\,dx$

$$= -xe^{-x} - e^{-x} + C \qquad \text{[integrating by parts]}$$

Dividing by e^{-x}, $\qquad y = -x - 1 + Ce^{x}$

which is the general solution.

EXAMPLE 2 Solve: $x\dfrac{dy}{dx} + (x+1)y = e^{x}$

Solution Dividing by x yields

$$\frac{dy}{dx} + \left(1 + \frac{1}{x}\right)y = \frac{e^{x}}{x}.$$

Here, $\qquad P = 1 + \dfrac{1}{x}, \quad \int P\,dx = x + \ln|x|$

So $\qquad e^{\int P\,dx} = e^{x+\ln|x|} = e^{x}\cdot e^{\ln x} = xe^{x} \qquad [\text{since } e^{\ln x} = x].$

Multiplying, $\qquad xe^{x}\left[\dfrac{dy}{dx} + \left(1 + \dfrac{1}{x}\right)y\right] = xe^{x}\cdot\dfrac{e^{x}}{x}$

or

$$e^{x}\left[x\frac{dy}{dx} + (x+1)y\right] = e^{2x}.$$

Then

$$\frac{d}{dx}(e^{x}xy) = e^{2x}.$$

Integrating, $$xye^x = \frac{1}{2}e^{2x} + C$$

or $$y = \frac{e^x}{2x} + \frac{C}{x}e^{-x},$$

the general solution.

Linear differential equations have the following characteristics. If we substitute in Eq. (1)

$$y = u + v, \qquad \frac{dy}{dx} = \frac{du}{dx} + \frac{dv}{dx},$$

where u and v are functions of x, then

$$\frac{du}{dx} + \frac{dv}{dx} + P(x)(u + v) = Q(x)$$

or $$\left(\frac{du}{dx} + Pu\right) + \left(\frac{dv}{dx} + Pv - Q\right) = 0.$$

Let's say that $dv/dx + Pv - Q = 0$ has the particular solution $y = v(x)$. Then $du/dx + Pu = 0$ remains to be solved. If the latter has the particular nonzero solution $y = u(x)$, then it can be shown that

$$\boxed{y = v(x) + Cu(x)} \tag{4}$$

is the general solution. In other words, if we can determine a particular solution of Eq. (1), we are left with the simpler equation (2) to solve. If the latter has a nonzero solution, the general solution is then recognized as a combination of the two solutions as shown in Eq. (4).

EXAMPLE 3 Solve: $\dfrac{dy}{dx} + y = 2.$

Solution By inspection, $y = v(x) = 2$ is a particular solution, since $0 + 2 = 2$. It now remains to solve $dy/dx + y = 0$. Then

$$\begin{aligned} \frac{dy}{y} &= -dx, \\ \ln|y| &= -x + C \\ y &= e^{-x+C} = e^{-x} \cdot e^C \\ &= C_1e^{-x}, \qquad C_1 = e^C \end{aligned}$$

Thus, $y = C_1e^{-x}$ represents the nonzero solution $y = u(x)$ in the discussion above. So

$$y = 2 + C_1e^{-x}$$

is the general solution according to Eq. (4).

Let us consider a real-world application of the ideas of this section.

EXAMPLE 4 A firm is asked by the local citizenry to do something about the pollution in a nearby reservoir. The initial pollution level is found to be l_0 grams per gallon. The firm begins piping water out of the reservoir at the constant rate of r gallons per hour and replaces it with less noxious water whose pollution level is l_1 grams per gallon. If the capacity (volume) of the reservoir is V gallons, how long should it take the firm to decrease the pollution level to l grams per gallon?

Solution We let $T(t)$ be the total number of grams of pollution in the reservoir at any time t and we wish to determine t so that

$$\frac{T(t)}{V} = l.$$

The rate at which polluted matter leaves the reservoir is given by

$$\frac{rT(t)}{V} \qquad \text{(grams per hour).}$$

The rate at which pollutants enter the reservoir is defined by

$$rl_1 \qquad \text{(grams per hour).}$$

Then the rate of change of the number of grams of pollution with respect to time t is

$$\frac{dT}{dt} = rl_1 - \frac{rT}{V}$$

or

$$\frac{dT}{dt} + \frac{r}{V}T = rl_1. \tag{5}$$

We recognize Eq. (5) as a first-order linear differential equation where $P = r/V$. The integrating factor is

$$e^{\int P\,dt} = e^{\int r/V\,dt} = e^{(r/V)t}$$

which is the multiplier for solving Eq. (5). Thus,

$$e^{(r/V)t}\left(\frac{dT}{dt} + \frac{r}{V}T\right) = e^{(r/V)t}rl_1$$

$$\frac{d}{dt}(e^{(r/V)t}T) = e^{(r/V)t}rl_1$$

Integrating,

$$e^{(r/V)t}T = Vl_1e^{(r/V)t} + C$$

$$T = Vl_1 + Ce^{-(r/V)t}$$

$$\frac{T}{V} = l_1 + \frac{C}{V}e^{-(r/V)t}.$$

For the initial condition l_0 at $t = 0$ we have

$$l_0 = l_1 + \frac{C}{V}$$

or

$$C = V(l_0 - l_1).$$

Substituting, we get

$$\frac{T}{V} = l_1 + (l_0 - l_1)e^{-(r/V)t}$$

which defines the pollution level at some t. The pollution level will be decreased to l grams when

$$l_1 + (l_0 - l_1)e^{-(r/V)t} = l.$$

Solving for t, we obtain

$$e^{-(r/V)t} = \frac{l - l_1}{l_0 - l_1}$$

$$-\frac{r}{V}t = \ln\frac{l - l_1}{l_0 - l_1}$$

$$t = -\frac{V}{r}\ln\frac{l - l_1}{l_0 - l_1}$$

or

$$t = \frac{V}{r}\ln\frac{l_0 - l_1}{l - l_1} \tag{6}$$

The firm next decides to use this formula to determine how long it would take to decrease the pollution level by 25%, knowing that the reservoir contains 10^8 gallons and that the water being piped in at the rate of 10^4 gallons per hour is free of pollution. Thus, the known quantities are

$$l = \frac{3}{4}l_0, \quad l_1 = 0, \quad V = 10^8, \quad \text{and} \quad r = 10^4.$$

Substituting in Eq. (6) yields

$$\begin{aligned} t &= \frac{10^8}{10^4}\ln\frac{l_0}{3l_0/4} \\ &= 10^4 \ln\frac{4}{3} \\ &= 10^4(\ln 4 - \ln 3) \\ &\cong 10^4(0.287) = 2870 \text{ hours.} \end{aligned}$$

PROBLEM SET 8.5

Solve problems 1–8 by applying the integrating factor $e^{\int P\,dx}$:

1. $\dfrac{dy}{dx} + y = e^x$

2. $\dfrac{dy}{dx} - y = x - 1$

3. $\dfrac{dy}{dx} + 2xy = x$

4. $x\dfrac{dy}{dx} + 3x^2y = x^2$

5. $x\dfrac{dy}{dx} + (x + 1)y = x$

6. $\dfrac{dy}{dx} + \dfrac{2}{x}y = e^{x^3}$

7. $2\frac{dy}{dx} - 8xy = x$ **8.** $\frac{dy}{dx} + \frac{y}{x} = 6$

Find the general solution $y = v(x) + Cu(x)$ in problems 9–14:

9. $\frac{dy}{dx} - y = 3$ **10.** $\frac{dy}{dx} + 3y = 3$

11. $\frac{dy}{dx} + y = -1$ **12.** $\frac{dy}{dx} + 2y = 4$

13. $2\frac{dy}{dx} - 3y = 12$ **14.** $3\frac{dy}{dx} - 8y = 12$

15. The initial pollution level of a lake is found to be b grams per gallon. Water is withdrawn from the lake at the constant rate of w gallons per hour and replaced with less polluted water whose pollution level is r grams per gallon. If the capacity of the lake is g gallons, find:

(a) the time t it will take to reduce the pollution level to m grams per gallon;
(b) the time t it will take to reduce the pollution level by 50% if $g = 10^5$, $r = 0$, and $w = 10^2$.

8.6 HOMOGENEOUS EQUATIONS

Certain equations whose variables are not separable can be transformed by a suitable substitution into equations whose variables are separable. A case in point occurs when the coefficients of the equation are homogeneous functions of the same degree.

DEFINITION 1 *A function $f(x, y)$ is said to be* ***homogeneous of degree*** *k in a region G if*

$$\boxed{f(tx, ty) = t^k f(x, y)}$$

for all (x, y) in G and all $t > 0$.

For example, if $f_1(x, y) = x^2 + y^2$ and $f_2(x, y) = xy$, then

$$\begin{aligned} f_1(tx, ty) &= (tx)^2 + (ty)^2 \\ &= t^2(x^2 + y^2) = t^2 f_1(x, y) \end{aligned}$$

and

$$\begin{aligned} f_2(tx, ty) &= (tx)(ty) \\ &= t^2(xy) = t^2 f_2(x, y). \end{aligned}$$

Thus, f_1 and f_2 are homogeneous of degree 2.

We see that the ratio

$$\frac{f_1(tx, ty)}{f_2(tx, ty)} = \frac{t^2 f_1(x, y)}{t^2 f_2(x, y)} = \frac{t^0 f_1(x, y)}{f_2(x, y)}$$

is of degree 0. This leads us to statements of two theorems which we give without proof.

In an equation of order one such as

$$\boxed{M(x, y)\,dx + N(x, y)\,dy = 0} : \tag{1}$$

THEOREM 1 *If M and N are both homogeneous of the same degree, then the ratio* $\dfrac{M}{N}$ *is homogeneous of degree* 0.

THEOREM 2 *If* $\dfrac{M}{N}$ *is homogeneous of degree* 0, *then* $\dfrac{M}{N}$ *is a function of* $\dfrac{y}{x}$ *alone.*

From this we conclude that Eq. (1) may be expressed as

$$\frac{dy}{dx} + h\left(\frac{y}{x}\right) = 0, \tag{2}$$

which suggests that the use of a new variable $v = y/x$ may be practical. Substituting v in Eq. (2) yields (since $y = vx$ and $dy/dx = v + x(dv/dx)$)

$$x\frac{dv}{dx} + v + h(v) = 0, \tag{3}$$

where the variables are separable. Equation (3) can be solved by the method of separation of variables. Upon substitution of the original variables, we get the solution for Eq. (1).

EXAMPLE 1 Solve: $(2x^2 + y^2)\,dx - 2xy\,dy = 0$.

Solution The coefficients of dx and dy are both homogeneous of degree 2, so that their ratio is of degree 0. Substituting $y = vx$ and $dy = v\,dx + x\,dv$ in the given equation, we obtain

$$(2x^2 + v^2x^2)\,dx - 2vx^2(v\,dx + x\,dv) = 0.$$

Simplifying, $$\frac{dx}{x} - \frac{2v\,dv}{2 - v^2} = 0.$$

Integrating, $$\ln|x| + \ln|2 - v^2| = C$$

Then
$$\ln|x(2 - v^2)| = C$$
$$x(2 - v^2) = e^C = C_1$$
$$x\left(2 - \frac{y^2}{x^2}\right) = C_1 \qquad \left[v = \frac{y}{x}\right]$$
$$\frac{2x^2 - y^2}{x} = C_1, \quad \text{the desired solution.}$$

When solving differential equations having homogeneous coefficients, it is sometimes more convenient to find the solution by using $x = vy$ instead of $y = vx$. The choice is determined by the relative complexity of the coefficients of dx and dy. In Example 1, the coefficient of dy is simpler than that of dx. So we used the substitution $y = vx$. In the next example we use $x = vy$ since the coefficient of dx is simpler than that of dy.

EXAMPLE 2 Solve: $xy\,dx - (x + 2y)^2\,dy = 0$.

Solution The coefficients of dx and dy are both homogeneous of degree 2, so their ratio is of degree 0. Let $x = vy$. Then $dx = v\,dy + y\,dv$. Substituting, we get

$$vy^2(v\,dy + y\,dv) - (vy + 2y)^2\,dy = 0.$$

Then

$$(v^2y^2 - v^2y^2 - 4vy^2 - 4y^2)\,dy + vy^3\,dv = 0$$

$$vy^3\,dv - 4y^2(v + 1)\,dy = 0$$

$$\frac{v\,dv}{v + 1} - 4\frac{dy}{y} = 0$$

$$\left(1 - \frac{1}{v + 1}\right) dv - 4\frac{dy}{y} = 0$$

$$v - \ln|v + 1| - 4\ln|y| = \ln|C|$$

$$v = \ln|Cy^4(v + 1)|$$

$$Cy^4(v + 1) = e^v$$

$$Cy^4\left(\frac{x}{y} + 1\right) = e^{x/y} \qquad \left[v = \frac{x}{y}\right]$$

$$y^3(x + y) = ke^{x/y} \qquad \left[k = \frac{1}{C}\right]$$

which is the desired solution.

PROBLEM SET 8.6

Find the general solution in problems 1–12:

1. $(x^2 + y^2)\,dx - xy\,dy = 0$

2. $(3y - 4x)\,dx + x\,dy = 0$

3. $(3x^2 + y^2)\,dx - 2xy\,dy = 0$

4. $(2x^2 - y^2)\,dx + 3xy\,dy = 0$

5. $(xy + 2y^2)\,dx + x^2\,dy = 0$

6. $(x^4 + 2y^4)\,dx - xy^3\,dy = 0$

7. $2xy\,dx - (y^2 - x^2)\,dy = 0$

8. $xy\,dx - (x + y)^2\,dy = 0$

9. $xy\,dx + (x^2 + y^2)\,dy = 0$

10. $2\,dx - \left(\frac{2x}{y} + \frac{x^2}{y^2}\right) dy = 0$

11. $3\,dx - \left(\frac{3x}{y} - \frac{4x^2}{y^2}\right) dy = 0$

12. $x^2y\,dx - x(x^2 - y^2)\,dy = 0$

Solve the given equations in problems 13–19 subject to the stated conditions.

13. $(x^2 + xy + y^2)\,dx - xy\,dy = 0$; $y = 1$ when $x = 1$

14. $(x^3 + y^3)\,dx - xy^2\,dy = 0$; $y = 3$ when $x = 1$

15. $(xy - y^2)\,dx - x^2\,dy = 0$; $y = 1/2$ when $x = 1/2$

16. $x^2y\,dx - (x^3 + y^3)\,dy = 0$; $y = e$ when $x = 3$

17. $x^3y\,dx - (x^4 - y^4)\,dy = 0;\quad y = \sqrt{2}$ when $x = 4$
18. $y\,dx - (x + ye^{x/y})\,dy = 0;\quad y = 1$ when $x = 0$. Use $y = vx$.
19. $xy\,dy - (y^2 - \sqrt{x^2 - y^2})\,dx = 0;\quad y = 1$ when $x = 1$.

8.7 REVIEW PROBLEMS

1. Show that $y^2 = Ce^{3x}$ is a general solution of $2\,dy/dx - 3y = 0$.
2. Determine whether $y = e^{2x} + e^{-2x}$ is a solution of $y'' + 2y' - 4e^{2x} = 0$.
3. Verify that $y = Cxe^{-x^2}$ is a general solution of $dy/dx + (2x - 1/x)y = 0$.
4. Solve: $x(y - 1)\,dx - y^2\,dy = 0$.
5. Solve: $x\sqrt{9 - y^2}\,dx + y\sqrt{16 - x^2}\,dy = 0;\quad y = \sqrt{5}$ when $x = \sqrt{7}$.
6. The growth rate of a certain population is proportional to its size P and inversely proportional to $(t + 1)$, t is time in years. Find a formula for $P(t)$.
7. A certain amount is deposited in a credit union paying $6\frac{1}{2}\%$ compounded continuously. How long will it take to triple the amount deposited? Use $A = Pe^{it}$.
8. The GNP (gross national product) of a country grows at an annual rate proportional to 5% of the GNP. Find the GNP in 1980 if it was 10^6 dollars in 1970.
9. Solve: $xy\,dy + (2x^2 + 1)\,dx = 0$
10. A thermometer is taken from indoors where the temperature is 75° to the outdoors where the temperature is 30°. Ten minutes later the thermometer reads 40°. What will it read after it has been outdoors for 20 minutes?
11. Solve: $3(3y + xy)\,dy - (y^2 - 2)\,dx = 0$.
12. Solve: $e^{2x - y^2}\,dx + y\,dy = 0$.
13. The growth rate of a population is proportional to its size y and also to the remaining amount of growth $1 - y$. Find the pertinent differential equation and obtain its general solution.
14. Solve: $\dfrac{dy}{dx} + 3y = e^{-x}$.
15. Solve: $\dfrac{dy}{dx} - 2y = 2x - 1$.
16. Find the particular solution determined by the given initial condition: $\dfrac{dy}{dx} + y = x;\quad y(0) = 1$.
17. If $y = 1$ when $x = 0$, find the particular solution of $y' - y = e^{2x}$.
18. Find the general solution $y = v(x) + Cu(x)$ if $\dfrac{dy}{dx} - 2y = 6$.
19. If $\dfrac{dy}{dx} + \dfrac{y}{3} = \dfrac{2}{3}$, find the general solution $y = v(x) + Cu(x)$.
20. The initial pollution level of a reservoir is found to be k grams per gallon. Water is piped out of the reservoir at the constant rate of r gallons per hour and is replaced with less noxious water whose pollution level is p grams per

gallon. If the capacity of the lake is V gallons, find:

(a) The time t it will take to reduce the pollution level to q grams per gallon;
(b) The time t it will take to reduce the pollution level by 60% if the replacement water is pollution-free, $V = 10^6$ gallons, and $r = 10^3$ gallons per hour.

21. In problem 20 find the time it will take to reduce the pollution level by 80%.

8.8 ENRICHMENT TOPICS AND PROBLEMS

1. Show that $y = Cx^{-3/4}$ is a solution of $3xy^3 + 4x^2y^2\dfrac{dy}{dx} = 0$.
2. Find a differential equation having the general solution $y^2 = e^{Cx}$, C is constant.
3. Solve the linear equation: $3x\dfrac{dy}{dx} + 3y = x^2$
4. Solve: $x\,dy + y\,dx = x\ln x\,dx$
5. Solve: $x\dfrac{dy}{dx} + y = x\ln x$
6. Solve: $\dfrac{dy}{dx} + \dfrac{2xy}{1 + x^2} = \dfrac{1}{x(1 + x^2)}$
7. Solve: $\dfrac{dy}{dx} + (1 + 2x)y = e^{3x - x^2}$
8. Solve the homogeneous equation: $(x\sqrt{x^2 + y^2} - y^2)\,dx + xy\,dy = 0$.
9. Solve: $\ln\dfrac{y}{x}\,dy - \dfrac{y}{x}\ln\dfrac{y}{x}\,dx = \dfrac{x}{y}\,dx$.
10. Solve: $4xy\,dx + (2x - y)^2\,dy = 0$.
11. If $y = 1$ when $x = 1$, find the particular solution of

$$(y^2 + 8xy + 8x^2)\,dx - xy\,dy = 0.$$

12. A tank initially holds V_0 gallons of brine which contains m pounds of salt. A second solution of brine containing n pounds of salt per gallon is piped into the tank at the rate of a gallons per minute. This input replaces the brine being piped out at b gallons per minute. Consider the solution in the tank as thoroughly mixed. To determine the amount of salt in the tank at any time t, we let P be that amount so that dP/dt is the rate at which salt pours in (na pounds per minute) minus the rate at which it pours out. The latter rate is found as follows.

The volume of brine is $V_0 + at - bt$ at any t.

The concentration of salt is

$$\frac{P}{V_0 + at - bt}$$

at any t. So, salt pours out at the rate

$$\frac{bP}{V_0 + at - bt}$$

pounds per minute. It follows that

$$\frac{dP}{dt} = na - \frac{bP}{V_0 + at - bt}$$

or

$$\frac{dP}{dt} + \frac{b}{V_0 + at - bt} \cdot P = na.$$

Use this equation to solve the following problem: A tank initially holds 1000 gallons of a brine solution which contains 200 pounds of salt. Fresh water enters the tank at 20 gallons per minute while the thoroughly mixed solution pours out of the tank at the same rate. Determine the amount of salt in the tank at any t; that is, find the particular solution. [*Hint*: $V_0 = 1000$, $m = 200$, $n = 0$, $a = b = 20$; at $t = 0$, $P = m = 200$.]

13. In problem 12, find the solution when $V_0 = 500$, $m = 100$, $n = 0$, $a = b = 15$; at $t = 0$, $P = m = 100$.

14. In the equation $M(x, y)\,dx + N(x, y)\,dy = 0$, if M and N have continuous partial derivatives and if $\partial M/\partial y = \partial N/\partial x$, then the equation is said to be **exact**. Thus, the equation

$$(2xy + y^2)\,dx + (x^2 + 2xy)\,dy = 0$$

is exact since

$$\frac{\partial M}{\partial y} = \frac{\partial}{\partial y}(2xy + y^2) = 2x + 2y$$

and

$$\frac{\partial N}{\partial x} = \frac{\partial}{\partial x}(x^2 + 2xy) = 2x + 2y = \frac{\partial M}{\partial y}.$$

Use these ideas to show that the following differential equations are exact:

(a) $(2xy^3 + 4x^3)\,dx + 3x^2y^2\,dy = 0$
(b) $\ln(y^2 + 1)\,dx + 2xy(y^2 + 1)^{-1}\,dy = 0$
(c) $ye^{xy}\,dx + xe^{xy}\,dy = 0$
(d) $(x - x^{-1}y + \ln y)\,dx + (xy^{-1} - \ln x)\,dy = 0$

15. A differential equation $M(x, y)\,dx + N(x, y)\,dy = 0$ which is exact has the solution $H(x, y) = C$, where C is a constant and H is such that $H_x = M$ and $H_y = N$. To solve

$$(1 + ye^{xy})\,dx + (y + xe^{xy})\,dy = 0,$$

we first show that $M_y = xye^{xy} + e^{xy} = N_x$. Thus, the equation is exact and $H_x = M$ and $H_y = N$. We integrate H_x with respect to x and obtain

$$H(x, y) = x + e^{xy} + g(y).$$

To find the constant of integration $g(y)$ in the above equation, we first find

$$H_y = xe^{xy} + g'(y).$$

Since $H_y = N$,

$$xe^{xy} + g'(y) = y + xe^{xy}$$

or

$$g'(y) = y$$

and

$$g(y) = \frac{y^2}{2} + C_1$$

Substituting: $H(x, y) = x + e^{xy} + \dfrac{y^2}{2} + C_1 = C_2$

Finally, $H(x, y) = x + e^{xy} + \dfrac{y^2}{2} = C \qquad (C = C_2 - C_1)$

Using the above discussion as a guide, solve the following equations:

(a) $(1 + 3x^2y^2 + y)\,dx + (x + 2x^3y)\,dy = 0$
(b) $(y^2 + 2xy - x^2)\,dx - (y^2 - 2xy - x^2)\,dy = 0$
(c) $(x + \sqrt{y^2 + 1})\,dx + [xy(y^2 + 1)^{-1/2} - y]\,dy = 0$
(d) $(2 + e^{2y})\,dx + 2(xe^{2y} - y)\,dy = 0$

16. If the equation $M(x, y)\,dx + N(x, y)\,dy = 0$ is multiplied by a function $\phi(x, y)$ and the resulting equation is exact, then $\phi(x, y)$ is said to be an **integrating factor** of the given equation. For instance, to solve

$$(3xy - 4y^2)\,dx + (2x^2 - 6xy)\,dy = 0,$$

we first discover that

$$M_y = 3x - 8y \quad \text{and} \quad N_x = 4x - 6y \neq M_y.$$

So the equation is not exact. It can be made exact if an integrating factor could be found. In this case it can be shown that $\phi(x, y) = xy$ is a suitable integrating factor. Multiplying by xy, we get

$$(3x^2y^2 - 4xy^3)\,dx + (2x^3y - 6x^2y^2)\,dy = 0.$$

Then
$$M_y = 6x^2y - 12xy^2 = N_x$$

and the equation is exact. Solving in the manner of problem 15, we obtain the solution

$$x^3y^2 - 2x^2y^3 = C.$$

With the above ideas in mind, solve the following equations by use of the given integrating factor:

(a) $(y^2 + 2xy)\,dx + (2x^2 - xy)\,dy = 0;\quad \phi(x, y) = x^{-2}y^{-1}$.
(b) $(2y^2 - 3xy)\,dx + (3x^2 + 2xy)\,dy = 0;\quad \phi(x, y) = x^{-1}y^{-2}$.

17. Refer to problem 16 and solve the equation $\dfrac{dy}{dx} = \dfrac{3x^2y}{x^3 + y^4}$, using $\phi(y) = y^{-2}$

18. Refer to problem 16 and solve the equation $(y + 2xy^2)\,dx + (x - 2x^2y)\,dy = 0$ using $\phi(x, y) = (2xy)^{-2}$.

19. The differential equation

$$\frac{dy}{dx} + P(x)y = Q(x)y^n, \qquad n \neq 0, \quad n \neq 1,$$

is called a **Bernoulli equation**. A special case of this equation is

$$\frac{dy}{dx} - 2xy = xy^2 \tag{1}$$

where $P = -2x$, $Q = x$, and $n = 2$. In order to solve, we substitute

$$u = y^{1-n},$$

which results in a linear equation that can be solved by previous methods.

Thus,

$$u = y^{1-2} = y^{-1} = \frac{1}{y} \quad \text{or} \quad y = \frac{1}{u}$$

and

$$\frac{dy}{dx} = -\frac{1}{u^2}\frac{du}{dx}.$$

Substituting in Eq. (1):

$$-\frac{1}{u^2}\frac{du}{dx} - \frac{2x}{u} = \frac{x}{u^2}$$

or

$$\frac{du}{dx} + 2xu = -x.$$

Then

$$e^{\int P dx} = e^{\int 2x dx} = e^{x^2}.$$

Multiplying: $e^{x^2}\left(\frac{du}{dx} + 2xu\right) = -xe^{x^2}$

Solving:

$$\frac{d}{dx}(e^{x^2}u) = -xe^{x^2}$$

$$e^{x^2}u = -\frac{1}{2}e^{x^2} + C$$

$$u = -\frac{1}{2} + Ce^{-x^2}$$

$$\frac{1}{y} = \frac{-1 + 2Ce^{-x^2}}{2} \qquad \left(u = \frac{1}{y}\right).$$

Therefore,

$$y = \frac{2}{2Ce^{-x^2} - 1}$$

Use the above ideas as a guide to solve:

(a) $\frac{dy}{dx} + 4xy = 2xy^2$

(b) $\frac{dy}{dx} + \frac{y}{1 + x} = (1 + x)y^2$

APPENDIX A

A Short Review of Trigonometry and Its Applications to Calculus

IMPORTANT EQUATIONS

$$\sin^2 \theta + \cos^2 \theta = 1$$

$$1 + \tan^2 \theta = \sec^2 \theta$$

$$\sin(x + y) = \sin x \cos y + \cos x \sin y$$

$$\sin 2x = 2 \sin x \cos x$$

$$\cos 2x = \cos^2 x - \sin^2 x$$

$$a^2 = b^2 + c^2 - 2bc \cos A$$

$$\frac{d}{dx}(\sin u) = \cos u \cdot \frac{du}{dx}$$

$$\frac{d}{dx}(\cos u) = -\sin u \cdot \frac{du}{dx}$$

$$\int \sin u \, du = -\cos u + C$$

$$\int \cos u \, du = \sin u + C$$

A.1 BASIC TOPICS

In this appendix we will review some basic ideas in trigonometry, together with some important formulas and their applications.

Consider a circle of radius r and a cartesian coordinate system with its origin at the center of the circle. We exhibit angle θ as shown in Figure A.1, where θ is zero on the positive x-axis and is positive when measured counterclockwise (negative when measured clockwise). Angle θ is shown with its terminal side in quadrants I, II, III, and IV, respectively.

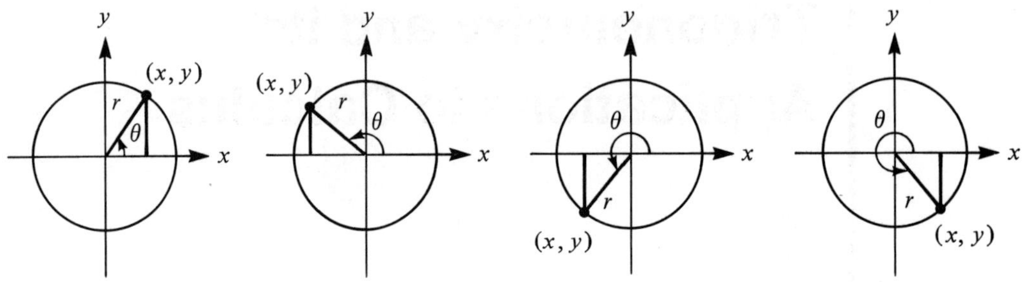

Figure A.1

Basic definitions of the trigonometric functions are as follows:

$$\sin\theta = \frac{y}{r}, \qquad \cos\theta = \frac{x}{r}, \qquad \tan\theta = \frac{y}{x},$$

$$\csc\theta = \frac{r}{y}, \qquad \sec\theta = \frac{r}{x}, \qquad \cot\theta = \frac{x}{y}.$$

The sin, cos, tan, csc, sec, and cot are abbreviations for sine, cosine, tangent, cosecant, secant, and cotangent, respectively. Note that (x, y) is the point where the terminating side of angle θ meets the circle of radius r. These basic definitions are dependent only on angle θ. The size of the circle is immaterial and radius r is considered to be positive. The signs of x and y depend on the quadrant (or axis) where point (x, y) occurs.

The following formulas are often useful:

$$\sec\theta = \frac{1}{\cos\theta}, \qquad \tan\theta = \frac{y/r}{x/r} = \frac{\sin\theta}{\cos\theta},$$

$$\csc\theta = \frac{1}{\sin\theta}, \qquad \cot\theta = \frac{1}{\tan\theta} = \frac{\cos\theta}{\sin\theta}.$$

Angle θ is most commonly measured (a) in degrees and minutes (chiefly used in the solution of triangles), and (b) in radians (mainly useful in calculus and in the analytic approach to trigonometry). A **degree** is the measure of an angle which, when its vertex is at the center of a circle, subtends on the circumference an arc equal to 1/360 of the circumference of the circle. A minute is 1/60 of one degree and a second is 1/60 of one minute. For instance, 44°47.3′ is read "forty-four degrees, forty-seven and three tenths minutes." The whole circumference is subtended by four right angles, each right angle containing 90°. Two right angles contain 180°, called a straight angle (see Figure A.2(a)). A **radian** is the measure of an angle which, when its vertex is at the center of a circle, subtends on the circumference an arc equal to the radius of the circle (see Figure A.2(b)). The radian is independent of the length of the radius of the circle.

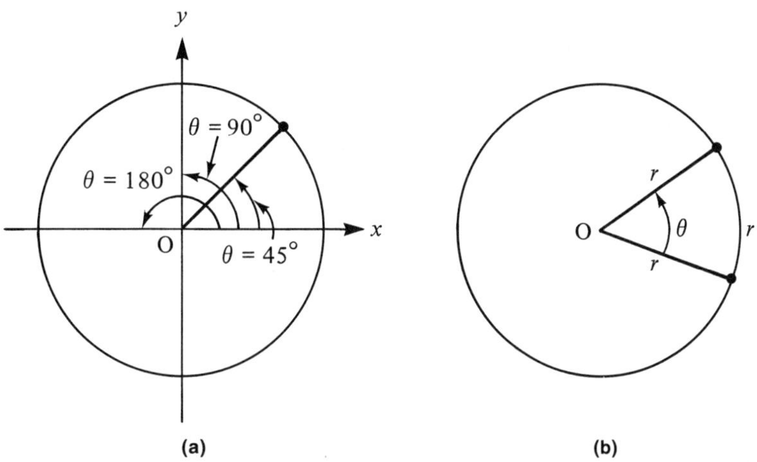

Figure A.2

The circumference of a circle is given by $C = 2\pi r$, or $2\pi = C/r$ defines one revolution. One revolution is also defined by 360°. Thus,

$$2\pi \text{ radians} = 360^\circ$$

or

$$\pi \text{ radians} = 180^\circ,$$

a precise relation between degree and radian measurements. From this relation we have

$$1 \text{ radian} = \frac{180^\circ}{\pi} \cong 57.296^\circ \qquad (\pi \cong 3.1416)$$

and

$$1^\circ = \frac{\pi}{180} \text{ rad} \cong 0.01745 \text{ rad}$$

where "rad" means radians. When writing the radian measures of angles, the "rad" is usually omitted. Thus,

$$15^\circ = \frac{\pi}{12}, \quad 30^\circ = \frac{\pi}{6}, \quad 45^\circ = \frac{\pi}{4}, \quad 60^\circ = \frac{\pi}{3}, \quad 90^\circ = \frac{\pi}{2}, \quad \text{etc.}$$

The following table exhibits some special values of the three main trigonometric functions.

θ	0	$\frac{\pi}{6}$	$\frac{\pi}{4}$	$\frac{\pi}{3}$	$\frac{\pi}{2}$	π	$\frac{3}{2}\pi$
$\sin\theta$	0	$\frac{1}{2}$	$\frac{\sqrt{2}}{2}$	$\frac{\sqrt{3}}{2}$	1	0	-1
$\cos\theta$	1	$\frac{\sqrt{3}}{2}$	$\frac{\sqrt{2}}{2}$	$\frac{1}{2}$	0	-1	0
$\tan\theta$	0	$\frac{\sqrt{3}}{3}$	1	$\sqrt{3}$	undefined	0	undefined

From Figure A.1 we see, by the Pythagorean Theorem, that

$$x^2 + y^2 = r^2.$$

Dividing both sides of this equation by r^2, we have

$$\left(\frac{x}{r}\right)^2 + \left(\frac{y}{r}\right)^2 = 1$$

or

$$\boxed{\cos^2\theta + \sin^2\theta = 1}$$

for all real values of θ. Since $\cos^2\theta$ and $\sin^2\theta$ are nonnegative, we see that

$$\cos^2\theta \le 1$$

and

$$\sin^2\theta \le 1.$$

For all real values of θ it is clear that

$$-1 \le \cos\theta \le 1$$

and

$$-1 \le \sin\theta \le 1.$$

The trigonometric functions repeat for angles greater than 2π. For θ any real number, θ and $\theta \pm 2\pi$ denote the same point (x, y). It then follows that

$$\sin(\theta \pm 2\pi) = \sin\theta$$

and

$$\cos(\theta \pm 2\pi) = \cos\theta.$$

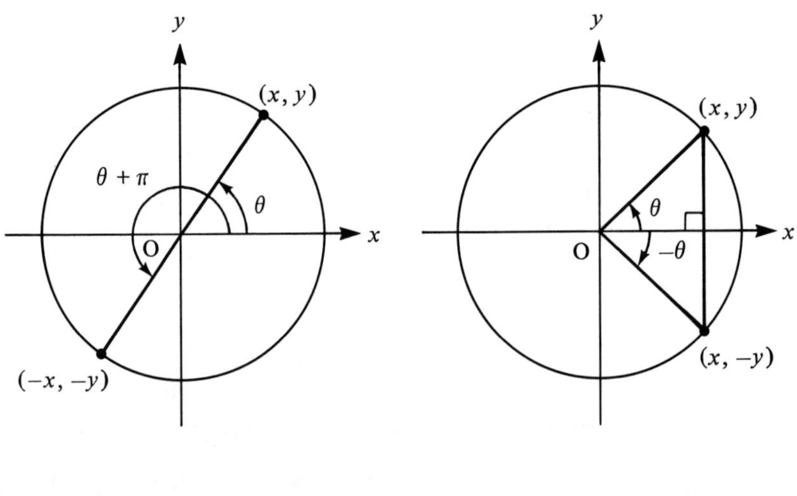

Figure A.3

Functions with this property are called **periodic** functions with period 2π. In general,

$$\sin(\theta \pm n \cdot 2\pi) = \sin \theta,$$
$$\cos(\theta \pm n \cdot 2\pi) = \cos \theta, \qquad n = 0, \pm 1, \pm 2, \ldots.$$

Let us consider Figure A.3. If (x, y) is a point on the circle that corresponds to θ, then $(-x, -y)$ is a point on the circle that corresponds to $\theta + \pi$. Accordingly, we have

$$\sin(\theta + \pi) = -\sin \theta$$

and

$$\cos(\theta + \pi) = -\cos \theta.$$

EXAMPLES

$$\sin 210^\circ = \sin(30^\circ + 180^\circ) = -\sin 30^\circ.$$
$$\cos \frac{4}{3}\pi = \cos\left(\frac{\pi}{3} + \pi\right) = -\cos \frac{\pi}{3}.$$

In the second circle of Figure A.3, θ and $-\theta$ correspond to the points (x, y) and $(x, -y)$, which are images of each other when reflected in the x-axis. Therefore,

$$\sin(-\theta) = -\sin \theta$$

and

$$\cos(-\theta) = \cos \theta.$$

When $f(-\theta) = -f(\theta)$, then f is called an **odd** function. When $f(-\theta) = f(\theta)$, then f is called an **even** function. Thus, the sine function is odd and the cosine function is even. It can be shown that

$$\sec(-\theta) = \sec \theta; \qquad \csc(-\theta) = -\csc \theta;$$
$$\tan(-\theta) = -\tan \theta; \qquad \cot(-\theta) = -\cot \theta.$$

EXAMPLES

$$\sin(-45^\circ) = -\sin 45^\circ$$
$$\cos(-30^\circ) = \cos 30^\circ$$
$$\tan(-60^\circ) = -\tan 60^\circ.$$

If θ and $\frac{\pi}{2} - \theta$ correspond to two points $P(x, y)$ and $Q(y, x)$, as in Figure A.4, then P and Q are reflections of each other in the line $y = x$.

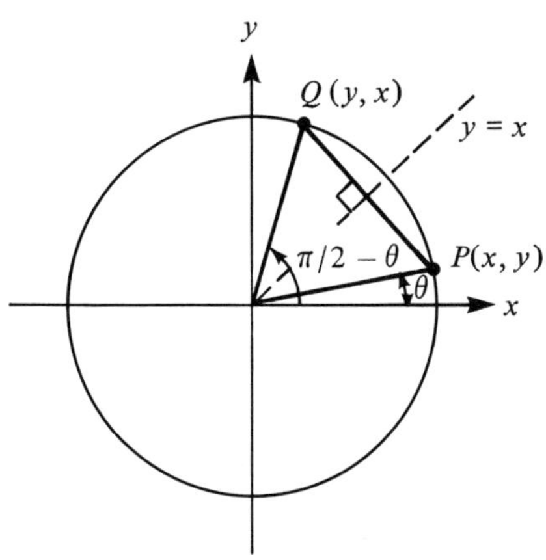

Figure A.4

Because of this property, we have

$$\sin\left(\frac{\pi}{2} - \theta\right) = \cos\theta; \qquad \cos\left(\frac{\pi}{2} - \theta\right) = \sin\theta.$$

The two angles $\frac{\pi}{2} - \theta$ and θ are said to be complementary angles since their sum is $\pi/2$.

EXAMPLES

$$\sin(90^\circ - 30^\circ) = \cos 30^\circ$$
$$\cos\left(\frac{\pi}{2} - \frac{\pi}{3}\right) = \sin\frac{\pi}{3}.$$

The sine and cosine of complementary angles are equal. That is,

$$\sin 30^\circ = \cos 60^\circ; \qquad \sin 40^\circ = \cos 50^\circ; \qquad \sin 45^\circ = \cos 45^\circ, \quad \text{etc.}$$

We can use the preceding formulas to find two more important formulas:

$$\sin\left(\theta + \frac{\pi}{2}\right) = \sin\left[\frac{\pi}{2} - (-\theta)\right] = \cos(-\theta) = \cos\theta$$
$$\cos\left(\theta + \frac{\pi}{2}\right) = \cos\left[\frac{\pi}{2} - (-\theta)\right] = \sin(-\theta) = -\sin\theta$$

EXAMPLES

$$\sin\left(\frac{\pi}{4} + \frac{\pi}{2}\right) = \cos\frac{\pi}{4}$$
$$\cos(30^\circ + 90^\circ) = -\sin 30^\circ.$$

From the identity $\sin^2\theta + \cos^2\theta = 1$, we obtain, upon dividing both sides by $\cos^2\theta$,

$$\boxed{\tan^2\theta + 1 = \sec^2\theta.}$$

When dividing both sides by $\sin^2\theta$ we get

$$1 + \cot^2\theta = \csc^2\theta.$$

These two identities are important and useful in the study of trigonometry.

To find the value of a trigonometric function of an angle in a quadrant other than the first, we initially determine the sign of the function from the quadrant in which the terminal side of the angle lies.

EXAMPLES If $\theta = 120^\circ$, then θ is in quadrant II where x is negative and y is positive. So $\sin\theta > 0$, $\cos\theta < 0$, $\tan\theta < 0$, etc. Thus,

$$\sin 120^\circ = \sin(180^\circ - 60^\circ) = \sin 60^\circ;$$
$$\cos 120^\circ = \cos(180^\circ - 60^\circ) = -\cos 60^\circ; \quad \text{etc.}$$

We define the 60°-angle in the preceding discussion to be the **related angle**. It then follows that the numerical (absolute) value of any trigonometric function of an angle is equal to the value of that function of the related angle.

EXAMPLES

$$\begin{aligned}\sin 150^\circ &= \sin(180^\circ - 30^\circ) = \sin 30^\circ\\ \sec 240^\circ &= \sec(180^\circ + 60^\circ) = -\sec 60^\circ\\ \csc 300^\circ &= \csc(360^\circ - 60^\circ) = -\csc 60^\circ\\ \cot 315^\circ &= \cot(360^\circ - 45^\circ) = -\cot 45^\circ\end{aligned}$$

where the 30°, 60°, and 45°-angles are the related angles.

We state without proof some basic **addition formulas**.

$$\boxed{\begin{aligned}\sin(x + y) &= \sin x \cos y + \cos x \sin y\\ \cos(x + y) &= \cos x \cos y - \sin x \sin y\\ \tan(x + y) &= \frac{\tan x + \tan y}{1 - \tan x \tan y}\end{aligned}}$$

If we replace y by $-y$ in the preceding equations we have formulas for functions of $(x - y)$:

$$\begin{aligned}\sin(x - y) &= \sin[x + (-y)]\\ &= \sin x \cos(-y) + \cos x \sin(-y)\\ &= \sin x \cos y - \cos x \sin y.\end{aligned}$$

The other two formulas are found similarly:

$$\begin{aligned}\cos(x - y) &= \cos x \cos y + \sin x \sin y\\ \tan(x - y) &= \frac{\tan x - \tan y}{1 + \tan x \tan y}\end{aligned}$$

EXAMPLES

$$\begin{aligned}\sin 75^\circ &= \sin(30^\circ + 45^\circ)\\ &= \sin 30^\circ \cos 45^\circ + \cos 30^\circ \sin 45^\circ\\ &= \frac{1}{2}\cdot\frac{\sqrt{2}}{2} + \frac{\sqrt{3}}{2}\cdot\frac{\sqrt{2}}{2} = \frac{\sqrt{2} + \sqrt{6}}{4}\\ \cos 15^\circ &= \cos(45^\circ - 30^\circ)\\ &= \cos 45^\circ \cos 30^\circ + \sin 45^\circ \sin 30^\circ\\ &= \frac{\sqrt{2}}{2}\cdot\frac{\sqrt{3}}{2} + \frac{\sqrt{2}}{2}\cdot\frac{1}{2} = \frac{\sqrt{6} + \sqrt{2}}{4}\end{aligned}$$

If we let $x = y$ in the addition formulas, we can find the so-called **double-angle formulas**. Thus,

$$\sin(x + x) = \sin x \cos x + \cos x \sin x$$

or

$$\boxed{\sin 2x = 2 \sin x \cos x}$$

Similarly,

$$\begin{aligned}\cos 2x &= \cos^2 x - \sin^2 x\\ &= 2\cos^2 x - 1\end{aligned}$$

$$\boxed{\cos 2x = 1 - 2\sin^2 x} \quad [\text{since } \sin^2 x + \cos^2 x = 1]$$

and

$$\tan 2x = \frac{2 \tan x}{1 - \tan^2 x}$$

EXAMPLES

$$\begin{aligned}
\sin\frac{2\pi}{3} &= 2\sin\frac{\pi}{3}\cos\frac{\pi}{3}\\
&= 2\cdot\frac{\sqrt{3}}{2}\cdot\frac{1}{2} = \frac{\sqrt{3}}{2}\\
\cos 480^\circ &= 2\cos^2 240^\circ - 1\\
&= 2(-\cos 60^\circ)^2 - 1\\
&= 2\left(\frac{1}{4}\right) - 1 = -\frac{1}{2}\\
\tan 240^\circ &= \frac{2\tan 120^\circ}{1-\tan^2 120^\circ}\\
&= \frac{2(-\tan 60^\circ)}{1-(-\tan 60^\circ)^2}\\
&= \frac{-2\sqrt{3}}{1-3} = \sqrt{3}\\
8\sin 2x\cos 2x &= 4(2\sin 2x\cos 2x)\\
&= 4\sin 4x\\
2 - 4\sin^2\frac{x}{2} &= 2\left(1 - 2\sin^2\frac{x}{2}\right)\\
&= 2\cos x
\end{aligned}$$

Half-angle formulas can be found from the double-angle formulas as follows. If we substitute $\theta/2$ for x in $\cos 2x$, we obtain

$$\cos 2\cdot\frac{\theta}{2} = \cos\theta = 2\cos^2\frac{\theta}{2} - 1$$

or

$$2\cos^2\frac{\theta}{2} = 1 + \cos\theta$$

or

$$\cos^2\frac{\theta}{2} = \frac{1+\cos\theta}{2}$$

Then

$$\cos\frac{\theta}{2} = \pm\sqrt{\frac{1+\cos\theta}{2}}$$

Similarly,

$$\sin\frac{\theta}{2} = \pm\sqrt{\frac{1-\cos\theta}{2}}$$

and

$$\tan\frac{\theta}{2} = \pm\sqrt{\frac{1-\cos\theta}{1+\cos\theta}}$$

EXAMPLES

$$\sin 22\tfrac{1}{2}^\circ = \sqrt{\frac{1-\cos 45^\circ}{2}} = \sqrt{\frac{1-\frac{\sqrt{2}}{2}}{2}} = \frac{1}{2}\sqrt{2-\sqrt{2}}$$

$$\cos 15^\circ = \sqrt{\frac{1+\cos 30^\circ}{2}} = \sqrt{\frac{1+\frac{\sqrt{3}}{2}}{2}} = \frac{1}{2}\sqrt{2+\sqrt{3}}$$

$$\tan 112.5^\circ = \sqrt{\frac{1 - \cos 225^\circ}{1 + \cos 225^\circ}} = -\sqrt{\frac{1 + \frac{\sqrt{2}}{2}}{1 - \frac{\sqrt{2}}{2}}} = -\sqrt{\frac{2 + \sqrt{2}}{2 - \sqrt{2}}}$$

Product formulas may be obtained from the addition formulas. We introduce them without proof.

$$\sin x + \sin y = 2 \sin \frac{1}{2}(x + y) \cos \frac{1}{2}(x - y)$$

$$\sin x - \sin y = 2 \cos \frac{1}{2}(x + y) \sin \frac{1}{2}(x - y)$$

$$\cos x + \cos y = 2 \cos \frac{1}{2}(x + y) \cos \frac{1}{2}(x - y)$$

$$\cos x - \cos y = -2 \sin \frac{1}{2}(x + y) \sin \frac{1}{2}(x - y)$$

EXAMPLES

$$\begin{aligned}\sin 50^\circ + \sin 10^\circ &= 2 \sin \frac{1}{2}(60^\circ) \cos \frac{1}{2}(40^\circ)\\ &= 2 \sin 30^\circ \cos 20^\circ\\ \cos 2\theta - \cos 4\theta &= -2 \sin \frac{1}{2}(6\theta) \sin \frac{1}{2}(-2\theta)\\ &= -2 \sin 3\theta \sin(-\theta)\\ &= 2 \sin 3\theta \sin \theta\end{aligned}$$

It is essential to be able to transform certain trigonometric expressions into other forms in order to solve certain types of problems. One of the best ways to learn to make these transformations is to practice verifying or proving the validity of given identities. To verify an identity we usually show that the right side (RS) of the equation equals the left side (LS) of the equation, or vice versa. That is, we operate with one side, leaving the other unchanged.

EXAMPLE 1 Verify: $\sin \theta \tan \theta + \cos \theta = \sec \theta$.

Solution

$$\begin{aligned}\text{LS} &= \sin \theta \cdot \frac{\sin \theta}{\cos \theta} + \cos \theta\\ &= \frac{\sin^2 \theta + \cos^2 \theta}{\cos \theta}\\ &= \frac{1}{\cos \theta} = \sec \theta = \text{RS}.\end{aligned}$$

EXAMPLE 2 Prove: $\tan \theta = \dfrac{1 - \cos 2\theta}{\sin 2\theta}$

Solution

$$\text{RS} = \frac{1 - (1 - 2\sin^2\theta)}{2\sin\theta\cos\theta}$$

$$= \frac{\sin\theta}{\cos\theta} = \tan\theta = \text{LS}.$$

EXAMPLE 3 Verify: $\dfrac{\sin 3x}{\sin x} - \dfrac{\cos 3x}{\cos x} = 2.$

Solution

$$\text{LS} = \frac{\sin 3x \cos x - \cos 3x \sin x}{\sin x \cos x}$$

$$= \frac{\sin(3x - x)}{\frac{1}{2}\sin 2x} = 2 = \text{RS}.$$

EXAMPLE 4 Prove: $\sec x + \tan x = \dfrac{1 + \tan\frac{x}{2}}{1 - \tan\frac{x}{2}}$

Solution

$$\text{RS} = \frac{1 + \dfrac{\sin\frac{x}{2}}{\cos\frac{x}{2}}}{1 - \dfrac{\sin\frac{x}{2}}{\cos\frac{x}{2}}}$$

$$= \frac{\cos\frac{x}{2} + \sin\frac{x}{2}}{\cos\frac{x}{2} - \sin\frac{x}{2}} \cdot \frac{\cos\frac{x}{2} + \sin\frac{x}{2}}{\cos\frac{x}{2} + \sin\frac{x}{2}}$$

$$= \frac{\cos^2\frac{x}{2} + 2\sin\frac{x}{2}\cos\frac{x}{2} + \sin^2\frac{x}{2}}{\cos^2\frac{x}{2} - \sin^2\frac{x}{2}}$$

$$= \frac{1 + \sin x}{\cos x}$$

$$= \frac{1}{\cos x} + \frac{\sin x}{\cos x} = \sec x + \tan x = \text{LS}.$$

Identities are useful in solving trigonometric equations. A knowledge of factoring is also essential.

EXAMPLE 5 Given $\sin 2\theta + \sin\theta = 0$, solve for θ where $0 \le \theta < 2\pi$.

Solution

$$2 \sin \theta \cos \theta + \sin \theta = 0$$
$$\sin \theta(2 \cos \theta + 1) = 0$$
$$\sin \theta = 0, \qquad \theta = 0, \pi.$$
$$2 \cos \theta + 1 = 0, \qquad \cos \theta = -\frac{1}{2},$$
$$\theta = \frac{2\pi}{3}, \frac{4\pi}{3}.$$

EXAMPLE 6 Solve for x if $\tan 2x - \cot x = 0, \quad 0° \leq x < 360°$.

Solution

$$\frac{\sin 2x}{\cos 2x} - \frac{\cos x}{\sin x} = 0$$
$$\frac{\sin 2x \sin x - \cos 2x \cos x}{\cos 2x \sin x} = 0$$
$$-(\cos 2x \cos x - \sin 2x \sin x) = 0$$
$$-\cos(2x + x) = -\cos 3x = 0$$
$$\cos 3x = 0$$
$$3x = 90°, 270°, 450°, 630°, 810°, 990°$$
$$x = 30°, 90°, 150°, 210°, 270°, 330°.$$

EXAMPLE 7 Solve: $\cos \frac{\theta}{2} = 1 + \cos \theta$

Solution

$$\sqrt{\frac{1 + \cos \theta}{2}} = 1 + \cos \theta$$
$$\frac{1 + \cos \theta}{2} = 1 + 2 \cos \theta + \cos^2 \theta$$
$$2 \cos^2 \theta + 3 \cos \theta + 1 = 0$$
$$(2 \cos \theta + 1)(\cos \theta + 1) = 0$$
$$2 \cos \theta + 1 = 0, \quad \cos \theta = -\frac{1}{2}, \quad \theta = 120° \pm n \cdot 360° \quad \text{or}$$
$$\theta = 240° \pm n \cdot 360°$$
$$\cos \theta + 1 = 0, \quad \cos \theta = -1, \quad \theta = 180° \pm n \cdot 360°, \quad n \text{ is any integer.}$$

If $\sin y = x$, what is y? By convention, we say that y is the angle (or number) whose sine is x. Symbolically, we write

$$y = \text{Arcsin } x$$

and read it as "y equals the principal inverse sine of x." When written as

$$y = \text{arcsin } x$$

then the inverse function is infinitely many-valued. We will emphasize the *principal inverse* functions. Some of the more important ones are as follows:

$$-\frac{\pi}{2} \le \text{Arcsin } x \le \frac{\pi}{2}$$
$$0 \le \text{Arccos } x \le \pi$$
$$-\frac{\pi}{2} < \text{Arctan } x < \frac{\pi}{2}$$
$$0 < \text{Arccot } x < \pi.$$

EXAMPLES

$$\text{Arcsin} \frac{1}{2} = 30^\circ$$
$$\text{Arccos} \frac{1}{2} = 60^\circ$$
$$\text{Arctan } 1 = \frac{\pi}{4}$$
$$\text{Arcsin}\left(-\frac{\sqrt{3}}{2}\right) = -\frac{\pi}{3}$$
$$\text{Arccos}\left(-\frac{1}{2}\right) = \frac{2\pi}{3}$$
$$\text{Arctan}\left(-\frac{\sqrt{3}}{3}\right) = -\frac{\pi}{6}$$
$$\text{Arccot}\,(-1) = 135^\circ.$$

PROBLEM SET A.1

1. Find:

   (a) $\sin 135^\circ$ **(b)** $\tan \frac{11\pi}{3}$ **(c)** $\sec 480^\circ$ **(d)** $\cos\left(-\frac{\pi}{4}\right)$

   (e) $\cot 22^\circ 10'$ **(f)** $\csc 98^\circ 25'$ **(g)** $\sin 269^\circ 50.5'$

2. Use tables to find θ:

   (a) $\sin \theta = 0.5432$ **(b)** $\cos \theta = 0.1357$
   (c) $\tan \theta = 1.7654$ **(d)** $\cot \theta = 1.3579$

3. A store manager places a ladder 17 meters long against his store, reaching a window 15 meters above the level ground. Find the six trigonometric functions of the angle that the ladder makes with the ground.

4. Express as a function of θ:

   (a) $\sin(270^\circ - \theta)$ **(b)** $\cos(540^\circ - \theta)$ **(c)** $\tan\left(\frac{7}{2}\pi - \theta\right)$

5. If $\theta = 45^\circ$, evaluate:

   (a) $\sin^2 \theta$ **(b)** $\cos^2 2\theta$ **(c)** $\tan 3\theta$
   (d) $\sec^4 3\theta$ **(e)** $\cos^2 \theta + \sec^2 \theta$

6. Evaluate:
   (a) $\sin(x + y)$ if $\sin x = 3/5$, x is in quadrant 1, and $\cos y = -5/13$, y is in quadrant 2.
   (b) $\cos(x - y)$ if $\tan x = 3/4$, x is in quadrant 3, and $\sin y = -12/13$, y is in quadrant 4.
   (c) $\tan(x + y)$ if $\cos x = 7/25$, x is in quadrant 4, and $\cos y = -4/5$, y is in quadrant 2.

7. Compute:
   (a) $\sin 2x$ if $\sin x = 2/3$ and $0 < x < \pi/2$
   (b) $\cos 2y$ if $\cot y = -5/12$ and $\pi/2 < y < \pi$
   (c) $\tan 2z$ if $\tan z = 4/3$ and $180° < z < 270°$

8. Compute:
   (a) $\sin x/2$ if $\sin x = 1$ and $0° < x < 360°$
   (b) $\cos x/2$ if $\sec x = -2$ and $180° < x < 270°$
   (c) $\tan x/2$ if $\csc x = 13/12$ and $\pi/2 < x < \pi$

9. Express as products:
   (a) $\sin 75° - \sin 45°$
   (b) $\cos \frac{3}{2}\theta + \cos \frac{1}{2}\theta$
   (c) $\sin 10° + \sin 80°$
   (d) $\cos 20°30' - \cos 40°30'$

10. Verify the following identities:
   (a) $\dfrac{\sec A}{\sin A} - \dfrac{\sin A}{\cos A} = \cot A$
   (b) $\dfrac{\sin \theta}{1 + \cos \theta} + \dfrac{1 + \cos \theta}{\sin \theta} = 2 \csc \theta$
   (c) $\dfrac{\tan 2x}{1 + \tan 2x} = \dfrac{\sin 2x}{\sin 2x + \cos 2x}$
   (d) $\cos^2 x - \sin^2 x = \cos(x + y)\cos(x - y) - \sin(x + y)\sin(x - y)$
   (e) $\tan \dfrac{\theta}{2} = \dfrac{1 + \sin \theta - \cos \theta}{1 + \sin \theta + \cos \theta}$
   (f) $\cos 3x = 4 \cos^3 x - 3 \cos x$
   (g) $\dfrac{\sin \theta + \sin 2\theta}{\cos \theta + \cos 2\theta + 1} = \tan \theta$

11. Solve the following equations for $0° \leq x < 360°$:
   (a) $\cos^2 x = \dfrac{3}{4}$
   (b) $\tan^2 x = 1$
   (c) $(\sin x + 1)(2 \cos x - 1) = 0$
   (d) $2 \sin^2 x - 3 \sin x + 1 = 0$
   (e) $2 \cos^2 x - 3 \sin x - 3 = 0$
   (f) $\tan^2 x - 2 \sec^2 x + 3 = 0$
   (g) $\cos 2x + \sin x = 0$
   (h) $4 \sin \dfrac{x}{2} = 3 - 2 \cos x$

12. Find:
   (a) Arcsin 1
   (b) Arccos (-1)
   (c) Arctan $\sqrt{3}$
   (d) Arccot $(-\sqrt{3})$
   (e) Arcsin $\left(-\dfrac{\sqrt{2}}{2}\right)$
   (f) Arccos 0.5055
   (g) Arctan 1.3806
   (h) Arccos π

13. Find:

(a) $\sin\left(\text{Arctan}\,\frac{3}{4}\right)$ (b) $\cos\left[\text{Arcsin}\left(-\frac{\sqrt{3}}{2}\right)\right]$

(c) $\sec\left(\text{Arccos}\,\frac{1}{5}\right)$ (d) $\tan 2[\text{Arcsin}\,(-1)]$

A.2 GRAPHS

The graph of the function $y = \sin x = f(x)$ is the totality of all points (x, y) whose coordinates satisfy the function. To graph the sine curve, which is periodic with period 2π, we need only know the shape of the curve for the interval $[0, 2\pi]$. This is so because the graph repeats itself to the right and left of this interval (see Figure A.5). We get the needed points from the following tabular form:

x	0	$\pi/2$	π	$3\pi/2$	2π
$f(x)$	0	1	0	-1	0

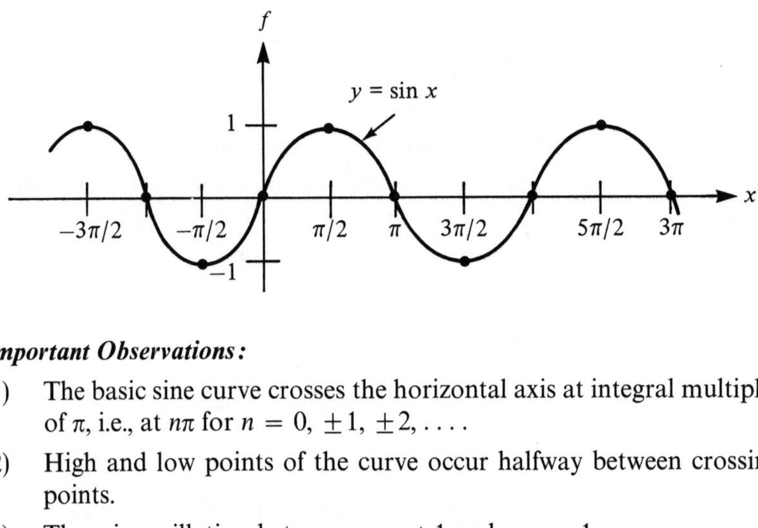

Figure A. 5

Important Observations:

(1) The basic sine curve crosses the horizontal axis at integral multiples of π, i.e., at $n\pi$ for $n = 0, \pm 1, \pm 2, \ldots$.

(2) High and low points of the curve occur halfway between crossing points.

(3) There is oscillation between $y = +1$ and $y = -1$.

The graph of $y = \cos x = f(x)$ is similar to that of the sine curve except that the cosine curve is displaced by $\pi/2$ units, as shown in Figure A.6.

Important Observations:

(1) The basic cosine curve crosses the horizontal axis at points which are odd multiples of $\pi/2$, i.e., at $(2n + 1)\pi/2$ for $n = 0, \pm 1, \pm 2, \ldots$.

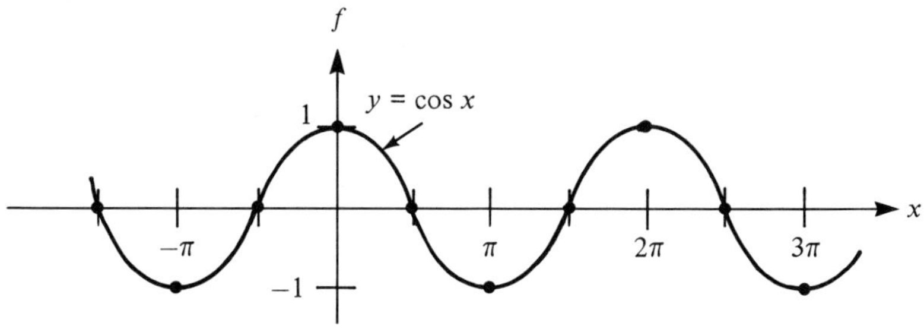

Figure A. 6

(2) High and low points of the curve occur halfway between crossing points.

(3) There is oscillation between $y = +1$ and $y = -1$.

EXAMPLES If $y = 2 \sin 3x$, the number 2 is called the **amplitude**, so that oscillation occurs between $y = 2$ and $y = -2$. To get crossing points we set

$$3x = n\pi$$

and solve for x. Then

$$\begin{aligned} x &= \frac{n\pi}{3} \\ &= 0, \frac{\pi}{3}, \frac{2\pi}{3} \qquad \text{(for } n = 0, 1, 2) \end{aligned}$$

which are the crossing points.

If $y = \frac{2}{3} \cos \frac{x}{2}$, then the amplitude is 2/3 and crossing points are found as follows:

$$\begin{aligned} \frac{x}{2} &= (2n + 1)\frac{\pi}{2} \\ x &= (2n + 1)\pi \\ &= \pi, 3\pi, 5\pi \qquad \text{(for } n = 0, 1, 2). \end{aligned}$$

If $y = 3 \sin(x - \pi/4)$, then the amplitude is 3 and the curve is displaced $\pi/4$ units (in the positive x-direction) from the basic sine curve. Crossing points are found as follows:

$$\begin{aligned} x - \frac{\pi}{4} &= n\pi \\ x &= n\pi + \frac{\pi}{4} \\ &= \frac{\pi}{4}, \frac{5\pi}{4}, \frac{9\pi}{4} \qquad \text{(for } n = 0, 1, 2). \end{aligned}$$

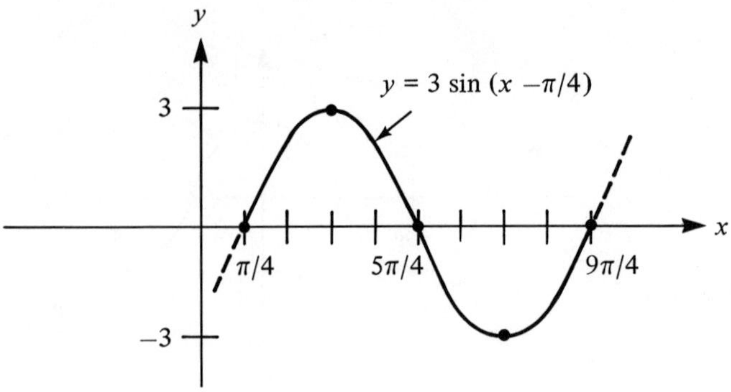

Figure A.7

We now have sufficient information to graph this curve, shown in Figure A.7.

Let us consider $y = 4\cos(x + \pi/3)$. The amplitude is 4 and the displacement from the basic cosine curve is $\pi/3$ units in the negative x-direction. The crossing points are as follows:

$$x + \frac{\pi}{3} = (2n + 1)\frac{\pi}{2}$$

$$x = (2n + 1)\frac{\pi}{2} - \frac{\pi}{3}$$

$$= \frac{\pi}{2} - \frac{\pi}{3} = \frac{\pi}{6} \quad \text{for } n = 0;$$

$$= \frac{3\pi}{2} - \frac{\pi}{3} = \frac{7\pi}{6} \quad \text{for } n = 1;$$

$$= \frac{5\pi}{2} - \frac{\pi}{3} = \frac{13\pi}{6} \quad \text{for } n = 2.$$

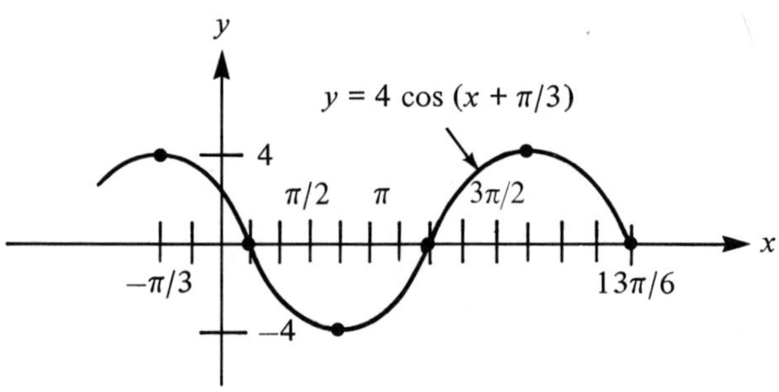

Figure A.8

We now graph $y = \tan x$. When $x = 0$, $y = \tan 0 = 0$; when $x = \pi/4$, $y = \tan \pi/4 = 1$; when $x = -\pi/4$, $y = \tan(-\pi/4) = -1$. If $x = \pi/2$, y is infinite positively, and for $x = -\pi/2$, y is infinite negatively. The graph repeats itself to the right and left, as shown in Figure A.9.

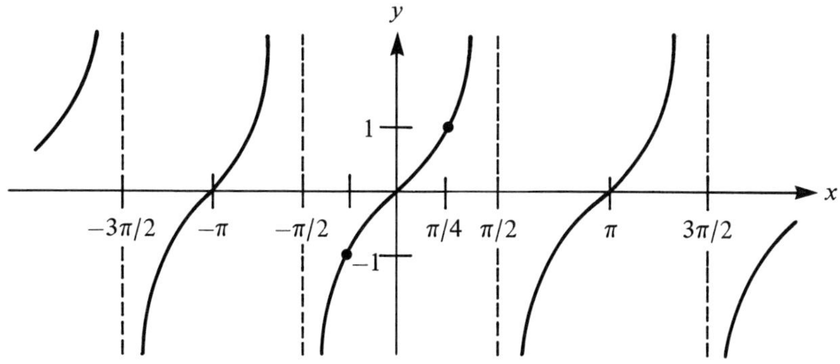

Figure A.9

Important Observations:

(1) The basic tangent curve crosses the horizontal axis at integral multiples of π (as does the sine).

(2) Vertical asymptotes occur at points which are odd multiples of $\pi/2$, or halfway between crossing points.

(3) The curve has period π.

The graph of $y = \cot x$ is shown in Figure A.10.

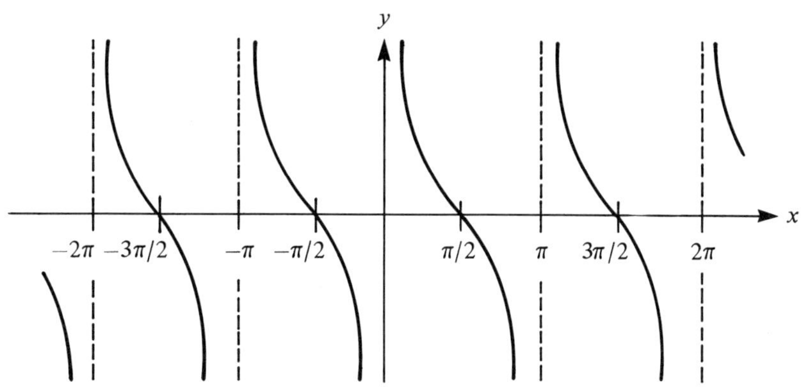

Figure A.10

Important Observations:

(1) The basic cotangent curve crosses the horizontal axis at odd multiples of $\pi/2$ (as does the cosine).

(2) Vertical asymptotes occur at integral multiples of π, or halfway between crossing points.

(3) The curve has period π.

If a function is given as a sum or difference of two other functions, we can graph by a method known as **addition of ordinates**.

EXAMPLES $y = x - \sin x$ may be graphed by first letting $y_1 = x$ and $y_2 = \sin x$. Then the graph of y is the graph of $y_1 - y_2$. When $x = 0$, $y_1 = 0$ and

$y_2 = 0$, so $y = 0 - 0 = 0$. When $x = \pi$, $y_1 = \pi$ and $y_2 = 0$, so $y = \pi - 0 = \pi$. Other convenient values of x may be located for use in computing values for y (see Figure A.11).

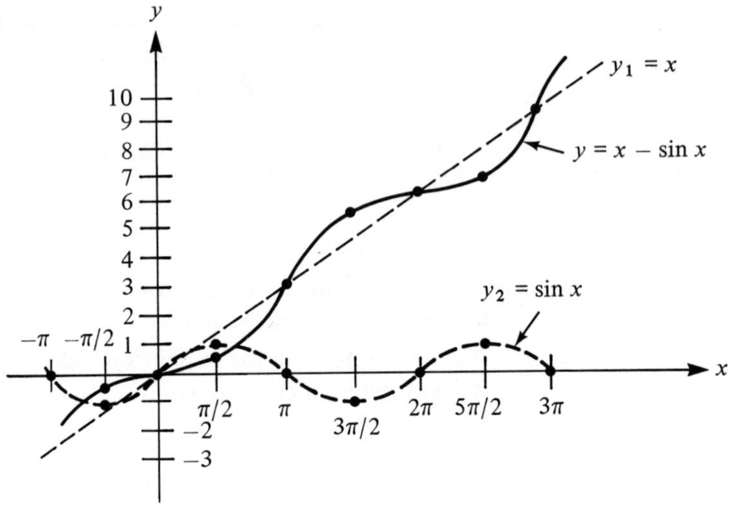

Figure A.11

Trigonometric functions enjoy a certain degree of compatibility with algebraic functions, especially those having applications in business problems. As illustrations, consider the three basic trigonometric functions: sine, cosine, and tangent.

The graph of $y = \sin x$ is shown in Figure A.12(a). Notice the close resemblance to the graph of the algebraic function $y = 3x - x^2$ in Figure A.12(b). The algebraic function often plays the role of a profit function, where profit increases with production up to $x = 3/2$, then decreases to zero at $x = 3$.

In Figure A.13(a) we see the graph of $y = \cos x$. Observe the similarity to the graph of the algebraic function $y = 1 - (x^2/9)$ in Figure A.13(b). The dotted portions of the two graphs are shown only for purposes of comparison. The function in Figure A.13(b) may occur as a demand function (in quadrant I), where the price decreases as the quantity of goods increases.

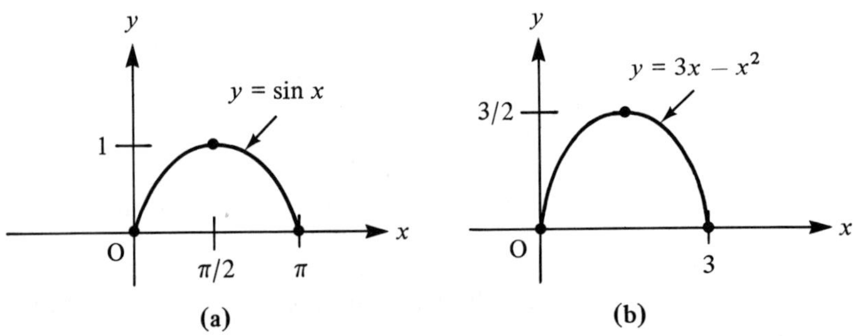

Figure A.12

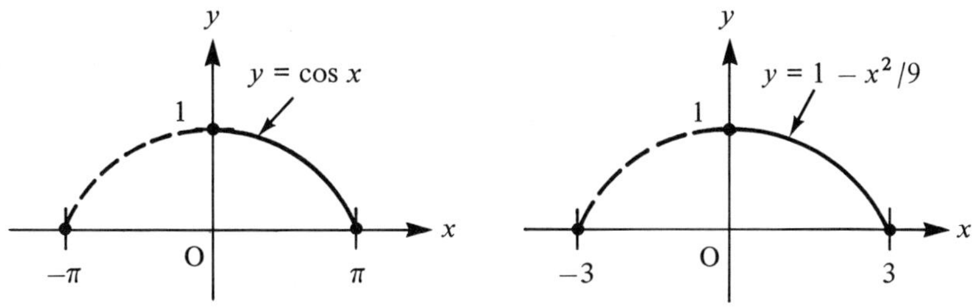

Figure A.13 (a) (b)

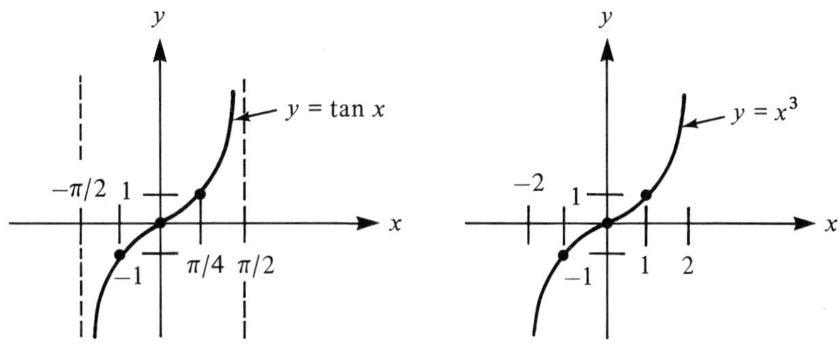

Figure A.14 (a) (b)

The graph of $y = \tan x$ appears in Figure A.14(a). Its resemblance to the graph of the algebraic function $y = x^3$ is remarkable (Figure A.14(b)). The latter may occur as (1) a cost function (cost increases with production); (2) a supply function (price increases as the supply of a commodity increases); or (3) an exponential growth function such as $y = e^x - 1$ (in quadrant I), where the similarity is not as good as in (1) and (2).

PROBLEM SET A.2

Sketch graphs of the following functions:

1. $y = 2 \sin 2x$ **2.** $y = \frac{3}{2} \cos \frac{x}{2}$

3. $y = \tan 2x$ **4.** $y = 2 \cot \frac{3x}{2}$

5. $y = 3 \cos \frac{\pi x}{2}$ **6.** $y = 4 \sin \frac{\pi x}{3}$

7. $y = \tan(x - \pi)$ **8.** $y = \cot\left(x + \frac{\pi}{2}\right)$

9. $y = 2 \sin\left(x - \frac{\pi}{3}\right)$ **10.** $y = \frac{2}{3} \cos\left(2x + \frac{\pi}{2}\right)$

11. $y = x + 2 \sin x$ **12.** $y = \cos x - \sin 2x$

A.3 APPLICATIONS

A right triangle has six important parts: three angles, one of which is a right angle, and three sides, one of which is the hypotenuse. The solution of a right triangle consists of solving for certain angles or sides, given sufficient conditions.

EXAMPLE 1 Solve the right triangle ABC, given $\angle A$ (angle A) $= 33^\circ$ and side $c = 400$. Make use of the tables in back of the text.

Solution See Figure A.15.

$$\sin A = \frac{a}{c}, \quad a = c \sin A = 400 \sin 33^\circ = 400(0.5446) = 217.84;$$

$$\cos A = \frac{b}{c}, \quad b = c \cos A = 400 \cos 33^\circ = 400(0.8387) = 335.48.$$

$$\text{Angle } C = 90^\circ, \quad \angle B = \angle C - \angle A = 90^\circ - 33^\circ = 57^\circ.$$

Thus, sides $a, b, c = 217.84, 335.48, 400$, and angles $A, B, C = 33^\circ, 57^\circ, 90^\circ$, respectively.

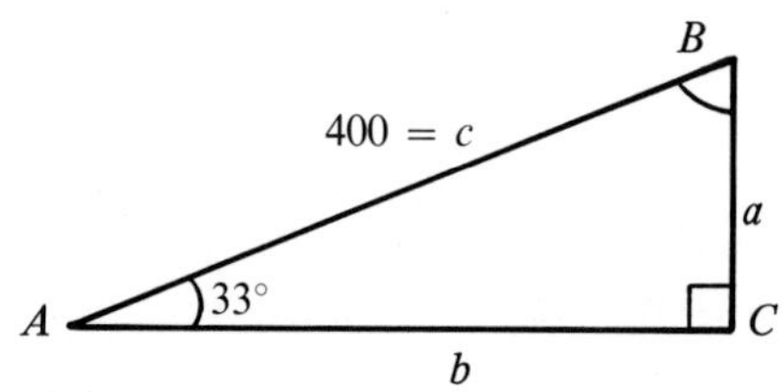

Figure A.15

EXAMPLE 2 A freighter sails from port A to port B in 3 hours and 30 minutes. Port C is 20 miles due west of port B and 60 miles due north of port A. Find the average speed of the freighter in knots (nautical miles per hour) and the direction of its course (neglect the curvature of the earth).

Solution See Figure A.16.

$$\tan A = \frac{a}{b} = \frac{20}{60} \cong 0.3333$$

Therefore,

$$A = \tan^{-1} 0.3333 \cong 18^\circ 26'.$$

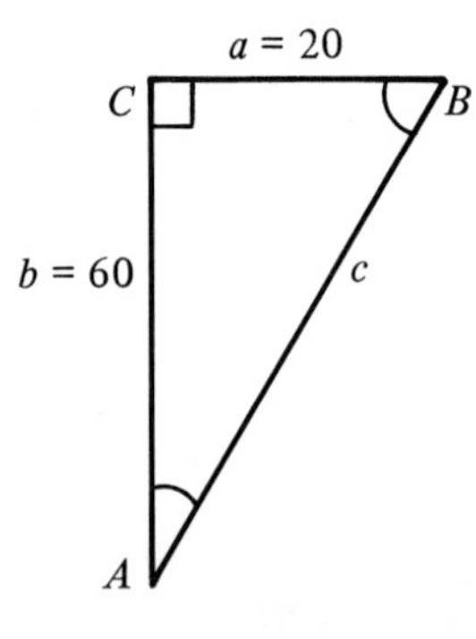

Figure A.16

That is, its course is about 18°26′ east of north. The side c is found as follows.

$$c = \frac{a}{\sin A} = \frac{20}{\sin 18°26'} = \frac{20}{0.3162} \cong 63.2.$$

Therefore, the freighter travels 63.2 nautical miles in $3\frac{1}{2}$ hours, resulting in an average speed of 63.2/3.5 or about 18.1 knots.

EXAMPLE 3 At point A the angle of elevation of the top of a bank building is 40°. At point B, 100 feet closer to the building, the angle of elevation is 50°. Find the height of the building.

Solution From Figure A.17 we see that $y/x = \tan 50°$, $y = x \tan 50°$ where y is the height of the building. Similarly, $y/(x + 100) = \tan 40°$, $y = (x + 100) \tan 40°$. Then

$$\begin{aligned} x \tan 50° &= x \tan 40° + 100 \tan 40° \\ (\tan 50° - \tan 40°)x &= 100 \tan 40° \\ (1.192 - 0.8391)x &= 100(0.8391) \\ 0.3529x &= 83.91 \\ x &= \frac{83.91}{0.3529} \cong 237.8 \text{ feet.} \end{aligned}$$

Then

$$\begin{aligned} y &= x \tan 50° = 237.8\,(1.192) \\ &\cong 283.5 \text{ feet, the height of the building} \end{aligned}$$

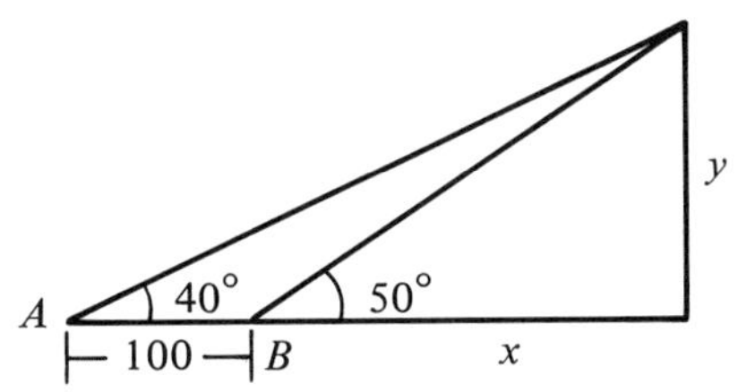

Figure A.17

For any triangle ABC with sides a, b, c, we can use the Law of Cosines to solve (a) for a side when two sides and the included angle are given, and (b) for an angle if the three sides are given. The Law of Cosines is given by

$$a^2 = b^2 + c^2 - 2bc \cos A$$

or

$$b^2 = a^2 + c^2 - 2ac \cos B$$

or

$$c^2 = a^2 + b^2 - 2ab \cos C.$$

EXAMPLE 4 Given side $a = 4$, side $b = 6$, $\angle C = 120°$, find side c.

Solution By the Law of Cosines,

$$\begin{aligned} c^2 &= 4^2 + 6^2 - 2(4)(6) \cos 120° \\ &= 16 + 36 - 48\left(-\frac{1}{2}\right) = 76 \\ c &= \sqrt{76} = 2\sqrt{19} \cong 8.7. \end{aligned}$$

Another property of any triangle ABC with sides a, b, c, is given by the Law of Sines:

$$\frac{a}{\sin A} = \frac{b}{\sin B} \quad \text{and} \quad \frac{b}{\sin B} = \frac{c}{\sin C}$$

Summarily, we have

$$\frac{a}{\sin A} = \frac{b}{\sin B} = \frac{c}{\sin C}$$

EXAMPLE 5 Solve the triangle ABC, given side $c = 100$, $\angle B = 31°$, and $\angle C = 120°$.

Solution See Figure A.18. By the Law of Sines,

$$\frac{b}{\sin 31°} = \frac{100}{\sin 120°},$$

$$b = \frac{100 \sin 31°}{\sin 60°} = \frac{100(0.5150)}{\frac{\sqrt{3}}{2}} \cong 59.46$$

$$\angle A = 180° - (31° + 120°) = 29°$$

$$\frac{a}{\sin 29°} = \frac{100}{\sin 120°},$$

$$a = \frac{100 \sin 29°}{\sin 60°} = \frac{100(0.4848)}{\frac{\sqrt{3}}{2}} \cong 55.98$$

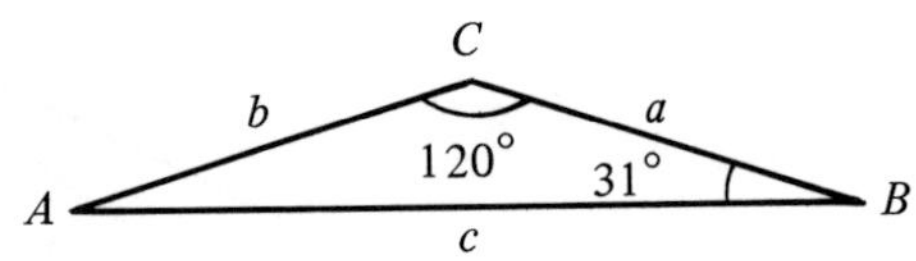

Figure A.18

PROBLEM SET A.3

1. Solve the right triangle ABC, given that
   (a) side $a = 100$, $\angle A = 38°40'$, $\angle B = 90°$
   (b) side $c = 1000$, $\angle B = 38°20'$, $\angle A = 90°$
   (c) side $b = 30$, side $a = 40$, $\angle C = 90°$.
2. The angle of elevation of a jet plane from a point P on level ground is $49°10'$. If the height of the plane is 10,000 feet, find the distance from P to a point on the ground directly beneath the plane.
3. The bottom of a 10-foot ladder is 6 feet from the foot of a wall. How high up the side of the wall is the top of the ladder?
4. Solve the triangle ABC, given that
   (a) side $a = 6$, $\angle A = 20°$, $\angle B = 50°$
   (b) side $a = 200$, side $b = 100$, $\angle B = 29°$
   (c) side $c = 100$, $\angle A = 45°$, $\angle B = 85°$.
5. To measure the length AB of a projected ramp between two floors of two hotels, a distance of 20 meters is measured from point A to a point C on the

same side as A. Angles ACB and BAC are found to be 43° and 82°, respectively. Find the length of the ramp.

6. Solve the triangle ABC, given that

(a) side $a = 20$, side $b = 12$, side $c = 30$
(b) side $a = 30$, side $c = 40$, $\angle B = 80°$
(c) side $b = 12$, side $c = 16$, $\angle A = 125°$.

7. To find the length AB of a proposed tunnel through a mountain from A to B, a point C is found from which A and B are both visible. If $AC = 500$ meters, $BC = 400$ meters, and $\angle ACB = 78°$, find the length of the tunnel.

A.4 DERIVATIVES

If $y = \sin u$ and u is a differentiable function of x, then

$$\boxed{\frac{dy}{dx} = \cos u \cdot \frac{du}{dx}}.$$

EXAMPLE 1

$$\begin{aligned} y &= \sin(3x - 1) \\ \frac{dy}{dx} &= \cos(3x - 1) \cdot \frac{d}{dx}(3x - 1) \\ &= 3\cos(3x - 1) \end{aligned}$$

If $y = \cos u$ and u is a differentiable function of x, then

$$\boxed{\frac{dy}{dx} = -\sin u \cdot \frac{du}{dx}}.$$

EXAMPLE 2

$$\begin{aligned} y &= \cos\sqrt{x} \\ \frac{dy}{dx} &= -\sin\sqrt{x} \cdot \frac{d}{dx}\sqrt{x} \\ &= -\sin\sqrt{x} \cdot \frac{1}{2\sqrt{x}} \\ &= -\frac{1}{2} \cdot \frac{\sin\sqrt{x}}{\sqrt{x}} \end{aligned}$$

The next example exhibits the *product rule* for finding a derivative.

EXAMPLE 3 If $f(t) = \sin^3 2t \cos^4 2t$, find $f'(t)$.

Solution

$$\begin{aligned} f'(t) &= \sin^3 2t \cdot D_t \cos^4 2t + \cos^4 2t \cdot D_t \sin^3 2t \\ &= \sin^3 2t \cdot 4\cos^3 2t(-\sin 2t \cdot 2) + \cos^4 2t \cdot 3\sin^2 2t \cdot \cos 2t \cdot 2 \\ &= -8\sin^4 2t \cos^3 2t + 6\cos^5 2t \sin^2 2t. \end{aligned}$$

Next we apply the *quotient rule* for finding a derivative.

EXAMPLE 4 If $g(x) = \dfrac{\sin 3x}{\cos^2 3x}$, find $g'(x)$.

Solution

$$g'(x) = \frac{\cos^2 3x \cdot D_x \sin 3x - \sin 3x \cdot D_x \cos^2 3x}{(\cos^2 3x)^2}$$

$$= \frac{\cos^2 3x \cdot \cos 3x \cdot 3 - \sin 3x \cdot 2 \cos 3x(-\sin 3x \cdot 3)}{\cos^4 3x}$$

$$= \frac{3 \cos^3 3x + 6 \cos 3x \sin^2 3x}{\cos^4 3x}$$

Let us now consider an application of the *chain rule* for finding a derivative.

EXAMPLE 5 If $x = \cos 2t$ and $y = \sin^3 2t$, then by the chain rule,

$$\frac{dy}{dx} = \frac{dy}{dt} \cdot \frac{dt}{dx} = \frac{dy/dt}{dx/dt}$$

$$= \frac{3 \sin^2 2t \cos 2t \cdot 2}{-\sin 2t \cdot 2}$$

$$= -3 \sin 2t \cos 2t \quad \text{or} \quad -\frac{3}{2} \sin 4t.$$

To find d^2y/dx^2, we write

$$\frac{d}{dx}(dy/dx) = \frac{d(dy/dx)/dt}{dx/dt} \qquad \text{[by the chain rule]}$$

$$= \frac{-\frac{3}{2} \cos 4t \cdot 4}{-\sin 2t \cdot 2} = \frac{3 \cos 4t}{\sin 2t}$$

We use the equation

$$y - y_1 = m(x - x_1)$$

to find an equation of the *line tangent to a curve*. Here (x_1, y_1) is the point of tangency and m is the slope of the tangent line at (x_1, y_1).

EXAMPLE 6 If $f(x) = 2 \sin \frac{x}{2}$, then $f'(x) = 2 \cos \frac{x}{2} \cdot \frac{1}{2} = \cos \frac{x}{2}$. At a given point such as $\left(\frac{\pi}{3}, 1\right)$,

$$m = f'\left(\frac{\pi}{3}\right) = \cos \frac{\pi}{6} = \frac{\sqrt{3}}{2}$$

Then

$$y - 1 = \frac{\sqrt{3}}{2}\left(x - \frac{\pi}{3}\right)$$

$$2y - 2 = \sqrt{3}x - \frac{\sqrt{3}\pi}{3}$$

$$6y - 6 = 3\sqrt{3}x - \sqrt{3}\pi$$

or

$$3\sqrt{3}x - 6y = \sqrt{3}\pi - 6, \quad \text{equation of tangent line.}$$

We now find *extrema* for certain trigonometric functions.

EXAMPLE 7 If $f(x) = x - \sin 2x$, then $f'(x) = 1 - 2\cos 2x$. $f'(x) = 0$ when $\cos 2x = 1/2$, or $2x = \pm\pi/3 + 2n\pi$, $x = \pm\pi/6 + n\pi$. When $n = 0$, $x = \pm\pi/6$; when $n = 1$, $x = 5\pi/6$ or $7\pi/6$.

Applying the second derivative test, we have

$$f''(x) = 4\sin 2x;$$

$$f''\left(\frac{\pi}{6}\right) = 4\sin\frac{\pi}{3} > 0 \to \text{rel min at } \left(\frac{\pi}{6}, f\left(\frac{\pi}{6}\right)\right) = \left(\frac{\pi}{6}, \frac{\pi}{6} - \frac{\sqrt{3}}{2}\right).$$

$$f''\left(-\frac{\pi}{6}\right) = 4\sin\left(-\frac{\pi}{3}\right) < 0 \to \text{rel max at } \left(-\frac{\pi}{6}, f\left(-\frac{\pi}{6}\right)\right) = \left(-\frac{\pi}{6}, -\frac{\pi}{6} + \frac{\sqrt{3}}{2}\right),$$

$$f''\left(\frac{5\pi}{6}\right) = 4\sin\frac{5\pi}{3} < 0 \to \text{rel max at } \left(\frac{5\pi}{6}, \frac{5\pi}{6} + \frac{\sqrt{3}}{2}\right),$$

$$f''\left(\frac{7\pi}{6}\right) = 4\sin\frac{7\pi}{3} > 0 \to \text{rel min at } \left(\frac{7\pi}{6}, \frac{7\pi}{6} - \frac{\sqrt{3}}{2}\right).$$

EXAMPLE 8 Find maximum profit P on $[0, \pi]$ if $P(x) = \sin 2x + \cos 2x$, where x is the amount (in hundreds) of units produced and sold and P is in thousands of dollars.

Solution

$$P'(x) = 2\cos 2x - 2\sin 2x = 0 \quad \text{when}$$

$$\cos 2x = \sin 2x$$

$$2x = \frac{\pi}{4} \quad \text{or} \quad x = \frac{\pi}{8}: \quad \text{cv}$$

Also, $$2x = \frac{5\pi}{4} \quad \text{or} \quad x = \frac{5\pi}{8}: \quad \text{cv}$$

$$P''(x) = -4\sin 2x - 4\cos 2x$$

$$P''\left(\frac{\pi}{8}\right) = -4\sin\frac{\pi}{4} - 4\cos\frac{\pi}{4} < 0 \to \text{rel max at } x = \frac{\pi}{8}.$$

Then $$P\left(\frac{\pi}{8}\right) = \sqrt{2}.$$

$$P''\left(\frac{5\pi}{8}\right) > 0 \to \text{rel min at } x = \frac{5\pi}{8}.$$

Since $P(0) = 1$, $P(\pi) = 1$, and $P(\pi/8) = \sqrt{2}$, we have $\sqrt{2} \cong 1.414$ or \$1414 is the max profit on $[0, \pi]$.

Let us now consider a *related-rate* problem, where we deal with the rate of change of a function with respect to time t.

EXAMPLE 9 Given the supply function $S(x) = \sin x \cos x$, where S denotes the amount of a product supplied and x is the price per unit of product. Find the rate at which the supply is changing at the instant when $x = \pi/6$ and x is increasing at the rate of $4 per day.

Solution

$$\begin{aligned}\frac{dS}{dt} &= \sin x\left(-\sin x\frac{dx}{dt}\right) + \cos x \cdot \cos x\frac{dx}{dt}\\ &= (\cos^2 x - \sin^2 x)\frac{dx}{dt}\\ &= \cos 2x\frac{dx}{dt}\end{aligned}$$

When $x = \pi/6$ and $\dfrac{dx}{dt} = 4$, we have

$$\begin{aligned}\frac{dS}{dt} &= \cos\frac{\pi}{3}(4)\\ &= \frac{1}{2}(4) = 2 \text{ units supplied per day.}\end{aligned}$$

This example typifies a possible occurrence during inflationary times.

PROBLEM SET A.4

Perform the indicated differentiations:

1. $\dfrac{d}{dx}\cos(x^2)$

2. $\dfrac{d}{dt}(\sin^3 2t)$

3. $\dfrac{d}{d\theta}(\cos^4 3\theta)$

4. $\dfrac{d}{dy}(4\tan\sqrt{y})$

5. $\dfrac{d}{dt}\left(\sin\dfrac{3t}{2}\tan\dfrac{3t}{2}\right)$

6. $\dfrac{d}{d\theta}(\cos 2\theta\sqrt{\sin 2\theta})$

7. $\dfrac{d}{dx}\left(\dfrac{1-2\cos 3x}{1+2\cos 3x}\right)$

8. $\dfrac{d}{dt}\left(\dfrac{1+\cos t}{t\sin t}\right)$

Use the chain rule to find the indicated derivative for the given functions:

9. $\dfrac{dy}{dx}$ if $y = \sin\dfrac{\theta}{3}$ and $\theta = \cos\dfrac{x}{6}$

10. $\dfrac{dG}{du}$ if $G(x) = x - \cos x$ and $x = \sin u - u$

11. $\dfrac{du}{dt}$ if $u(x) = \dfrac{x}{\sin x}$ and $x = \dfrac{\cos t}{t}$

12. $\dfrac{dy}{dx}$ if $y = t\sin t$ and $x = t\cos t$

13. Determine an equation of the line tangent to the given curve at the indicated point:

(a) $f(x) = x - \cos 2x$; $(0, -1)$; **(b)** $g(t) = \sin 3t + \cos 3t$; $(\pi/3, -1)$
(c) $h(y) = y\sin(\pi y/2)$; $(3, -3)$; **(d)** $H(x) = \tan 2x$; $(\pi/6, \sqrt{3})$.

14. Find extrema for the following functions on the interval $[0, 2\pi]$:

(a) $g(t) = \cos 2t - t$

(b) $f(x) = 2\cos x + \cos 2x$

(c) $F(u) = \sin u + \dfrac{\sin 2u}{2}$

(d) $G(z) = \sin\dfrac{z}{2} + \cos\dfrac{z}{2}$.

15. A firm wishes to build on a lot in the shape of a right triangle with floor dimensions as shown in the figure. Maximize the area of the triangle and thus find the largest floor space.

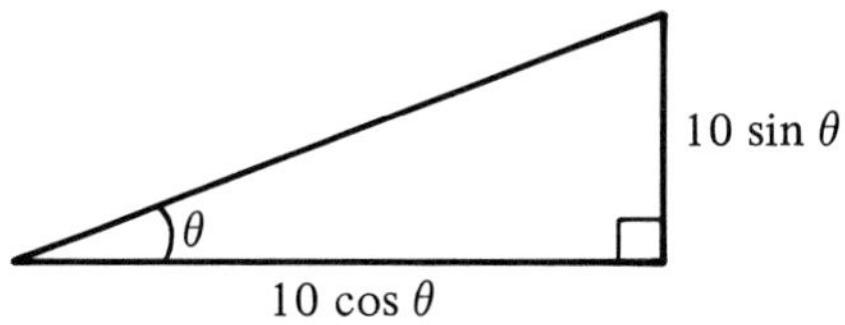

16. A cost function is given by $Q(x) = \cos x + (x/2)$ where x is the number of units produced. For what value of x is the cost a minimum?

17. In problem 16, find the rate at which x is increasing at the instant when $dQ/dt = 2$ and $x = \pi/2$.

18. In problem 15, find the rate at which the area is changing at the instant when $d\theta/dt = \pi/5$ and $\theta = \pi/2$.

A.5 INTEGRALS

Since $D_x \cos u = -\sin u \cdot D_x u$, then

$$\boxed{\int \sin u\, du = -\cos u + C}.$$

Since $D_x \sin u = \cos u \cdot D_x u$, then

$$\boxed{\int \cos u\, du = \sin u + C}$$

where u is a differentiable function of x.

EXAMPLES

$$\int \sin 3x\, dx = \frac{1}{3}\int \sin 3x \cdot 3\, dx$$
$$= -\frac{1}{3}\cos 3x + C$$

$$\int \cos(1 - 2t)\, dt = -\frac{1}{2}\int \cos(1 - 2t)(-2\, dt)$$
$$= -\frac{1}{2}\sin(1 - 2t) + C$$

$$\int \sin^3 \frac{x}{2} \cos \frac{x}{2}\, dx = 2\int \sin^3 \frac{x}{2} \cos \frac{x}{2} \cdot \frac{1}{2}\, dx$$
$$= 2\frac{\sin^4}{4}\frac{x}{2} + C$$
$$= \frac{1}{2}\sin^4 \frac{x}{2} + C$$

$$\int \frac{2\sin 3z\,dz}{\sqrt{1+2\cos 3z}} = 2\left(-\frac{1}{6}\right)\int (1+2\cos 3z)^{-1/2}(-6\sin 3z)\,dz$$
$$= -\frac{1}{3}(1+2\cos 3z)^{1/2}\cdot 2 + C$$
$$= -\frac{2}{3}\sqrt{1+2\cos 3z} + C$$

We continue integration with functions involving e^u and $\ln u$.

EXAMPLES

$$\int e^x \sin e^x\,dx = -\cos e^x + C \quad \text{where } u = e^x$$
$$\int \frac{e^t \cos e^t\,dt}{1+\sin e^t} = \ln|1+\sin e^t| + C \quad \text{where } u = 1+\sin e^t$$
$$\int e^{\sin 3x}\cos 3x\,dx = \frac{1}{3}\int e^{\sin 3x}\cos 3x\cdot 3\,dx$$
$$= \frac{1}{3}e^{\sin 3x} + C \quad \text{where } u = \sin 3x.$$

Let us apply the method of integration by parts.

EXAMPLES

$$\int x\sin x\,dx = uv - \int v\,du$$

where
$$\left.\begin{aligned} u &= x,\ du = dx,\\ dv &= \sin x\,dx,\\ v &= -\cos x \end{aligned}\right\} \begin{aligned} &= -x\cos x - \int -\cos x\,dx\\ &= -x\cos x + \sin x + C \end{aligned}$$

$$\int e^{3t}\cos 2t\,dt = uv - \int v\,du$$

where
$$\left.\begin{aligned} u &= e^{3t},\ du = 3e^{3t}\,dt,\\ dv &= \cos 2t\,dt,\\ v &= \frac{1}{2}\sin 2t \end{aligned}\right\} = \frac{1}{2}e^{3t}\sin 2t - \frac{3}{2}\int e^{3t}\sin 2t\,dt$$

To evaluate the second integral we let

$$\left.\begin{aligned} u &= e^{3t},\\ du &= 3e^{3t}\,dt,\\ dv &= \sin 2t\,dt,\\ v &= -\frac{1}{2}\cos 2t \end{aligned}\right\} \begin{aligned} \int e^{3t}\cos 2t\,dt &= \frac{1}{2}e^{3t}\sin 2t\\ &\quad -\frac{3}{2}\left[-\frac{e^{3t}}{2}\cos 2t + \frac{3}{2}\int e^{3t}\cos 2t\,dt\right]\\ &= \frac{1}{2}e^{3t}\sin 2t + \frac{3}{4}e^{3t}\cos 2t\\ &\quad - \frac{9}{4}\int e^{3t}\cos 2t\,dt \end{aligned}$$

Adding $\frac{3}{4}\int e^{3t}\cos 2t\,dt$ to both sides of the equation, we get

$$\frac{7}{4}\int e^{3t}\cos 2t\,dt = \frac{1}{2}e^{3t}\sin 2t + \frac{3}{4}e^{3t}\cos 2t$$

$$= \frac{e^{3t}}{4}(2\sin 2t + 3\cos 2t)$$

Therefore, $$\int e^{3t}\cos 2t\,dt = \frac{e^{3t}}{7}(2\sin 2t + 3\cos 2t)$$

Certain trigonometric substitutions are available for performing certain integrations.

EXAMPLES $$\int\sqrt{4-x^2}\,dx = \int\sqrt{4-4\sin^2\theta}\cdot 2\cos\theta\,d\theta \qquad [\text{where } x = 2\sin\theta,\ dx = 2\cos\theta\,d\theta]$$

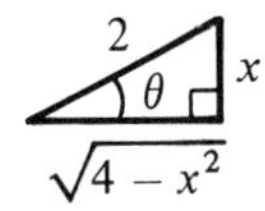

$$= 4\int\sqrt{\cos^2\theta}\cdot\cos\theta\,d\theta$$

$$= 4\int\cos^2\theta\,d\theta$$

$$= 4\int\frac{1+\cos 2\theta}{2}\,d\theta$$

$$= 2\left[\theta + \frac{1}{2}\sin 2\theta\right] + C$$

$$= 2[\theta + \sin\theta\cos\theta] + C$$

$$= 2\left[\text{Arcsin}\,\frac{x}{2} + \frac{x}{2}\cdot\frac{\sqrt{4-x^2}}{2}\right] + C$$

$$= 2\,\text{Arcsin}\,\frac{x}{2} + \frac{x}{2}\sqrt{4-x^2} + C$$

$$\int\frac{x+1}{4x^2+9}\,dx = \int\frac{x\,dx}{4x^2+9} + \int\frac{dx}{4x^2+9} \qquad [\text{where } 2x = 3\tan\theta,\ dx = \frac{3}{2}\sec^2\theta\,d\theta]$$

$$= \frac{1}{8}\int\frac{8x\,dx}{4x^2+9} + \int\frac{\frac{3}{2}\sec^2\theta\,d\theta}{9\tan^2\theta + 9}$$

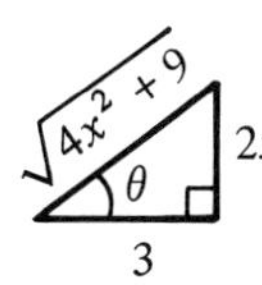

$$= \frac{1}{8}\ln|4x^2+9| + \frac{1}{6}\int\frac{\sec^2\theta\,d\theta}{\sec^2\theta}$$

$$= \frac{1}{8}\ln|4x^2+9| + \frac{1}{6}\theta + C$$

$$= \frac{1}{8}\ln|4x^2+9| + \frac{1}{6}\text{Arctan}\,\frac{2x}{3} + C$$

The next two examples are concerned with finding *areas under a curve* and *areas between curves.*

EXAMPLE 1 Find the area under one arch of $f(x) = \sin 2x$.

Solution sin $2x$ crosses the horizontal axis where

$$\begin{aligned} 2x &= n\pi, \\ x &= \frac{n\pi}{2} \\ &= 0, \frac{\pi}{2}, \qquad \text{for } n = 0, 1. \end{aligned}$$

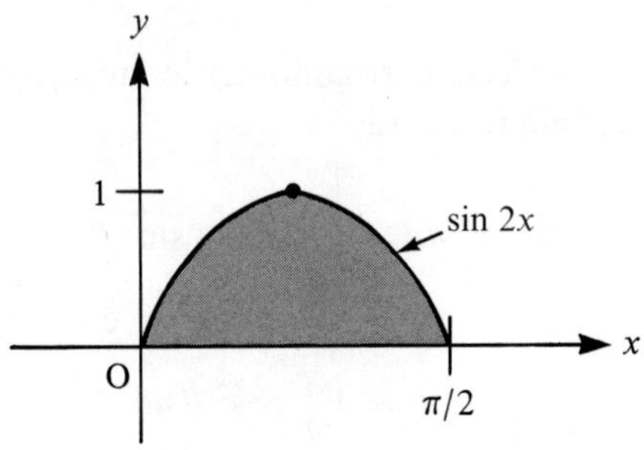

Then the area desired is

$$\begin{aligned} A &= \int_0^{\pi/2} \sin 2x \, dx \\ &= \frac{1}{2}\int_0^{\pi/2} \sin 2x \cdot 2\, dx \\ &= \frac{1}{2}(-\cos 2x)\Big]_0^{\pi/2} \\ &= \frac{1}{2}[-\cos \pi + \cos 0] \\ &= \frac{1}{2}(1 + 1) = 1 \text{ unit}^2. \end{aligned}$$

EXAMPLE 2 Find the area of the region bounded by the curves $f(x) = 2 \sin x$, $g(x) = 2 \cos x$, and $x = 0$.

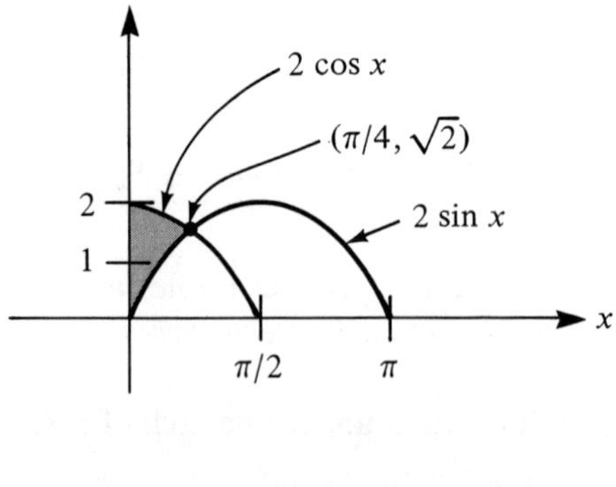

Solution $2 \cos x = 2 \sin x$ when $x = \pi/4$. Therefore,

$$\begin{aligned}
\text{area } A &= \int_0^{\pi/4} (2 \cos x - 2 \sin x)\, dx \\
&= 2 \sin x + 2 \cos x \Big]_0^{\pi/4} \\
&= 2 \sin \frac{\pi}{4} + 2 \cos \frac{\pi}{4} - 2 \sin 0 - 2 \cos 0 \\
&= \frac{2\sqrt{2}}{2} + \frac{2\sqrt{2}}{2} - 0 - 2 = (2\sqrt{2} - 2) \text{ units}^2.
\end{aligned}$$

In the next (and final) example we illustrate how to find maximum profit when marginal revenue, marginal cost, and fixed cost are given.

EXAMPLE 3 Given marginal revenue $MR = 6$, marginal cost $MQ = 12 \cos x$, and fixed cost = \$10. Find maximum profit (assume answer in thousands of dollars).

Solution

$$R = \int 6\, dx = 6x + C.$$

When $x = 0$ we assume that $R = 0$; then $R = 6x$.

$$Q = \int 12 \cos x\, dx = 12 \sin x + C_1.$$

When $x = 0$, $Q = 10$, so

$$\begin{aligned}
10 &= 12 \sin 0 + C_1 = 0 + C_1; \quad \text{then } C_1 = 10 \\
Q(x) &= 12 \sin x + 10
\end{aligned}$$

$$\begin{aligned}
\text{Profit} &= \text{Revenue minus cost} \\
&= R - Q
\end{aligned}$$

or

$$\begin{aligned}
P(x) &= 6x - 12 \sin x - 10 \\
P'(x) &= 6 - 12 \cos x = 0 \quad \text{when} \\
12 \cos x &= 6, \cos x = \frac{1}{2}, \quad x = \frac{\pi}{3} \text{ or } \frac{5\pi}{3}: \quad \text{cv} \\
P''(x) &= 12 \sin x \\
P''\left(\frac{\pi}{3}\right) &= 12 \sin \frac{\pi}{3} > 0 \rightarrow \min P \text{ at } x = \frac{\pi}{3}.
\end{aligned}$$

But we want max P. So we find

$$\begin{aligned}
P''\left(\frac{5\pi}{3}\right) &= 12 \sin \frac{5\pi}{3} \\
&= -12 \sin \frac{\pi}{3} < 0 \rightarrow \max P \text{ at } x = \frac{5\pi}{3}.
\end{aligned}$$

Then

$$\begin{aligned}\max P &= P\left(\frac{5\pi}{3}\right) \\ &= 6\left(\frac{5\pi}{3}\right) - 12 \sin\frac{5\pi}{3} - 10 \\ &= 10(3.1416) - 12\left(\frac{\sqrt{3}}{2}\right) - 10 \\ &= 21.416 + 6(1.732) \\ &= 31.808 \quad \text{or} \quad \$31{,}808 \text{ profit.}\end{aligned}$$

PROBLEM SET A.5

1. Integrate:

(a) $\int \cos^3 x \sin x \, dx$ **(b)** $\int \frac{\sin 2t \, dt}{\sqrt{\cos 2t}}$

(c) $\int \frac{\cos 3\theta \, d\theta}{3 + \sin 3\theta}$ **(d)** $\int \frac{\sin(2x + 1)}{1 + \cos(2x + 1)} dx$

(e) $\int e^{2t} \cos e^{2t} \, dt$ **(f)** $\int \tan(1 - 4z) \, dz$

2. Compute:

(a) $\int_0^\pi \cos \frac{x}{2} \, dx$ **(b)** $\int_0^1 (t + 2 \cos t) \, dt$

(c) $\int_{\pi/6}^{\pi/3} \sin^2 \theta \, d\theta$ **(d)** $\int_0^2 \cos^2 \frac{\pi x}{2} \, dx$

3. Integrate by parts:

(a) $\int x \cos x \, dx$ **(b)** $\int t \sin 2t \, dt$

(c) $\int e^{2x} \sin x \, dx$ **(d)** $\int e^z \cos \frac{z}{2} \, dz$

4. Integrate by an appropriate trigonometric substitution:

(a) $\int \sqrt{9 - x^2} \, dx$ **(b)** $\int \frac{dt}{\sqrt{1 - 4t^2}}$

(c) $\int \frac{dx}{(x^2 + 1)^2}$ **(d)** $\int \frac{t + 2}{9t^2 + 1} dt$

5. Find the area of the region described as follows:

(a) Under one arch of $y = \cos 2x$
(b) Under one arch of $y = 2 \sin(3x/2)$
(c) Bounded by $y = \sin 2x$, $y = \cos 2x$, and $x = 0$
(d) Bounded by $y = \sin(x/2)$, $y = \cos(x/2)$, x-axis, on $[0, \pi]$.

6. Given marginal revenue, marginal cost, and fixed cost as follows, find maximum profit. Assume the answer is in thousands of dollars.

(a) $MR = 8$, $MQ = 16 \cos x$, fixed cost = \$20.
(b) $MR = 2$, $MQ = -4 \cos x$, fixed cost = \$6.
(c) $MR = 8$, $MQ = 8 \tan x$, fixed cost = \$2.

7. Given marginal revenue MR, marginal cost MQ, and fixed cost as shown, find maximum profit (in thousands of dollars):

(a) $MR = 4$, $MQ = 8 \sin x$, fixed cost = \$4

(b) $MR = 4$, $MQ = 8 \cos x$, fixed cost = \$6.

8. Evaluate:

(a) $\int_0^{\pi/2} \cos 4t \, dt$

(b) $\int_{\pi/2}^{3\pi/4} (\sin x + \cos x)^2 \, dx$

APPENDIX B

Some Examples of the Use of Logarithms

EXAMPLE 1 Determine the logarithm of (a) 543; (b) 0.00309.

Solution (a) We first write 543 as 5.43×10^2 (scientific notation). Then

$$\begin{aligned}\log 543 &= \log 5.43 \times 10^2 \\ &= \log 5.43 + \log 10^2 && \text{[property of logs]} \\ &= 0.7348 + 2 \log 10 && \text{[from common log table]} \\ &= 0.7348 + 2 \text{ or } 2.7348.\end{aligned}$$

(b) Write 0.00309 as 3.09×10^{-3}. Then

$$\begin{aligned}\log 0.00309 &= \log 3.09 \times 10^{-3} \\ &= 0.4900 + (-3 \log 10) && \text{[from common log table]} \\ &= 0.4900 - 3.\end{aligned}$$

If $\log N = A$, then the number N is called the **antilogarithm** of the number A.

EXAMPLE 2 Given $\log N = 4.2041$, find N.

Solution We write 4.2041 as $0.2041 + 4$. Then

$$\begin{aligned}\log N &= 0.2041 + 4 \\ N &= (\text{antilog } 0.2041) \times 10^4 && \text{[by definition]} \\ &= 1.60 \times 10^4 = 16{,}000. && \text{[from common log table]}\end{aligned}$$

EXAMPLE 3 Use logarithms to compute $(1.99)^4(2.01)^{1/2}/(8.02)^{1/3}$

Solution Let N be the value of the given fraction. Then

$$\begin{aligned}\log N &= \log(1.99)^4(2.01)^{1/2} - \log(8.02)^{1/3} && \text{[property of logs]} \\ &= 4 \log 1.99 + \frac{1}{2} \log 2.01 - \frac{1}{3} \log 8.02 && \text{[property of logs]} \\ &= 4(0.2989) + \frac{1}{2}(0.3032) - \frac{1}{3}(0.9042) && \text{[common log table]} \\ &= 1.1956 + 0.1516 - 0.03014 \\ &= 1.0458 \text{ or } 0.0458 + 1.\end{aligned}$$

Therefore, $N = (\text{antilog } 0.0458) \times 10^1$.

Using the table of common logs, we can set up the following tabular form:

$$0.0039 \left\{ \begin{array}{l} 0.0005 \left\{ \begin{array}{l} 0.0453 \to \log 1.110 \\ 0.0458 \to \log ? \end{array} \right\} x \\ \qquad\quad 0.0492 \to \log 1.120 \end{array} \right\} 0.010$$

By the principle of proportional parts, we obtain

$$\frac{x}{0.010} = \frac{0.0005}{0.0039} = \frac{5}{39} \tag{1}$$

Here we have assumed that changes in the logarithm are proportional to the corresponding changes in the number. This isn't exactly true, but is sufficiently valid for our purposes. Solving Eq. (1) for x, we get

$$x = \frac{5}{39}(0.010) = \frac{0.050}{39} \cong 0.0013.$$

Adding 0.0013 to 1.110, we obtain 1.1113; therefore,

$$\text{antilog } 0.0458 = 1.1113.$$

Then

$$N = 1.1113 \times 10^1 = 11.113,$$

the desired answer. We have found antilog 0.0458 by a process called **linear interpolation**. A quick check on our answer for reasonableness yields

$$\frac{(1.99)^4(2.01)^{1/2}}{(8.02)^{1/3}} \cong \frac{(16)(1.4)}{2} = 8(1.4) = 11.2.$$

Thus, our answer is "within the ball park."

EXAMPLE 4 Evaluate ln 12.

Solution We are asked to find the natural log of 12. Then

$$\begin{aligned} \ln 12 &= \ln(2^2)(3) \\ &= 2 \ln 2 + \ln 3 && \text{[property of logs]} \\ &= 2(0.6931) + 1.0986 && \text{[from natural log tables]} \\ &= 1.3862 + 1.0986 = 2.4848. \end{aligned}$$

EXAMPLE 5 Evaluate $\ln \frac{3}{7}$.

Solution

$$\begin{aligned} \ln \tfrac{3}{7} &= \ln 3 - \ln 7 && \text{[property of logs]} \\ &= 1.0986 - 1.9459 && \text{[natural log table]} \\ &= -0.8473 \text{ or } 0.1527 - 1. \end{aligned}$$

It is preferable to express the answer with a positive decimal. In this case the answer was obtained by adding and subtracting 1:

$$\begin{array}{r} 1.0000 - 1 \\ -0.8473 \\ \hline 0.1527 - 1. \end{array}$$

EXAMPLE 6 Solve for k if $e^{-2k} = 10^{-4}$.

Solution

$$\begin{aligned} -2k \ln e &= -4 \ln 10 && \text{[property of logs]} \\ -2k &= -4 \ln 10 && [\ln e = 1] \\ k &= 2 \ln 10 && \text{[dividing both sides by } -2] \\ &= 2(2.3026) = 4.6052. && \text{[common log table]} \end{aligned}$$

TABLE I: COMMON LOGARITHMS*

x	0	1	2	3	4	5	6	7	8	9
1.0	.0000	.0043	.0086	.0128	.0170	.0212	.0253	.0294	.0334	.0374
1.1	.0414	.0453	.0492	.0531	.0569	.0607	.0645	.0682	.0719	.0755
1.2	.0792	.0828	.0864	.0899	.0934	.0969	.1004	.1038	.1072	.1106
1.3	.1139	.1173	.1206	.1239	.1271	.1303	.1335	.1367	.1399	.1430
1.4	.1461	.1492	.1523	.1553	.1584	.1614	.1644	.1673	.1703	.1732
1.5	.1761	.1790	.1818	.1847	.1875	.1903	.1931	.1959	.1987	.2014
1.6	.2041	.2068	.2095	.2122	.2148	.2175	.2201	.2227	.2253	.2279
1.7	.2304	.2330	.2355	.2380	.2405	.2430	.2455	.2480	.2504	.2529
1.8	.2553	.2577	.2601	.2625	.2648	.2672	.2695	.2718	.2742	.2765
1.9	.2788	.2810	.2833	.2856	.2878	.2900	.2923	.2945	.2967	.2989
2.0	.3010	.3032	.3054	.3075	.3096	.3118	.3139	.3160	.3181	.3201
2.1	.3222	.3243	.3263	.3284	.3304	.3324	.3345	.3365	.3385	.3404
2.2	.3424	.3444	.3464	.3483	.3502	.3522	.3541	.3560	.3579	.3598
2.3	.3617	.3636	.3655	.3674	.3692	.3711	.3729	.3747	.3766	.3784
2.4	.3802	.3820	.3838	.3856	.3874	.3892	.3909	.3927	.3945	.3962
2.5	.3979	.3997	.4014	.4031	.4048	.4065	.4082	.4099	.4116	.4133
2.6	.4150	.4166	.4183	.4200	.4216	.4232	.4249	.4265	.4281	.4298
2.7	.4314	.4330	.4346	.4362	.4378	.4393	.4409	.4425	.4440	.4456
2.8	.4472	.4487	.4502	.4518	.4533	.4548	.4564	.4579	.4594	.4609
2.9	.4624	.4639	.4654	.4669	.4683	.4698	.4713	.4728	.4742	.4757
3.0	.4771	.4786	.4800	.4814	.4829	.4843	.4857	.4871	.4886	.4900
3.1	.4914	.4928	.4942	.4955	.4969	.4983	.4997	.5011	.5024	.5038
3.2	.5051	.5065	.5079	.5092	.5105	.5119	.5132	.5145	.5159	.5172
3.3	.5185	.5198	.5211	.5224	.5237	.5250	.5263	.5276	.5289	.5302
3.4	.5315	.5328	.5340	.5353	.5366	.5378	.5391	.5403	.5416	.5428
3.5	.5441	.5453	.5465	.5478	.5490	.5502	.5514	.5527	.5539	.5551
3.6	.5563	.5575	.5587	.5599	.5611	.5623	.5635	.5647	.5658	.5670
3.7	.5682	.5694	.5705	.5717	.5729	.5740	.5752	.5763	.5775	.5786
3.8	.5798	.5809	.5821	.5832	.5843	.5855	.5866	.5877	.5888	.5899
3.9	.5911	.5922	.5933	.5944	.5955	.5966	.5977	.5988	.5999	.6010
4.0	.6021	.6031	.6042	.6053	.6064	.6075	.6085	.6096	.6107	.6117
4.1	.6128	.6138	.6149	.6160	.6170	.6180	.6191	.6201	.6212	.6222
4.2	.6232	.6243	.6253	.6263	.6274	.6284	.6294	.6304	.6314	.6325
4.3	.6335	.6345	.6355	.6365	.6375	.6385	.6395	.6405	.6415	.6425
4.4	.6435	.6444	.6454	.6464	.6474	.6484	.6493	.6503	.6513	.6522
4.5	.6532	.6542	.6551	.6561	.6571	.6580	.6590	.6599	.6609	.6618
4.6	.6628	.6637	.6646	.6656	.6665	.6675	.6684	.6693	.6702	.6712
4.7	.6721	.6730	.6739	.6749	.6758	.6767	.6776	.6785	.6794	.6803
4.8	.6812	.6821	.6830	.6839	.6848	.6857	.6866	.6875	.6884	.6893
4.9	.6902	.6911	.6920	.6928	.6937	.6946	.6955	.6964	.6972	.6981
5.0	.6990	.6998	.7007	.7016	.7024	.7033	.7042	.7050	.7059	.7067
5.1	.7076	.7084	.7093	.7101	.7110	.7118	.7126	.7135	.7143	.7152
5.2	.7160	.7168	.7177	.7185	.7193	.7202	.7210	.7218	.7226	.7235
5.3	.7243	.7251	.7259	.7267	.7275	.7284	.7292	.7300	.7308	.7316
5.4	.7324	.7332	.7340	.7348	.7356	.7364	.7372	.7380	.7388	.7396
x	0	1	2	3	4	5	6	7	8	9

* From Table I in *Modern College Algebra and Trigonometry*, Second Edition, by Edwin F. Beckenbach and Irving Drooyan. Copyright 1968 by Wadsworth Publishing Company, Inc.

TABLE I (*continued*)

x	0	1	2	3	4	5	6	7	8	9
5.5	.7404	.7412	.7419	.7427	.7435	.7443	.7451	.7459	.7466	.7474
5.6	.7482	.7490	.7497	.7505	.7513	.7520	.7528	.7536	.7543	.7551
5.7	.7559	.7566	.7574	.7582	.7589	.7597	.7604	.7612	.7619	.7627
5.8	.7634	.7642	.7649	.7657	.7664	.7672	.7679	.7686	.7694	.7701
5.9	.7709	.7716	.7723	.7731	.7738	.7745	.7752	.7760	.7767	.7774
6.0	.7782	.7789	.7796	.7803	.7810	.7818	.7825	.7832	.7839	.7846
6.1	.7853	.7860	.7868	.7875	.7882	.7889	.7896	.7903	.7910	.7917
6.2	.7924	.7931	.7938	.7945	.7952	.7959	.7966	.7973	.7980	.7987
6.3	.7993	.8000	.8007	.8014	.8021	.8028	.8035	.8041	.8048	.8055
6.4	.8062	.8069	.8075	.8082	.8089	.8096	.8102	.8109	.8116	.8122
6.5	.8129	.8136	.8142	.8149	.8156	.8162	.8169	.8176	.8182	.8189
6.6	.8195	.8202	.8209	.8215	.8222	.8228	.8235	.8241	.8248	.8254
6.7	.8261	.8267	.8274	.8280	.8287	.8293	.8299	.8306	.8312	.8319
6.8	.8325	.8331	.8338	.8344	.8351	.8357	.8363	.8370	.8376	.8382
6.9	.8388	.8395	.8401	.8407	.8414	.8420	.8426	.8432	.8439	.8445
7.0	.8451	.8457	.8463	.8470	.8476	.8482	.8488	.8494	.8500	.8506
7.1	.8513	.8519	.8525	.8531	.8537	.8543	.8549	.8555	.8561	.8567
7.2	.8573	.8579	.8585	.8591	.8597	.8603	.8609	.8615	.8621	.8627
7.3	.8633	.8639	.8645	.8651	.8657	.8663	.8669	.8675	.8681	.8686
7.4	.8692	.8698	.8704	.8710	.8716	.8722	.8727	.8733	.8739	.8745
7.5	.8751	.8756	.8762	.8768	.8774	.8779	.8785	.8791	.8797	.8802
7.6	.8808	.8814	.8820	.8825	.8831	.8837	.8842	.8848	.8854	.8859
7.7	.8865	.8871	.8876	.8882	.8887	.8893	.8899	.8904	.8910	.8915
7.8	.8921	.8927	.8932	.8938	.8943	.8949	.8954	.8960	.8965	.8971
7.9	.8976	.8982	.8987	.8993	.8998	.9004	.9009	.9015	.9020	.9025
8.0	.9031	.9036	.9042	.9047	.9053	.9058	.9063	.9069	.9074	.9079
8.1	.9085	.9090	.9096	.9101	.9106	.9112	.9117	.9122	.9128	.9133
8.2	.9138	.9143	.9149	.9154	.9159	.9165	.9170	.9175	.9180	.9186
8.3	.9191	.9196	.9201	.9206	.9212	.9217	.9222	.9227	.9232	.9238
8.4	.9243	.9248	.9253	.9258	.9263	.9269	.9274	.9279	.9284	.9289
8.5	.9294	.9299	.9304	.9309	.9315	.9320	.9325	.9330	.9335	.9340
8.6	.9345	.9350	.9355	.9360	.9365	.9370	.9375	.9380	.9385	.9390
8.7	.9395	.9400	.9405	.9410	.9415	.9420	.9425	.9430	.9435	.9440
8.8	.9445	.9450	.9455	.9460	.9465	.9469	.9474	.9479	.9484	.9489
8.9	.9494	.9499	.9504	.9509	.9513	.9518	.9523	.9528	.9533	.9538
9.0	.9542	.9547	.9552	.9557	.9562	.9566	.9571	.9576	.9581	.9586
9.1	.9590	.9595	.9600	.9605	.9609	.9614	.9619	.9624	.9628	.9633
9.2	.9638	.9643	.9647	.9652	.9657	.9661	.9666	.9671	.9675	.9680
9.3	.9685	.9689	.9694	.9699	.9703	.9708	.9713	.9717	.9722	.9727
9.4	.9731	.9736	.9741	.9745	.9750	.9754	.9759	.9763	.9768	.9773
9.5	.9777	.9782	.9786	.9791	.9795	.9800	.9805	.9809	.9814	.9818
9.6	.9823	.9827	.9832	.9836	.9841	.9845	.9850	.9854	.9859	.9863
9.7	.9868	.9872	.9877	.9881	.9886	.9890	.9894	.9899	.9903	.9908
9.8	.9912	.9917	.9921	.9926	.9930	.9934	.9939	.9943	.9948	.9952
9.9	.9956	.9961	.9965	.9969	.9974	.9978	.9983	.9987	.9991	.9996
x	0	1	2	3	4	5	6	7	8	9

TABLE II: EXPONENTIAL FUNCTIONS*

x	e^x	e^{-x}	x	e^x	e^{-x}
0.00	1.0000	1.0000	1.5	4.4817	0.2231
0.01	1.0101	0.9901	1.6	4.9530	0.2019
0.02	1.0202	0.9802	1.7	5.4739	0.1827
0.03	1.0305	0.9705	1.8	6.0496	0.1653
0.04	1.0408	0.9608	1.9	6.6859	0.1496
0.05	1.0513	0.9512	2.0	7.3891	0.1353
0.06	1.0618	0.9418	2.1	8.1662	0.1225
0.07	1.0725	0.9324	2.2	9.0250	0.1108
0.08	1.0833	0.9331	2.3	9.9742	0.1003
0.09	1.0942	0.9139	2.4	11.023	0.0907
0.10	1.1052	0.9048	2.5	12.182	0.0821
0.11	1.1163	0.8958	2.6	13.464	0.0743
0.12	1.1275	0.8869	2.7	14.880	0.0672
0.13	1.1388	0.8781	2.8	16.445	0.0608
0.14	1.1503	0.8694	2.9	18.174	0.0550
0.15	1.1618	0.8607	3.0	20.086	0.0498
0.16	1.1735	0.8521	3.1	22.198	0.0450
0.17	1.1853	0.8437	3.2	24.533	0.0408
0.18	1.1972	0.8353	3.3	27.113	0.0369
0.19	1.2092	0.8270	3.4	29.964	0.0334
0.20	1.2214	0.8187	3.5	33.115	0.0302
0.21	1.2337	0.8106	3.6	36.598	0.0273
0.22	1.2461	0.8025	3.7	40.447	0.0247
0.23	1.2586	0.7945	3.8	44.701	0.0224
0.24	1.2712	0.7866	3.9	49.402	0.0202
0.25	1.2840	0.7788	4.0	54.598	0.0183
0.30	1.3499	0.7408	4.1	60.340	0.0166
0.35	1.4191	0.7047	4.2	66.686	0.0150
0.40	1.4918	0.6703	4.3	73.700	0.0136
0.45	1.5683	0.6376	4.4	81.451	0.0123
0.50	1.6487	0.6065	4.5	90.017	0.0111
0.55	1.7333	0.5769	4.6	99.484	0.0101
0.60	1.8221	0.5488	4.7	109.95	0.0091
0.65	1.9155	0.5220	4.8	121.51	0.0082
0.70	2.0138	0.4966	4.9	134.29	0.0074
0.75	2.1170	0.4724	5.0	148.41	0.0067
0.80	2.2255	0.4493	5.5	244.69	0.0041
0.85	2.3396	0.4274	6.0	403.43	0.0025
0.90	2.4596	0.4066	6.5	665.14	0.0015
0.95	2.5857	0.3867	7.0	1096.6	0.0009
1.0	2.7183	0.3679	7.5	1808.0	0.0006
1.1	3.0042	0.3329	8.0	2981.0	0.0003
1.2	3.3201	0.3012	8.5	4914.8	0.0002
1.3	3.6693	0.2725	9.0	8103.1	0.0001
1.4	4.0552	0.2466	10.0	22026	0.00005

TABLE III: NATURAL LOGARITHMS†

n	$\log_e n$	n	$\log_e n$	n	$\log_e n$
	*	4.5	1.5041	9.0	2.1972
0.1	7.6974	4.6	1.5261	9.1	2.2083
0.2	8.3906	4.7	1.5476	9.2	2.2192
0.3	8.7960	4.8	1.5686	9.3	2.2300
0.4	9.0837	4.9	1.5892	9.4	2.2407
0.5	9.3069	5.0	1.6094	9.5	2.2513
0.6	9.4892	5.1	1.6292	9.6	2.2618
0.7	9.6433	5.2	1.6487	9.7	2.2721
0.8	9.7769	5.3	1.6677	9.8	2.2824
0.9	9.8946	5.4	1.6864	9.9	2.2925
1.0	0.0000	5.5	1.7047	10	2.3026
1.1	0.0953	5.6	1.7228	11	2.3979
1.2	0.1823	5.7	1.7405	12	2.4849
1.3	0.2624	5.8	1.7579	13	2.5649
1.4	0.3365	5.9	1.7750	14	2.6391
1.5	0.4055	6.0	1.7918	15	2.7081
1.6	0.4700	6.1	1.8083	16	2.7726
1.7	0.5306	6.2	1.8245	17	2.8332
1.8	0.5878	6.3	1.8405	18	2.8904
1.9	0.6419	6.4	1.8563	19	2.9444
2.0	0.6931	6.5	1.8718	20	2.9957
2.1	0.7419	6.6	1.8871	25	3.2189
2.2	0.7885	6.7	1.9021	30	3.4012
2.3	0.8329	6.8	1.9169	35	3.5553
2.4	0.8755	6.9	1.9315	40	3.6889
2.5	0.9163	7.0	1.9459	45	3.8067
2.6	0.9555	7.1	1.9601	50	3.9120
2.7	0.9933	7.2	1.9741	55	4.0073
2.8	1.0296	7.3	1.9879	60	4.0943
2.9	1.0647	7.4	2.0015	65	4.1744
3.0	1.0986	7.5	2.0149	70	4.2485
3.1	1.1314	7.6	2.0281	75	4.3175
3.2	1.1632	7.7	2.0412	80	4.3820
3.3	1.1939	7.8	2.0541	85	4.4427
3.4	1.2238	7.9	2.0669	90	4.4998
3.5	1.2528	8.0	2.0794	100	4.6052
3.6	1.2809	8.1	2.0919	110	4.7005
3.7	1.3083	8.2	2.1041	120	4.7875
3.8	1.3350	8.3	2.1163	130	4.8676
3.9	1.3610	8.4	2.1282	140	4.9416
4.0	1.3863	8.5	2.1401	150	5.0106
4.1	1.4110	8.6	2.1518	160	5.0752
4.2	1.4351	8.7	2.1633	170	5.1358
4.3	1.4586	8.8	2.1748	180	5.1930
4.4	1.4816	8.9	2.1861	190	5.2470

* Subtract 10 for $n < 1$. Thus $\log_e 0.1 = 7.6974 - 10 = -2.3026$.

† From Table IV in *Modern College Algebra and Trigonometry*, Second Edition, by Edwin F. Beckenbach and Irving Drooyan. Copyright 1968 by Wadsworth Publishing Company, Inc.

TABLE IV: POWERS AND ROOTS*

n	n^2	$\sqrt{n}$	n^3	$\sqrt[3]{n}$	n	n^2	$\sqrt{n}$	n^3	$\sqrt[3]{n}$
1	1	1.000	1	1.000	**51**	2,601	7.141	132,651	3.708
2	4	1.414	8	1.260	**52**	2,704	7.211	140,608	3.733
3	9	1.732	27	1.442	**53**	2,809	7.280	148,877	3.756
4	16	2.000	64	1.587	**54**	2,916	7.348	157,464	3.780
5	25	2.236	125	1.710	**55**	3,025	7.416	166,375	3.803
6	36	2.449	216	1.817	**56**	3,136	7.483	175,616	3.826
7	49	2.646	343	1.913	**57**	3,249	7.550	185,193	3.849
8	64	2.828	512	2.000	**58**	3,364	7.616	195,112	3.871
9	81	3.000	729	2.080	**59**	3,481	7.681	205,379	3.893
10	100	3.162	1,000	2.154	**60**	3,600	7.746	216,000	3.915
11	121	3.317	1,331	2.224	**61**	3,721	7.810	226,981	3.936
12	144	3.464	1,728	2.289	**62**	3,844	7.874	238,328	3.958
13	169	3.606	2,197	2.351	**63**	3,969	7.937	250,047	3.979
14	196	3.742	2,744	2.410	**64**	4,096	8.000	262,144	4.000
15	225	3.873	3,375	2.466	**65**	4,225	8.062	274,625	4.021
16	256	4.000	4,096	2.520	**66**	4,356	8.124	287,496	4.041
17	289	4.123	4,913	2.571	**67**	4,489	8.185	300,763	4.062
18	324	4.243	5,832	2.621	**68**	4,624	8.246	314,432	4.082
19	361	4.359	6,859	2.668	**69**	4,761	8.307	328,509	4.102
20	400	4.472	8,000	2.714	**70**	4,900	8.367	343,000	4.121
21	441	4.583	9,261	2.759	**71**	5,041	8.426	357,911	4.141
22	484	4.690	10,648	2.802	**72**	5,184	8.485	373,248	4.160
23	529	4.796	12,167	2.844	**73**	5,329	8.544	389,017	4.179
24	576	4.899	13,824	2.884	**74**	5,476	8.602	405,224	4.198
25	625	5.000	15,625	2.924	**75**	5,625	8.660	421,875	4.217
26	676	5.099	17,576	2.962	**76**	5,776	8.718	438,976	4.236
27	729	5.196	19,683	3.000	**77**	5,929	8.775	456,533	4.254
28	784	5.292	21,952	3.037	**78**	6,084	8.832	474,552	4.273
29	841	5.385	24,389	3.072	**79**	6,241	8.888	493,039	4.291
30	900	5.477	27,000	3.107	**80**	6,400	8.944	512,000	4.309
31	961	5.568	29,791	3.141	**81**	6,561	9.000	531,441	4.327
32	1,024	5.657	32,768	3.175	**82**	6,724	9.055	551,368	4.344
33	1,089	5.745	35,937	3.208	**83**	6,889	9.110	571,787	4.362
34	1,156	5.831	39,304	3.240	**84**	7,056	9.165	592,704	4.380
35	1,225	5.916	42,875	3.271	**85**	7,225	9.220	614,125	4.397
36	1,296	6.000	46,656	3.302	**86**	7,396	9.274	636,056	4.414
37	1,369	6.083	50,653	3.332	**87**	7,569	9.327	658,503	4.431
38	1,444	6.164	54,872	3.362	**88**	7,744	9.381	681,472	4.448
39	1,521	6.245	59,319	3.391	**89**	7,921	9.434	704,969	4.465
40	1,600	6.325	64,000	3.420	**90**	8,100	9.487	729,000	4.481
41	1,681	6.403	68,921	3.448	**91**	8,281	9.539	753,571	4.498
42	1,764	6.481	74,088	3.476	**92**	8,464	9.592	778,688	4.514
43	1,849	6.557	79,507	3.503	**93**	8,649	9.644	804,357	4.531
44	1,936	6.633	85,184	3.530	**94**	8,836	9.695	830,584	4.547
45	2,025	6.708	91,125	3.557	**95**	9,025	9.747	857,375	4.563
46	2,116	6.782	97,336	3.583	**96**	9,216	9.798	884,736	4.579
47	2,209	6.856	103,823	3.609	**97**	9,409	9.849	912,673	4.595
48	2,304	6.928	110,592	3.634	**98**	9,604	9.899	941,192	4.610
49	2,401	7.000	117,649	3.659	**99**	9,801	9.950	970,299	4.626
50	2,500	7.071	125,000	3.684	**100**	10,000	10.000	1,000,000	4.642

* From Table 2 in *Fundamentals of College Algebra*, Second Edition, by Earl Swokowski. Copyright 1971 by Prindle, Weber & Schmidt, Inc.

REFERENCE LIST OF MATHEMATICAL EQUATIONS

All letters represent real numbers.

1. $a + b = b + a,$ $ab = ba.$

2. $a + (b + c) = (a + b) + c,$ $a(bc) = (ab)c.$

3. $a(b + c) = ab + ac.$

4. $a + 0 = 0 + a = a,$ $a \cdot 1 = 1 \cdot a = a.$

5. $a + (-a) = -a + a = 0,$ $a \cdot a^{-1} = a^{-1} \cdot a = 1,$ $a \neq 0.$

6. If $a + c = b + c$, then $a = b.$

7. $a \cdot 0 = 0 \cdot a = 0.$

8. If $a + b = 0$, then $a = -b.$

9. $-(-a) = a.$

10. If $ab = 1$, then $b = a^{-1}.$

11. $a^{-1} \cdot b^{-1} = (ab)^{-1}.$

12. $(-a)b = -(ab),$ $a(-b) = -(ab).$

13. $(-a)(-b) = ab.$

14. $a - b = a + (-b).$

15. $a(b - c) = ab - ac.$

16. $-(a + b) = -a - b.$

17. $a \div b = ab^{-1}, \quad b \neq 0.$

18. $a^{-1} = \dfrac{1}{a}, \quad a \neq 0.$

19. $\dfrac{a}{b} \cdot \dfrac{c}{d} = \dfrac{ac}{bd}, \quad b, d \neq 0.$

20. $\dfrac{ad}{bd} = \dfrac{a}{b}, \quad b, d \neq 0.$

21. $\dfrac{a}{b} = \dfrac{c}{d}$ iff $ad = bc, \quad b, d \neq 0.$

22. $\dfrac{a}{b} + \dfrac{c}{d} = \dfrac{ad + bc}{bd}, \quad b, d \neq 0.$

23. If $ab = 0$, then $a = 0$ or $b = 0.$

24. If $ac = bc$, $c \neq 0$, then $a = b.$

25. One and only one of the following is true:
$a < b, \quad a = b, \quad a > b.$

26. If $a \neq 0$, then $a^2 > 0.$

27. If $a > 0$, then $-a < 0.$

28. If $a > b$, then $a - b > 0.$

29. If $a < b$, then $a - b < 0.$

30. If $a < b$, then $b > a.$

31. If $a > 0$ and $b > 0$, then $a + b > 0.$

32. If $a > b$ and $b > c$, then $a > c.$

33. If $a > b$, then $a + c > b + c.$

34. If $a > b$ and $c > 0$, then $ac > bc.$

35. If $a > b$ and $c < 0$, then $ac < bc.$

36. The product (quotient) of two nonzero real numbers is positive if the two numbers agree in sign, and negative if they differ in sign.

37. If $a = bc$, then b is a factor of a and c is a factor of $a.$

38. $|a| = a$ if $a \geq 0$; $= -a$ if $a < 0.$

39. $|ab| = |a|\,|b|.$

40. $|a| <$ iff $-b < a < b.$

41. $|a| > b$ iff $a > b$ or $a < -b.$

42. $|a| = b$ iff $a = b$ or $a = -b.$

43. $x^n = x \cdot x \cdot x \cdot x \cdots$ to n factors, n is a natural number.

44. $x^a \cdot x^b = x^{a+b}$.

45. $(x^a)^b = x^{ab}$.

46. $(xy)^a = x^a y^a; \left(\dfrac{x}{y}\right)^a = \dfrac{x^a}{y^a}$.

47. $\dfrac{x^a}{x^b} = x^{a-b}, \quad a > b; \qquad = \dfrac{1}{x^{b-a}}, \quad b > a; \qquad = 1, \quad a = b.$

48. $x^0 = 1, \qquad x \neq 0.$

49. $\sqrt{x^2} = |x|$.

50. $\sqrt[n]{x^m} = x^{m/n} = (\sqrt[n]{x})^m$.

51. $(x + y)(x - y) = x^2 - y^2$.

52. $(x + y)(x^2 - xy + y^2) = x^3 + y^3$.

53. $(x - y)(x^2 + xy + y^2) = x^3 - y^3$.

54. $(ax + b)(cx + d) = acx^2 + (ad + bc)x + bd$.

55. $(x \pm y)^2 = x^2 \pm 2xy + y^2$.

56. $(x \pm y)^3 = x^3 \pm 3x^2y + 3xy^2 \pm y^3$.

APPENDIX C

Answers to Odd-Numbered Problems

Note: Numerical answers given here may differ slightly from answers obtained using different electronic calculators or five-place logarithm tables.

CHAPTER 1

Problem Set 1.1 (page 11)

1. **(a)** $|-\sqrt{2}| = \sqrt{2}$ **(b)** $|3 - (-4)| = 7$ **(c)** $\left|\dfrac{-1}{-2}\right| = \dfrac{1}{2}$

   (d) $|(-2)^3| = 8$ **(e)** $|-2(-3)^2| = |-2(9)| = 18$

   (f) $\left|\dfrac{-3 + 1}{-1 + 3}\right| = \left|\dfrac{-2}{2}\right| = 1$

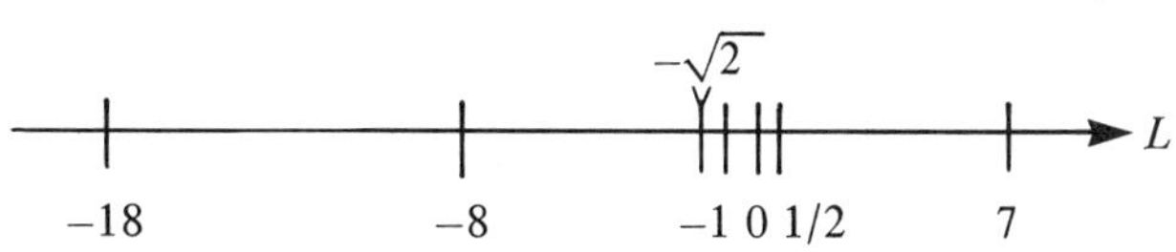

3. **(a)** $b > c$ **(b)** $c < d$ **(c)** $a \le b$ **(d)** $a \le c < b$
   (e) $x_i < t < x_{i-1}$ **(f)** $a \le x \le b$ **(g)** $|x| < 2$
5. **(a)** 1 or -3 **(b)** $-2 < k < 6$ **(c)** $-4 \le k \le -2$
   (d) $k > 9$ or $k < -9$

7.

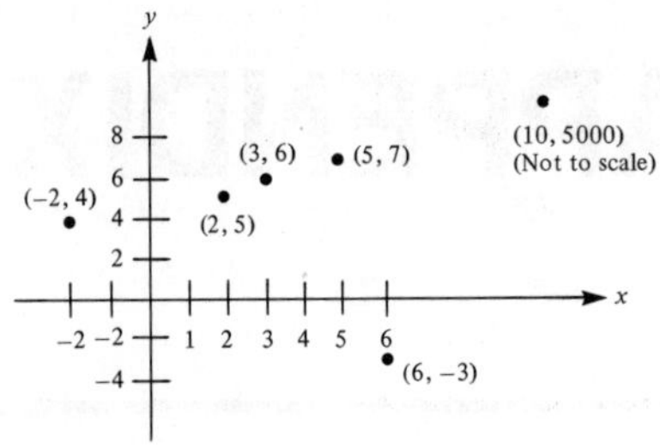

9. **(a)** $m = 4/3$ **(b)** $m = 3/2$ **(c)** $m = -1/2$
(d) $m =$ undefined **(e)** $m = -1/12$ **(f)** $m = -1.7/2.8$

11. **(a)** $y = \frac{3}{2}x + \frac{1}{2}$ **(b)** $y = -\dfrac{1.7}{2.8}x + 3$ **(c)** $y = \frac{4}{3}x + \frac{1}{12}$

Problem Set 1.2, 1.3 (page 20)

1. **(a)** The vertical line $x = 3/2$ **(b)** Horizontal line $y = 6$
(c) Oblique line $y = 5x$ **(d)** Oblique line $y = 3x + 7$
(e) Parabola opening up, $y = x^2 + 6$
(f) Parabola opening down, $y = 1 - x^2$
(g) Parabolic curve $u = 2\sqrt{t}$, in quadrant 1.
(h) Hyperbolic curve $p = 1/2q$

3. **(a)** $g(9) = \sqrt{9} = 3$ **(b)** $g(16/25) = 4/5$ **(c)** $g(0.04) = 0.2$
(d) 0.1 **(e)** 10/21

5. $$\frac{F(1.1) - F(1)}{1.1 - 1} = \frac{3.63 - 3}{0.1} = 6.3$$

7. **(a)** Domain: all reals except -1; Range: $y \neq 0$
(b) $\dfrac{1}{a}$ **(c)** $\dfrac{x + 1}{x + 2}$

9. **(a)**

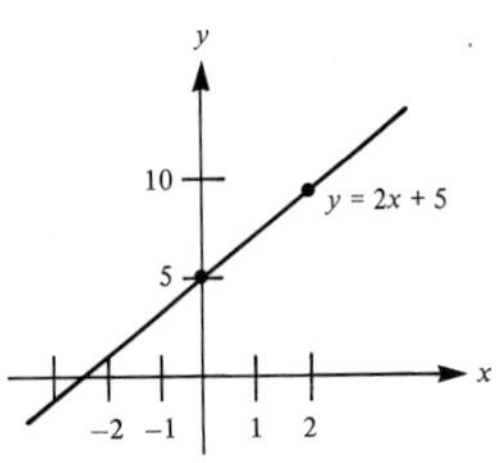

(b)

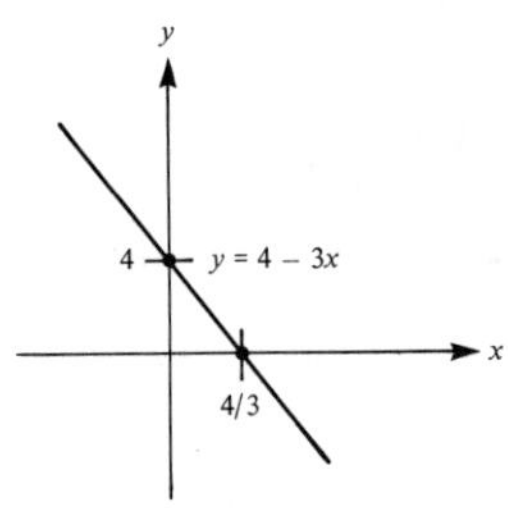

(c)

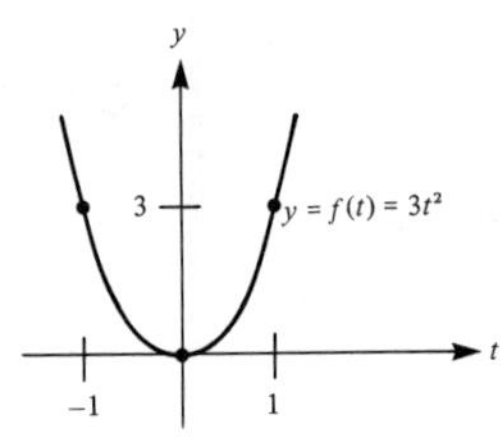

(d)

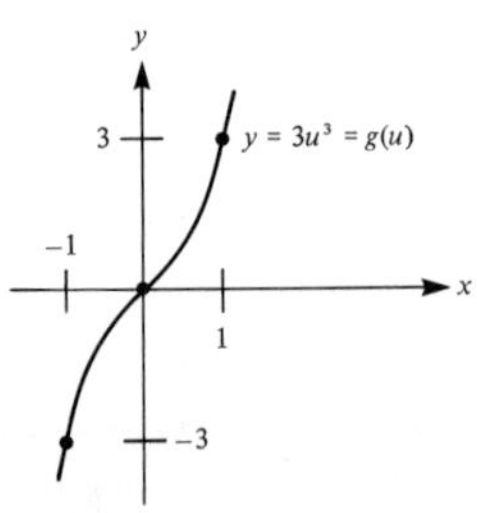

(e)

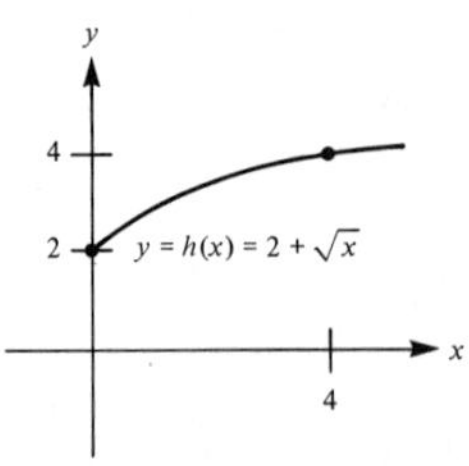

(f)

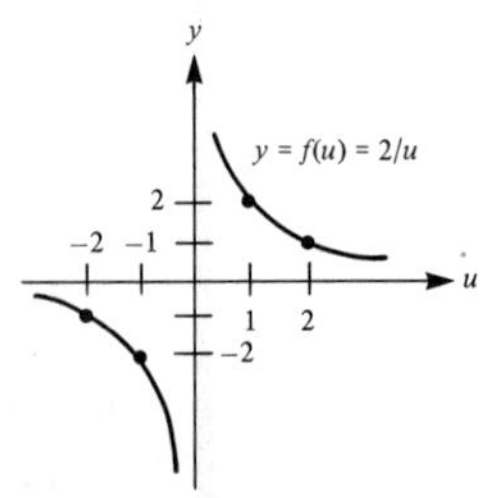

11. (a)

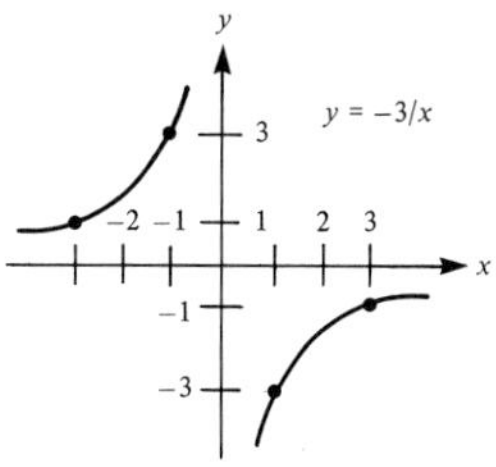

Domain: all reals except 0
Range: all reals except 0
Asymptotes: $x = 0, y = 0$

(b)

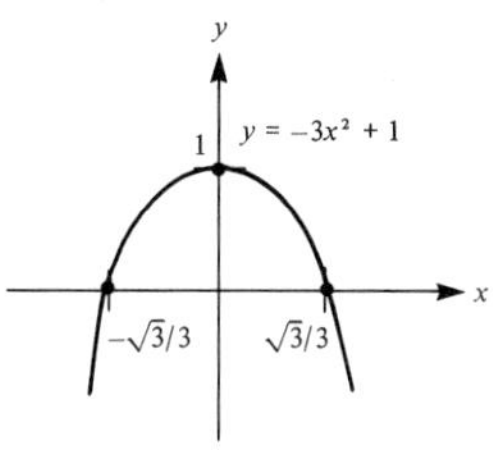

Domain: all reals
Range: $y \leq 1$
Asymptotes: none

(c)

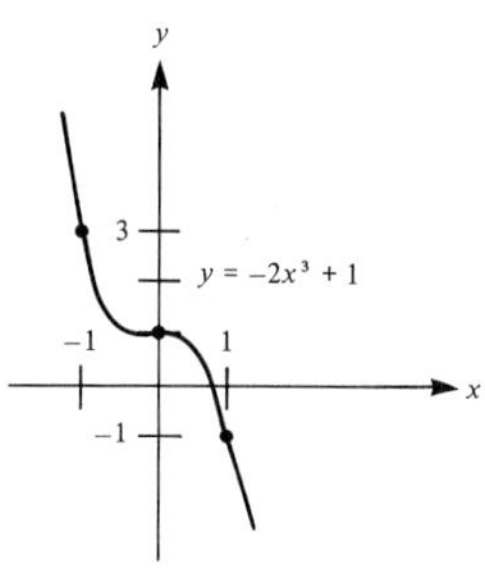

Domain: all reals
Range: all reals
Asymptotes: none

(d)

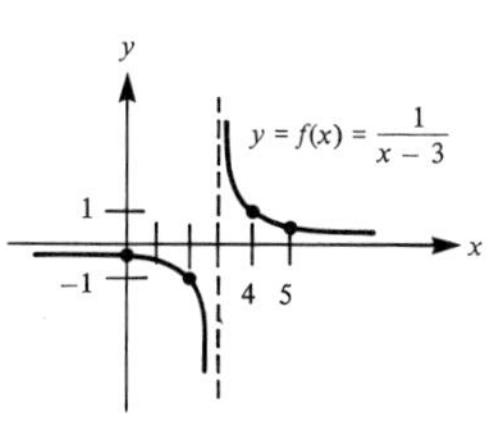

Domain: all reals $\neq 3$
Range: all reals $\neq 0$
Asymptotes: $x = 3, y = 0$

(e)

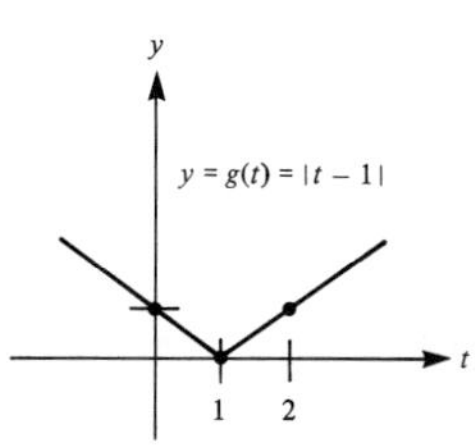

Domain: all reals
Range: all reals ≥ 0
Asymptotes: none

(f)

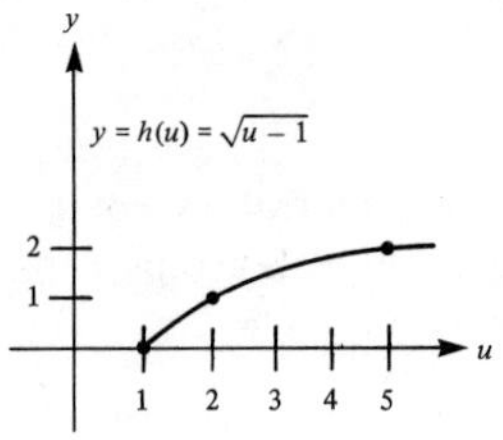

Domain: $x \geq 1$
Range: $y \geq 0$
Asymptotes: none

13. **(a)** (8, 5) **(b)** (4, 3) **(c)** (5, 2) **(d)** (6, 6)

Problem Set 1.4 (page 30)

1. **(a)** $2 + 4x - x^2$ **(b)** $1 + 4x - x^2 - 4x^3$ **(c)** $5 - 4x^2$

3. **(a)** $t^2 + 4$ **(b)** $3t^2 - 4$ **(c)** $32 - 16t^2 + 2t^4$

(d) 3(a) Domain: all reals; Range: $y \geq 4$
3(b) Domain: all reals; Range: $y \geq -4$
3(c) Domain: all reals; Range: $y \geq 0$

5. **(a)** x **(b)** x

7. $f(a + b) = 3(a + b) = 3a + 3b$; $f(a) = 3a$, $f(b) = 3b$, $f(a) + f(b) = 3a + 3b = f(a + b)$.

9. **(I)** $f^{-1}(x) = (x + 2)/3$
Domain: all reals
Range: all reals

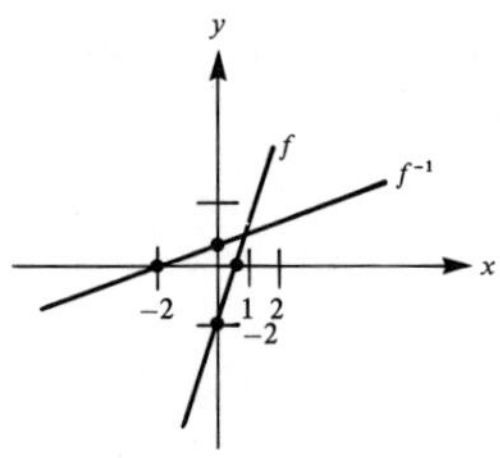

(II) $f^{-1}(x) = x^2, x \geq 0$
Domain: $x \geq 0$
Range: $y \geq 0$

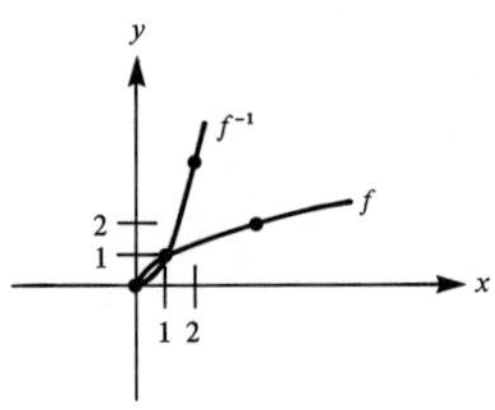

(III) $f^{-1}(x) = x^{1/3}$
Domain: $x \geq 0$
Range: $y \geq 0$

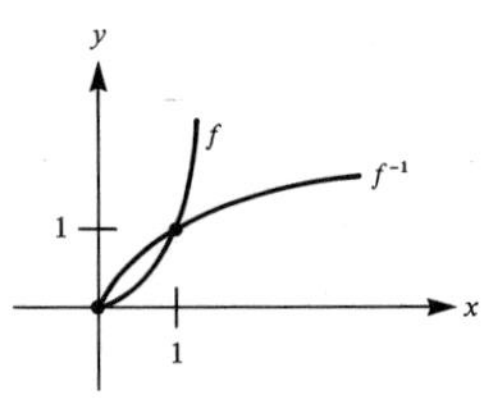

(IV) $f^{-1}(x) = \sqrt{x - 2}$
Domain: $x \geq 2$
Range: $y \geq 0$

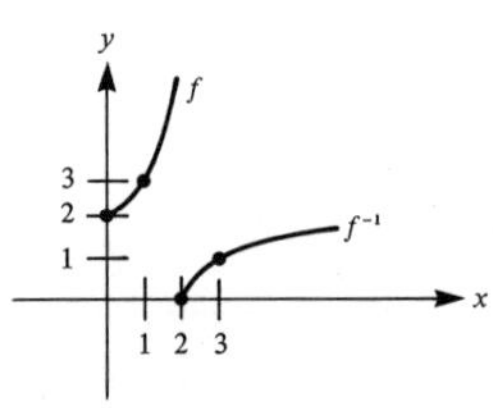

(V) $f^{-1}(x) = -\sqrt{1 - x}$
Domain: $x \leq 1$
Range: $y \leq 0$

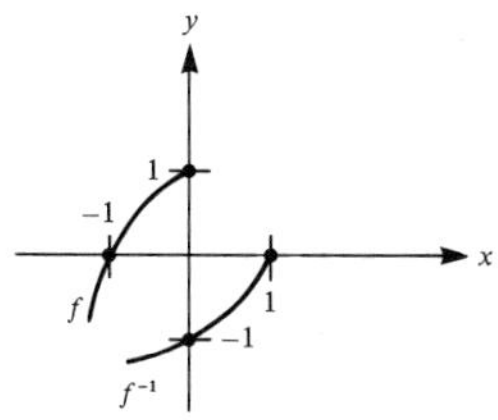

11. **(a)** $g^{-1}(x) = (2x + 1)/4$ **(b)** $g^{-1}(x) = x^2 + 2, x \geq 0$
(c) $g^{-1}(x) = 1/x, x > 0$ **(d)** $g^{-1}(x) = 2 + (1/x)$

13. **(a)** **(b)** **(c)**

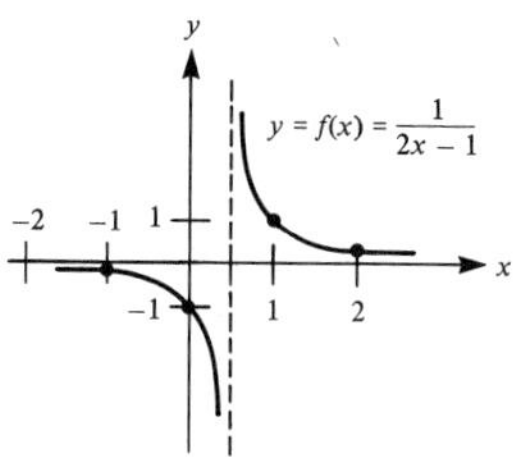

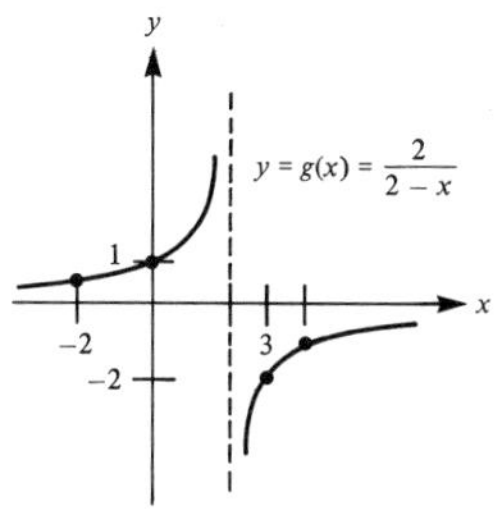

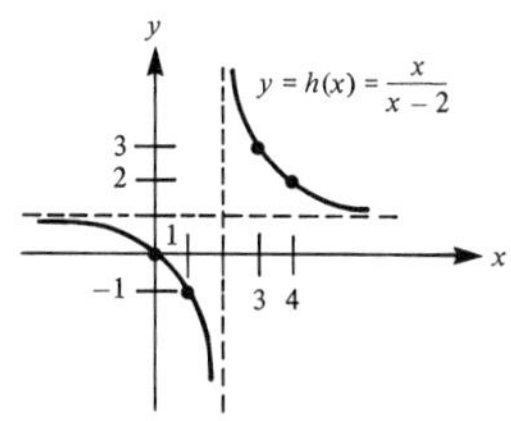

(d) **(e)**

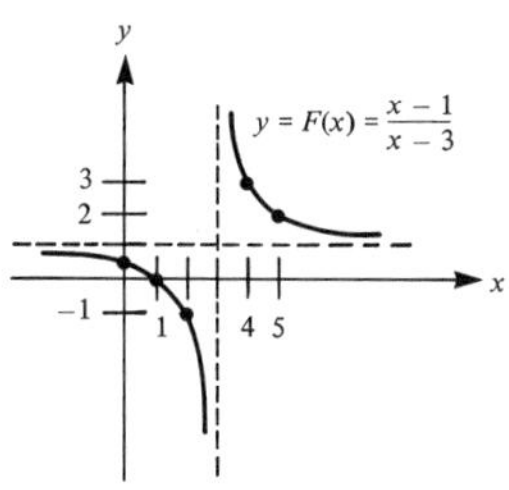

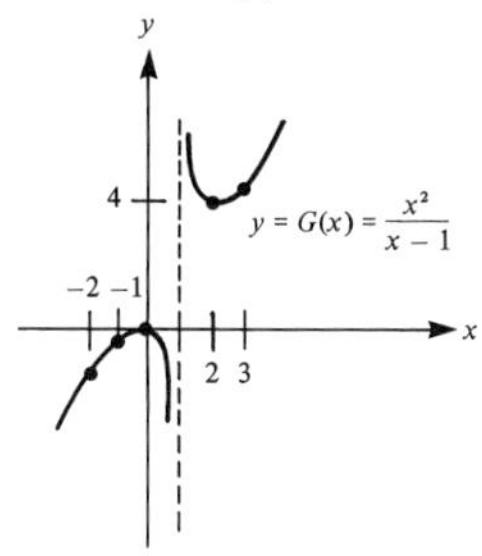

15. **(a)** ± 4 **(b)** ± 5 **(c)** 2, 3

17. **(a)** 1/4 **(b)** ± 3 **(c)** 4
(d) 2/3 **(e)** 5 **(f)** $x = \pm 3$

1.5 Review Problems (page 31)

1. **(a)** $C > \$2000$ **(b)** $P \leq \$30{,}000$ **(c)** $1.2 < \text{GNP} < 1.4$

3. **(a)** $x = \pm 2$ **(b)** $-1/2 < x < 1/2$ **(c)** $x > 3/2$ or $x < -3/2$

5. **(a)** 19 **(b)** 29 **(c)** $4(x + h)^2 + 2(x + h) - 1$
(d) $8x + 4h + 2$

7. **(a)** $\sqrt{2t}$ **(b)** 1 **(c)** 1

9. **(a)** $-6/7$ **(b)** $-1/2$

11. **(a)** **(b)** **(c)**

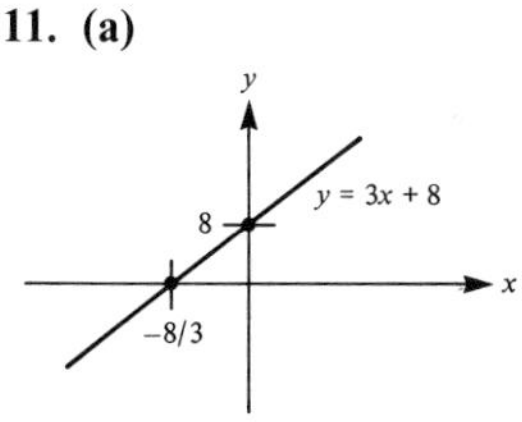

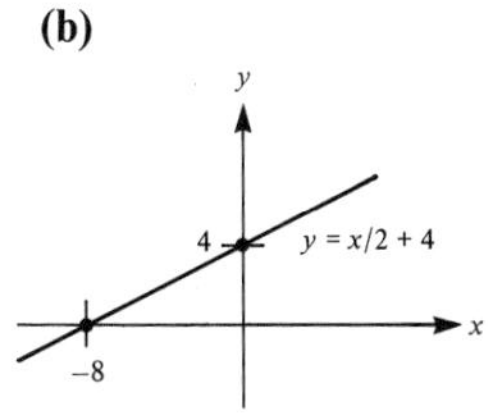

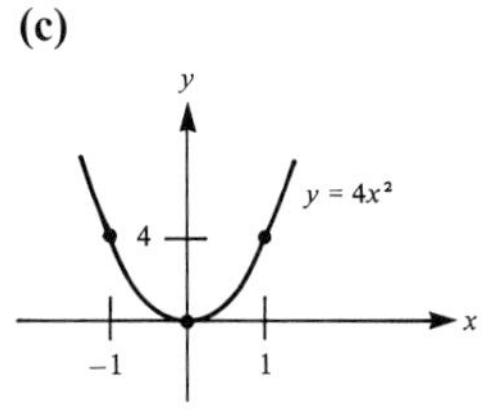

(d)

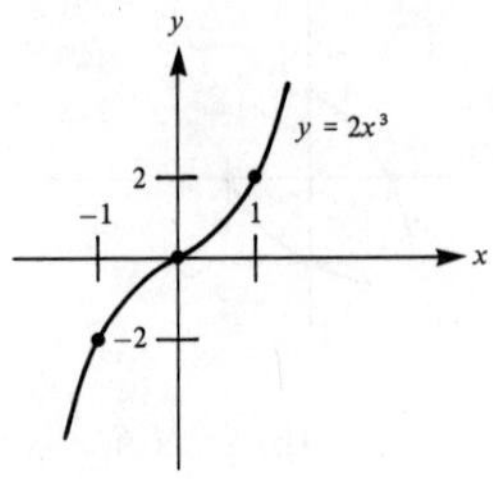

(e)

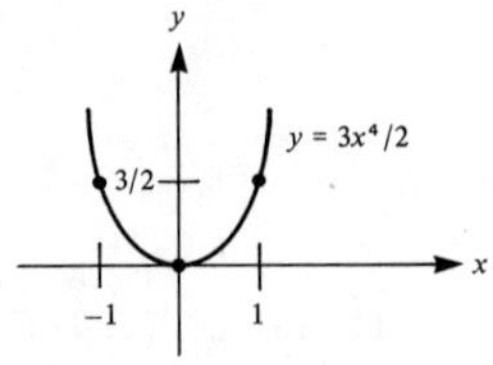

13. (a)

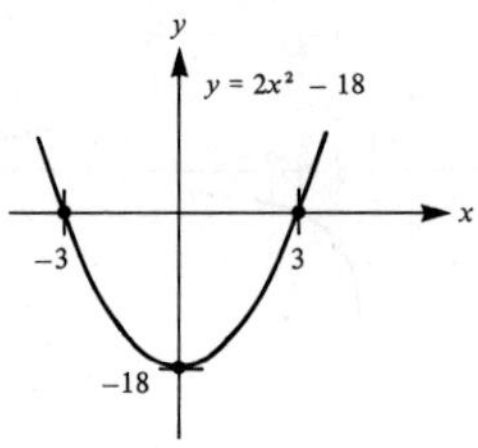

(b)

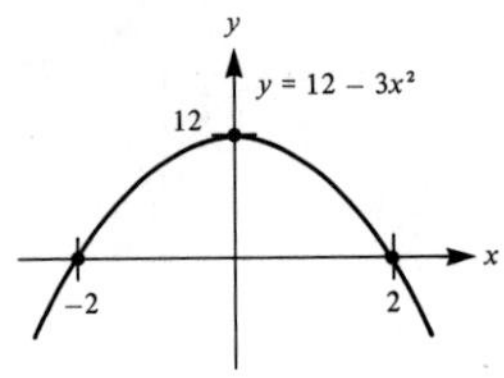

(c)

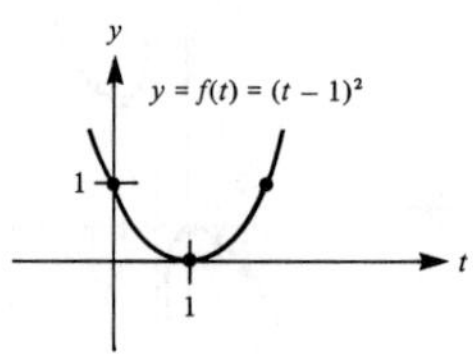

(d)

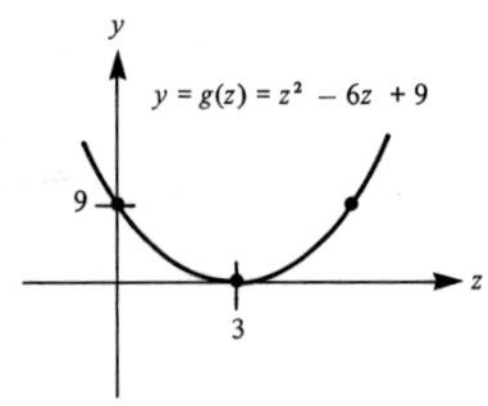

15. (a)

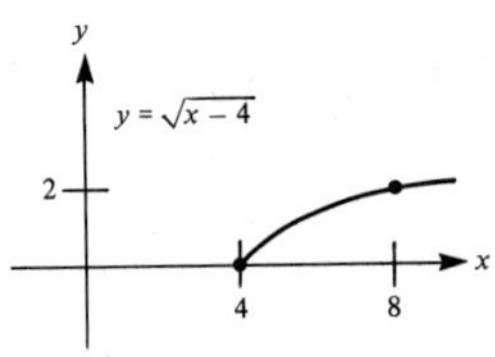

(b)

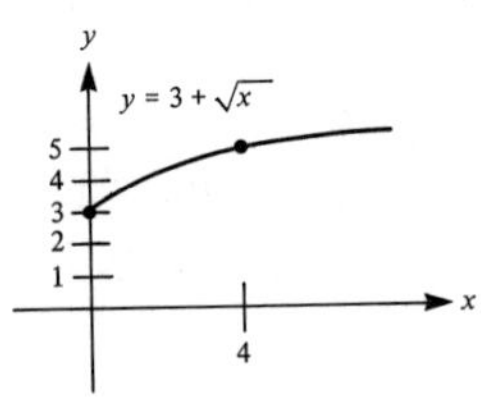

(c)

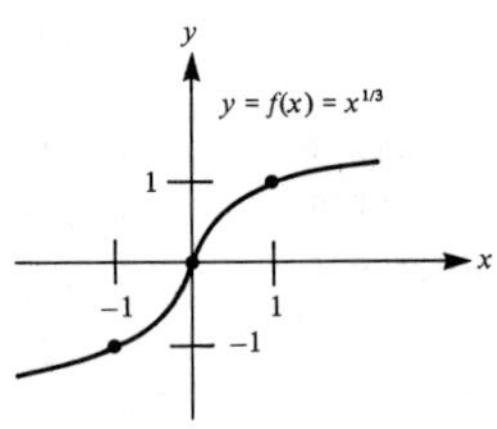

(d)

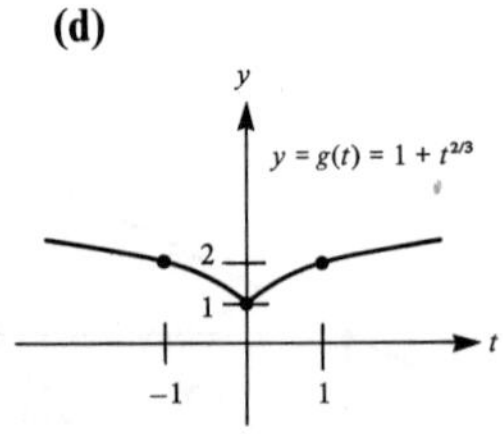

(e)

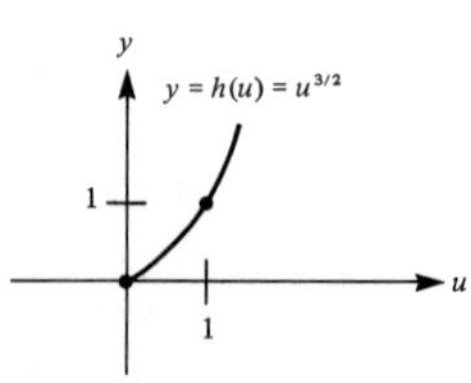

17. (a)

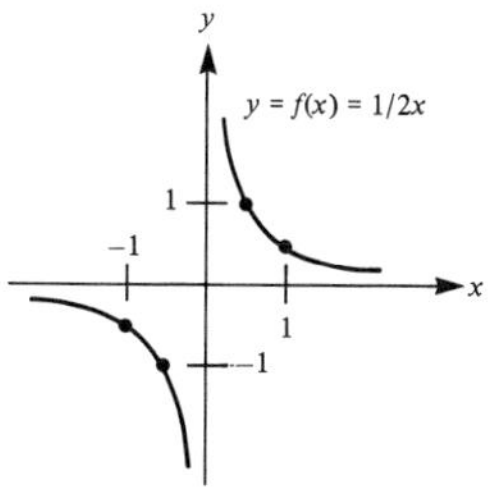

(b)

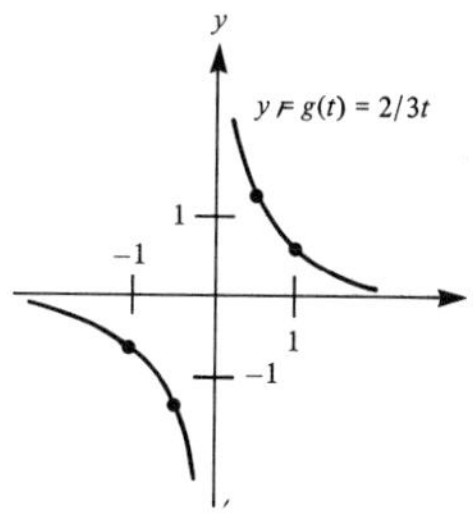

(c)

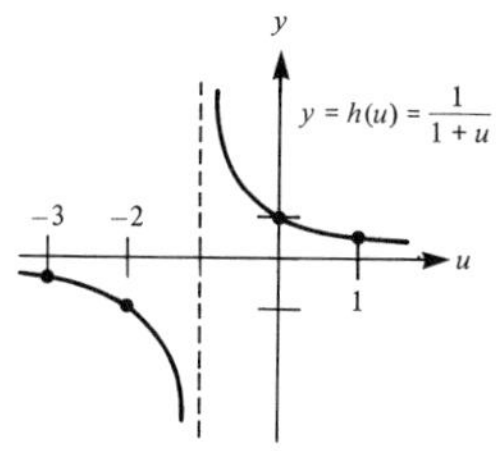

(d)

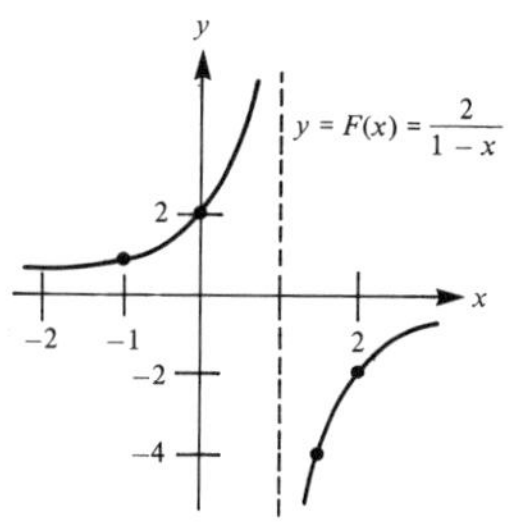

(e)

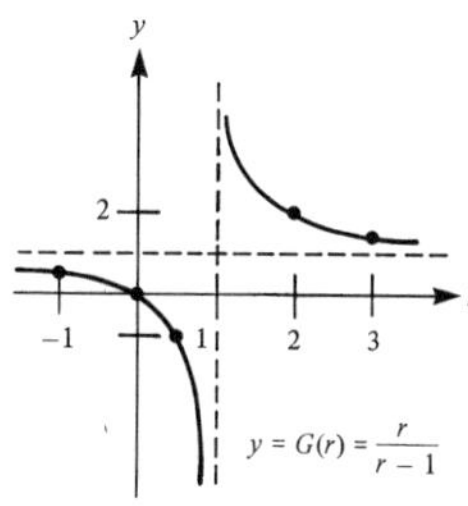

19. $f^{-1}(x) = (x - 1)^2$

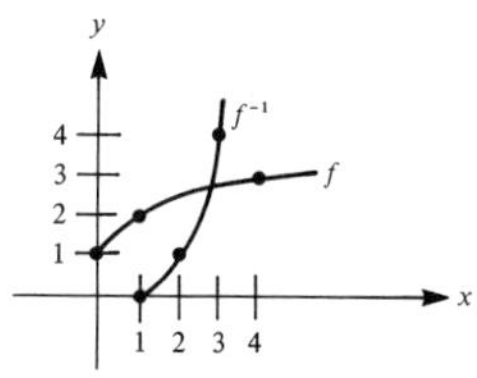

21. **(a)** $f(x) = 1 + \dfrac{1}{x - 3}$ **(b)** $g(x) = x + 1 - \dfrac{3}{x - 1}$

(c) $h(x) = 2x - 2 + \dfrac{2}{x + 1}$ **(d)** $f(x) = 3x + 12 + \dfrac{48}{x - 4}$

1.6 Enrichment Topics and Problems (page 33)

1. **(a)** $A \cup B = \{0, 1, 2, 3, 4, 5\}$ **(b)** $A \cap B = \{2, 3\}$ **(c)** $A' \cap B' = 6$

3. **(a)** $x^3 + x^2 - 2x$ **(b)** $2a^2 + 5ab - 3b^2$

5. **(a)** $6xy^2$ **(b)** $2a + 3b$ **(c)** $x^2 - 2x + 1$
(d) $x^2 + 6 + 11/(x^2 - 2)$

7. **(a)** 16 **(b)** 3 **(c)** $c - d$
(d) $2 - x$

9. **(a)** 0 or 3 **(b)** 1/3

11. $E = \$1180$, $F = \$1250$

13. 565 at 5¢ each, 425 at 10¢ each

15. $\$40 < C < \$60$

17. $A = 2x^2 + (4V/x)$

19. $\sqrt{2h^2 - 1}$

21. $V = x(12 - 2x)^2$

23. **(a)** $V(1, 2)$

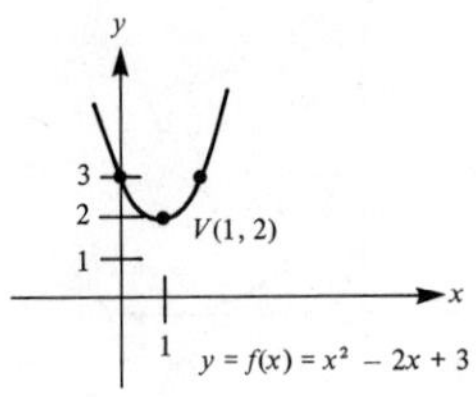

(b) $V(1/2, 2)$

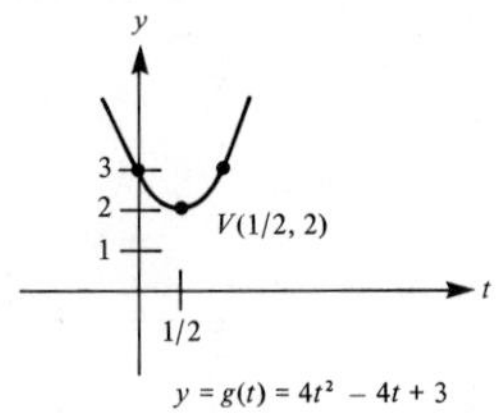

25. **(a)**

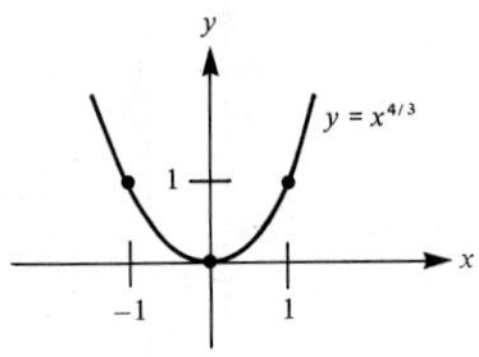

(b)

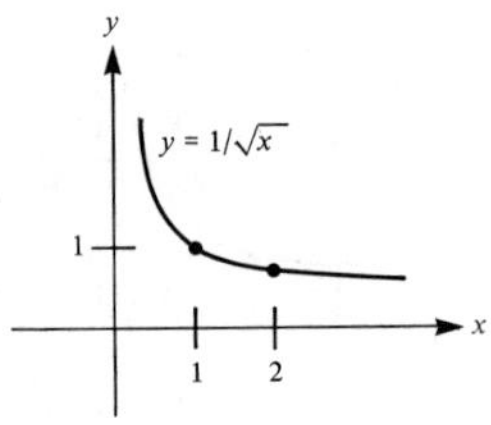

(c)

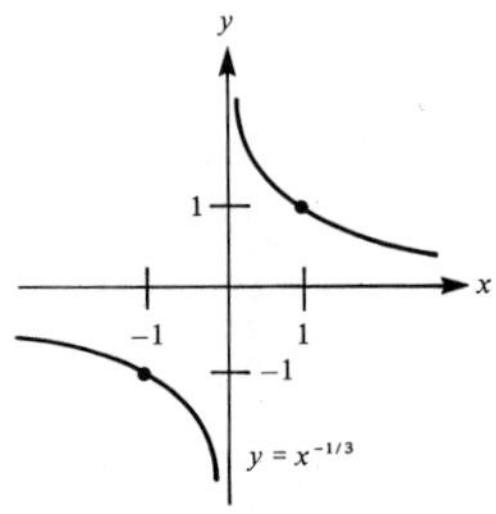

(d)

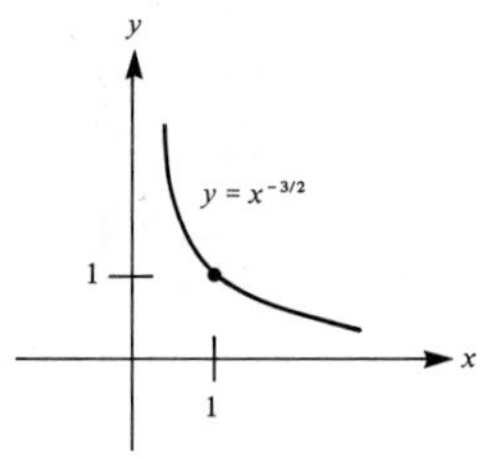

(e)

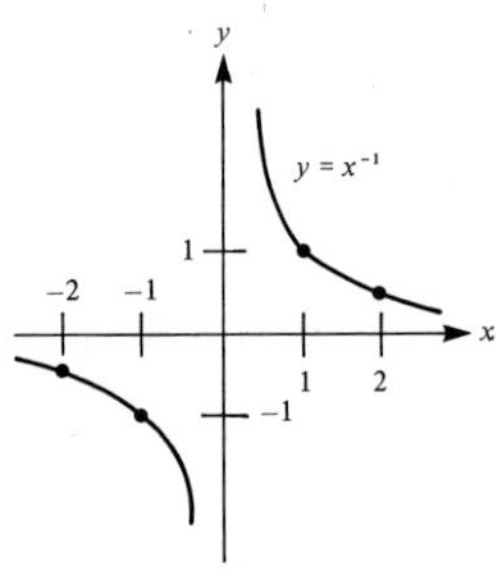

27. **(a)** 3 or $-3/2$ **(b)** 4

29. $(2, 5)$

31. \$325 for $6\frac{1}{2}\%$; \$362.50 for $7\frac{1}{4}\%$

33. **(a)** $\sqrt{2}$ **(b)** $\sqrt[3]{3}$

35. **(a)** $V(2, 3) \rightarrow \text{Max } P = \$3$; **(b)** $V(3, 7) \rightarrow \text{Max } P = \$7$

37. **(a)** $A = 2000(1 + 0.02)^{16} \cong \$2744.96$, using four terms
(b) $A = 5000(1 + 0.01)^{60} \cong \$9080.48$, using five terms

39. **(a)** $0.9744 - 1$ **(b)** $0.9829 - 1$ **(c)** 2.4592

41. **(a)** 35/18 **(b)** 97/60

43. **(a)** $x = 1$ **(b)** $x = 2, -1$ **(c)** $x = 3, -2$
(d) $x = -6, 4$

45. **(a)** $(x/2) + 2$ **(b)** $2x^2 - x$ **(c)** $t - 2 + (3/t)$

47. $C(x) = 15x^2 + (400/x)$

49. **(a)** $x^3 + 6x^2 + 12x + 8$ **(b)** $4x^2 - 20x + 25$
(c) $1 - 4x + 6x^2 - 4x^3 + x^4$

CHAPTER 2

Problem Set 2.1 (page 47)

1. (a) $\lim_{x\to 1}(x^2-2x+1) = \lim_{x\to 1} x^2 + \lim_{x\to 1}(-2x) + \lim_{x\to 1} 1$
$= \lim_{x\to 1} x \cdot \lim_{x\to 1} x + \lim_{x\to 1}(-2)\cdot \lim_{x\to 1} x + 1$
$= 1\cdot 1 - 2(1) + 1 = 0$

(b) $\lim_{x\to 3}\sqrt{\frac{x}{x^2+1}} = \sqrt{\lim_{x\to 3}\frac{x}{x^2+1}} = \sqrt{\frac{\lim_{x\to 3} x}{\lim_{x\to 3}(x^2+1)}}$
$= \sqrt{\frac{3}{\lim_{x\to 3} x^2 + \lim_{x\to 3} 1}} = \sqrt{\frac{3}{\lim_{x\to 3} x \cdot \lim_{x\to 3} x + 1}}$
$= \sqrt{\frac{3}{3\cdot 3+1}} = \sqrt{\frac{3}{10}}$

3. 1/2 5. 3 7. 999/10 9. 10
11. 1 13. Undefined 15. $\sqrt{2}/4$ 17. Undefined
19. 12 21. 1/7

23. (a)

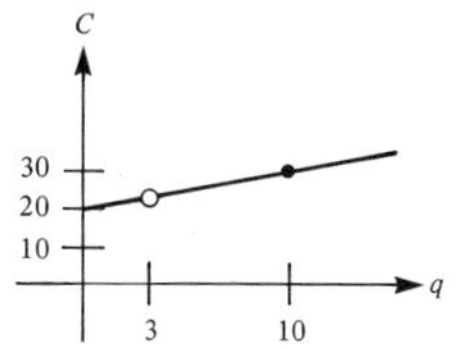

(b) 23

25. (a)

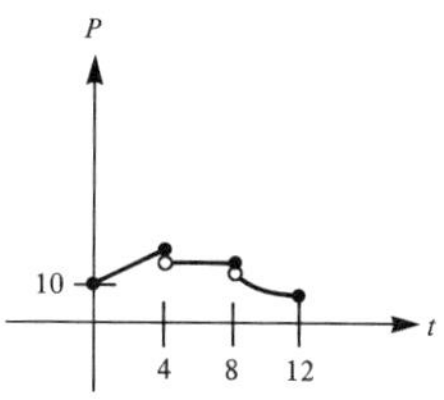

(b) $\lim_{t\to 4^-}(10+t) = 14$

$\lim_{t\to 4^+} 13 = 13$

Problem Set 2.2 (page 52)

1. $f(0)$ is undefined; therefore f is not continuous at $x = 0$
$f(50) = 10$; $\lim_{x\to 50} 500/x = 10$; $f(50) = \lim_{x\to 50} f(x)$. Therefore f is continuous at $x = 50$
3. $h(10) = 5 - 200 + 1000 = 805$; $\lim_{x\to 10} h(x) = 805$;
$h(10) = \lim_{x\to 10} h(x) \to h$ is continuous
5. $G(3)$ is undefined; therefore G is not continuous.
7. $P(2)$ is 9, by hypothesis; $\lim_{q\to 2} P = 2(2) + 5 = 9$;
$P(2) = \lim_{q\to 2} P \to P$ is continuous.
9. $g(\pm 2)$ is undefined; therefore g is not continuous
11. $f(1) = 3$; $\lim_{x\to 1} f(x) = \lim_{x\to 1^-}(x+2) = 3$
or $= \lim_{x\to 1^+} -(x+1)/2 = -1$. Limit does not exist since $3 \neq -1$.
Therefore f is not continuous at $x = 1$ and similarly for $x = 3$.

13. h is not continuous.

15.

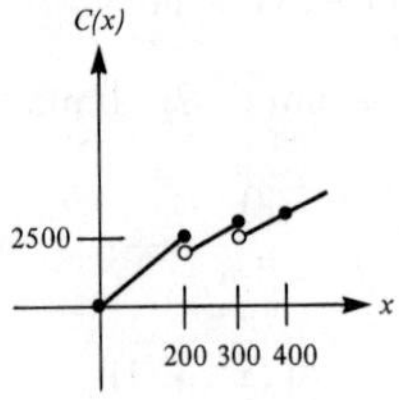

$\lim_{x\to 200^-}(12.5x) = 2500$; $\lim_{x\to 200^+} 10x = 2000$
Therefore, limit does not exist, since $2500 \neq 2000$, and similarly for $\lim_{x\to 300} C(x)$

17. (a)

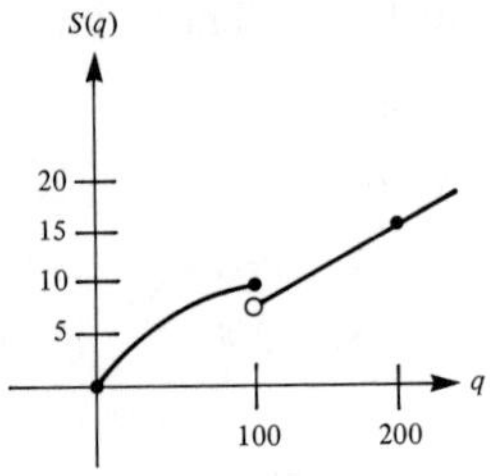

(b) $\lim_{q\to 100} S(q) = \lim_{q\to 100^-} \sqrt{q} = 10$
or $= \lim_{q\to 100^+} q/12 = 100/12 = 25/3$
Therefore, S is not continuous since $10 \neq 25/3$

Problem Set 2.3 (page 59)

1. $f(x) = 3x + 4$; $\dfrac{\Delta y}{\Delta x} = \dfrac{f(x + \Delta x) - f(x)}{\Delta x}$

$$= \frac{3(x + \Delta x) + 4 - f(x)}{\Delta x} = 3;$$

$f'(x) = 3$

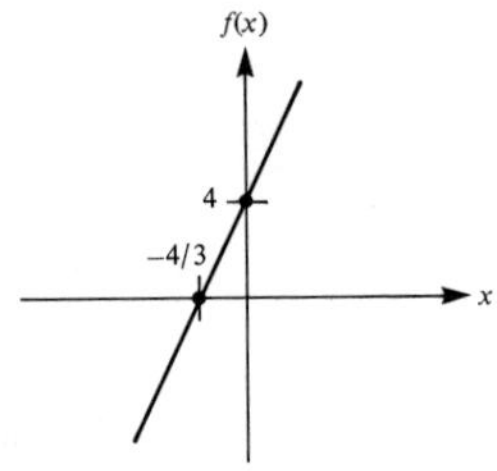

3. $\dfrac{\Delta y}{\Delta x} = \dfrac{(x + \Delta x)^2 - f(x)}{\Delta x} = 2x + \Delta x$;

$f'(x) = 2x$

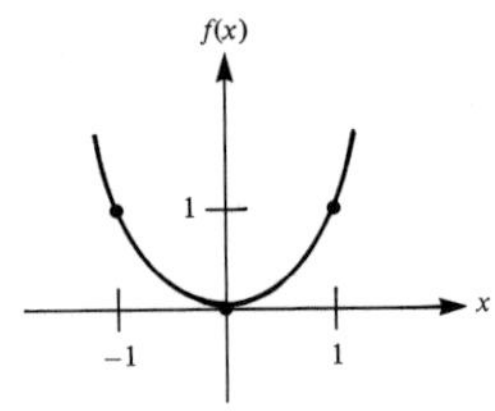

5. $\dfrac{\Delta y}{\Delta x} = 2x + \Delta x + 1$;

$f'(x) = 2x + 1$

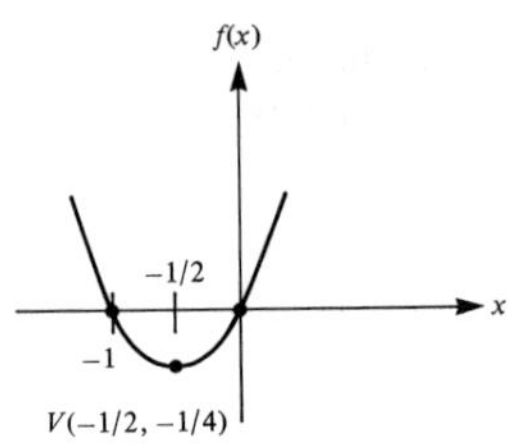

7. $\dfrac{\Delta y}{\Delta x} = 1 - 4x - 2\,\Delta x$;

$f'(x) = 1 - 4x$

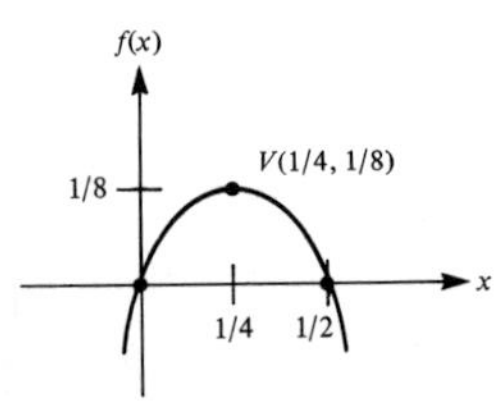

9. $\dfrac{\Delta y}{\Delta x} = 3x^2 + 3x\,\Delta x + \Delta x^2;$

$f'(x) = 3x^2$

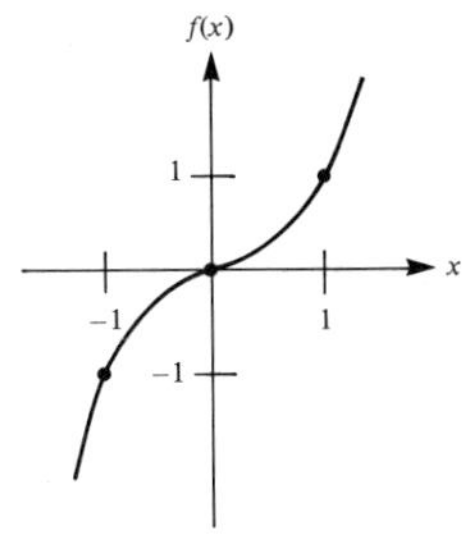

11. $\dfrac{\Delta y}{\Delta x} = \dfrac{1}{\sqrt{x + \Delta x} + \sqrt{x}};$

$f'(x) = \dfrac{1}{2\sqrt{x}}$

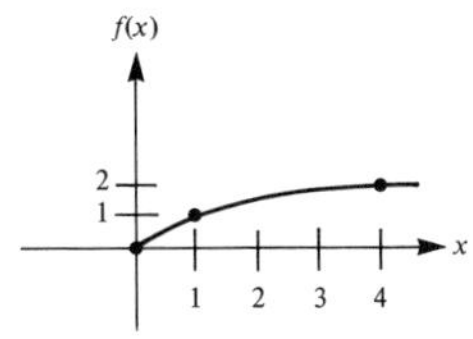

13. $\dfrac{\Delta y}{\Delta x} = \dfrac{-1}{x(x + \Delta x)};$

$f'(x) = -\dfrac{1}{x^2}$

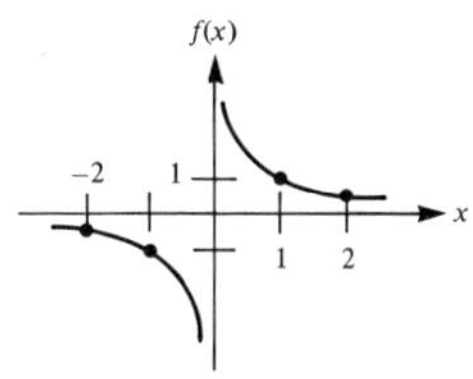

15. $\dfrac{\Delta p}{\Delta d} = 30 - 6d - 3\,\Delta d;\quad p'(d) = 30 - 6d$

17. $f'(x) = -6x;\quad f'(1) = -6$

19. $f'(x) = 8x - 3;\quad f'(3) = 21$

21. $f'(x) = \dfrac{1}{2\sqrt{x + 1}};\quad f'(3) = 1/4$

23. **(a)** $y = x + 4$ at $(2, 6)$; $y = -x + 7$ at $(1, 6)$
(b) $y = 3x + 6$ at $(2, 12)$
(c) $y = 2$ at $(2, 2)$
(d) $y = 0$ at $(0, 0)$; $y = -16$ at $(\pm 2, -16)$

25. **(a)**

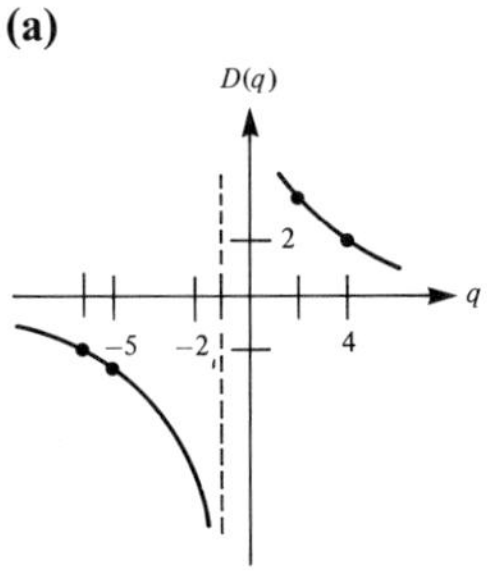

(b) $D'(q) = -10/(q + 1)^2$
$D'(9) = -1/10$

Problem Set 2.4.1 (page 62)

1. $y' = 0$

3. $y' = 0$

5. $f'(x) = -3$

7. $L'(x) = 2x + 2$

9. $f'(x) = 3x - 2/3$

11. $H'(x) = 6x^2 - 6x + 4$

13. $g'(x) = -\dfrac{12}{5x^5} - \dfrac{2}{3x^{3/2}} - \dfrac{5}{6x^{4/3}}$

15. $y' = 4x + 1$

17. $f'(x) = 1 + (1/\sqrt{x})$

19. $h'(t) = 4t^3 + 2t$

21. $G'(x) = 4x - 3 + \dfrac{5}{x^2} - \dfrac{12}{x^3}$ **23.** $C'(x) = 3 - \dfrac{100}{x^2}$; $C'(20) = \dfrac{11}{4}$

25. (a) $C'(x) = 7.68 - 0.03x^2$ **(b)** $C'(x) = 0$ for $x = 16$

Problem Set 2.4.2 (page 67)

1. $f'(x) = (2x - 1)(2x - 1) + (x^2 - x + 1) \cdot 2$; $f'(1) = 3$

3. $f'(x) = x^2 \cdot 2(2x - 3) \cdot 2 + (2x - 3)^2 \cdot 2x$; $f'(2) = 20$

5. $h'(u) = \sqrt{u} \cdot 3(u - 3)^2 \cdot 1 + (u - 3)^3 \cdot 1/(2\sqrt{u})$; $h'(4) = 25/4$

7. $G'(x) = (x^2 + 1) \cdot 3(2x - 1)^2 \cdot 2 + (2x - 1)^3 \cdot 2x$; $G'(1) = 14$

9. $y' = -10/(x + 2)^2$; $y'(8) = -1/10$

11. $g'(t) = \dfrac{(4t + 1) \cdot 2 - (2t - 1) \cdot 4}{(4t + 1)^2}$; $g'(0) = 6$

13. $h'(z) = \dfrac{(3 + z^2)(-2z) - (5 - z^2)(2z)}{(3 + z^2)^2}$; $h'(1) = -1$

15. $\dfrac{dy}{dt} = \dfrac{dy}{dx} \cdot \dfrac{dx}{dt} = 2 \cdot 3 = 6$

17. $\dfrac{dy}{dx} = \dfrac{dy}{du} \cdot \dfrac{du}{dx} = \dfrac{1}{2\sqrt{u}} \cdot \dfrac{1}{3x^{2/3}} = \dfrac{1}{6\sqrt{x^{1/3}} \cdot x^{2/3}} = \dfrac{1}{6x^{5/6}}$

19. $\dfrac{dy}{dz} = \dfrac{dy}{dx} \cdot \dfrac{dx}{dz} = (3x^2 + 3)(-1) = -3x^2 - 3$

21. $\dfrac{du}{dx} = \dfrac{du}{dt} \cdot \dfrac{dt}{dx} = -\dfrac{1}{t^2}(-1) = \dfrac{1}{t^2} = \dfrac{1}{(1 - x)^2}$

$\dfrac{dP}{dv} = \dfrac{dP}{ds} \cdot \dfrac{ds}{dv} = 0.03s^2 \cdot \dfrac{2}{3v^{1/3}} = \dfrac{0.02s^2}{v^{1/3}}$

25. $g(1) = 0 \rightarrow (1, 0)$; $g'(u) = \dfrac{1}{\sqrt{u}} + \dfrac{1}{u^{3/2}}$; $g'(1) = 2 = m$

$G(u) = 2u - 2$, equation of tangent line.

27. $F(x) = x/2 + 2$

29. (a) $R'(8) = \dfrac{1}{6}$ **(b)** $R(x) = \dfrac{x}{4} + \dfrac{1}{4}$

Problem Set 2.4.3 (page 71)

1. $f'(x) = 3x^2 - 4$; $f''(x) = 6x$; $f^{(3)}(x) = 6$

3. $h''(x) = 3(1 - x/2)^2 = 0$ when $1 - x/2 = 0$ or $x = 2$

5. $F(p) = 2 + \frac{3}{2}p^{-1}$; $F'(p) = -\frac{3}{2}p^{-2}$; $F''(p) = 3/p^3$

7. $g''(-2) = -124$; $g^{(5)}(0) = 120$

9. (a) $R'(x) = 15 + 3x - 2x^2$ **(b)** $R''(x) = 3 - 4x$ **(c)** $R''(4) = -13$

11. (a) $M'(x) = -1600/x^2 + 10^{-6}$ **(b)** $x = 4(10^4)$ **(c)** $M''(10) = 16/5$

13. $f''(x) = 12x^2 - 6x$; $f''(1/2) = 0$

15. $a(t) = s''(t) = 4(t^2 - t - 1) + 2(2t - 1)^2$; $a(1) = -2$

Problem Set 2.5 (page 75)

1. $dy = 2\,dx$ **3.** $dy = \dfrac{3}{2\sqrt{x}}\,dx$ **5.** $dy = 4(2x - 3)\,dx$

7. $dy = -\dfrac{4}{(x - 2)^2}\,dx$ **9.** $dy = \dfrac{dx}{\sqrt{2x + 1}}$ **11.** $dy = \left(-\dfrac{3}{t^2} - \dfrac{1}{3}\right)dt$

13. $dy = (3z^2 - 1)\,dz$ **15.** $dy = (5x^4 + 20x^3 + 60x^2 + 120x + 120)\,dx$

17. 8.0625 **19.** 9.98333 **21.** 1.96875 **23.** 0.000102
25. 5/9 of 1° **27.** 5.56% **29.** About 3.704%

Problem Set 2.6 (page 78)

1. $c = 3/2$ **3.** $c = \pm 2\sqrt{3}/3$ **5.** $c = 1$ **7.** $c = 5/2$
9. $c = 3$ **11.** 1.992 **13.** 4.033 **15.** 33.600

Problem Set 2.7 (page 82)

1. 1.91 **3.** 1.78 **5.** 2.73 **7.** 0.31
9. 1.15 **11.** 2.732

2.8 Review Problems (page 82)

1. 7 **3.** 6 **5.** 1 **7.** 2 **9.** 2 **11.** -2
13. Discontinuous at $x = \pm 2$ **15.** Discontinuous at $x = 2$
17. $y = 12x - 17$
19. **(a)** $S'(x) = 2x$ **(b)** $D'(x) = -3x^2$ **(c)** $C'(x) = 1/2\sqrt{x}$
21. $2x(1 - 3x)(1 - 6x)$ **23.** $2x/(x^2 + 1)^2$ **25.** $2/(1 - 2x)^2$
27. **(a)** $v = 3t^2 + 2t;\quad a = 6t + 2$
(b) $v = \dfrac{3}{\sqrt{t}} + \dfrac{3}{2}\sqrt{t};\quad a = -\dfrac{3}{2t^{3/2}} + \dfrac{3}{4\sqrt{t}}$
29. $(x^2 - 2x)/(x - 1)^2$ **31.** $-2/(x^2 + 2x - 1)^{3/2}$
33. $-10/27r^{4/3}$ **35.** $-6(1 - 2z)^2$
37. **(a)** $R'(t) = 10^6/t^2;\quad M'(t) = 1 + 1000t$ **(b)** About 10 years
39. **(a)** $C'(100) = \$40$ **(b)** $x = 0$ or $x \cong 67$
41. **(a)** $r'(s) = 6/s^2$
$r'(2) = 3/2$ or \$150,000
(b)

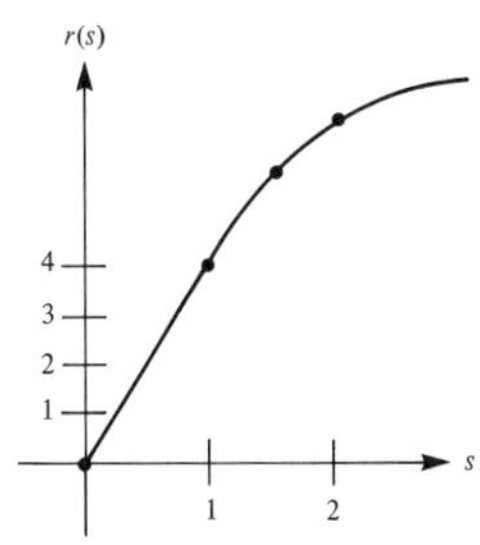

43. **(a)** $C'(1) = 0$ **(b)** $C''(q) = -4 + 12q^2$ **(c)** About 577 units
45. **(a)** $S'(64) = 3$ **(b)** $x = 64$ units
47. $16\frac{2}{3}\%$ **49.** 9.0556 **51.** 3.0093
53. 0.9867 **55.** 7.80 **57.** 3/2
59. $2\sqrt{3}/3$ **61.** $1 + \sqrt{2}$ **63.** 10.954
65. 5.013 **67.** 2.991 **69.** 1.988

2.9 Enrichment Topics and Problems (page 85)

1. Graph of $y = [x]$ for $\$0 \le x \le \$4$

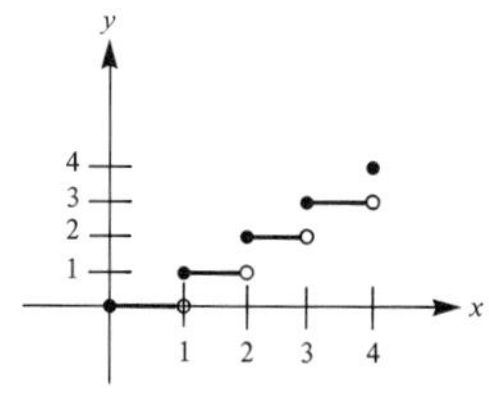

3. $y^2 = (x^2 + x)^2, \dfrac{dy^2}{du} = \dfrac{dy^2/dx}{du/dx} = (x + 1)(2x + 1)$

5. $\lim_{n\to\infty}(\sqrt{n^2+1}-n)=\infty-\infty$: meaningless.

$$\lim_{n\to\infty}(\sqrt{n^2+1}-n)\left[\frac{\sqrt{n^2+1}+n}{\sqrt{n^2+1}+n}\right]=\lim_{n\to\infty}\frac{1}{\sqrt{n^2+1}+n}=0$$

7. 0

9. **(a)** 7.06% for 1971; 8.7% for 1972; 10.71% for 1974.
(b) 11.54% for 1971; 36.36% for 1972; 53.93% for 1974.
(c) $16\frac{2}{3}$% for 1971; 21.05% for 1972; 75.8% for 1974.

11. **(a)** $-1/6$ **(b)** Decrease of 2%

13. 0.000137

15. 8.89%

17. **(a)** 3/2 **(b)** $(3+2\sqrt{3})/3$

19. **(a)** 11.979 **(b)** 32.400 **(c)** 1.037

CHAPTER 3

Problem Set 3.1 (page 97)

1. rel. min at $(2, -8)$

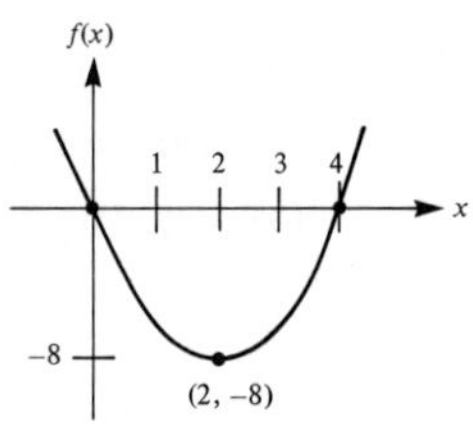

3. rel. max at $(4, 36)$

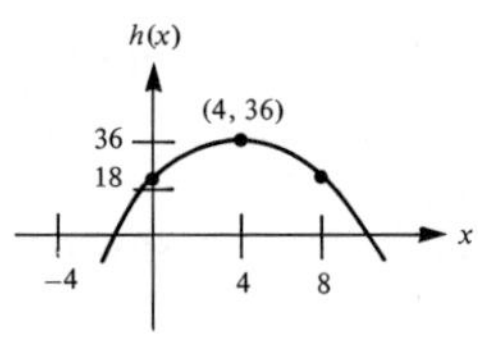

5. rel. max at $(0, 0)$;
rel. min at $(4, -32)$

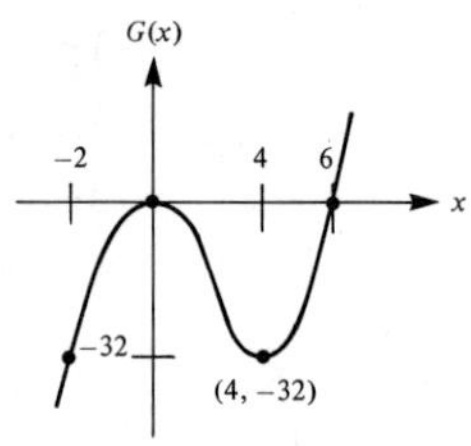

7. rel. min at $(3, -82)$;
rel. max at $(-3, 242)$

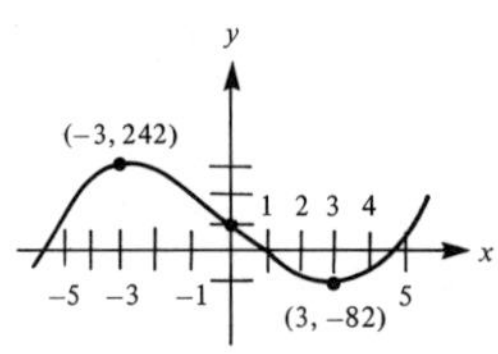

9. rel. min at $(0, 0)$;
rel. max at $(5, 125)$

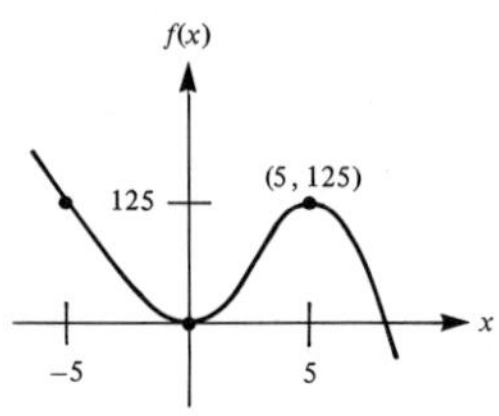

11. rel. max at $(3, 243)$

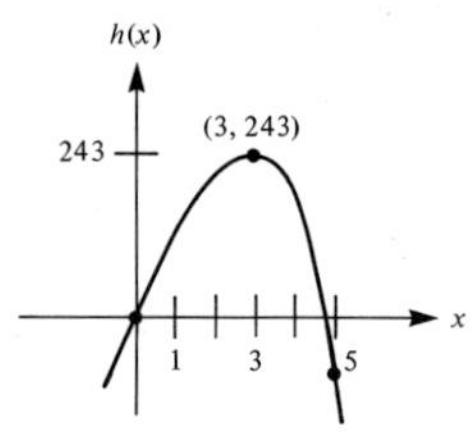

13. rel. min at $(1, 2)$
rel. max at $(-1, -2)$

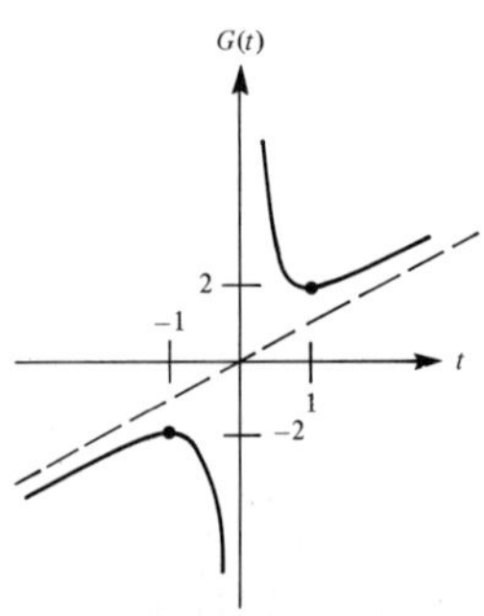

15. No rel. max or rel. min

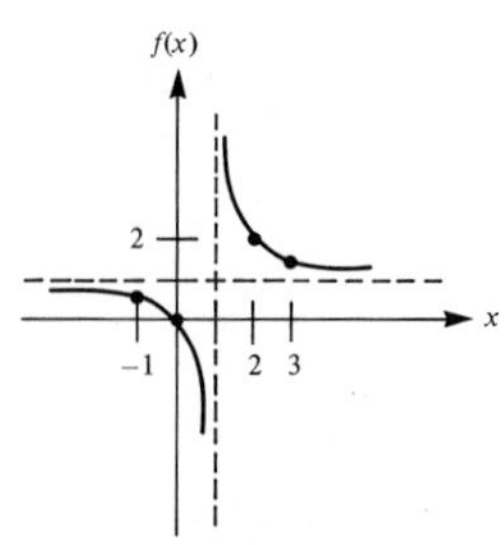

17. rel. max at (0, 1)

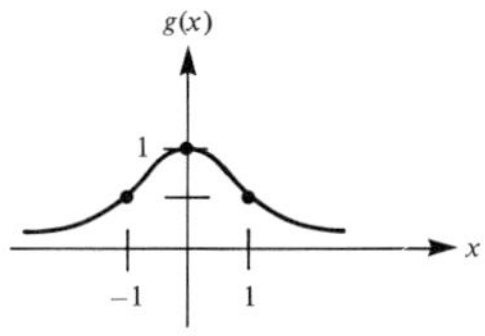

19. rel. max at (0, 3)

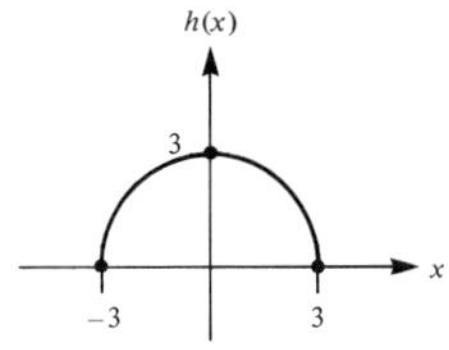

Problem Set 3.2 (page 103)

1. (a) $x \geq 1$ **(b)** No **(c)** Q is min at $x = 1$; q is min at $x = 2\sqrt{2}$

(d) $Q'(x) = 2x - 2$, $Q'(2\sqrt{2}) = 4\sqrt{2} - 2$
$q(x) = x - 2 + 8/x$, $q(2\sqrt{2}) = 4\sqrt{2} - 2 = \text{min}$

(e)

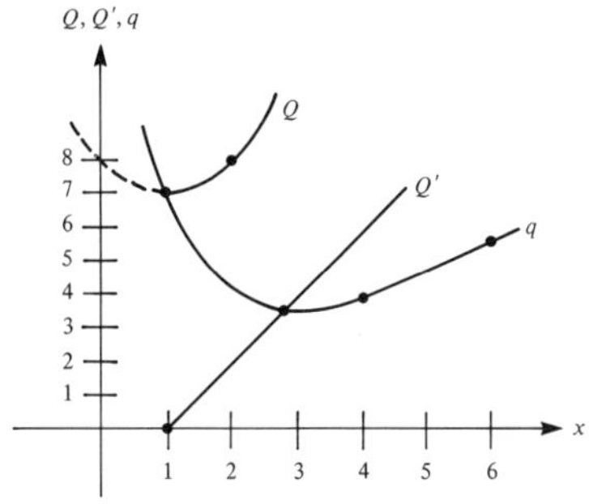

3. $q = \dfrac{z}{4} + \dfrac{9}{z}$

min q at (6, 3)
$Q'(z) = z/2$
$Q'(6) = 3 = \text{min } q$

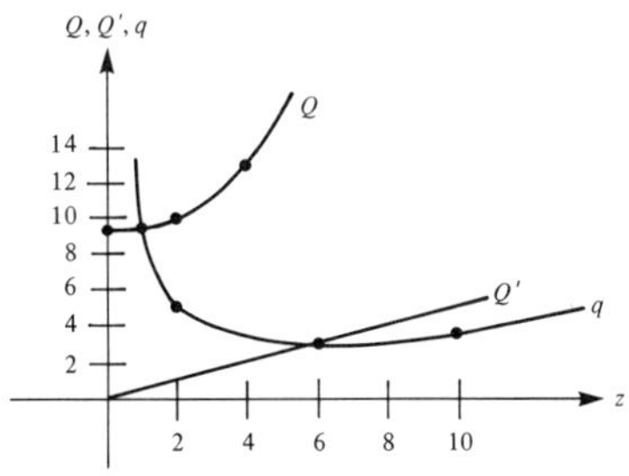

5. $r = 27 - x^2$
max r at (0, 27)
$R'(x) = 27 - 3x^2$
$R'(0) = 27 = \text{max } r$

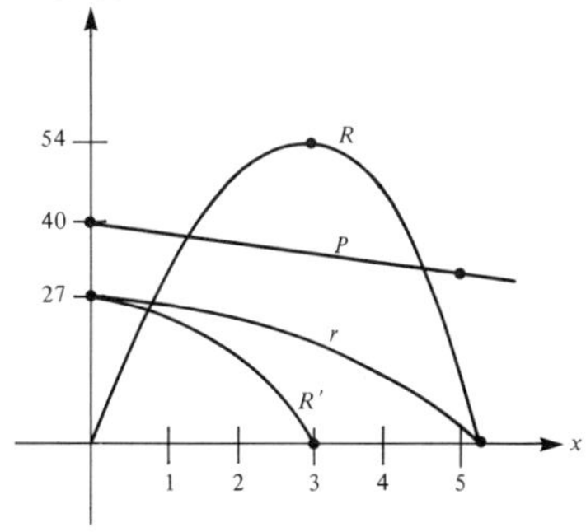

7. min q at $x = 4$

9. max r at $q = 2$

Problem Set 3.3 (page 107)

1. $x = 4000$; min cost = \$4000 **3.** $x = 300\sqrt{10}$ **5.** $x = 1000$
7. $z = 200$ **9.** $x = 1010$ **11.** $x = 50{,}025$

Problem Set 3.4 (page 113)

1. 200 meters by 200 meters **3.** 40 employees **5.** 4 m by 4 m by 4 m
7. 3 m by 3 m by 3 m **9.** $R' = Q' \rightarrow 60 - q = 2, q = 58$ units

11. min $C = \$400$

13. 60 m by 25 m

15. **(a)** $P(x) = -40 + 8x - 0.02x^2$ **(b)** 200 units
(c) $R' = Q' \to 10 = 2 + 0.04x$, $x = 200$ units

3.5 Review Problems (page 115)

1. rel. max at $(-1, 5/3)$;
rel. min at $(3, -9)$;
inflection pt. at $(1, -11/3)$

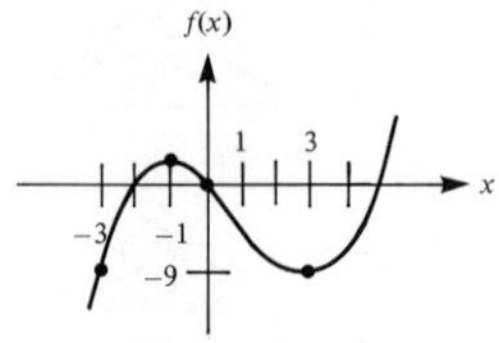

3. rel. max at $(0, 0)$;
rel. min at $(\pm 2, -24)$;
inflection pts. at $(\pm 2/\sqrt{3}, -40/3)$

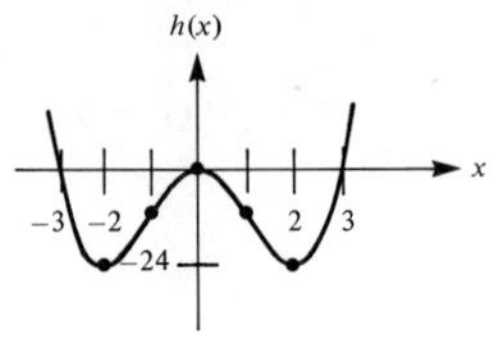

5. rel. max at $(0, 4)$;
inflection pts. at $(\pm 1/\sqrt{3}, 3)$

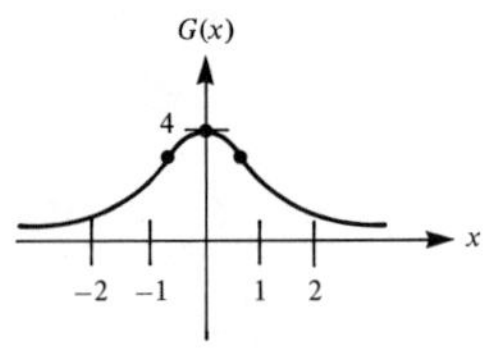

7. rel. min at $(1, 2)$;
rel. max at $(-1, -2)$;
no inflection pts.

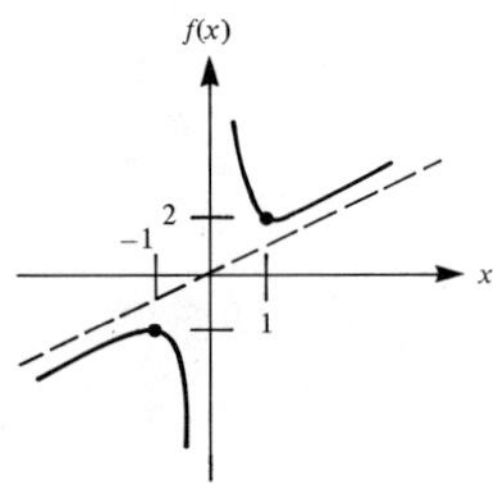

9. rel. max at $(0, 4)$;
no inflection pts.

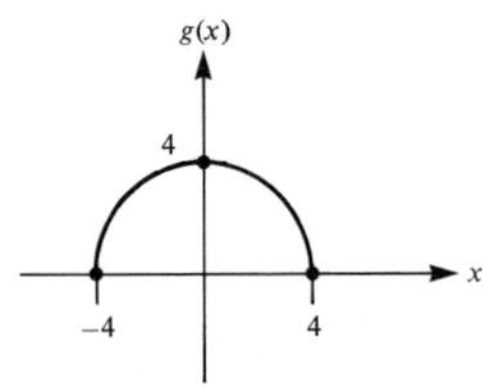

11. $\sqrt[3]{10}$ m $\times$ $\sqrt[3]{10}$ m $\times$ $30/10^{2/3}$ m

13. 150 passengers

15. min Q at $(1, 24)$;
min q at $(3, 12)$;
$Q'(3) = 12 = \min q$

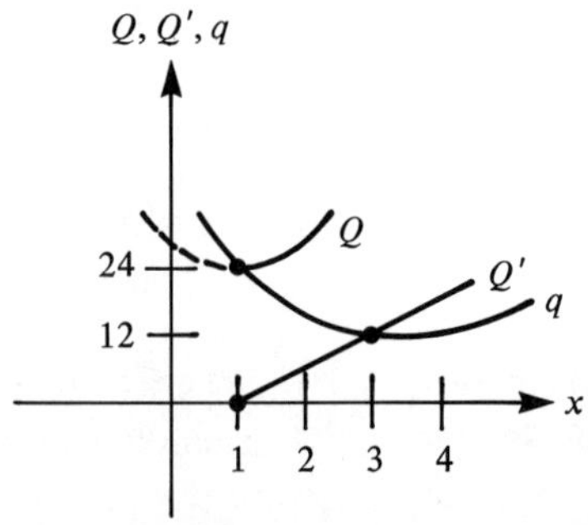

17. $x = \sqrt{\dfrac{2CD}{B + 2E + 2F}}$ 19. min A for radius $x = 2$ cm, ht. $y = 4$ cm.

3.6 Enrichment Topics and Problems (page 116)

1. rel. max at (0, 3); rel. min at (2, 3/5)
3. rel. max at $(-2, 22/3)$; rel. min at $(3, -27/2)$
5. rel. max at $(-3, -6)$; rel. min at (3, 6)
7. Absolute min $= -2/3$; absolute max $= 98$
9. (2, 1) 11. \$1,719,600
13. (a) 26 cm^2/sec (b) 128π cm^3/min (c) $-1/12\pi$ m/hr

CHAPTER 4

Problem Set 4.1 (page 126)

1. (a)

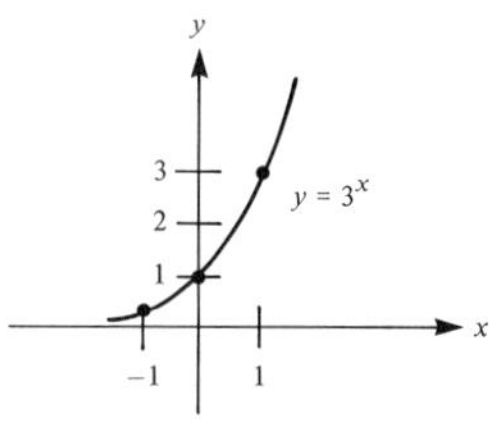

(b)

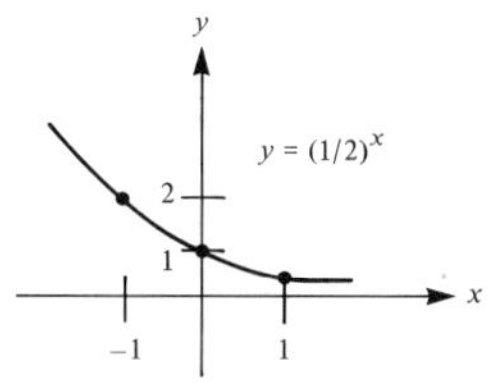

(c)

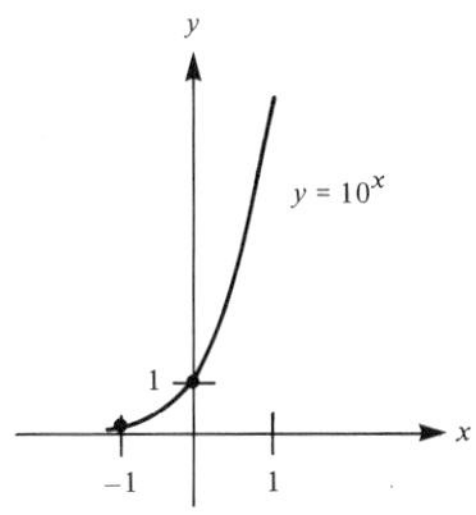

(d)

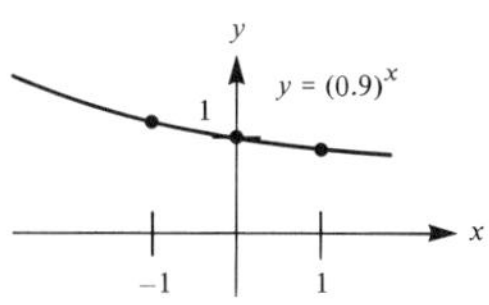

(e)

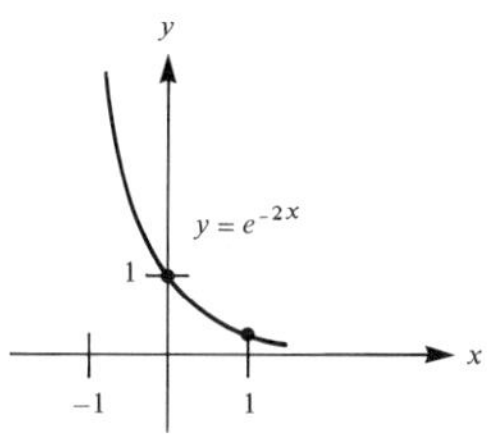

(f)

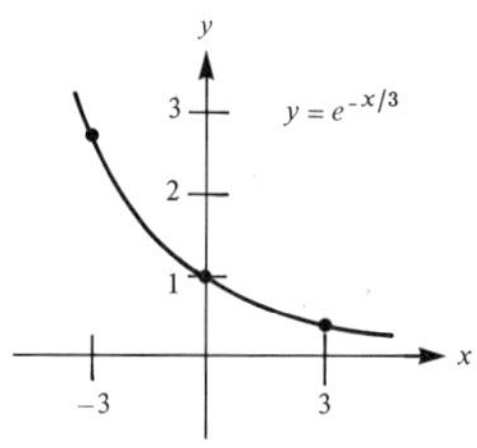

3. \$3466.60 5. \$4852.00 7. Optimum = 200; about 86.5%

9. $S(10) = 190.4$;
$S(20) = 362.6$

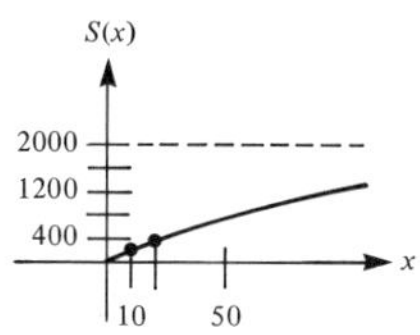

11. \$22,408.50
13. (a) $2e^x(1 - 2e^x)$ (b) $3xe^{-3x}(2xe^x + 3)$
(c) $(e^x - 4)(e^x + 3)$ (d) $(e^t + 5)(e^t - 3)$

Problem Set 4.2 (page 134)

1. (a) $x = \log_{10} y$ (b) $1/2 = \log_{16} 4$ (c) $-2 = \log_2(1/4)$
(d) $2/3 = \log_{64} 16$ (e) $1/2 = \log_{81} 9$ (f) $-3 = \log_{10} 0.001$
3. (a) $y = \frac{1}{2} \ln x$ (b) $y = \ln 2x$ (c) $y = 2^x$
(d) $y = -\ln 2x$ (e) $y = 1 + \ln x$

5. **(a)** $\dfrac{\log 2}{\log 3}$ **(b)** $\dfrac{\log 5}{\log 4}$ **(c)** $\dfrac{\log 27}{\log 2}$ **(d)** $\dfrac{\log 57}{\log 5}$

7. **(a)** 2.1303 **(b)** 0.9667 − 1 **(c)** 0.1900 − 1 **(d)** 0.0629

9. $x = 2/5$ 11. $x = 2$

13. **(a)** 1.847 **(b)** 12.189 **(c)** 0.2575

15. 7.27 years 17. 3.8 years 19. $x = 0$

21. **(a)** 64 **(b)** 1/2 **(c)** 2
(d) $-4/3$ 23. 15.26 years

Problem Set 4.3 (page 138)

1. $2/x$ 3. $\dfrac{4u^2+2}{u^3+u}$ 5. $\dfrac{t}{1+t^2}$ 7. $\dfrac{1}{3}\left(\ln\dfrac{3}{x} - 1\right)$

9. $\dfrac{dy}{dt} = f(t)\left[\dfrac{2}{t} + \dfrac{1}{1-2t} - \dfrac{9}{3t-1}\right]$ 11. $2h(x)\left[1 - \dfrac{e^x}{1-e^x} - \dfrac{e^x}{1+e^x}\right]$

13. Min C at (0, 0) 15. Max P at $x = e$, or about 2718 items.

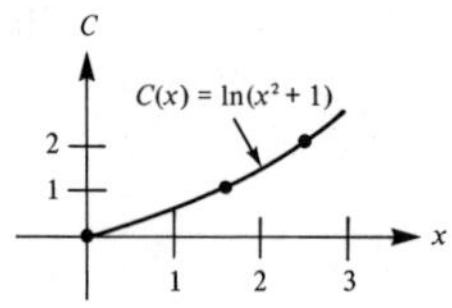

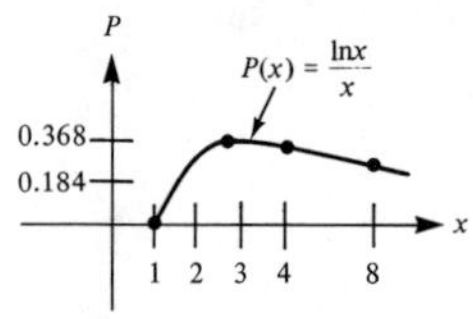

Problem Set 4.4 (page 141)

1. $9e^{3x}$ 3. $-2e^{1-2x}$ 5. $-2e^{2-t}$ 7. $e^w/(1-e^w)^2$

9. 3 11. $3e^u(e^u-1)^2$ 13. $3e^{3t} - e^{1-t}$ 15. $e^x(2+x)$

17. $e^{-x}\left(\ln x - \dfrac{2}{x} - \dfrac{1}{x^2}\right)$ 19. $9e^{3v} - e^v$

21. **(a)** $R(p) = 1000pe^{-0.2p}$ **(b)** $R'(p) = 1000e^{-0.2p}(1-0.2p)$
(c) $R''(10) = 0$

23. **(a)** rel. min at $(-1, -e^{-1})$ **(b)** rel. min at (1, 1)
(c) rel. max at (0, 1)

25. **(a)** **(b)** min C at $x = e^{-1}$ millions

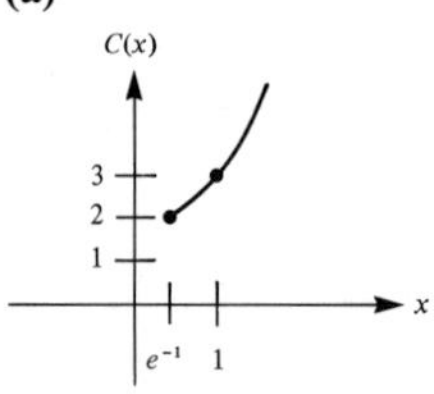

27. max R at $q = 200$ items

4.5 Review Problems (page 142)

1. **(a)** **(b)** **(c)**

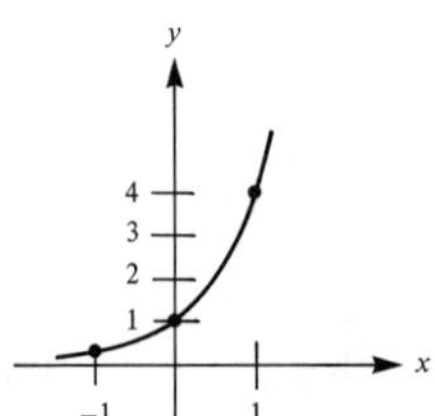

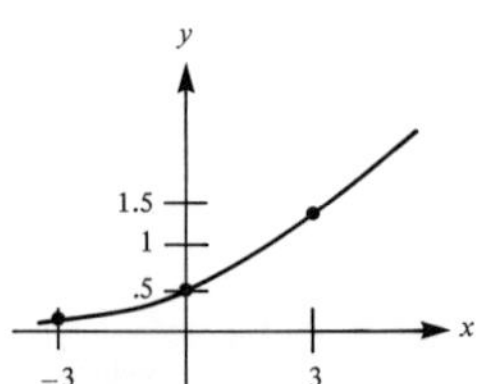

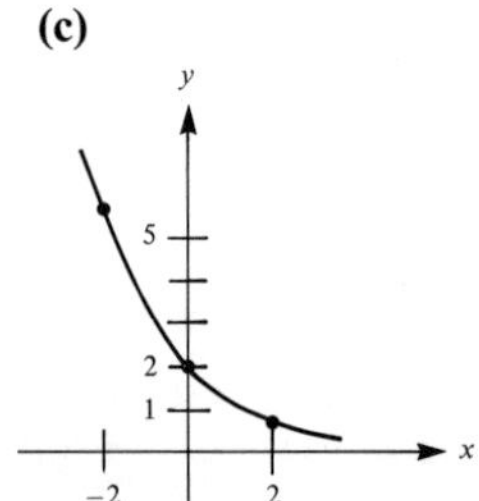

(d)

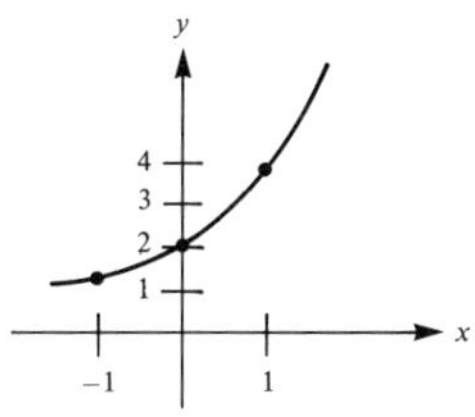

(e)

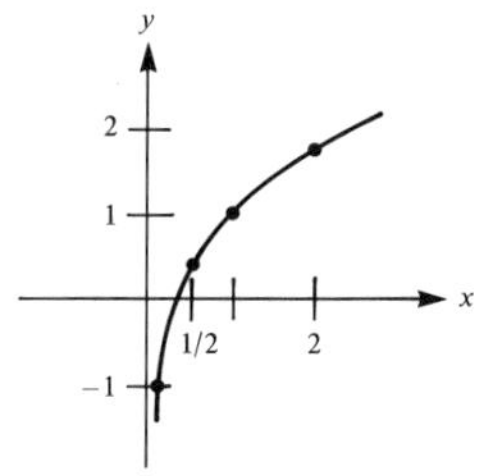

(f)

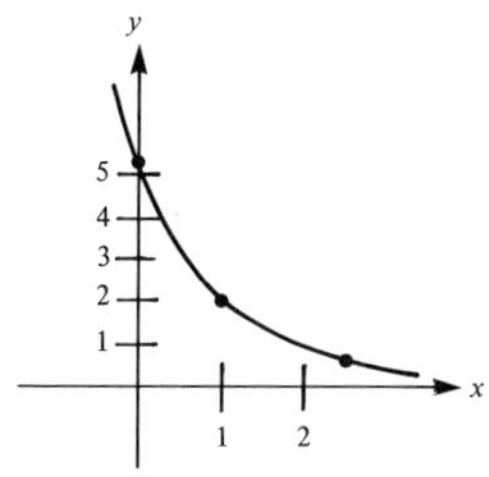

3. 12.6 years **5.** \$1986.40 **7.** Optimum = 300; 95% in 15 days

9. **(a)** $\dfrac{40e^{-x} + 45e^{2x} - 12e^{-2x}}{60}$ **(b)** $\dfrac{e^x + xe^x + 2 - x - x^2}{e^{2x}}$

11. **(a)** $x = \ln y - 1$ **(b)** $x = \ln(1 + \sqrt{2})$ **(c)** $x = 1$
(d) $x = 2e^2$ **(e)** $x = e^3 - 1$ **(f)** $x = 0$ or 1

13. **(a)** $10^x = 0.01$ **(b)** $10^{0.3010} = 2$ **(c)** $e^{0.03} = 1.03$
(d) $5^{2/3} = \sqrt[3]{25}$ **(e)** $2^{-4} = x$

15. **(a)** 99.65 **(b)** 130.6 **(c)** 9.982

17. **(a)** $x = 3$ **(b)** $x = 2$ or 5 **(c)** $x = 1 + \sqrt{1 + e}$
(d) $x = -4$ or 2

19. 9.01 years **21.** About 22.4 years **23.** $f'(x) = \dfrac{2x + 1}{x^2 + x}$

25. $g'(t) = \dfrac{2t}{1 - t^2}$ **27.** $F'(x) = (e^{3x}/x) + 3e^{3x} \ln 4x$

29. $H'(u) = 4(e^u + e^{-u})^{-2}$ **31.** $f'(x) = \dfrac{-e^{1/x}}{x^2}\left(x + \ln \dfrac{1}{x}\right)$

33. $h'(u) = \dfrac{2e^u}{(1 - e^u)^2}$ **35.** $G'(v) = 1 + \ln v$

37. $y' = 1 + \dfrac{x}{x^2 - 1} - \dfrac{2x}{x^2 + 1}$ **39.** $y' = y\left[\dfrac{6}{2x - 1} - \dfrac{12}{1 - 3x} - \dfrac{1}{2x}\right]$

41. **(a)** $e^{2x-1}\left(\dfrac{2}{x^2} - \dfrac{8}{x} - 8 \ln x\right)$ **(b)** $2x^{-3}$
(c) $-\frac{1}{4}x^{-3/2}$ **(d)** $-\frac{1}{4}(x - 1)^{-3/2}$

4.6 Enrichment Topics and Problems (page 145)

1. **(a)** P_{max} at $x = e$, or about 2718 units **(b)** Graph of P, $x \geq 1$

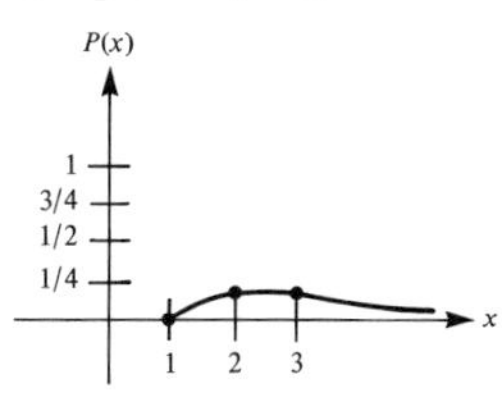

3. About 17.3 years **5.** 4%

7. (a) $y = 3^{|x|}$

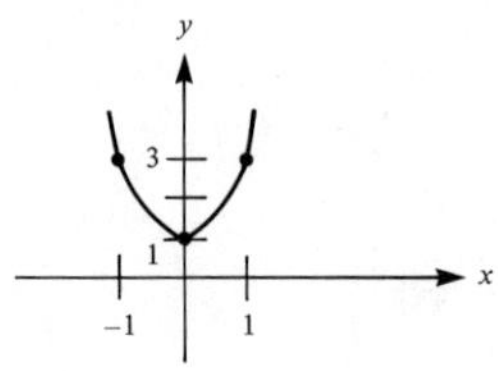

(b) $y = (\frac{1}{2})^{|x|}$

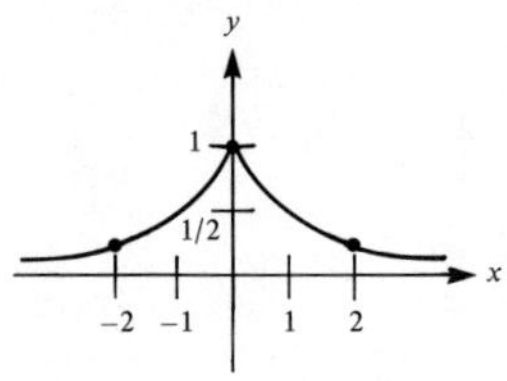

(c) $y = e^{-x^2}$

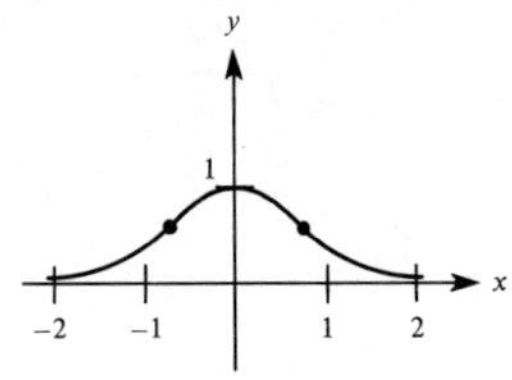

(d) $y = \ln(1/x)$

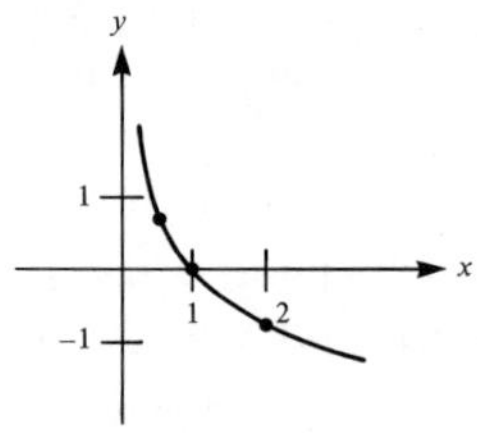

(e) $y = 10 - 2e^{-x}$

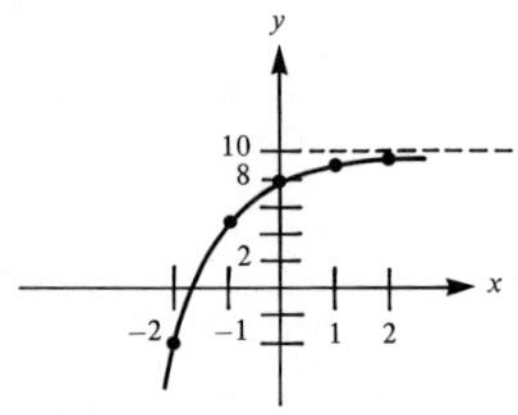

9. \$2983.60 **11.** \$3679 **13.** About 11 years **15.** Max C at $p = \$200$

CHAPTER 5

Problem Set 5.1, 5.2 (page 153)

1. (a)

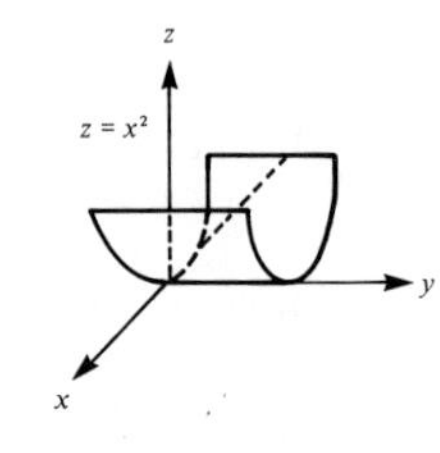

(b)

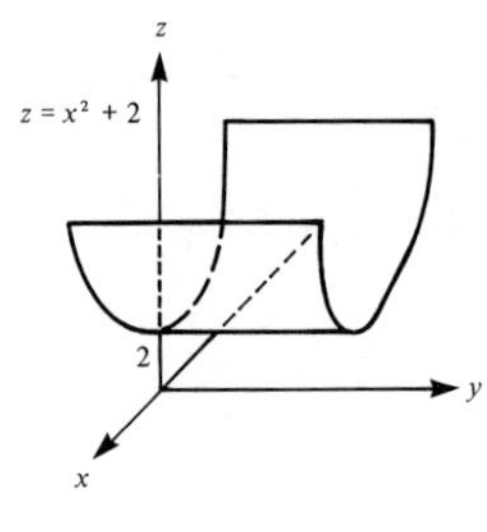

(c)

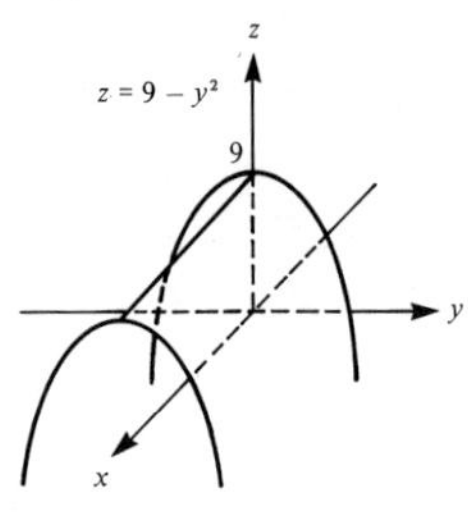

3. (a)

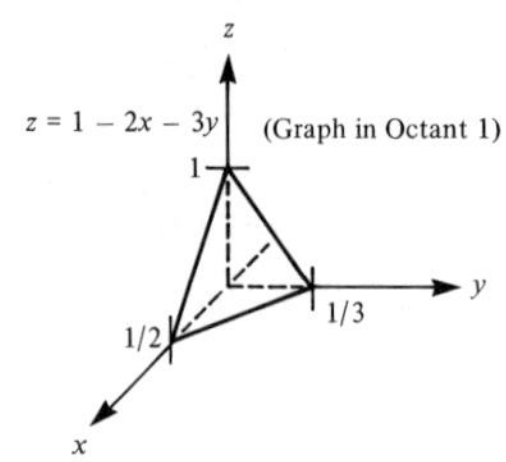

(b)

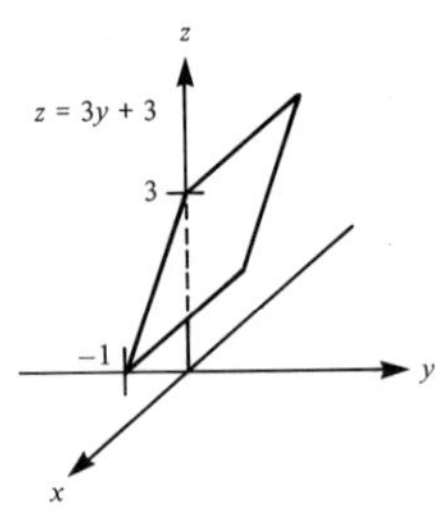

(c)

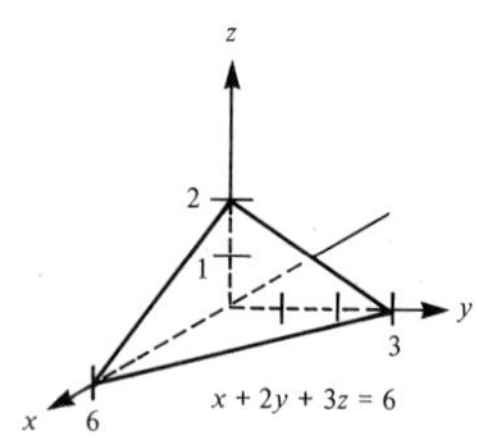

5. $f_x = 4xy + 3y^2; \quad f_y = 2x^2 + 6xy$ **7.** $w_x = \dfrac{y}{xy + 1}; \quad w_y = \dfrac{x}{xy + 1}$

9. $F_x = \dfrac{1}{y} - \dfrac{y}{x^2}; \quad F_y = -\dfrac{x}{y^2} + \dfrac{1}{x}$ **11.** $h_x = 2xy^2e^{x^2}; \quad h_y = 2ye^{x^2}$

13. $V_x = \dfrac{y + e^x - (x + e^y)e^x}{(y + e^x)^2}$; $V_y = \dfrac{(y + e^x)e^y - (x + e^y)}{(y + e^x)^2}$

15. $u_x = ze^{xz}$; $u_y = -ze^{yz}$; $u_z = xe^{xz} - ye^{yz}$

17. $P_y = x + 3z - 4y$ indicates the rate of change of profits with respect to the number of gasoline pumps when the number of employees and the value of other supplies are held fixed; $P_z = 2x + 3y - 2z$ has a similar interpretation.

Problem Set 5.3 (page 157)

1. $f_x = 1 + y, f_{xx} = 0$; $f_y = x, f_{yy} = 0$; $f_{xy} = 1, f_{yx} = 1$

3. $u_x = 6x^2y^2 - 6xy, u_{xx} = 12xy^2 - 6y$; $u_y = 4x^3y - 3x^2, u_{yy} = 4x^3$; $u_{xy} = 12x^2y - 6x = u_{yx}$

5. $w_x = \frac{1}{2}(x + y)^{-1/2}, w_{xx} = -\dfrac{1}{4(x + y)^{3/2}} = w_{xy} = w_{yx}$

7. $G_x = \ln y \cdot e^x = G_{xx}$; $G_y = e^x/y, G_{yy} = -e^x/y^2$; $G_{xy} = e^x/y = G_{yx}$

9. $u_x = e^y, u_{xx} = 0$; $u_y = (x - 1)e^y = u_{yy}$; $u_{xy} = e^y = u_{yx}$

11. $z_x = 2e^{2x-3y}, z_{xx} = 4e^{2x-3y}$; $z_y = -3e^{2x-3y}, z_{yy} = 9e^{2x-3y}$; $z_{xy} = -6e^{2x-3y} = z_{yx}$

13. $w_x = -\dfrac{1}{x^2} \cdot \ln \dfrac{1}{y}, w_{xx} = \dfrac{2}{x^3} \cdot \ln \dfrac{1}{y}$; $w_y = -\dfrac{1}{xy}$; $w_{yy} = \dfrac{1}{xy^2}$; $w_{xy} = \dfrac{1}{x^2y} = w_{yx}$

15. $z_{xyy} = 4$; $z_{yxx} = 6$ 17. $z_{xyy} = xe^{xy}(xy + 2)$; $z_{yxx} = ye^{xy}(xy + 2)$

19. **(a)** $P_l(30, 4) = -30$; $P_k(30, 4) = 182$
(b) $P_{ll}(6, 2) = -2$; $P_{kk}(6, 2) = -2$; $P_{lk}(6, 2) = 5 = P_{kl}(6, 2)$

21. $du/dx = 5x - 5y$ 23. $\frac{5}{2}xe^{x^2y}$

25. $e^{x+2y-z}(1 - 4x - 1/x)$ 27. $dP/da = 1/9$

Problem Set 5.4 (page 161)

1. $-\dfrac{2x + 2}{2y - 3}$ 3. $-\dfrac{x - xy^2}{y + 1 - x^2y}$

5. $-\dfrac{e^x(1 - \ln y)}{y^{-1}(1 - e^x)}$ 7. $-\dfrac{2x + 2y - 1}{2x + 2y + 1}$

9. **(1)** -6 **(2)** $4/5$ **(3)** $-6/5$ **(4)** -15

11. $dr/ds = -12/11$ 13. $3y = x - 1$ 15. $2/(y - 1)$

Problem Set 5.5 (page 165)

1. rel. min at (2, 2) 3. No extrema

5. rel. max at $(1, -2)$ 7. rel. min at $(e, 1)$

9. $C_{\min} = \$28{,}000$ at $P(10, 8)$ 11. 5 m × 5 m × 5 m; min $A = 150$ m²

13. max $P = \$10{,}000$ 15. CP at (4, 4, 3); min $C = \$720$

Problem Set 5.6 (page 171)

1. min cost at $(k, w) = (1, 0)$

3. max sales at $(c, w) = (5, 4)$; $S(5, 4) = \$689$

5. min $E = E(6, 6) = \$90$ 7. $P_{\max} = P(8, 7) \cong \$6{,}666{,}667$

9. $CP(1/2, 1, -1/2)$; $d_{\min} = \sqrt{6}/2$ 11. 3 ft × 3 ft × 3 ft

13. max P at (2, 3, 4); $P(2, 3, 4) = 1{,}000{,}000$ units.

Problem Set 5.7 (page 176)

1. **(a)** $y = \frac{3}{2}t + \frac{1}{3}$ **(b)** About \$6.33

3. $y = \frac{10}{11}x + 30$ 5. $y = \frac{65}{2}x + \frac{80}{3}$; for $x = 4$ (1975), $y \cong \$156.67$

7. $y = 0.35x + 2.3/3$; for $x = 4$ (1973), $y \cong \$2.17$

9. $y = 3x + 22/3$; for $x = 4$ (1980), $y = 19\frac{1}{3}$ million

11. $y = 3x + 242/3$; for $x = 4$ (Test 4), $y \cong 93$

13. $y = -0.01x + 1.1$; for $x = 5$ ($A = 500{,}000$), $y = \$1{,}095{,}000$.

5.8 Review Problems (page 177)

1. (a)

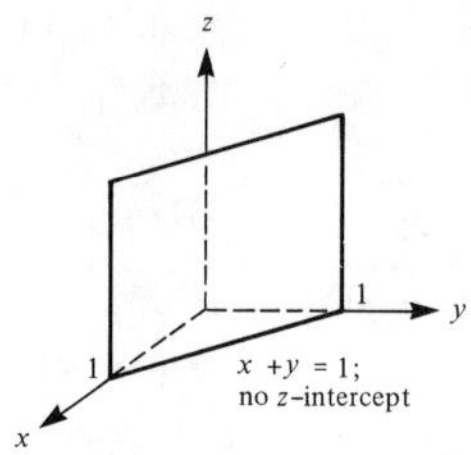

(b)

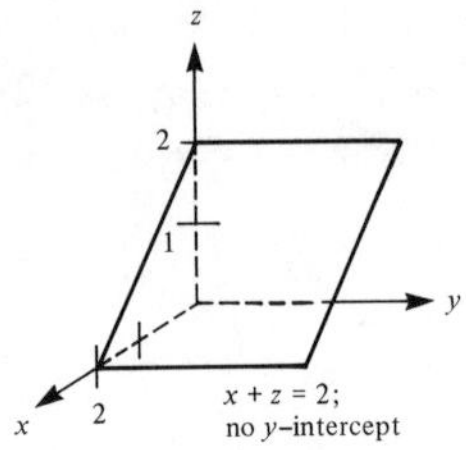

(c)

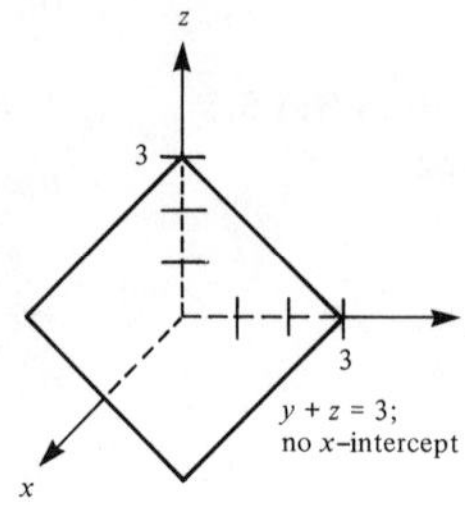

(d)

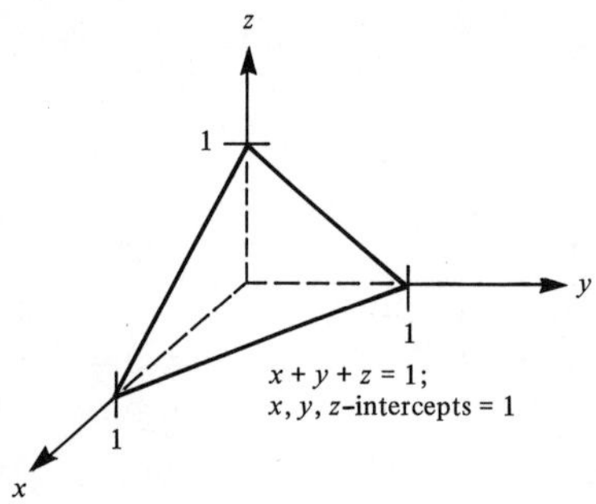

(e)

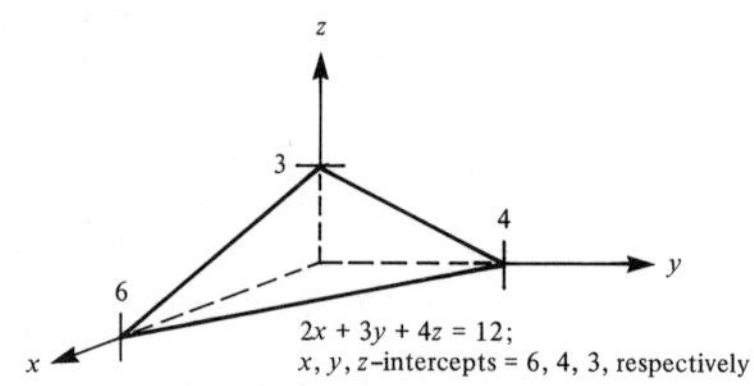

3. (a) $P_k(2, 1) = 6$; $P_l(2, 1) = 2$ **(b)** $P_{kk} = 4$; $P_{ll} = 6$; $P_{lk} = -2$

5. (a) $P(w, s) = w^2 + s^2 - ws - 8w - 2s + 10$
(b) $P_s = 2s - w - 2$; $P_s(2, 4) = 4$ **(c)** $P_{ww} = 2$; $P_{ss} = 2$; $P_{ws} = -1$

7. (a) $G_{uu} = ve^u$; $G_{vv} = ue^v$; $G_{uv} = e^v + e^u$
(b) $G_{uu} = e^{u-v} + 2v/u^3$; $G_{vv} = e^{u-v}$; $G_{uv} = -e^{u-v} - 1/u^2$
(c) $G_{uu} = 24(2u + 3v)$; $G_{vv} = 54(2u + 3v)$; $G_{uv} = 36(2u + 3v)$
(d) $G_{uu} = -v^2/(1 - uv)^2$; $G_{vv} = -u^2/(1 - uv)^2$; $G_{uv} = -1/(1 - uv)^2$

9. (a) $y = -x + 2$ **(b)** $y = 2x - 5$
(c) $y = -2x + 10$ **(d)** $y = -x/2 + 9/2$

11. $\dfrac{ds}{dt} = -\dfrac{-2e^{x^2+y^2}}{t^3}(xt + 2y)$

13. max $P = \$127$ at $(u, v) = (6, 5)$ **15.** rel. max at $(2/3, 4/3)$

17. (a) $2x - 4xy$ **(b)** $\dfrac{1 + 2x}{2\sqrt{x + y}}$

(c) $e^{x-2y}(1 - 2/x)$ **(d)** $\dfrac{x + 2}{x(x + 2y)}$

19. 3 ft × 3 ft × 3 ft

21. $x = \$20{,}000$, $y = \$10{,}000$, $s(20, 10) = 20{,}000$ units

23. Critical point $(4, -4)$ **25.** $x = 44$, $y = 22$

27. $y = 0.2x + 2.8/3$; for $x = 4$ (1975), $y \cong 1.73 \times 10^8$

29. $y = 1.9x + 4$; for $x = 5$ (1975), $y = 13.5$ thousands; for $x = 6$ (1976), $y = 15.4$ thousands.

5.9 Enrichment Topics and Problems (page 179)

1. **(a)** 0 **(b)** 18 **(c)** 0 **(d)** 1/8 **(e)** 1/2

3. max P at (3, 2); $P(3, 2) = \$45$

5. $(x, y, z) = (1/2, 0, 1/2)$

7. Critical point (2, 1, 7/2); max $P = P(2, 1, 7/2) = \$8000$

9. $P_{max} = P(2, 4, 4) = \$9600$ **11.** About 101

13. $y = 0.38x - 0.1$; for $x = 5$ (1975), $y = 1.80$

15. $|\Delta V| = 0.162008$; $dV = 0.16$

17. About 217.53 **19.** 24¢

21. relative error = 0.01; percentage error = 1%

23. relative error = 0.0024; percentage error = 0.24%

25. **(a)** 0 **(b)** 1 **(c)** ∞

CHAPTER 6

Problem Set 6.1 (page 189)

1. **(a)** $\frac{3x^2}{2} - 4x + C$ **(b)** $\frac{x^2}{6} + \frac{5x}{3} + C$

(c) $\frac{x^3}{3} + 2x + C$ **(d)** $x - \frac{2x^3}{3} + C$

(e) $\frac{4x^3}{3} - \frac{x^2}{2} + x + C$ **(f)** $x + x^2 - 2x^3 + C$

3. $Q(x) = 10x - 6x^2 + 5x^3 + 100$ **5.** $P(q) = q^4 + q^3 + 20$

7. $2u - \ln|u| + C$ **9.** $2t - t^3 + t^4 + C$

11. $\frac{x^2}{2} - x + 3\ln|x| + C$ **13.** $(x - 2)^3/3 + C$

15. $-\frac{1}{8}(1 - 2v)^4 + C$ **17.** $\frac{x^5}{5} - \frac{x^4}{4} + 6\sqrt{x} + C$

19. $S(q) = \frac{2}{9}(3q - 2)^{3/2} + 74/9$ **21.** $D(x) = 2\ln|5x + 1| + 1$

23. $S(q) = \frac{q^3}{6} + q + 10$

Problem Set 6.2 (page 194)

1. $y = (x^3/3) - 2x + C$ **3.** $y^2 = 4t^2 + C$

5. $\sqrt{y} = \frac{1}{3}x^{3/2} + C$ **7.** $G(s) = ks^2/2 + 2k$

9. $S(q) = \frac{Mq^2}{2} + (1 - M)q + 2$ **11.** $P(c) = \frac{200}{4c + 10} + 396$

13. $y + \frac{3}{2}y^2 = t + \frac{3}{2}t^2 + \frac{5}{2}$ **15.** $y = 1 - 1/x^2$

17. $y - 2y^2 = t^2 - t - 3$

Problem Set 6.3 (page 198)

1. $y = x^2 - x + 1$ **3.** $y = \frac{1}{6}(x^4 - x^3 + 5x + 1)$

5. $y = -6\ln|x| + 7x - 6$ **7.** $P(x) = \frac{x^3}{3} - \frac{x^4}{4} + x - \frac{1}{12}$

9. $S(x) = \ln|x + 1| + 2$

11. $P(t) = -36\sqrt{t} + 40t - 80$

13. $R(t) = \dfrac{t^4}{4} + \dfrac{2t^3}{3} + \dfrac{5t^2}{2} + C_1 t + C_2$

15. $R(y) = (3y - 2)^6/270 + C_1 y + C_2$

17. $H(q) = (1 - 3q)^{10}/30 + C_1 q + C_2$

Problem Set 6.4 (page 201)

1. $P(x) = \dfrac{(x^2 - 1)^4}{8} + C$

3. $R(x) = 2\left[\dfrac{(x + 1)^{5/2}}{5} - \dfrac{(x + 1)^{3/2}}{3}\right] + C$

5. $T(x) = -2[\sqrt{x} + \ln|\sqrt{x} - 1|] + C$

7. $V(t) = \dfrac{(t + 1)^7}{7} - \dfrac{(t + 1)^6}{6} + C$

9. $X(u) = \frac{1}{2}\ln(u^2 + 1) + C$

11. $F(x) = (x^2 - 6x)^3/6 + C$

13. $H(t) = (t^3 - 3t)^4/12 + C$

15. $P(x) = 2\sqrt{x} + x + C$

Problem Set 6.5.1 (page 208)

1. (a, b)

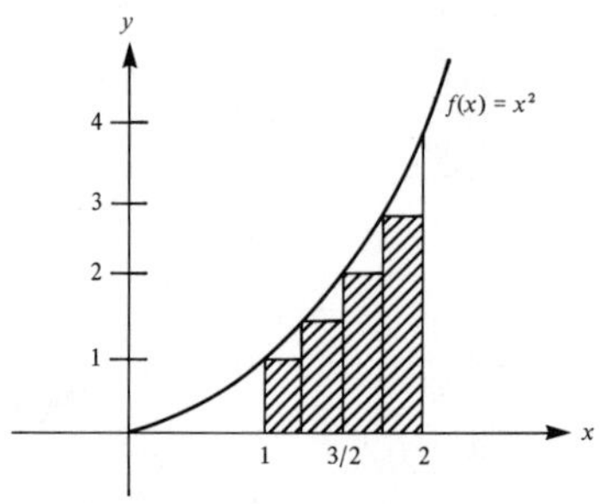

(c) $A = \frac{63}{32}$ units^2
(d) $\frac{63}{32} = \frac{35}{96}$ less than $\frac{7}{3}$

3. (a, b)

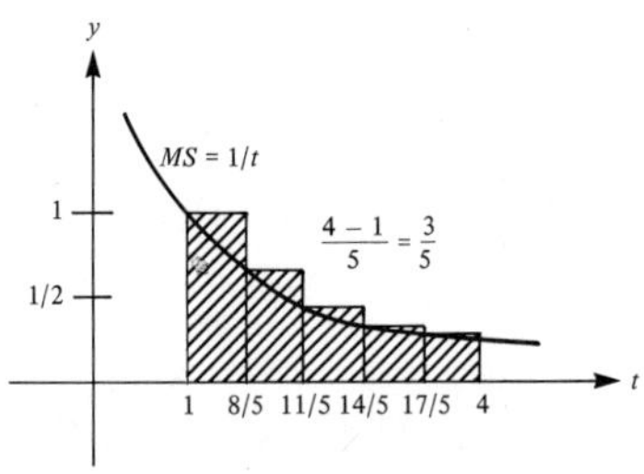

(c) $A \cong 1.64$ units^2
(d) 1.64 overestimates by 0.26 units^2

5. (a, b)

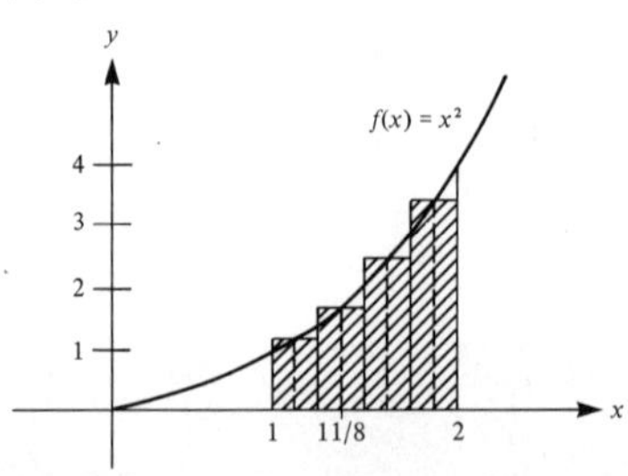

(c) $A = \frac{149}{64}$ units^2
(d) $\frac{149}{64}$ underestimates by $\frac{1}{192}$ units^2

7. **(a, b)**

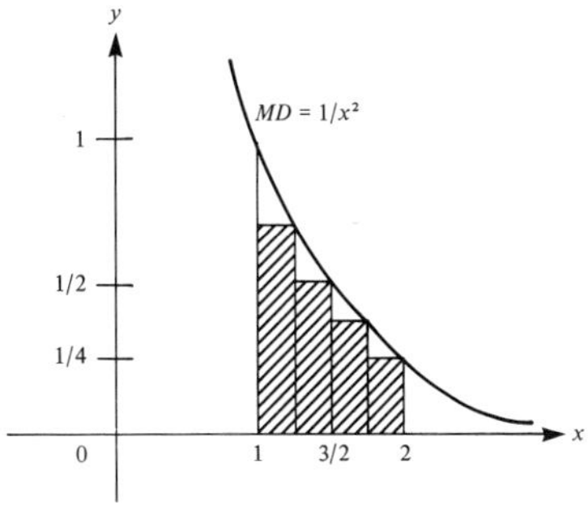

(c) $A \cong 0.415$ units^2

(d) 0.415 underestimates by 0.085 units^2

9. **(a)**

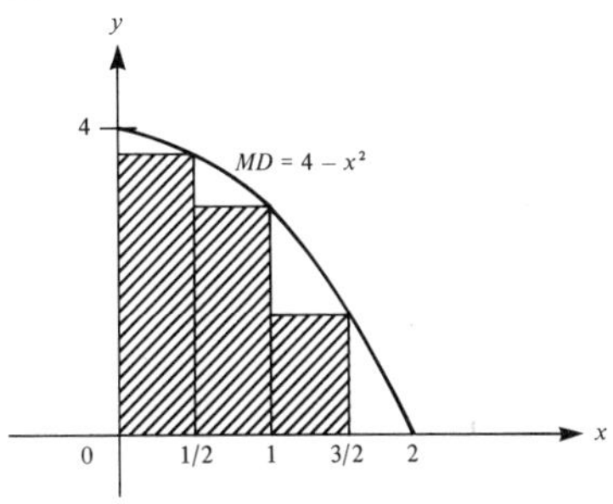

(b) $A = \frac{17}{4}$ units^2

(c) $\frac{17}{4}$ underestimates $\frac{16}{3}$ by $\frac{13}{12}$ units^2

11. $A = \frac{1}{2}[f(\frac{1}{4}) + f(\frac{3}{4}) + f(\frac{5}{4}) + f(\frac{7}{4})] = \frac{43}{8}$ units^2. This answer overestimates $\frac{16}{3}$ by $\frac{1}{24}$ units^2

13. **(a)** Area $\cong$ 15.740 units^2

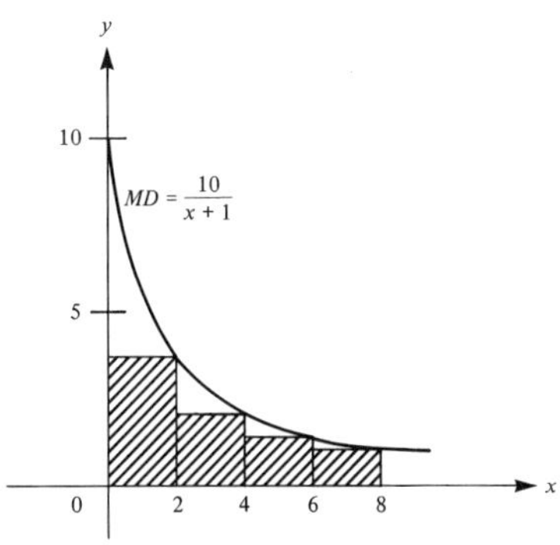

(b) Area by midpoint method $\cong$ 20.833 units

True area = 21.972 units^2

Problem Set 6.5.2 (page 216)

1. 4

3. 15

5. 32/3

7. **(a)**

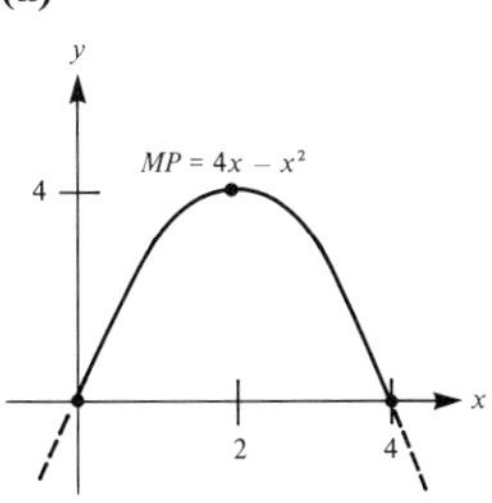

(b) $A = \int_0^4 (4x - x^2)\,dx = \dfrac{32}{3}$ units^2

(c) $P \cong \$10.67$

9. \$64 **11.** \$2048 **13.** 4/3

15. 3/16 **17.** $6.8x$ **19.** \$4.80

21. **(a)** $P(x) = -2x^2 + 16x - 8;\quad \max P = P(4) = \$24$
(b) $P(x) = -2.4x^2 + 24x - 8;\quad \max P = P(5) = \$52$
(c) $P(x) = -\frac{9}{8}x^2 + 18x - 8;\quad \max P = P(8) = \$64$

Problem Set 6.6 (page 226)

1. $\sum_{i=1}^{n} (2i - 1) = n^2$ **3.** $\sum_{i=1}^{n} 4i = 2n(n + 1)$

5. $Q(x) = \frac{3}{2}x^2\Big]_0^2 = 6$ **7.** $7 + 9 + 11 + \cdots + (2n + 5)$

9. $2 + 18 + 52 + \cdots + (n^3 - 3n)$ **11.** \$9

13. $\sum_{k=1}^{n} y_k \, \Delta x_k$ **15.** $\frac{1}{1!} + \frac{1}{2!} + \frac{1}{3!} + \frac{1}{4!} + \frac{1}{5!}$ **17.** 10 units

19. $x^{2n-1};\ \sum_{i=1}^{n} x^{2i-1}$ **21.** \$5,637.08 **23.** 16

25. $(-1)^{n+1} \cdot 2^{n-1};\ \sum_{i=1}^{\infty} (-1)^{i+1} \cdot 2^{i-1}$ **27.** 1

29. $\sum_{r=0}^{6} \binom{6}{r} x^{6-r} y^r$ **31.** $\sum_{r=0}^{4} \binom{4}{r} (3a)^{4-r}(2b)^r$

33. $-\frac{1}{(x + 1)^2} + \frac{1}{(x + 1)^3} - \frac{1}{(x + 1)^4} + \cdots$

35. **(a)** 4/9 **(b)** 5/3 **(c)** 28/3

37. **(a)** 0.961 **(b)** 1.104

6.7 Review Problems (page 228)

1. $2x^2 - 5x + C$ **3.** $\frac{2t^3}{3} - \frac{3t^2}{2} + C$ **5.** $\frac{x^4}{4} - \frac{3x^2}{2} + x + C$

7. $\frac{t^5}{5} - \frac{2t^3}{3} + 2t + C$ **9.** $\frac{1}{3}(2q + 1)^{3/2} - 6$

11. **(a)** $-4u^{-1/2} - 3 \ln|u| + C$ **(b)** $-\frac{1}{9}(1 - 3t)^3 + C$
(c) $\frac{1}{8}(2\theta + 1)^4 + C$ **(d)** $-\frac{4}{3}\sqrt{1 - 3v} + C$

13. $\frac{1}{2}(2q + 1)^3 - 25/2$ **15.** $y^2 + 2y = x^2 + 2x + C$

17. $\ln|y| = 4\sqrt{t - 1} + C$ **19.** $u = 1 - \frac{e^C}{x + 1}$

21. $P(x) = 10x^5 + 80$ **23.** $P(x) = \frac{(1 + 2x)^4}{12} - 17x - \frac{1}{12}$

25. **(a)** $T(x) = \frac{(x - 1)^4}{12} + C_1x + C_2$ **(b)** $T(x) = \frac{2\sqrt{2}}{3}x^{3/2} + C_1x + C_2$
(c) $T(x) = \frac{4}{27}(3x + 2)^{3/2} + C_1x + C_2$
(d) $T(x) = -\frac{1}{4}\ln|2x - 1| + C_1x + C_2$

27. $S(x) = \frac{1}{10}(1 + x^2)^5 + C$ **29.** $P(q) = \frac{1}{180}(1 + 3q)^3 + C$

31. **(a)**

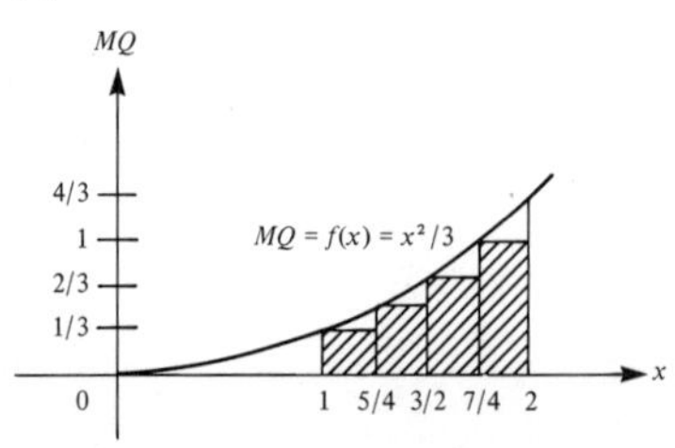

(b) Sum of areas $= \frac{189}{288}$ or $\frac{21}{32}$
(c) Approximate area underestimates true area by $\frac{35}{288}$ units^2

33. (a)

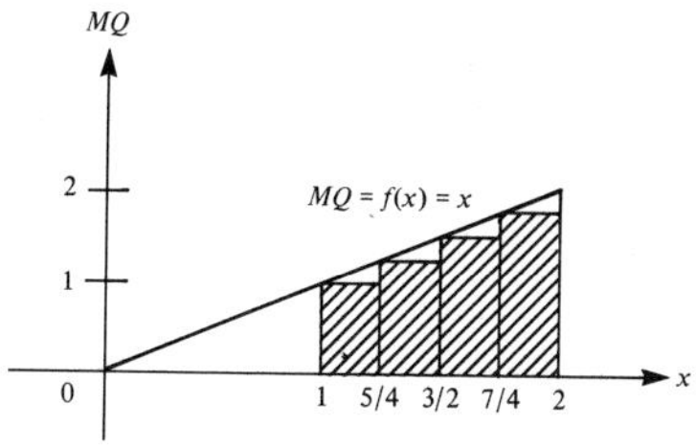

(b) Sum of areas = 11/8
(c) Underestimates by $\frac{1}{8}$ units^2

35. (30) (a)

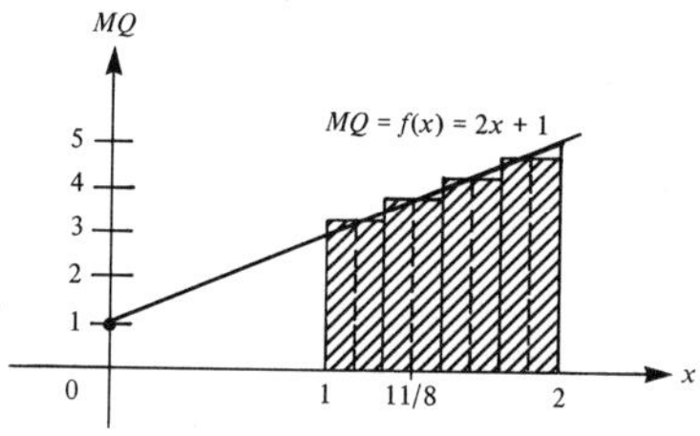

(b) Sum of areas = 4
(c) No underestimate

35. (32) (a)

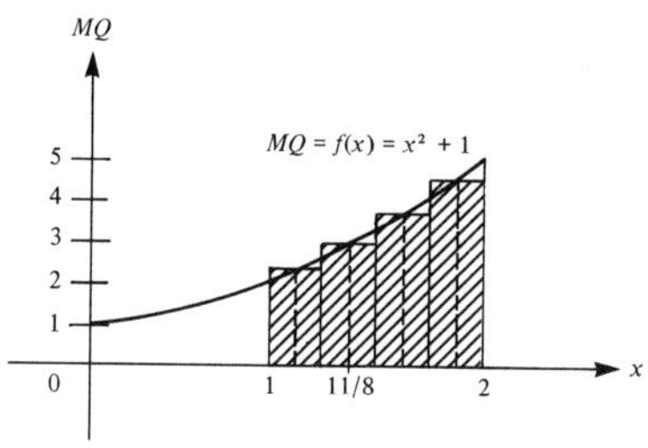

(b) Sum of areas = $\frac{213}{64}$ or $\frac{639}{192}$
(c) Underestimates by $\frac{1}{192}$ units^2

37. 16 **39.** 256/3 **41.** 4/3 **43.** 14/3

45. $P(q) = 2q^2 - \dfrac{2q^3}{3} + 3$ **47.** \$100 **49.** \$62.40

51. $$S_n = f(0)\cdot\frac{3}{n} + f\left(\frac{3}{n}\right)\cdot\frac{3}{n} + f\left(\frac{6}{n}\right)\cdot\frac{3}{n} + \cdots + f\left(\frac{3(n-1)}{n}\right)\cdot\frac{3}{n} = \cdots = \frac{27}{2}\left(1 + \frac{1}{n}\right)$$

$\lim_{n\to\infty} S_n = 27/2$;
check: $\int_0^3 3x\,dx = 27/2$

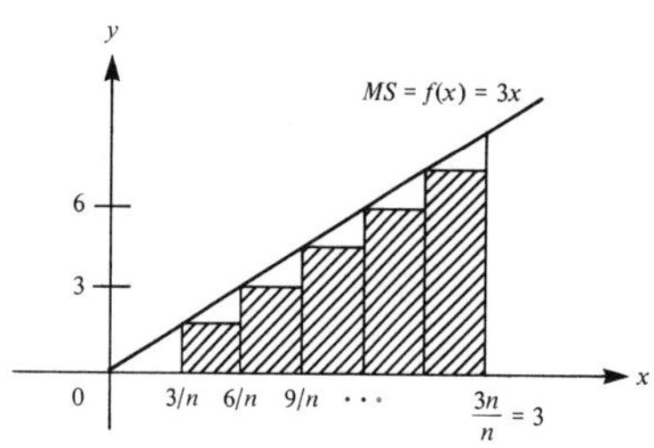

53. \$5,976.68

55. **(a)** About 6.80 **(b)** About 10.46 **(c)** 63/64

57. **(a)** 1 **(b)** 0.972

59. **(a)** 2 **(b)** 4 **(c)** 8

61. **(a)** $\sum_{k=1}^{n} H(q_k)\,\Delta q_k$ **(b)** $\sum_{i=0}^{n} a_i z^i$

63. **(a)** $G(u_1) + G(u_2) + G(u_3) + \cdots + G(u_n)$
(b) $1(1 + 1) + 2(2 + 1) + 3(3 + 1) = 20$

65. **(a)** $\frac{1}{9}(3x + 4)^3 + C$ **(b)** $\frac{x^2}{2} + \frac{4}{3}x^{3/2} + x + C$

6.8 Enrichment Topics and Problems (page 231)

1. **(a)** $f(x) = kx$, $5 = k \cdot \frac{1}{2}$, $k = 10$, $f(x) = 10x$, $W = \int_0^2 10x\,dx = 20$ inch-lb
(b) $W = \int_0^2 9x\,dx = 18$ ft-lb **(c)** 45,000 inch-lb

3. $21{,}060\pi$ ft-lb **5.** 6 years **7.** \$10,737,418.23

9. **(a)** \$12,880.88 **(b)** \$48,808.80 **11.** \$19,680

13. **(a)** $\left.\frac{t^2}{2} - \frac{4t^{5/2}}{5} + \frac{t^3}{3}\right]_0^1 = \frac{1}{30}$ **(b)** $\left.3\sqrt{u^2 - 15}\,\right]_4^8 = 18$
(c) Area of 1/4 of a circle is $\frac{1}{4}\pi a^2$

15. **(a)** $2\sqrt{x} + x + C$ **(b)** $3t^{1/3} + 3t^{2/3}/2 + C$

17. ln 3 **19.** 1/2 **21.** 7260 **23.** 51

25. **(a)** Sum of the first n integers
(b) Sum of the squares of the first n integers

27. **(a)** $\frac{1}{40}(2t^4 + 1)^5 + C$ **(b)** $\frac{1}{6}(3x^2 - x^3)^2 + C$
(c) $(x^4 + 8x^2 + 9)^2/8 + C$ **(d)** $(4x^3 - 3x^2 - 2)^2/6 + C$

29. **(a)** $(x + 1)^8 = x^8 + 8x^7 + 28x^6 + 56x^5 + 70x^4 + 56x^3 + 28x^2 + 8x + 1$
(b) $(1 - 2x)^5 = 1 - 10x + 40x^2 - 80x^3 + 80x^4 - 32x^5$
(c) $(ab + c)^3 = a^3b^3 + 3a^2b^2c + 3abc^2 + c^3$

31. **(a)** 1.219 **(b)** 0.990 **(c)** 117.576
(d) 0.172 **(e)** 100.130 **(f)** 0.099

CHAPTER 7

Problem Set 7.1 (page 243)

1. Area = 1/6 units^2

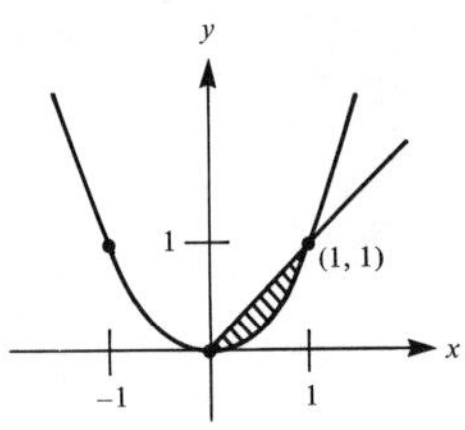

3. Area = 32/3 units^2

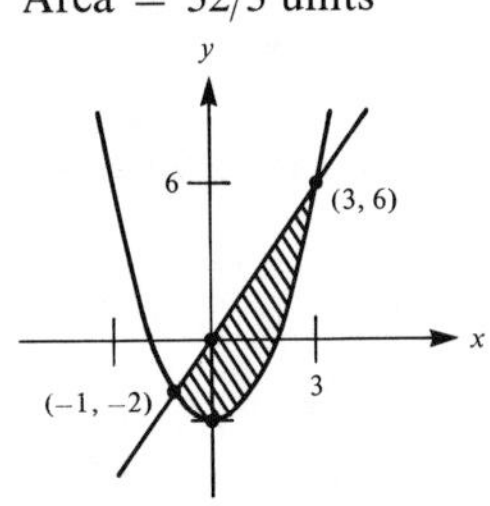

5. Area = 1/6 units^2

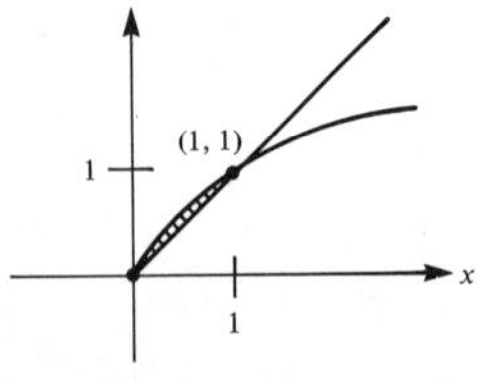

7. Area = 8 units^2

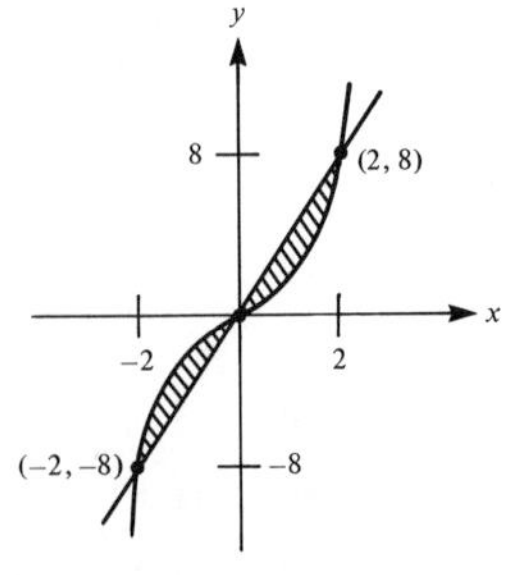

9. Area = $(2 \ln 2 - 1)$ units^2

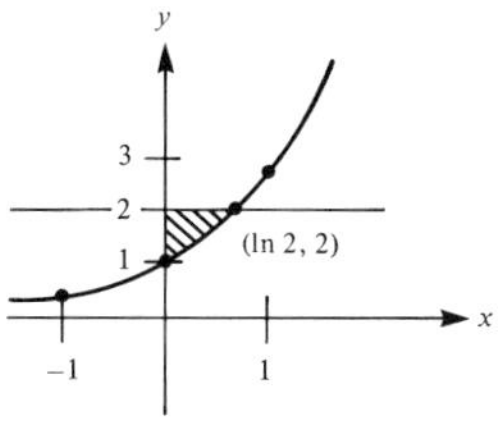

11. Area = $(1 - \ln 2)$ units^2

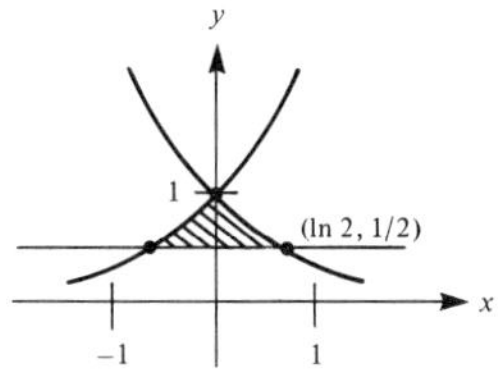

13. \$4,500 **15.** \$167 **17.** \$17,333.33 **19.** \$647.90

21. \$4000 **23.** 3 units^2 **25.** 64/3 units^2

Problem Set 7.2 (page 247)

1.

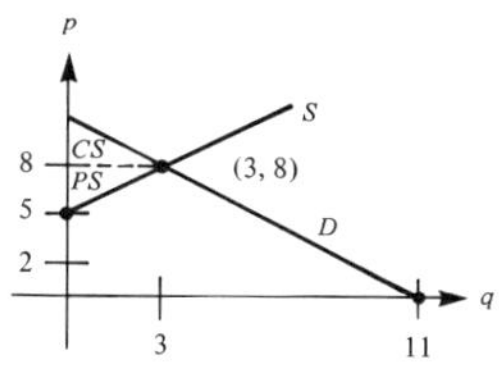

CS = \$4.50, PS = \$4.50

3.

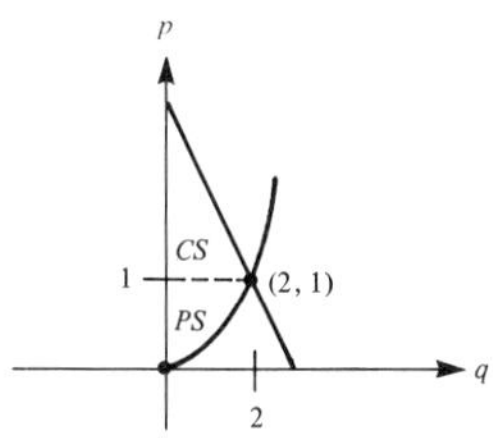

CS = \$2, PS = \$4/3

5.

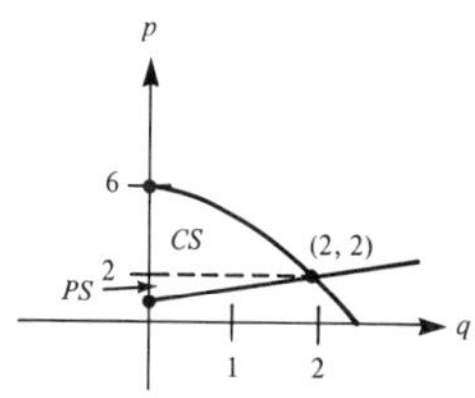

CS = \$16/3, PS = \$1

7.

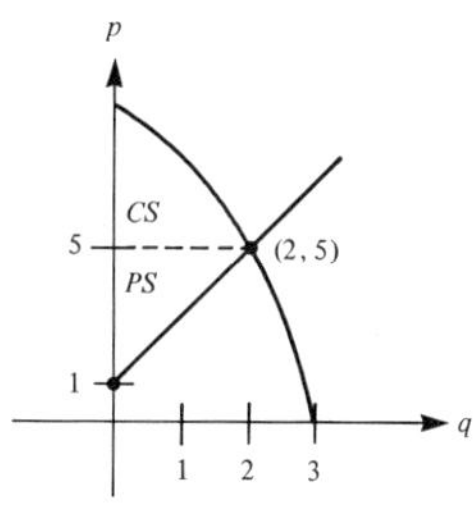

CS = \$16/3, PS = \$16/3

9.

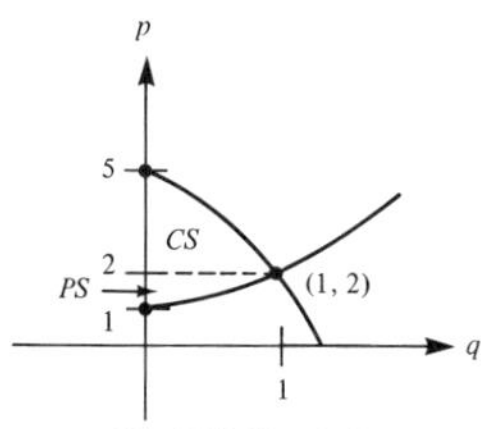

CS = \$5/3, PS = \$2/3

11.

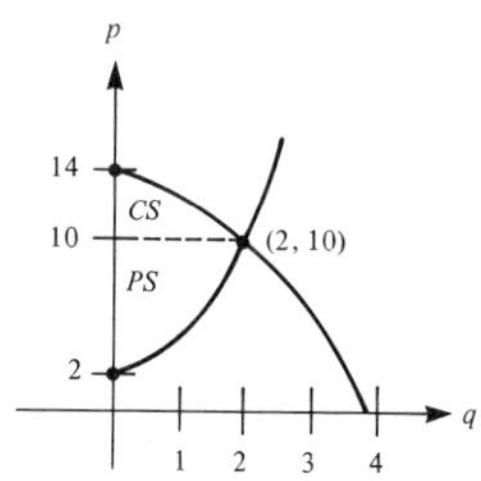

CS = \$16/3, PS = \$32/3

13.

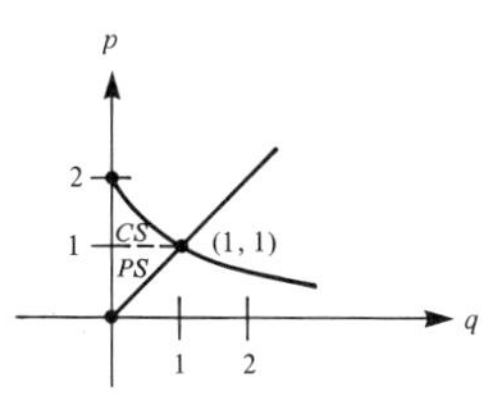

CS = \$0.39, PS = \$0.50

15.

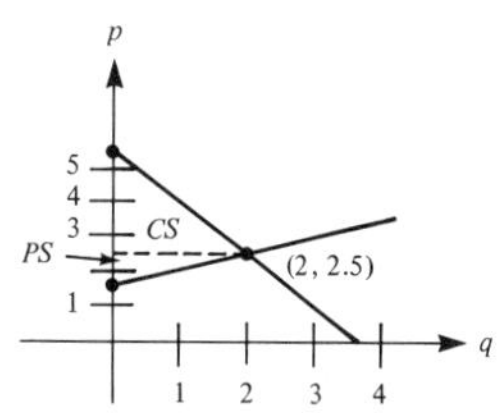

CS = \$3, PS \$3

Problem Set 7.3 (page 250)

1. Average $Q = 35/3$; $x = 5/2$ **3.** Average $D = 45$; $x = 10$

5. Average $R = 50/3$; $x = 10\sqrt{3}/3$ **7.** Average $P = 12$; $x \cong 1.4$

9. 700 **11.** 156 **13.** About 53 **15.** 15/8 thousands

17. 25.00 **19.** About 4.58

Problem Set 7.4 (page 254)

1. $-xe^{-x} - e^{-x} + C$

3. $\frac{1}{2}te^{2t} - \frac{1}{4}e^{2t} + C$

5. $\frac{1}{2}y^2 \ln y - \frac{1}{4}y^2 + C$

7. $\dfrac{x(x-1)^4}{4} - \dfrac{(x-1)^5}{20} + C$

9. $\frac{1}{3}ze^{3z} - \frac{1}{9}e^{3z} + C$

11. 9 units^2

13. $(\frac{7}{2} + \ln 2)$ units^2

15. 2500

17. $9 + 2 \ln 2$

19. **(a)** $3(x \ln x - x) + C$ **(b)** $xe^{2x-1} - \frac{1}{2}e^{2x-1} + C$
(c) $-0.2e^{-1} + 0.1$ **(d)** $16 \ln 2 - 4$

Problem Set 7.5 (page 259)

1. $\dfrac{1}{3} \ln \left|\dfrac{x-1}{x+2}\right| + C$

3. $x^2/2 + 12 \ln|x+5| + 13 \ln|x-5| + C$

5. $\ln \left|\dfrac{x-1}{x}\right| + C$

7. $-\ln x + \ln(x-3)^{3/2} - \ln \sqrt{x+1} + C$

9. $\ln \dfrac{16}{9} - \dfrac{1}{6}$

11. $1 + \dfrac{2}{3} \ln 5$

13. $\dfrac{x^2}{2} + x + \dfrac{3}{2} \ln(x^2 + 3) + C$

15. $\ln|x-1| - \dfrac{2}{x-1} + C$

Problem Set 7.6 (page 269)

1. 1/2 **3.** Diverges **5.** 1/4

7. 1/2 **9.** Diverges **11.** 0

13. 2 **15.** Diverges **17.** Diverges

19. 4 **21.** 6 units^2

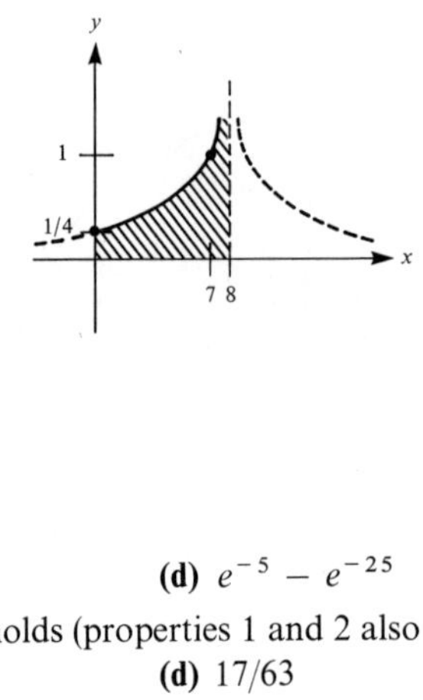

23. Diverges **25.** 5

Problem Set 7.7 (page 276)

1. **(a)** All three properties hold
(b) e^{-50} **(c)** e^{-25} **(d)** $e^{-5} - e^{-25}$

3. **(a)** $\int_0^1 \frac{4}{7}(6x^2 - x^3)\,dx = 1$ so property 3 holds (properties 1 and 2 also hold)
(b) 15/112 **(c)** 97/112 **(d)** 17/63

5. $k = 1$ **7.** $k = 3$

9. **(a)** All three properties hold
(b) 7/48 **(c)** 41/48 **(d)** 23/81

11. **(a)** 3/4 **(b)** 1/9 **13.** 1/8

15. **(a)** 2/3 **(b)** 7/9 **(c)** 1/9

Section 7.8 (page 278)

Exercise I (page 279)

(a) $-1/4x + \dfrac{3}{16} \ln \left|\dfrac{4+3x}{x}\right| + C$

(b) $\dfrac{1}{3(3+4t)} - \dfrac{1}{9} \ln \left|\dfrac{3+4t}{t}\right| + C$

Exercise II (page 279)

(a) $2\sqrt{4+5x} + 2\ln\left|\frac{\sqrt{4+5x}-2}{\sqrt{4+5x}+2}\right| + C$

(b) $\frac{2(32-8z+3z^2)\sqrt{2+z}}{15} + C$

Exercise III (page 279)

(a) $-\frac{1}{10}\ln\left|\frac{w^2}{w^2-5}\right| + C$

(b) $\frac{1}{18}\ln\left|\frac{t^2}{t^2+9/4}\right| + C$

Exercise IV (page 280)

(a) $-\frac{\sqrt{6-\theta^2}}{6\theta} + C$

(b) $\frac{x}{2}\sqrt{x^2+9} + \frac{9}{2}\ln|x+\sqrt{x^2+9}| + C$

(c) $\left(\frac{z^2}{5} - \frac{8}{15}\right)(z^2+4)^{3/2} + C$

(d) $\frac{(t^2-10)^{5/2}}{5} + \frac{10}{3}(t^2-10)^{3/2} + C$

(e) $-\frac{\sqrt{v^2+16}}{32v^2} + \frac{1}{128}\ln\left|\frac{4+\sqrt{v^2+16}}{v}\right| + C$

7.9 Review Problems (page 280)

1. 6 **3.** 1/3 **5.** 3/4

7. 4/3 **9.** 13/3 **11.** $CS = \$128/3$, $PS = \$8$

13. $CS = \$686/3$, $PS = \$686/3$ **15.** 154/3 **17.** 240

19. 96 **21.** $\frac{t}{4(1-2t)^2} - \frac{1}{8(1-2t)} + C$

23. About 346 **25.** $2\sqrt{t}\ln\sqrt{t} - 2\sqrt{t} + C$

27. $-\frac{2}{3}(1-e^t)^{3/2} + C$ **29.** $\ln\left|\frac{t}{t-4}\right| - \frac{4}{t-4} + C$

31. $2x + 3\ln|x-2| + 5\ln|x+2| + C$ **33.** $\frac{1}{2}(1-e^{-1})$

35. (a) $\frac{1}{27}(9t^2+1)^{3/2} + C$ **(b)** $\frac{1}{128}(16x^2-9)^4 + C$

37. 0.01 **39. (a)** e^{-3} **(b)** $e^{-4} - e^{-5}$

7.10 Enrichment Topics and Problems (page 282)

1. $\pi/7$ units^3 **3.** $\pi/30$ **5.** $\pi/2\,(e-1)$

7. $\frac{376}{15}\pi$ **9.** $\frac{128}{5}\pi$

11. (a) $\mu_1 = 4/3, \mu_2 = 4/3, \mu_3 = 8/5$ **(b)** $\mu_1 = 3/4, \mu_2 = 8/15, \mu_3 = 5/12$
(c) $\mu_1 = 31/3, \mu_2 = 105/4, \mu_3 = 1013/15$

13. $\frac{D_0}{RE-K}\lim_{h\to\infty} e^{(RE-K)t}\Big]_0^h = \frac{D_0}{RE-K}\lim_{h\to\infty}\left[\frac{1}{e^{(K-RE)h}} - 1\right] = \frac{D_0}{K-RE}$

15. \$2.75 **17. (a)** f is a pdf **(b)** 9/16

19. (a) \$1 **(b)** \$32 **(c)** \$2000

21. $CS = \$2$, $PS = \$16/3$ **23.** $x + 9\ln|x-3| + \frac{18}{x-3} + C$

25. **(a)** 4 **(b)** $2(2 \ln 2 - 1)$ **(c)** $4 \ln 2 - 3/2$
(d) $32/3$

27. $-89/5$

29. **(a)** $x^2e^x - 2xe^x + 2e^x + C$ **(b)** $\frac{1}{2}t^2e^{2t} - \frac{1}{2}te^{2t} + \frac{1}{4}e^{2t} + C$
(c) $z(\ln z)^2 - 2z \ln z + 2z + C$
(d) $\dfrac{x^2(x+1)^5}{5} - \dfrac{x(x+1)^6}{15} + \dfrac{(x+1)^7}{105} + C$

31. **(a)** $3 - 4 \ln 2$ **(b)** $3/2 + 3 \ln 3/2$ **(c)** $\ln 2 - 5/8$
(d) $\ln 2 - \ln|1 + e^{-1}|$ **(e)** 6

CHAPTER 8

Problem Set 8.1, 8.2 (page 290)

1. $\dfrac{x-y}{xy} = \dfrac{1}{C}, \dfrac{1}{y} - \dfrac{1}{x} = \dfrac{1}{C}, -\dfrac{1}{y^2}y' + \dfrac{1}{x^2} = 0, y' = y^2/x^2;$
Therefore, $x^2 \cdot \dfrac{y^2}{x^2} - y^2 = 0$

3. $y' = -\dfrac{4x}{(x^2-1)^2}$; Therefore $-\dfrac{4x}{(x^2-1)^2} + x\left[\dfrac{4}{(x^2+1)^2}\right] = 0$

5. $y' = -xe^{-x}$; $y'' = xe^{-x} - e^{-x}$; substituting:
$xe^{-x} - e^{-x} + 2(-xe^{-x}) + (e^{-x} + xe^{-x} + 1) - 1 = 0$

7. $C_1 = -1, C_2 = 3$

9. $C = x^4/y^3, 0 = 4x^3y^3 - 3x^4y^2\dfrac{dy}{dx} = 0, \dfrac{dy}{dx} = \dfrac{4x^3y^3}{3x^4y^2} = \dfrac{4y}{3x}$

11. $y = \frac{1}{3}x^{3/2} - 7$

13. $\dfrac{dy}{dx} = C_1e^x + C_2(xe^x + e^x)$; $\dfrac{d^2y}{dx^2} = C_1e^x + C_2(xe^x + C_2xe^x)$
Therefore, $C_1e^x + C_2(xe^x + 2e^x) - 2[C_1e^x + C_2(xe^x + e^x)] + (C_1e^x + C_2xe^x) = 0$

15. $y' = 3e^x = y'''$; Therefore $3e^x - 3e^x = 0$
$y' = -\frac{1}{4}e^{-x}, y'' = \frac{1}{4}e^{-x}, y''' = -\frac{1}{4}e^{-x}$; Therefore $-\frac{1}{4}e^{-x} - (-\frac{1}{4}e^{-x}) = 0$
$y' = 0 = y'''$; $0 - 0 = 0$

17. $-4e^{-x}\,dx + 4e^{-4y}\,dy = 0, \dfrac{dy}{dx} = \dfrac{e^{-x}}{e^{-4y}} = e^{-x+4y}$
But $\dfrac{dy}{dx} = \dfrac{e^{x+y}}{e^{2x-3y}} = e^{-x+4y}$: checks.

Problem Set 8.3 (page 292)

1. $\dfrac{1+y}{2-x} = e^C$

3. $\ln|y| = x + \ln|1 - x| + C$

5. $\ln|y| = \frac{1}{2}\ln|3 + e^{2x}| + C$

7. $y \ln|y| = C - x \ln|x|$

9. $2\sqrt{y+1} + \ln|x| = 4$

11. $2 \ln|y| + e^x - x - \ln|1 + e^{-x}| = 1 - \ln 2$

13. $e^y + \ln|e^t - 2| = 1$

15. $3P^{1/3} = kt + C$

17. About 8.67 years

19. About \$14,918,000

21. **(a)** $P(t) = 2200e^{0.01352t}$; $P(100) \cong 8514$ million
(b) About 828 years or about the year 2758

Problem Set 8.4 (page 296)

1. $\ln|y - 100| = 2x - 2$
3. $\ln|3/4 - u| = 1 - \ln 2 - \frac{1}{2}t$
5. $\ln|2z^2 - 1| = \frac{8}{3}(t^{3/2} - 8)$
7. $2y^2 = x - \ln|1 + x| + 8$
9. $\dfrac{L - P}{P}\,dP = k\,dt;$

   $L \ln P - P = kt + C$
11. $H = 74 + 120e^{-0.077t}$
13. $\dfrac{dy}{dt} = \dfrac{ky}{y - 1}, \left(1 - \dfrac{1}{y}\right) dy = k\,dt, y - \ln y = kt + \ln C,$

   $\ln Cy = y - kt, Cy = e^y \cdot e^{-kt}, ye^{-y} = \dfrac{1}{C}e^{-kt}$
15. $\dfrac{dx}{x(T - x)} = k\,dt, \dfrac{1}{T}\ln|x| + \dfrac{1}{T}\ln|T - x| = kt + C$

   $\ln\left|\dfrac{x}{T - x}\right| = kTt + TC, \dfrac{x}{T - x} = e^{kTt} \cdot e^{TC} = De^{Bt}$

   where $D = e^{TC}$ and $B = Tk$. Then
   $x = (T - x)\,De^{Bt} = TDe^{Bt} - xDe^{Bt},$

   $x(1 + De^{Bt}) = TDe^{Bt}, x = \dfrac{TDe^{Bt}}{1 + De^{Bt}} = \dfrac{DT}{D + e^{-Bt}}$

   after multiplying numerator and denominator by e^{-Bt}

Problem Set 8.5 (page 302)

1. $y = \frac{1}{2}e^x + Ce^{-x}$
3. $y = \frac{1}{2} + Ce^{-x^2}$
5. $y = 1 - 1/x + \dfrac{C}{x}e^{-x}$
7. $y = Ce^{2x^2} - 1/8$
9. $y = -3 + C_1e^x$
11. $y = -1 + C_1e^{-x}$
13. $y = -4 + C_1e^{3x/2}$
15. **(a)** $t = \dfrac{g}{w}\ln\dfrac{b - r}{m - r}$

   (b) $m = \frac{1}{2}b, t = \dfrac{10^5}{10^2}\ln\dfrac{b - 0}{b/2} = 10^3 \ln 2 = 10^3(0.6931) \cong 693$ hours

Problem Set 8.6 (page 305)

1. $y^2 = 2x^2 \ln|x| - 2Cx^2$
3. $\dfrac{3x^2 - y^2}{x} = C_1$
5. $\dfrac{x^2y}{x + y} = C_1$
7. $\ln|y| + \dfrac{1}{3}\ln\left|\dfrac{3x^2 - y^2}{y^2}\right| = C$
9. $\ln|y| + \frac{1}{4}\ln\left|\dfrac{2x^2 + y^2}{y^2}\right| = C$
11. $4\ln|y| = 3y/x + C$
13. $\ln|x| - y/x + \ln\left|\dfrac{x + y}{x}\right| = \ln 2 - 1$
15. $\ln|x| - x/y + \ln 2 + 1 = 0$
17. $\dfrac{x^4}{4y^4} + \ln|y| = 16 + \ln\sqrt{2}$
19. $\sqrt{x^2 - y^2} + 1 = x$

8.7 Review Problems (page 306)

1. Let $y = C_1e^{3x/2}$. Then $\dfrac{dy}{dx} = \dfrac{3}{2}C_1e^{3x/2}$; substituting into the given equation, we get zero.

3. $\dfrac{dy}{dx} = -2Cx^2e^{-x^2} + Ce^{-x^2}$; substituting y and dy/dx into the given equation, we get zero.

5. $\sqrt{9 - y^2} + \sqrt{16 - x^2} = 5$

7. About 16.9 years

9. $y^2 + 2x^2 + 2\ln|x| = C$

11. $\frac{3}{2}\ln|y^2 - 2| - \ln|3 + x| = C$

13. $y = \dfrac{Ae^{kt}}{1 + Ae^{kt}}, A = e^C$

15. $y = Ce^{2x} - x$

17. $y = e^{2x}$

19. $y = 2 + C_1e^{-x/3}$

21. **(a)** $t = \dfrac{V}{r}\ln\left|\dfrac{k - P}{q - p}\right|$ **(b)** 1609 hours approx.

8.8 Enrichment Topics and Problems (page 307)

1. $\dfrac{dy}{dx} = -\dfrac{3}{4}Cx^{-7/4}$; substituting y and dy/dx in the equation, we get zero.

3. $y = \dfrac{x^2}{9} + \dfrac{C}{x}$

5. $y = \dfrac{x}{2}\ln x - \dfrac{1}{4}x + \dfrac{C}{x}$

7. $y = \dfrac{1}{4}e^{3x-x^2} + Ce^{-x-x^2}$

9. $2y^2\ln\left|\dfrac{y}{x}\right| - y^2 = 4x^2\ln|x| + 4Cx^2$

11. $\dfrac{y}{x} - \ln\left|\dfrac{y + x}{x}\right| = 8\ln|x| + C$

13. $P(t) = 100e^{-0.03t}$

15. **(a)** $H(x, y) = x + x^3y^2 + xy = C$

(b) $H(x, y) = xy^2 + x^2y - \dfrac{x^3}{3} - \dfrac{y^3}{3} = C$

(c) $H(x, y) = \dfrac{x^2}{2} + x\sqrt{y^2 + 1} - \dfrac{y^2}{2} = C$

(d) $H(x, y) = 2x + xe^{2y} - y^2 = C$

17. $H(x, y) = x^3y^{-1} - y^3/3 = C$

19. **(a)** $y = \dfrac{2}{C + e^{-2x}}$ **(b)** $y = \dfrac{2}{1 + 2Ce^{2x^2}}$

APPENDIX A

Problem Set A.1 (page A–12)

1. **(a)** $1/\sqrt{2}$ **(b)** $-\sqrt{3}$ **(c)** -2 **(d)** $\sqrt{2}/2$
(e) 2.4545 **(f)** 1.0109 **(g)** -1.000

3. $\sin\theta = 15/17$, $\cos\theta = 8/17$, $\tan\theta = 15/8$, $\csc\theta = 17/15$, $\sec\theta = 17/8$, $\cot\theta = 8/15$

5. **(a)** $\sin^2 45° = 1/2$; **(b)** $\cos^2 90° = 0$; **(c)** $\tan 135° = -1$
(d) 4 **(e)** 5/2

7. **(a)** $4\sqrt{5}/9$ **(b)** $-119/169$ **(c)** $-24/7$

9. **(a)** $2\cos 60° \sin 15°$ **(b)** $2\cos\theta\cos\frac{1}{2}\theta$
(c) $2\sin 45° \cos(-35°)$ **(d)** $-2\sin 30°30' \sin(-10°)$

11. **(a)** $x = 30°, 150°, 210°, 330°$ **(b)** $x = 45°, 135°, 225°, 315°$
(c) $x = 60°, 270°, 300°$ **(d)** $x = 30°, 90°, 150°$
(e) $x = 210°, 270°, 330°$ **(f)** $x = 45°, 135°, 225°, 315°$
(g) $x = 90°, 210°, 330°$ **(h)** $x = 60°, 300°$

13. **(a)** 3/5 **(b)** 1/2 **(c)** 5 **(d)** 0

Problem Set A.2 (page A–19)

1.

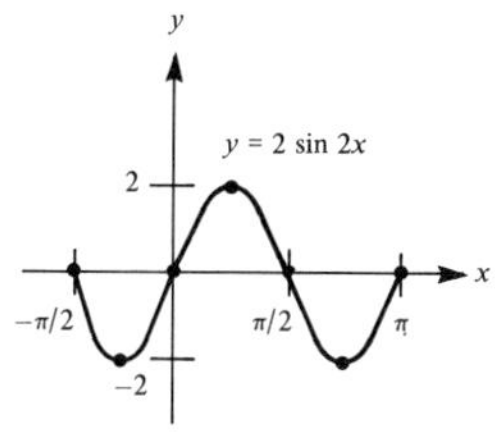

3.

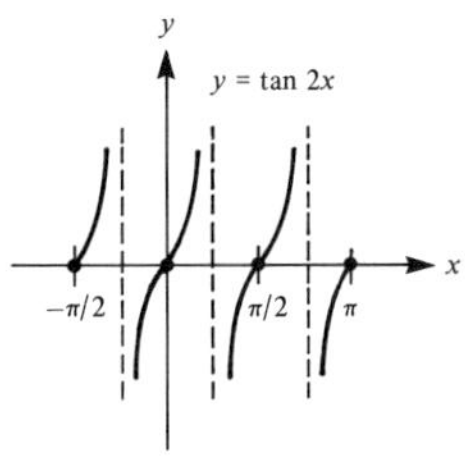

5.

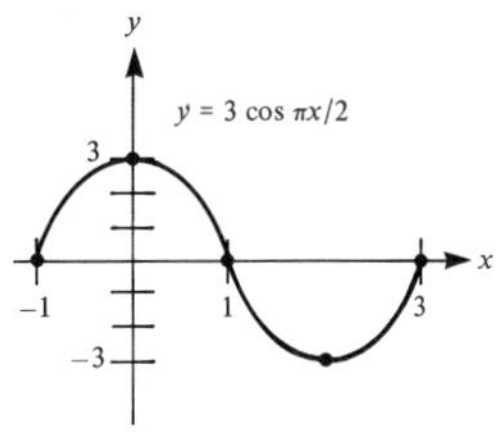

7.

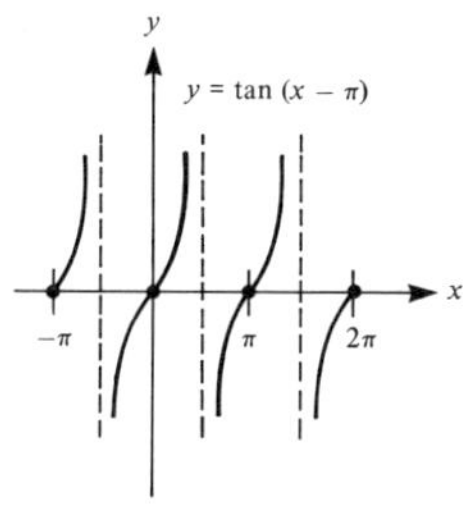

9.

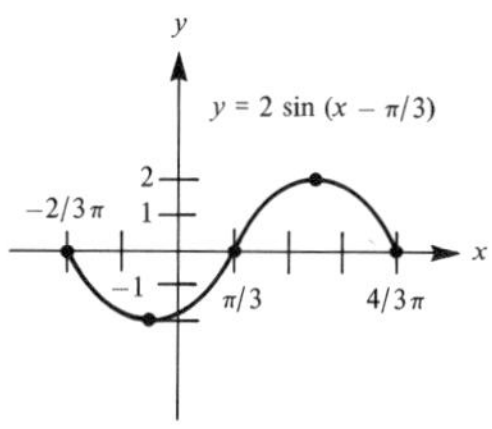

11.

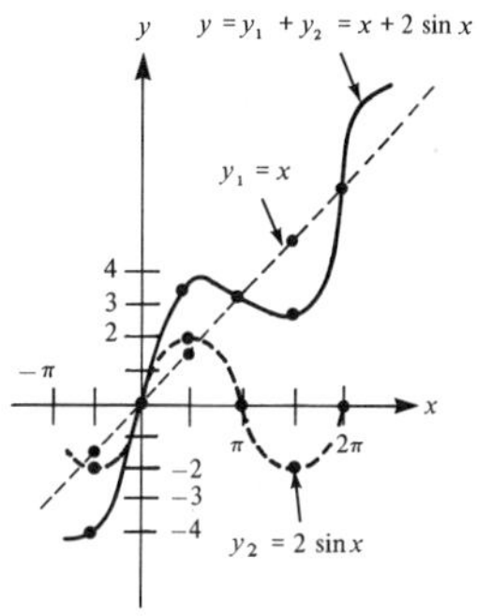

Problem Set A.3 (page A–22)

1. (a) $c = 124.97, b = 160.1; \quad \angle C = 51°20', \angle B = 90°$
(b) $a = 1275, b = 790.7; \quad \angle C = 51°40', \angle A = 90°$
(c) $c = 50; \quad \angle A \cong 53°08', \angle B \cong 36°52', \angle C = 90°$

3. 8 ft **5.** 16.65 meters **7.** 571.7 meters

Problem Set A.4 (page A–26)

1. $-2x \sin(x^2)$

3. $-12 \cos^3 3\theta \sin 3\theta$

5. $3/2\left[\dfrac{\sin \frac{3}{2}t}{\cos^2 \frac{3}{2}t} + \sin \frac{3}{2}t\right]$

7. $\dfrac{12 \sin 3x}{(1 + 2 \cos 3x)^2}$

9. $-\dfrac{1}{18} \cos \dfrac{\theta}{3} \sin \dfrac{x}{6}$

11. $-\dfrac{\sin x - x \cos x}{\sin^2 x}\left(\dfrac{t \sin t + \cos t}{t^2}\right)$

13. (a) $y = x - 1$ **(b)** $y = -3t + \pi - 1$

(c) $h(y) = -y$ **(d)** $y = 8x - \dfrac{4\pi}{3} + \sqrt{3}$

15. 50 units^2 **17.** $dx/dt = -4$

Problem Set A.5 (page A–32)

1. (a) $-\dfrac{\cos^4 x}{4} + C$ **(b)** $-\sqrt{\cos 2t} + C$

(c) $\frac{1}{3} \ln|3 + \sin 3\theta| + C$ **(d)** $-\frac{1}{2} \ln|1 + \cos(2x + 1)| + C$

(e) $\frac{1}{2} \sin e^{2t} + C$ **(f)** $\frac{1}{4} \ln|\cos(1 - 4z)| + C$

3. (a) $x \sin x + \cos x + C$ **(b)** $-\frac{1}{2}t \cos 2t + \frac{1}{4} \sin 2t + C$

(c) $\frac{e^{2x}}{5}\,[2\sin x - \cos x] + C$ **(d)** $\frac{2e^{z}}{5}\left[\sin\frac{z}{2} + 2\cos\frac{z}{2}\right] + C$

5. (a) 1 unit^2 **(b)** 8/3 **(c)** $\frac{\sqrt{2}}{2} - \frac{1}{2}$ **(d)** $4 - 2\sqrt{2}$

7. (a) max P at $P(\pi/6) \cong$ loss of \$2,977.60
(b) max P at $P(5\pi/3) \cong$ \$21,872

INDEX

Abscissa, 7
Absolute value, 6
Acceleration, 69, 196
Addition formulas, trigonometric, A-7
Addition of functions, 22
Addition of ordinates, A-17
Amplitude, A-15
Angle(s), A-2
  complementary, A-6
  of elevation, A-21
  related, A-5
  right, A-3
  straight, A-3
Antiderivative(s), 185, A-27
  formulas for, 187
Antidifferentiation, 183
Antilogarithm, B-2
Arcsin x, A-11
Area, 202, A-29
  approximation of, 203
  as a probability, 273
  between two curves, 236, A-30
  exact, 209
  of triangular region, 241
  of unbounded region, 264
  under a curve, 211, A-29
Arithmetic mean, 248, 284
Asymptote,
  horizontal, 17
  oblique, 30
  vertical, 17, 51
Average cost, 98
Average revenue, 102
Average value of a function, 248
Axis of symmetry, 18

Bernoulli equation, 309
Binomial coefficients, 226
Binomial expansion, 35, 225
Boundary conditions, 188
Boundedness, 260

Carrying costs, 104
Cartesian coordinate system, 7
Chain Rule, 63, 65, 139, 156, A-24
Circumscribed rectangles, 205
Closed interval, 6
Coefficients, equating of, 256
Common logarithms, 128
  table of, B-4
Composition of functions, 23
Compound amount, 123
Compound interest, 123
Conjugate of a binomial, 38
Constraints, 166
Consumers' surplus, 244

Continuous compounding, 124
Continuity, 48
  definition of, 50
Continuum, 272
Convergence, 262
Cosine function,
  definition of, A-2
  derivative of, A-23
  graph of, A-14
  integral of, A-27
Costs,
  carrying, 104
  reorder, 104
  setup, 104
  warehousing, 106
Critical point, 163
Critical values, 92
Cubic cost function, 99
Cylinders,
  elliptic, 149
  hyperbolic, 149
  parabolic, 149

Decay function, 121
Decomposition of functions, 30
Definite integral, 204, 209
  definition of, 211
  properties of, 215
Degree, A-2
Demand-supply system, 244
Dependent variable, 12
Depreciation, rate of, 144
Derivative(s),
  chain rule, 63
  of cos *u*, A-23
  definition of, 57, 151
  of e^u, 139
  first, 55
  first partial, 151
  of higher order, 68
  of ln *u*, 135, 136
  partial, 151
  of a product, 63
  of a quotient, 64
  second partial, 154
  of sin *u*, A-23
Deviation, 173
Difference quotient, 57
Differential equation(s), 190, 287
  applications of, 193, 195
  definition of, 190, 288
  degree of, 288
  exact, 308
  homogeneous, 303
  linear, 297
    first-order, 298
  order of, 288
  ordinary, 288
  partial, 288
  solutions of, 190, 289, 300
Differentials, 72
  approximations by, 73
  definition of, 72
  of a function of two variables, 180
Differentiation,
  implicit, 117, 158
  logarithmic, 138
  techniques of, 60, 63, 152
Directrix, 148
Discontinuous function, 45, 50
Distance between two points, 8
Divergence, 262
Domain, 12, 129, 147
Double-angle formulas, A-7
Dummy variables, 212

e, 121–123
Economic order quantity, 104
Elasticity of demand, 87
Ellipse, 37
Equation(s), 9
  logarithmic, 131
  of lines, 9
  point-slope form, 10
  quadratic, 16
  roots of, 28
  simultaneous solution, 10
  system of, 15
  trigonometric, A-10
Equilibrium point, 15
Error,
  percentage, 74, 182
  relative, 74, 182

Even function, 18
Event(s), 270
  mutually exclusive, 271
Experiment, 270
Exponential function, 120
Extrema, 92, 161, A-25

Factorial, 121
Factor of a polynomial, 28
First Derivative Test, 92
Forming new functions, 22
Function(s), 13
  absolute value, 13
  average cost, 98
  average revenue, 102
  average value of, 248, 285
  composition of, 23
  concave, 19
  continuity of, 48
  convex, 19
  cubic cost, 99
  decay, 121
  decreasing, 91
  definition of, 12
  demand, 15, 244
  derivative of, 54
  domain of, 12
  doubling, 13
  even, 18, A-5
  explicit, 158
  exponential, 120-122
    properties of, 122
  forming new, 22
  graphing of, 13-20
  growth, 120
  homogeneous, 303
  identity, 13
  image of, 24
  implicit, 159
  increasing, 91
  inverse, 24, 128, A-11
  limits of, 41
  linear, 14
  logarithmic, 119, 127
  marginal cost, 62
  marginal revenue, 61
  mirror image of, 129
  multivariable, 146
  odd, 18, A-5
  periodic, A-5
  polynomial, 26
  probability density, 269
  quadratic, 16
  quadratic cost, 98
  range of, 12
  rational, 26
  reciprocal, 13
  revenue, 102
  square root, 13
  squaring, 13
  supply, 15, 244
  tabular form, 13
  trigonometric, A-2
  zeros of, 28
Fundamental Theorem of Integral Calculus, 212-213

Generator, 149
Geometric progression, 223
  infinite, 224
  sum of, 223
Geometric series, 263
Graphs of functions, 8, 19, 147, A-14
Growth function, 121

Half-angle formulas, A-8
Homogeneous function, 303
Horizontal asymptote, 17
Horizontal element of area, 240
Hyperbola, 17, 37

Identical equality, 255
Identity, A-6
Image of a function, 24
Implicit differentiation, 117, 158
Improper integral(s),
  convergent, 262, 265
  definition of, 261
  divergent, 262, 265
Indefinite integral, 184, A-27
  definition of, 186

Independent variable, 12
Indeterminate form, 46, 182
Index of summation, 218
Inequalities, 6, 42
Infinite series, 262
Inflection point, 91
  test for, 96
Initial conditions, 193, 295
Inscribed rectangles, 203
Integers, 5
Integral(s),
  definite, 204, 209
  improper, 260, 265
  indefinite, 184, A-27
Integrand, 186
Integrating factor, 299, 309
Integration,
  applications of, 235
  constant of, 186
  of e^u, 238
  by parts, 251, A-28
  substitutions for, 198, A-29
  techniques of, 199
  by use of tables, 278
Intervals, 6, 49
Inventory, 104
  carrying, 105
Inverse function(s), 24
  definition of, 128
  principal, A-11
Irrational numbers, 5

Lagrange multiplier, 166, 168
Law of cosines, A-21
Law of sines, A-22
Law of the Mean, 75
Learning curve, 125
Least squares, 172
Leibniz, 2, 3, 212
Length of arc, 283
L'Hospital's Rule, 85, 182
Limit(s), 41, 86, 151, 182, 221
  definition of, 43
  properties of, 262
  theorems on, 43
Limits of integration, 212, 267
Linear differential equations, 297
Linear factors, 255
  repeated, 258
Linear function, 14
Linear regression, 172
Line(s),
  equation of, 9
  horizontal, 9
  oblique, 9
  parallel, 9
  of regression, 176
  secant, 55
  slope of, 9
  straight, 9
  tangent, 55
  vertical, 9
  y-intercept of, 9
ln $u(x)$, 135, 136
Logarithm(s),
  common, 128, B-4
  computations of, 131
  definition of, 127
  examples of use of, B-2
  natural, 128, B-7
  properties of, 129
Logarithmic differentiation, 138
Lowest common denominator (LCD), 36

Marginal analysis, 40
Marginal cost, 62, A-31
Marginal productivity, 155
Marginal revenue, 61, A-31
Maximum profit, 111
Maximum revenue, 102
Meaningless form, 45, 123
Mean value for integrals, 249
Mean Value Theorem, 76
  approximations by, 77
Mesh, 210
Midpoint method, 206
Minimum cost, 98, 170
Moment, 283
Motion, problems of, 118
Multiplication of functions, 22
Multivariable differentiation, 146

Natural numbers, 5
Newton, 1–3, 212
Newton's Law of Cooling, 295
Newton's method, 78
  finding roots by, 79
Null set, 270
Number line, 5

Oblique asymptote, 30
Octant, 150
Odd function, 18
Open interval, 6
Optimization analysis, 90
Optimum lot size, 104
Optimum performance, 126
Ordered pairs, 7
Ordered triple, 148
Ordinate, 8
Origin, 7

Parabola, 16
  vertex of, 16
Partial derivatives, 151
  of higher orders, 154
Partial fractions, 255
Partial sums, 223
Partition, 271
Pascal's triangle, 226
pdf. *See* Probability density function
Percentage error, 74
Percentage time rate, 87
Perimeter of a triangle, 8
Periodic function, A–5
Plane, 150
Point of equilibrium, 244
Point-slope form, 9
Polynomial(s),
  of degree *n*, 26
  division of, 29
  factors of, 28
Polynomial function, 26
Present value, 124
Probability density function, 269
  definition of, 274
Producers' surplus, 244
Product formulas, trigonometric, A–9
Product rule, 63, A–23

Quadrant, 7
Quadratic cost function, 98
Quadratic equation, 29
  roots of, 29
Quadratic Formula, 29
Quotient rule, 64, A–23

Radian, A–3
Random variable, 272
Range, 12, 129, 147
Rate of change,
  instantaneous, 55
  of growth, 293
Rational numbers, 5
Rationalizing the denominator, 38
Rationalizing the numerator, 38
Real numbers, 5
Rectangles,
  circumscribed, 205
  inscribed, 203
Rectangular coordinate system, 7
Reflection,
  in the origin, 19
  in the *y*-axis, 18
Regression analysis, 176
Related rates, 117, A–25
Relative error, 74
Relative maximum, 92, 163
Relative minimum, 92, 163
Relative time rate, 87
Reorder costs, 104
Repeating decimals, 225
Riemann, 2, 211
Riemann sum, 211
Roots, 28

Saddle point, 161, 163
Sample space, 270
  partition of, 271
  subset of, 270
Secant lines, 55

Second derivative test, 93, 163
Second partial derivative, 154
Separation of variables, 191, 291, 293
Series, infinite, 223
Set-builder notation, 5
Set(s), 5
  complement of, 271
  dense, 273
  elements of, 5
  empty, 271
  intersection of, 33
  union of, 33
Setup costs, 104
Sigma (Σ), 211
Sigma notation, 217
  applied, 217–226
  properties of, 218
Simultaneous solution, 10
Sine function,
  definition of, A–2
  derivative of, A–23
  graph of, A–14
  integral of, A–27
Single-variable substitution, 188
Slope, 9
  of tangent, 56
Straight line, 9
Subset, 5
Subinterval, 204
Sum, 210
  limit of, 211
  of infinite series, 224
  of integers, 217
  Riemann, 211
Summation (Σ), 211
Surface, 148
Symmetry,
  about the origin, 18
  axis of, 18

Table(s),
  of common logs, B–4
  of exponential functions, B–6
  of natural logs, B–7
  of powers and roots, B–8
Tabular form, 13
Tangent function,
  definition of, A–2
  graph of, A–14
Tangent line, 55, 152, A–24
Third derivative test, 96
Time rates, 87
Total derivative, 154, 156
Trace, 149
Trial, 270
Triangle, right, A–20
Trigonometric functions,
  applications in business, A–18
  defined, A–2
  extrema for, A–25
Trigonometric substitutions, A–29

Variable(s),
  dependent, 12
  independent, 12
  random, 272
Velocity, 69
Vertex, 16, 34
Vertical asymptote, 17
Vertical element of area, 239
Volume of a solid, 282

Work, 231

y-intercept, 9

Zeros of a polynomial, 28

INDEX OF IMPORTANT EQUATIONS

CHAPTER 1

1. $\{x | x \in \mathbf{R}\}$ page 5
2. $P_1P_2 = \sqrt{(x_2 - x_1)^2 + (y_2 - y_1)^2}$ page 8
3. $\{(x, y) | y = mx + b\}$ page 9
4. $y - y_1 = m(x - x_1)$ page 9
5. $[fg](x) = f(x) \cdot g(x)$ page 22
6. $[f \circ g](x) = f[g(x)]$ page 23
7. $P(x) = a_0 + a_1x + a_2x^2 + \cdots + a_nx^n, \qquad n \geq 0$ page 26

CHAPTER 2

8. $\lim_{x \to b} f(x) = L$ page 42
9. $f'(x) = \lim_{\Delta x \to 0} \frac{f(x + \Delta x) - f(x)}{\Delta x}$ page 57
10. $\frac{dy}{dt} = \frac{dy}{dx} \cdot \frac{dx}{dt}$ page 65
11. $dy = f'(x)\, dx$ page 72

CHAPTER 3

12. $f''(x_0) > 0 \to \min f \quad \text{at } [x_0, f(x_0)]$ page 95
13. $f''(x_0) < 0 \to \max f \quad \text{at } [x_0, f(x_0)]$ page 95
14. $q(x) = \frac{Q(x)}{x}$ page 98
15. $Q(x) = \frac{cx}{2} + \frac{sd}{x}$ page 105

CHAPTER 4

16. $f(x) = ae^{kx}$ page 121
17. $x = \log_b y$ page 127
18. $D_x \ln|u| = \frac{1}{u} \cdot D_x u$ page 136
19. $D_x e^u = e^u D_x u$ page 139

CHAPTER 5

20. $\frac{\partial P}{\partial x} = P_x = P_1(x, y)$ page 151
21. $\frac{\partial P}{\partial y} = P_y = P_2(x, y)$ page 151
22. $\frac{\partial}{\partial x}\left(\frac{\partial f}{\partial x}\right) = \frac{\partial^2 f}{\partial x^2} = f_{xx}$ page 154
23. $\frac{\partial}{\partial y}\left(\frac{\partial f}{\partial y}\right) = \frac{\partial^2 f}{\partial y^2} = f_{yy}$ page 154
24. $\frac{\partial}{\partial x}\left(\frac{\partial f}{\partial y}\right) = \frac{\partial^2 f}{\partial x\, \partial y} = f_{yx}$ page 154
25. $\frac{\partial}{\partial y}\left(\frac{\partial f}{\partial x}\right) = \frac{\partial^2 f}{\partial y\, \partial x} = f_{xy}$ page 154
26. $\frac{dy}{dx} = \frac{\partial y}{\partial x} \cdot \frac{dx}{dx} + \frac{\partial y}{\partial u} \cdot \frac{du}{dx}$ page 156
27. $\frac{dy}{dx} = -\frac{H_x}{H_y}$ page 159
28. $F(x, y, l) = f(x, y) + lg(x, y)$ page 168